Springer Series on Atomic, Optical, and Plasma Physics

Volume 84

The Springer Series on Atomic, Optical, and Plasma Physics covers in a comprehensive manner theory and experiment in the entire field of atoms and molecules and their interaction with electromagnetic radiation. Books in the series provide a rich source of new ideas and techniques with wide applications in fields such as chemistry, materials science, astrophysics, surface science, plasma technology, advanced optics, aeronomy, and engineering. Laser physics is a particular connecting theme that has provided much of the continuing impetus for new developments in the field, such as quantum computation and Bose-Einstein condensation. The purpose of the series is to cover the gap between standard undergraduate textbooks and the research literature with emphasis on the fundamental ideas, methods, techniques, and results in the field.

More information about this series at http://www.springer.com/series/411

Boris M. Smirnov

Theory of Gas Discharge Plasma

Springer

Boris M. Smirnov
Joint Institute for High Temperatures
Russian Academy of Sciences
Moscow
Russia

ISSN 1615-5653 ISSN 2197-6791 (electronic)
ISBN 978-3-319-11064-6 ISBN 978-3-319-11065-3 (eBook)
DOI 10.1007/978-3-319-11065-3

Library of Congress Control Number: 2014956686

Springer Cham Heidelberg New York Dordrecht London

Printed on acid-free paper

Springer is part of Springer Science+Business Media (www.springer.com)

Preface

According to its nature, gas discharge is the passage of electric current through a gas that is located under the action of external fields. The principal property of gas discharge, its self-maintenance, was formulated in the beginning of the twentieth century [1–5] and consists in the ionization balance inside gas discharge. This is provided by formation of an ionized gas called a gas discharge plasma, and the ionization equilibrium is supported in this plasma. Because of this character of the passage of electric current through a gas, the processes in the gas discharge plasma are of importance. Basic processes in gas discharge were studied long ago and contemporary information about them differs weakly from what we have 10 years ago. For example, the values of Townsend coefficients and ion mobilities [6–8] published more than 70 years ago differ weakly from contemporary values. In addition, the peculiarities of various schemes and regimes of gas discharge are studied and contained in various works on gas discharge (in particular, [9–23]), as well as the processes in a gas discharge plasma and its kinetics [24–40]. We add to this that the gas discharge plasma includes most part of plasma applications.

Thus, gas discharge plasma is the widespread type of plasma whose properties are determined by the processes in this plasma. Therefore, one can connect parameters of a certain gas discharge plasma with the parameters of processes in this plasma. However, in spite of understanding this connection in principle, it is impossible to give a universal algorithm to express properties of a gas discharge plasma through the rates of corresponding processes because of a variety of types of gas discharges, as well as a variety of their regimes, geometric constructions, and configurations of external fields that support this gas discharge. Therefore, considering this problem in a given book, we restrict by simple configurations and types of external fields as well as by helium and argon as a working discharge gas. Hence, this book has a methodology goal for fulfilling numerical calculations of various parameters of a gas discharge plasma of helium and argon. This restriction allows us to use the real rates and rate constants for the processes in the gas discharge plasma and to obtain certain simple algorithms for determining the plasma parameters and regimes of its evolution. This book is based on the Russian author's book [41] where the above concept is formulated and methodical approaches are developed. This book consists of two parts, so that the first part is a textbook on processes and properties of a gas discharge plasma, and the second part includes problems related to some aspects of a helium and argon gas discharge plasma.

The book is aimed at two groups of readers. The first group is students who first study the problem of the gas discharge plasma. They can understand general principles of a gas discharge plasma as well as the methods to analyze some aspects of this area up to numerical determination of plasma parameters. The second group of readers are users who can obtain methods and codes for computer solutions of some problems related to certain applications.

Moscow Boris M. Smirnov

Contents

1 **Introduction** 1

Part I Processes in Gas Discharge Plasma

2 **Properties of Gas Discharge Plasma** 13
2.1 Equilibria and Distributions of Particles in Gases and Plasmas 13
2.2 Basic Plasma Parameters 17
2.3 Transport Coefficients 19
2.4 Ionization Equilibrium in Gas Discharge Plasma 20
2.5 Thermoemission of Electrons from the Cathode 23
2.6 Parameters of Helium and Argon Atoms and Ions 25

3 **Elementary Processes in Gas Discharge Plasma** 31
3.1 Elastic Collisions of Classical Atomic Particles 31
3.2 Elastic Electron-Atom Scattering in Slow Collisions 34
3.3 Inelastic Electron Collisions with Atoms 40
3.4 Resonant Atom Transitions in Collisions with Electrons 43
3.5 Transitions Between Neighboring Atom States in Collisions With Electrons 47
3.6 Atom Ionization by Electron Impact 50
3.7 Recombination of Electrons and Ions in Plasma 58
3.8 Three Body Processes and Stepwise Ionization of Atoms 59
3.9 Collision Processes Involving Ions 62

4 **Radiative Processes in Gas Discharge Plasma** 65
4.1 Radiative Transitions in Atoms 65
4.2 Photoionization and Photorecombination Process 67
4.3 Broadening of Spectral Lines 69

4.4 Cross Section and Absorption Coefficient for Resonant Photons 75
4.5 Propagation of Resonant Radiation in the Gas 79
4.6 Resonant Radiation in Excited Helium and Argon 84
4.7 Block Model for Atom Levels 88

Part II Kinetics and Transport Phenomena in Gas Discharge Plasma

5 Collision Processes in Kinetics of Gas Discharge Plasma 93
5.1 Kinetic Equation for Charged Particles 93
5.2 Integral Relations for an Average Momentum and Energy of a Charged Particle in Gas in Electric Field 95
5.3 Integral of Electron-Atom Collisions 99
5.4 Landau Collision Integral 101
5.5 Kinetics of Fast Electrons in Plasma 105
5.6 Electron Regimes in Gas Discharge Plasma 110

6 Kinetic Processes in Gas Discharge Plasma 113
6.1 Distribution Function of Electrons Located in Rareness Gas in Electric Field 113
6.2 Regime of High Electron Density in Electron Kinetics 120
6.3 Atom Excitation as Electron Diffusion in Energy Space 125
6.4 Efficiency of Atom Excitation by Electron Impact 130
6.5 Electron Distribution Function Above the Atom Excitation Threshold 134
6.6 Electron Kinetics in Gas in Strong Field 141

7 Transport Processes in Gas Discharge Plasma 151
7.1 Transport Phenomena in Gases 151
7.2 Electron Drift in Gas in External Electric Field 156
7.3 Electrons in Gas of Hard Spheres 164
7.4 Conductivity of Weakly Ionized Gas 170
7.5 Electron Thermal Conductivity of Helium Arc Plasma 173
7.6 Ion Drift and Diffusion in Gas in External Electric Field 176
7.7 Ambipolar Diffusion of Plasma in Gas in Electric Field 184
7.8 Heat Processes in Gas Discharge Plasma 191
7.9 Plasma Transport in Magnetic Field 193

Part III Processes in Gas Discharge

8 Ionization Equilibrium in Gas Discharge Plasma 199
8.1 Townsend Scheme for Self-maintaining of Gas Discharge 199
8.2 Ionization Equilibrium in Positive Column of Gas Discharge 207
8.3 Stepwise Ionization of Atoms in Positive Column. 209
8.4 Plasma of Positive Column of Low Pressure Gas Discharge. . . 211
8.5 Heat Processes in Positive Column of Gas Discharge 214
8.6 Local Thermodynamic Equilibrium in Arc Plasma of High Pressure. 220
8.7 Ionization Equilibrium in Arc Plasma of High Pressure 225

9 Cathode and Wall Processes 229
9.1 Electric Breakdown of Gases 229
9.2 Electron Emission from Cathode in Ion Collisions 231
9.3 Properties of Cathode Region of Glow Discharge 234
9.4 Transition from Glow Discharge to Arc and Townsend Discharges 245
9.5 Plasma Sheath at Walls 248
9.6 Principles of Magnetron Discharge 252

Part IV Helium and Argon Gas Discharge Plasmas

10 Atom Excitation in Helium and Argon Uniform Plasma 259
10.1 Excitation of Metastable State in Helium Plasma at Low Electron Concentrations 259
10.2 Excitation of Metastable State in Helium Plasma at High Electron Concentrations 263
10.3 Inelastic Electron Collisions with Excited Helium Atoms 267
10.4 Excitation of Atoms in Argon Gas Discharge Plasma 278
10.5 Continuous Spectrum of Radiation of Equilibrium Plasma 285
10.6 Tail of the Energy Distribution Function of Electrons 288

11 Ionization in Helium and Argon Gas Discharge Plasma 291
11.1 Single Ionization of Atoms in Helium Gas Discharge Plasma. 291
11.2 Stepwise Ionization of Atoms in Helium Gas Discharge Plasma. 296
11.3 Single and Stepwise Ionization of Atoms in Argon Gas Discharge Plasma 303

11.4 Thermodynamic of Stepwise Ionization in Gas Discharge Plasma 307
11.5 Ionization in Argon Gas Discharge Plasma Involving Excited Atoms 311

12 Helium and Argon Plasma in Positive Column of Gas Discharge 315
12.1 Schottky Regime for Ionization Equilibrium in Positive Column 315
12.2 Stepwise Ionization in Helium Positive Column 321
12.3 Ionization in Argon Gas Discharge Plasma of Positive Column 331
12.4 Hot Gas Discharge Plasma of Positive Column 338
12.5 Capillary Discharge 347

13 Processes at Boundaries of Gas Discharge Plasma 357
13.1 Cathode Plasma of Glow Discharge Near Equilibrium 357
13.2 Cathode Layer of Abnormal Glow Discharge 364
13.3 Transition Region in Glow Discharge 368
13.4 The Wall Plasma Sheath for Positive Column in Helium and Argon 370
13.5 Magnetron Plasma in Helium and Argon 376

14 Principles of Gas Discharge Plasma 387
14.1 Gas Discharge Plasma as Complex Physical Object 387
14.2 Self-consistent Character of Processes Involving Excited Atom States 390
14.3 Regimes of Gas Discharge Plasma 395
14.4 Conclusion 398

Appendix 399

References 409

Index 421

Chapter 1
Introduction

According to the definition, gas discharge is the passing of an electric current through a gas. Correspondingly, a gas discharge plasma is a matter that provides this process. It is a weakly ionized gas that is supported by an external electric field. A gas discharge plasma as a physical object is an nonequilibrium system because of the character of energy transmission from an electric field to a gas through electrons. Indeed, electrons as more mobile charged atomic particles acquire an energy from the electric field and then transfer it to gas atoms or molecules in collisions with them. From this it follows that the kinetic theory is necessary for description a gas discharge plasma, and the velocity or energy electron distribution function (EEDF) is one of characteristics of a gas discharge plasma under consideration.

The first self-consistent scheme of a gas discharge plasma was represented by Townsend [42] and has the form

$$e + A \rightarrow 2e + A^+, \quad A^+ + M \rightarrow A + M^+ + e \tag{1.0.1}$$

Within the framework of this scheme the ionization equilibrium is established as a result of collisions of electrons (e) and atoms (A), and electrons are reproduced in collisions of ions (A^+) with the cathode surface (M). Such collisions lead to formation of secondary electrons. This scheme is working under certain conditions and is described by two parameters, namely, by α, the first Townsend coefficient, and by γ, the second Townsend coefficient (we exclude from the Townsend scheme ion-atom collisions). Here $1/\alpha$ is the mean free path of electrons in an external field with respect to atom ionization, and γ is the probability of formation of a secondary electron at the cathode as a result of collision with the cathode surface for an ion which is accelerated by an electric field. This scheme allowed one to explain the principal properties of gas discharge on the first stage of its study [1–5]. In addition, this understanding of physics of gas discharge gave an impetus for measurements the parameters of various processes involving electrons in gases located in an electric field [6, 7, 12, 43], such as the drift velocity of electrons, the transverse and longi-

B.M. Smirnov, *Theory of Gas Discharge Plasma*, Springer Series on Atomic, Optical, and Plasma Physics 84, DOI 10.1007/978-3-319-11065-3_1

tudinal diffusion coefficients of electrons, the effective electron temperature that is the ratio of the transverse diffusion coefficient to its mobility, the first and second Townsend coefficients and rates of various processes.

The principal property of a gas discharge plasma is its self-consistency. A space self-consistency of a gas discharge plasma follows from the Townsend scheme and leads to a nonuniform space distribution of a gas discharge plasma. If this plasma is located in a cylinder tube, this leads to separation of gas discharge in the cathode layer where electrons are generated, and the positive column with a low electric field strength, so that this plasma distribution provides the minimum voltage between electrodes. A kinetic self-consistency of a gas discharge plasma means that processes of atom excitation and ionization by electron impact lead to a drop of the electron distribution function above the thresholds of corresponding processes that in turn influences on the rates of these processes [44, 45]. Along with a self-consistency of a gas discharge plasma, its complexity may be connected with various schemes of the ionization equilibrium in a plasma if it is determined by different processes. As a result, many regimes and types of gas discharges may be realized each of these is supported by certain processes (for example [18]). Hence, various regimes of a gas discharge plasma may exist depending on used gases or gas mixtures, their pressures, electric currents and powers of gas discharge, temporary dependence for used fields and on the geometry of a gas discharge system. Due to the variety of these parameters, a gas discharge plasma cannot be described within an universal scheme of the ionization balance and basic processes as it takes place in the Townsend case. In addition, because of the complexity of this object, the description has usually a qualitative character and is based on simple models. In particular, tau-approximation for the kinetic equation of electrons is used, as well as simplified parameters of basic processes. Though this description does not allow one to obtain numerical parameters of a certain gas discharge plasma under given conditions, but only on the basis of this approach one can analyze many various discharge types simultaneously.

Hence different types of a gas discharge plasma as a physical object are not described within the framework of an universal scheme, and a family of gas discharge plasmas is analogous to a mosaic, where each type requires a specific scheme and used parameters. One can expect the contemporary development of the gas discharge plasma theory must be based on computer simulation that must be addressed to a certain gas discharge plasma and conditions which accompany this plasma.

The interest to a gas discharge plasma is connected with its various applications. In turn, development of applications induced a more deep study of a gas discharge plasma. Moreover, experimental investigation of gas discharge in a gas filled tube as a light source allowed one to formulate for Langmuir [46, 47] a plasma as a physical object. He exhibits that a matter inside a tube is a uniform quasineutral gas that includes electrons and ions. According to Langmuir [46] "we shall use the name plasma to describe this region containing balanced charges of ions and electrons". The plasma sheath is formed near electrodes and according to the Langmuir analysis of the positive column of mercury arc in 1923 [48] "Electrons are repelled from the negative electrode while positive ions are drawn towards it. Around each negative electrode there is thus a sheath of definite thickness containing only positive ions and neutral atoms".

The first application of a gas discharge plasma, gas discharge lamps, was started more than a century ago. These lamps were gas discharge tubes filled with an inert gas and mercury addition. The light color was determined by an inert gas type. From that time an efficacy of gas discharge lamps was increased up to 100 lm/W (the efficacy of an incandescent lamp is approximately 14 lm/W), as well as a number of their types. Along with inert gases with a mercury addition, gas discharge tubes may be filled with sodium, sulfur, hydrogen, deuterium, nitrogen, oxygen and halogens. In addition, various metal vapor additions may be used in gas discharge tube filled with inert gases. An example of a contemporary lamp of white color is represented in Fig. 1.1.

A subsequent development of technics of gas discharge lamps is connected with gas lasers as a source of coherent radiation. He-Ne and argon lasers are the most spread laser types. Recently, lasers in a vacuum ultrahigh radiation and X-ray lasers on transitions between ion levels were created on the basis of capillary discharge. This discharge provides high electron energies that allows one to generate short-wavelength radiation. As an example, Fig. 1.2 gives a scheme of transitions for Ne-like argon multicharge ions which allows one to generate laser radiation at the wavelength 46.9 nm [50–52].

Gas discharge is a simple method to create a plasma, and because gas discharge has many regimes and forms which are increased in time, a number of applications of a gas discharge plasma grows. Often new applications are based on old ideas, as

Fig. 1.1 Fluorescent lamp as a source of visible light [49]

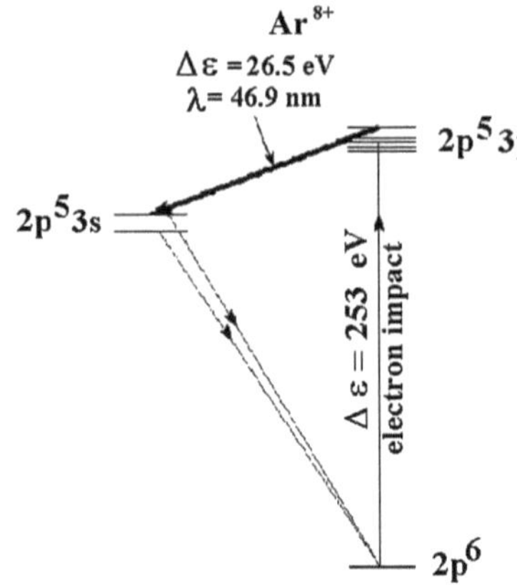

Fig. 1.2 Scheme of transitions in X-ray laser based on transitions between Ne-like levels of a multicharge argon ion in capillary gas discharge [50–52]

plasma displays. Its concept for television was described in 1936 by Tihanyi [53] as a system of single transmission points of a grid with cells arranged in a thin panel display, where these points are excited to different levels by varying the voltages at this point. But production of plasma displays started in eighties when they can compete with electron-beam tubes or liquid crystal displays. Roughly, the plasma display construction includes two parallel planes with parallel bus-bars in each plane. Bus-bars of each plane are surrounded by dielectric planes and bus-bars of different planes have perpendicular directions. As a result, the space inside the bas-bar plane is divided in some cells, and each cell relates to one intersection of bus-bars. The space is filled with a mixture of nitrogen and neon (or other inert gas), and gas discharges occur in each cell almost independently. In back to the cathode each cell is divided in three parts covered by colored luminophors, so that after absorption radiation of local gas discharge one of subcells gives a red color, second one gives a green color, and the third subcell transforms radiation of gas discharge in a blue color. These three colors are joined in the overall color of this element that depends on the intensity of local discharge, and this intensity in turn is operated through bus-bars. Competing with other types of displays, plasma displays or plasma panels are favorable for screens of large sizes and their applications are determined by possibilities of a new technology [54–57]. A subsequent development of the plasma panel technics allows one to use this system as a detector of radiation [58] and even for muon detection [59].

Another system where a plasma is used for generation of electric energy is the thermoemission converter. It contains two parallel metal plates with different work functions (the work function is the binding energy of an electron at a surface). One of these plates is heated and emits electrons which reach the other plate, i.e. an electric current occurs between electrodes. Connection of the plates through a load leads to release of the electrical energy in the load. It is evident that a plasma is not the underlying basis for this device. Nevertheless, using a plasma in the gap between the plates, one can overcome this trouble. If a plasma is absent in the gap, electric charge occurs in this region that creates a blocking voltage and locks a current. For typical energy fluxes in these systems $\sim 1\,\mathrm{W/cm^2}$ the distance between plates must be less than $10\,\mu\mathrm{m}$. It is difficult to combine a high temperature of the heated plate $\sim 2{,}000\,\mathrm{K}$ with small gap sizes, and a gap plasma allows one to overcome this trouble.

In various applications of a plasma with an intense input of energy, the plasma is generated in a moving medium. The plasma generator or plasmatrone is usually arc discharge in a flowing gas or vapor that produces an equilibrium thermal plasma [36]. As a result, plasma torches are formed with wide applications in various technological problems including incineration of waste and special medical uses. A moving gas discharge plasma is used in rocket-propelled vehicles [60–64]. Indeed, the velocity of combustion products of a rocket vehicle with a chemical fuel is of the order of the sound speed, or of the order of 10^6 cm/s. In reality one can accelerate ions in an electric field up to 10^8 cm/s, but ion currents are small because a space charge locks the ion current. To increase this current, one can use a gas discharge plasma which is accelerated in various configurations of electric and magnetic fields. Of course, the power of such engines is some orders of magnitude lower than that for engines with a chemical fuel, and hence such rocket-propelled vehicles may be used for the control of a spacecraft motion in a space.

Many plasma applications are based on the possibility to insert a high electrical energy in the plasma. This leads to the creation and maintenance of an ionized gas containing active atomic particles, electrons, ions, excited atoms, radicals. These particles may be analyzed by a variety of techniques, and therefore a plasma can used not only in energetic systems, but also in measuring instruments. In particular, plasma-based methods of spectral analysis are widely used in metallurgy. In these methods a small amount of metal in the form of a solution or powder is injected in a flowing arc plasma, and spectral analysis of the plasma makes it possible to determine the metal composition. The accuracy of this spectral determination of admixture concentration with respect to a primary component is of the order of 0.01–0.001 %. The optohalvanic method [65, 39] that is based on the connection between radiative and ionization properties of a gas discharge plasma allows one to detect small concentrations of gas and vapor admixtures up to 10^{-10}–10^{-9} g/g. In such an analysis gas discharge is burnt, and a laser signal is tuned to a resonant line of a given element that leads to variation of a gas discharge current due to subsequent ionization processes.

Plasma processing for environmental applications is developing in two directions. The first one is decomposition of toxic substances, explosive materials, and other hazardous wastes which are injected in an arc plasma and are decomposed in a plasma into simple chemical constituents. The second one is connected with an improvement of air quality by using corona discharge of a low power. This discharge generates active atomic particles, such as oxygen atoms. These atoms have an affinity for active chemical compounds in air and react with them. Such discharges also destroy microbes, but do not lead to hazards for humans because of low concentrations of these particles.

Plasma applications are widen in time and are included in new sides of the man activity. Plasma applications in medicine started several decades ago and are based on the mechanical or chemical action of a plasma on a living object. For example,

Fig. 1.3 Plasma knife (or plasma scalpel) for surgery [66]

a plasma knife that is a think plasma flux is used similar to a knife in surgery. An example of this instrument is represented in Fig. 1.3. As a matter, it is a flowing plasma which propagates through a nozzle or orifice in atmospheric or rareness air. Depending on tasks, different types of a gas discharge plasma may be used, and specific requirements relate to the wall material that must not be destroyed under the action of this plasma during a long time. In a simple construction, the basis of a plasma knife may be capillary discharge where plasma motion outside results from electron drift in an electric field.

The action of a plasma as an active media may be various because of existence of many plasma forms. In particular, destruction of medical waste including bandages, medicinal cotton and other disposable materials, with using an arc plasma proceeds similar to combustion of these materials in oven, but plasma set up is more compact and exclude danger products, though it is also more expensive. Nevertheless, plasma methods are more favorable in practice than combustion of medicine materials in oven. A not so power plasma intends for decontamination and sterilization of medicine tool. Because a plasma contains various active particles, as electrons, ions, metastable atoms and molecules, radicals, it is applied in wound healing, dermatology and dentistry where the plasma kills microbes and does not destroy a living tissue. Of course, a suitable plasma form and a certain current regime must be used for each case and follows from the corresponding study. In addition, plasma technology is used for production of specific medicine materials and devices.

The oldest applications of a plasma as a heat-transfer agent [36, 67] are in the welding or cutting of metals. Since the maximum temperature in chemical torches is about 3,000 K, they cannot be used for some materials. The arc discharge (electric arc) allows one to increase this temperature by a factor of three compared to chemical sources of energy, so that melting or evaporation of any material is possible by plasma methods. Therefore the electric discharge is used starting a century ago for welding and cutting of metals. Presently, plasma torches with power up to 10 MW are used for iron melting in cupolas, for scarp melting, for production of steel alloys, and for steel reheating in tundishes and landlies. Plasma processing is used for extraction of metals from ores. In some cases plasma methods compete with traditional ones which are based on chemical heating. One can conclude from comparison plasma and chemical methods that plasma methods provide a higher specific output, a higher quality of product, a smaller amount of waste, but require a larger energy expenditure and more expensive equipment. In particular, this relates to plasma chemistry [68–73] that allows one to produce various chemical compounds and some of them may be obtained on the basis of standard chemical methods. Especially plasmachemical processing is convenient for production fine chemical compounds including both inorganic materials and organic compounds. The latter includes the production of polymers and polymeric membranes, processes of fine organic synthesis in a cold plasma [70, 71, 74], etc. In a qualitative assessment of the technological applications of plasmas, we conclude that plasma technologies have a sound basis, and present promising prospects for important further improvements.

Returning to the theory of a gas discharge plasma, we note a lot of geometric constructions of gas discharges and configurations of electric and magnetic fields. Various gaseous components, different parameters of these components and external fields create various regimes of evolution of such a plasma and a variety of processes which determine the plasma properties and its evolution. In other words, if at a given construction of gas discharge such parameters as its power, the current strength, the pressure of a gas (and especially, its composition) vary, a lot of regimes of gas discharge is realized, and these regimes are determined by different processes in a gas discharge plasma. Therefore, though general principles of gas discharge are understood a century ago, new types of gas discharge for certain applications are created now. For this reason a universal description of a gas discharge plasma is cumbrous and is not practical. Being guided by certain plasma components and field configurations, one can restrict a group of process that is responsible for plasma properties and the character of its evolution. In this book we restrict ourselves by helium and argon as plasma components, by the cylinder and plane geometries of a gas discharge chamber, and also by constant electric and magnetic fields. Thus, the goal of this book is the modeling of a gas discharge plasma of helium and argon at a simple field configuration.

Let us formulate a general scheme for the model of a gas discharge plasma. Because of a non-equilibrium character of kinetics for electrons and ions in a gas discharge plasma [41, 75, 76], the model of a gas discharge plasma must be based on the concept of the distribution function of electrons and ions. In other words, hydrodynamic and thermodynamic description of a gas discharge plasma has a qualitative character. In addition, it is necessary to take into account a real dependence of the cross sections on the collision energy, in particular, the Ramsauer effect in the cross section of elastic electron-atom scattering. In this case the minimal cross section is two orders of magnitude less than the cross section at zero electron energy. It is clear that the tau-approximation that is used often in the analysis of kinetics of a gas discharge plasma and simplifies this analysis, is not applicable in this and other case. From this it follows that detailed information about the dominant processes is required for the kinetic analysis of a gas discharge plasma, and this information is analyzed in this book for processes in a helium and argon plasma.

Thus, violation the equilibrium in the gas discharge plasma under consideration is of importance. In particular, an electron energy changes weakly in single collisions with atoms and ions because of a large mass difference. Hence, an average electron energy is enough to ionize gas atoms at moderate electric field strengths, while an average ion energy is close to a thermal energy of gas atoms. Another example relates to propagation of a plasma in a gas on its boundary, in particular, near the walls of a chamber where this plasma is located. Then the rate of plasma motion is determined by electrons as a light plasma component, and along with diffusion of these electrons in a gas their displacement is determined by ion drift under the action of an electric field that is created by electrons which move away from ions and pull them. If the electron average energy exceeds the ion energy significantly, this field gives the main contribution to the rate of plasma propagation to its boundary. These examples exhibit the specifics of a gas discharge plasma as a non-equilibrium object.

Evidently, methods of computer simulation may be used for the analysis of a certain gas discharge plasma. In contrast to an universal analysis which has a descriptive character like to [18], computer simulation allows one in principle to describe a certain gas discharge plasma. For example, computer simulation of an argon gas discharge plasma is fulfilled in [77–81] with taking into account a large number of excited states. In particular, in the Vlacek model [82] that is the basis of evaluations [78–80] in which 64 levels are included. But in spite of the affinity of these evaluations, they are not reliable, i.e. their results do not correspond to real objects. Indeed, the authors are focused on computer aspects of the problem and cannot consider correctly peculiarity of processes which accompany a certain problem. We give one of reasons of the result invalidation. The atom excitation process by electron impact in a gas discharge plasma is the consistent process [44, 45, 83], i.e. the electron distribution function above the excitation threshold decreases sharply with an increasing electron energy, and this in turn influences on the process rate. This fact is not taken into account in the above papers [77–81]. These troubles are determined by the complexity of the problem where a certain nonuniform gas discharge plasma is an object of each paper. This means that though just computer simulation is the prospective method for the analysis of a nonuniform gas discharge plasma, the contemporary computer analysis of a certain gas discharge plasma is not reliable because of ignoring the peculiarities of processes in this plasma in this analysis.

We keep another position compared to computer specialists who use universal computer codes without a careful analysis of the processes and kinetics of gas discharge plasma. It is convenient to divide a general problem in some blocks, and each block solves a certain problem. In particular, in the cases of a helium and argon gas discharge plasma under consideration the first block includes the kinetics of a uniform gas discharge plasma with elastic and inelastic processes electron-atom collisions, electron-electron collisions and radiative processes. The second block relates to a nonuniform plasma and considers the kinetics, transport phenomena in this plasma and its ionization balance. Note that the first block may be identical for different problems of a gas discharge plasma if the same processes are its basis. This approach requires simple computer codes, but allows one to obtain reliable results. We also tend the accuracy of this approach to be in accordance with the accuracy of information used.

Practically, this book demonstrates the indicated approach based on the analysis of physics of a gas discharge plasma. Moreover, the most part of the book is devoted to the analysis of processes, kinetics and properties of a uniform gas discharge plasma, rather than computer simulation of problems of nonuniform gas discharge plasmas. This exhibits that the understanding of the problem and its details with treatment of information is of importance for solution of a certain problem. Moreover, computer codes may be simplified significantly compared to universal computer codes if the physics of a certain problem is clear. On contrary, replacing of the problem understanding by universal and complex computer codes makes the result to be unreliable.

This book consists of four parts, so that first three parts contain a description of processes in a gas discharge plasma and the analysis of properties of simple gas discharges. The fourth chapter is devoted to various aspects of gas discharge plasmas of helium an argon. General properties of a gas discharge plasma and processes in this plasma, mainly elementary ones, are analyzed in the first book part. Kinetics of a gas discharge plasma and transport phenomena in this plasma are the content of the second book part. The third part of the book is a description of elements of simple gas discharges. The four part is devoted to a uniform helium and argon plasma, and properties of such a plasma in the positive column, in the cathode layer and near the walls of a gas discharge chamber is considered in the fifth part of the book. Believing that properties of certain gas discharge plasmas and processes which are responsible for regimes of this plasma and its evolution, we give in a preface a list of books where physics of some gas discharges is analyzed.

Part I
Processes in Gas Discharge Plasma

Chapter 2
Properties of Gas Discharge Plasma

Abstract The distribution function of plasma particles as a function of coordinates and energies (velocities) is a basis for plasma description as a nonequilibrium system. The Debye-Hückel radius and plasma frequency are basic plasma parameters. Transport phenomena in a weakly ionized nonuniform plasma and ionization equilibrium through the processes of formation and decay of charged particles in a plasma are analyzed. Some parameters are represented which are of interest for a helium and argon gas discharge plasma.

2.1 Equilibria and Distributions of Particles in Gases and Plasmas

Processes of collisions of atomic particles in gases and plasmas as the behavior of these particles in external fields lead to certain distributions of particles. We consider below the simplest distribution of atomic particles in gases and plasmas. As a result of elastic collisions of atoms (or molecules) in gases

$$A + \widetilde{A} \to A + \widetilde{A}$$

the Maxwell distribution [84] of atoms on velocities is established. Then the velocity distribution function $f(v_x)$ of atoms is given by

$$f(v_x) = N\left(\frac{T}{2\pi m}\right)^{1/2} \exp\left(-\frac{mv_x^2}{2T}\right), \qquad (2.1.1)$$

where v_x is the velocity projection onto the x-axis, N is the number density of atoms, m is the atom mass, T is the gas temperature that is expressed through the book in energetic units. Here $f(v_x)dv_x$ is the number of atoms per unit volume which have the velocity projection between v_x and $v_x + dv_x$. This means that the distribution function is normalized as

B.M. Smirnov, *Theory of Gas Discharge Plasma*, Springer Series on Atomic, Optical, and Plasma Physics 84, DOI 10.1007/978-3-319-11065-3_2

$$\int_{-\infty}^{\infty} f(v_x) dv_x = N$$

In the case of an isotropic distribution on velocities v, the Maxwell distribution function has the form

$$f(v) = N \left(\frac{m}{2\pi T}\right)^{3/2} \exp\left(-\frac{mv^2}{2T}\right), \qquad (2.1.2)$$

for the following character of normalization

$$\int_0^{\infty} f(v) \cdot 4\pi v^2 dv = N$$

The Maxwell distribution function over the atom kinetic energies $\varepsilon = mv^2/2$ has the form

$$f(\varepsilon) = \frac{2N}{(\pi)^{1/2} T^{3/2}} \exp\left(-\frac{\varepsilon}{T}\right), \qquad (2.1.3)$$

and is normalized by the condition

$$\int_0^{\infty} f(\varepsilon) \cdot \varepsilon^{1/2} d\varepsilon = N$$

If the equilibrium of atomic particle is established in a weak external field, the space distribution function of atomic particles is determined by the Boltzmann distribution

$$N(\mathbf{r}) = N_o \exp\left[-\frac{U(\mathbf{r})}{T}\right], \qquad (2.1.4)$$

where $U(\mathbf{r})$ is the particle interaction potential with an external field, N_o is the particle number density at point where $U = 0$. The Boltzmann distribution may be used for distribution on states. If N_o is the number density of atomic particles in a given (ground) state, and the statistical weight (a number of degenerate states) for this state is g_o, the number density of atomic particles in a state i is given by

$$\frac{N_i}{N_o} = \frac{g_i}{g_o} \exp\left(-\frac{\Delta\varepsilon_i}{T}\right), \qquad (2.1.5)$$

where g_i is the statistical weight of ith state, $\Delta\varepsilon_i$ is the excitation energy of this state, and T is the temperature of particles which establish this distribution. In particular, if o, i are the ground and excited atom states, and atom excitation and quenching result from collisions with electrons according to the scheme

$$e + A \leftrightarrow e + A^i, \tag{2.1.6}$$

T in formula (2.1.5) means the electron temperature. In this case the equilibrium for the energy distribution of electrons is established in elastic collisions between electrons, and the energy change in electron-electron collision proceeds more effective than that in electron-atom collisions.

Let us consider an ionization equilibrium in a weakly ionized gas that is established as a result of atom ionization by electron impact and three-body electron-ion recombination according to the scheme

$$e + A \leftrightarrow 2e + A^+ \tag{2.1.7}$$

Then the relation between the number density of electrons N_e, ions N_i and atoms N_a is given by Saha formula [85]

$$\frac{N_e N_i}{N_a} = \frac{g_e g_i}{g_a}\left(\frac{m_e T_e}{2\pi\hbar^2}\right)^{3/2} \exp\left(-\frac{J}{T_e}\right) \tag{2.1.8}$$

Here g_e, g_i, g_a are statistical weights of electron, ion and atom correspondingly, m_e is the electron mass, T_e is the electron temperature, J is the atom ionization potential.

Let us apply the Saha formula (2.1.8) for the helium case, where it has the form

$$\frac{N_e^2}{N_a} = 3.6 \times 10^{22} T_e^{3/2} \exp\left(-\frac{24.59}{T_e}\right), \tag{2.1.9}$$

where the electron temperature T_e is expressed in eV, and the number density is given in cm^{-3}. Transferring to the electron concentration $c_e = N_e/N_a$, we rewrite this equation for the gas temperature $T = 300\,\text{K}$ as

$$c_e = \frac{1.2 \times 10^3 T_e^{3/4}}{p^{1/2}} \exp\left(-\frac{12.29}{T_e}\right), \tag{2.1.10}$$

where the gas pressure p is expressed in Torr. According to this formula total ionization of equilibrium helium at the helium pressure 10–100 Torr proceeds at the electron temperatures 2–3 eV. In reality, the degree of ionization of a plasma of glow gas discharge in helium with the electron temperature (or typical electron energy) 2–4 eV is less by several orders of magnitude.

The Saha distribution (2.1.8) is an example of equilibrium between the bound electron state and states of continuous spectrum, and the bound state respects to the ground atom state. In principle an atom can be found in excited states, but the number density of excited atoms is small compared to that of atoms in the ground state. But for the equilibrium between atomic and molecular ions according to the scheme

$$A + A^+ \leftrightarrow A_2^+ \tag{2.1.11}$$

a molecular ion may be found in excited rotation and vibration states. For this equilibrium the Saha formula has the form [86]

$$\frac{N_a N_{ia}}{N_{im}} = \frac{g_a g_i}{g_{mol}} \left(\frac{\mu T}{2\pi\hbar^2}\right)^{3/2} \exp\left(-\frac{D}{T}\right); \; g_{mol} = \left(\frac{1}{2}\right) \frac{T}{B} \left[1 - \exp\left(-\frac{\hbar\omega}{T}\right)\right]^{-1}, \tag{2.1.12}$$

where N_{ia}, N_{im} are the number densities of atomic and molecular ions correspondingly, T is the gas temperature, $\mu = M/2$ is the reduced ion-atom mass (M is the atom mass), D is the dissociation energy of the molecular ion, $\hbar\omega$ is the vibration energy of the molecular ion, B is the rotation constant for the molecular ion, and the factor $(\frac{1}{2})$ equals to one when nuclei are different isotopes, and is $1/2$ if nuclei are identical isotopes because under the condition $T \gg B$ one half of states is forbidden due to the state symmetry.

Let us use the dissociation equilibrium (2.1.12) for dissociation equilibrium between atomic and molecular argon ions in argon in accordance with the scheme

$$Ar^+ + Ar \leftrightarrow Ar_2^{\cdot} \tag{2.1.13}$$

where the Saha relation (2.1.12) between the number densities of atomic $N(Ar^+)$ and molecular $N(Ar_2^+)$ ions leads to the following relation for the equilibrium constant of dissociation equilibrium for argon ions

$$\chi_{dis} \equiv \frac{N(Ar^+)N_a}{N(Ar_2^+)} = \frac{g_a g_i}{g_{mol}} \left(\frac{\mu T}{2\pi\hbar^2}\right)^{3/2} \frac{B}{T} \left[1 - \exp\left(-\frac{\hbar\omega}{T}\right)\right] \exp\left(-\frac{D}{T}\right), \tag{2.1.14}$$

where χ_{dis} is the equilibrium constant for dissociation of molecular ions, g_a, g_i, g_{mol} are the statistical weights for electron states of the atom, atomic and molecular ions (for the ground electron states we have $g_a = 1, g_i = 6, g_{mol} = 2$ in the argon case, if a molecular ion consisting of nuclei-different isotopes), $\mu = M/2$ is the reduced mass of the atom and atomic ion, D is the dissociation energy of the molecular ion, $\hbar\omega$ is the energy of vibration excitation, and B is the rotation constant. We assume here the classical character of rotation degrees of freedom $T \gg B$. In considering the argon case, we use the following parameters of the molecular argon ion $Ar_2^+(^2\Sigma_g^+)$ formed from the argon atom Ar and ion Ar^+ in the ground states [87]: the dissociation energy of the molecular ion is $D = 1.23\,\text{eV}$, the vibration excitation energy is $\hbar\omega = 308.9\,\text{cm}^{-1}$, the rotation constant is $B = 0.143\,\text{cm}^{-1}$.

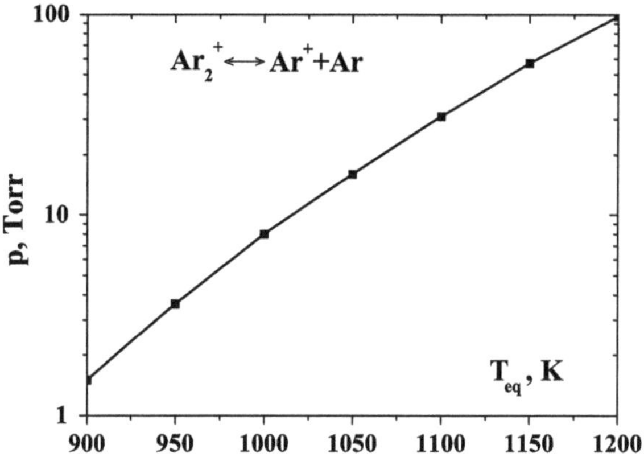

Fig. 2.1 The equilibrium argon temperature T_{eq} according to formula (2.1.15) where the number density of argon atomic and molecular ions are equal. The argon pressure is expressed in Torr

We below determine the equilibrium temperature T_{eq} of equality for the number density of atomic and molecular ions, i.e.

$$N(Ar^+, T_{eq}) = N(Ar_2^+, T_{eq}) \tag{2.1.15}$$

Figure 2.1 gives the dependence of the equilibrium temperature T_{eq} for ions on the argon pressure $p = N_a T_{eq}$.

2.2 Basic Plasma Parameters

Let us consider general plasma properties which are determined by a long-range Coulomb interaction between charged plasma particles (electrons and ions). Since this interaction remains in presence of neutral particles, these properties become apparent also in a weakly ionized gas. If a weak electric field is introduced in a plasma, it causes displacement of electrons and ions, that leads in the end to screening of this field. Correspondingly, the electric field strength E on a distance x from the plasma boundary is given by

$$E = E_o \exp\left(-\frac{x}{r_D}\right), \tag{2.2.1}$$

where E_o is the electric field strength on the plasma boundary, and r_D is the Debye-Hückel radius. In the same manner the interaction potential of two particles of a charge e at a distance r between them equals to

$$U(r) = \frac{e^2}{r} \exp\left(-\frac{r}{r_D}\right), \qquad (2.2.2)$$

the Debye-Hückel radius is given by formula [88]

$$r_D = \left[4\pi N_e e^2 \left(\frac{1}{T_e} + \frac{1}{T_i}\right)\right]^{-1/2} \qquad (2.2.3)$$

Here T_e is the electron temperature, T_i is the ion temperature, N_e is the number density of electrons or ions (a plasma is quasineutral). In the case of an equilibrium plasma $T_e = T_i = T$ this formula takes the form

$$r_D = \sqrt{\frac{T}{8\pi N_e e^2}}, \qquad (2.2.4)$$

$T_e = T_i = T$. As is seen, the Debye-Hückel radius r_D is an important plasma parameter.

Take a plasma that occupies an infinite space and displace plasma electrons on some distance. Then an electric field arises that tends to return the electrons in the initial positions. Because of inertia of electrons, their returning to the initial positions has the vibration character, and the oscillation frequency ω_p, the frequency of plasma oscillations, is equal to [47, 89, 90]

$$\omega_p = \sqrt{\frac{4\pi N_e e^2}{m_e}}, \qquad (2.2.5)$$

where m_e is the electron mass. The reciprocal value $1/\omega_p$ is a typical time of plasma reaction on the action of external fields.

The above parameters, the Debye-Hückel radius r_D and plasma frequency ω_p are basic parameters which characterize collective plasma properties. The presence of atoms and molecules in a plasma does not change collective plasma properties which are determined by Coulomb interaction of charged particles because neutral particles does not influence on this interaction. Note that the ratio of the above parameters r_D/ω_p equals to the thermal velocity of electrons, i.e.

$$\frac{r_D}{\omega_p} = \sqrt{\frac{2T}{m_e}} \qquad (2.2.6)$$

Since at first energy is transferred from an external field to electrons, parameters of electrons characterize a gas discharge plasma in the first turn. Figures 2.2 and 2.3 contain electron parameters of various types of nonequilibrium and equilibrium plasmas.

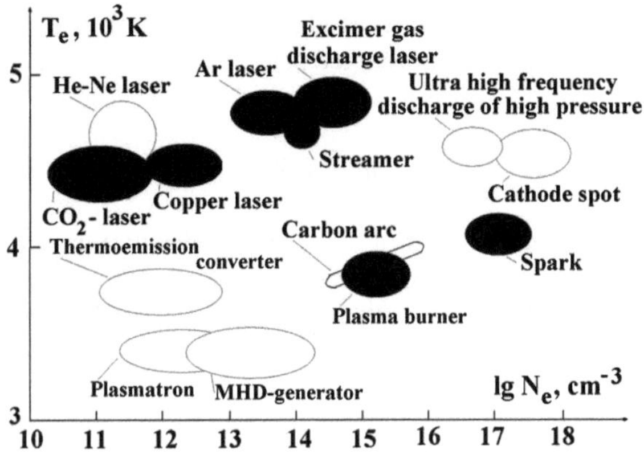

Fig. 2.2 Parameters of gas discharges plasmas of various types [91]

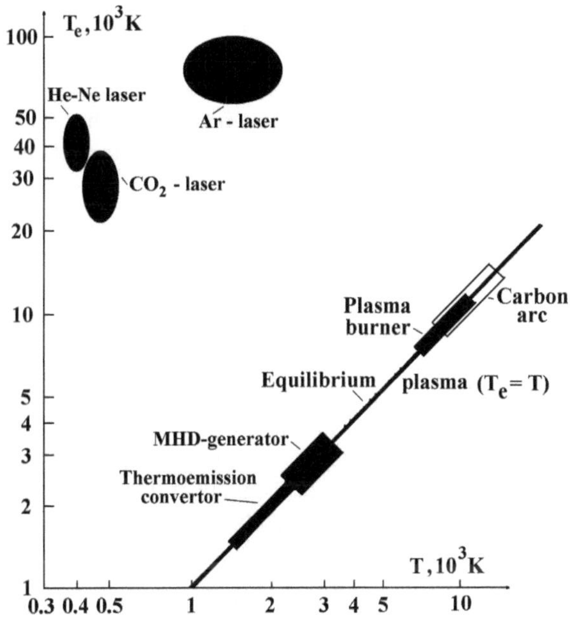

Fig. 2.3 The electron temperature of gas discharge plasmas via the gas temperature [83]

2.3 Transport Coefficients

Transport of a charged particle in a gas is characterized by the diffusion coefficient D and the mobility K are defined that the flux j of particles is given by

$$j = -D\nabla N + wN, \tag{2.3.1}$$

where N is the number density of charged particles, w is their drift velocity that is connected with the mobility K of charged particles at low electric field strengths E by the relation

$$w = EK \tag{2.3.2}$$

Let us consider an equilibrium of charged particles in a gas in a weak electric field and find from this the relation between transport coefficients. On the one hand, the space distribution of charged particles in a gas in a weak electric field is given by the Boltzmann formula (2.1.4) that has the form

$$N(x) = N_o \exp\left[-\frac{U(x)}{T}\right],$$

where x is the force direction, $U(x) = eEx$ is the interaction potential of a particle with an electric field, and E is the electric field strength. This gives the ratio between the number density of charged particles $N(x)$ and their gradient ∇N that has the form $\nabla N = -eEN/T$. Next, the total flux of charged particles (2.3.1) is zero. Substituting this expression for the gradient number density in (2.3.1) with accounting for (2.3.2), we find the following relation between transport coefficients of charged particles that is called the Einstein relation [92–94]

$$K = \frac{eD}{T} \tag{2.3.3}$$

Note that though this relation is called the Einstein relation, it was derived by Nernst [95] and Townsend [96, 97] several years before (see [98, 27]). Einstein used these results in the analysis of Brownian motion of particles [94, 99].

The criterion of validity of the Einstein relation relates to a weak field, so that the mean free path of a charged particle λ is small compared to a typical distance $l \sim N/\nabla N$ of a remarkable variation of the number density, and particle displacement on this distance results from many collisions of a test charged particles with gas neutral particles. Second, an external field is weak

$$eE\lambda \ll T \tag{2.3.4}$$

and does not change parameters of charged particles because their energy from an external field is small compared to a thermal energy.

2.4 Ionization Equilibrium in Gas Discharge Plasma

A gas discharge plasma is a self-consistent state of an ionized gas as a result of passing of an electric current through a gas under the action of external fields and is a wide type of a low-temperature plasma. Gas discharge consists of several parts with a specific destination. Usually gas discharge is burnt in a cylindrical tube where a gas under certain pressure is located, or between two parallel plates. These cases will be considered below. The basic elements of gas discharge are the electrodes, the cathode and anode, to which a voltage is applied, and a gas discharge tube which is restricted by walls and contains a gas. If gas discharge is maintained by high frequency fields including a laser field, gas discharge can have another construction. Nevertheless, below we will be guided by a stationary or quasi-equilibrium gas discharge plasma. Along with an electric field, a gas discharge plasma under consideration may be under the action of a magnetic field.

At variety of gas discharges and different regions of gas discharge with a certain destination we extract two basic regions of gas discharge, the cathode layer and positive column, that are of principle. Reproduction of electrons and ions proceeds in the cathode layer, i.e. this region is of importance for self-maintenance of gas discharge. In accordance with the processes in the cathode layer, the electric field strength drops at removal from the cathode, and a plasma of the cathode region is non-uniform. On contrary, a plasma of the positive column is uniform with respect to displacement along the discharge current. The ionization balance in the positive column is provided by n external uniform electric field. Hence a variation of the positive column length with conservation of the electric field strength inside it does not change parameters of the positive column plasma (Fig. 2.4).

Gas discharge is a widespread method to create and to apply a low temperature plasma. Therefore a variety of low temperature plasmas and their peculiarities are determined in a most degree by gas discharge plasmas. Figure 2.2 [91] contains the basic parameters of different types of gas discharge plasmas which include the number density of electrons and their temperature (or a typical electron energy). Because the energy is introduced in a gas discharge plasma through electrons, and a number of electrons in a typical gas discharge plasma is less significantly than the number of

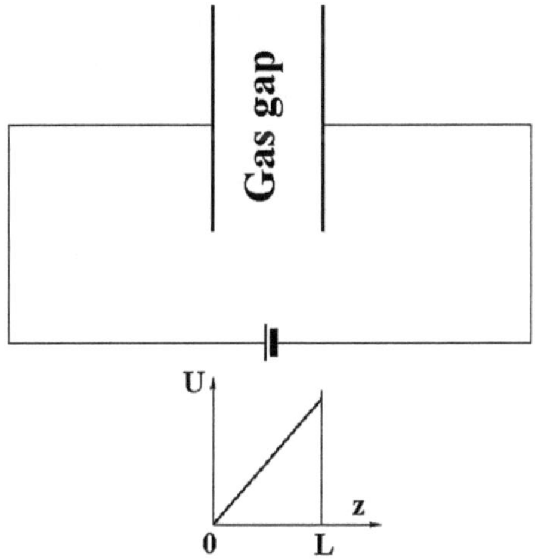

Fig. 2.4 Scheme of Townsend gas discharge in a gaseous gap between two parallel electrodes and the distribution of the electric voltage along this gap in absence of charged particles inside it

atoms, non-equilibrium conditions are realized often in a gas discharge plasma when the electron temperature exceeds the gaseous one. Figure 2.3 [91] compares the electron and gas temperatures for various plasma types. For an equilibrium gas discharge plasma the electron and gas temperatures are identical, while for a non-equilibrium plasma they are different.

A gas discharge plasma under consideration is formed in a gas located in a certain space where an electric current propagates through the gas under the action of an external electric field that means existence of gas discharge. We have many forms and regimes of gas discharge depending on a type and parameters of a gas, a geometry of a chamber where this gas located as well as positions of electrodes, and by parameters of external fields. We below restrict ourselves by a small part of these regimes using helium or argon as a working gas for gas discharge which is located in a gap between two large parallel electrodes or in a cylinder tube. In addition, an electric voltage between electrodes does not vary in time, i.e. this is d.c. (direct current) gas discharge. But even in this case we have several regimes of gas discharge which will be considered below.

For simplicity, we consider first gas discharge in a gaseous gap between two parallel electrodes. One can see that an ionization equilibrium in this gas discharge plasma is required for maintaining of an electric current between electrodes, and this equilibrium is determined by processes involving the gas discharge plasma. In the simplest and real case we include in consideration two processes following to Townsend [42]. The first process is ionization in a space in collisions of atoms with electrons which are accelerated by an external electric field. This process is characterized by the first Townsend coefficient α that is a number of forming electrons per unit length of a test electron. The second process in this ionization equilibrium is described by the second Townsend coefficient γ that is the probability to form a secondary electron at the cathode as a result of its bombardment by a plasma ion. The ionization equilibrium in this case means reproduction of electrons and ions attached to electrodes as a result of gas ionization in a space. We have two regimes of self-consistent gas discharge for this ionization equilibrium. Townsend gas discharge in a gaseous gap between two parallel electrodes is realized at low electric current of gas discharge if a small charge density does not influence on the voltage distribution inside the gaseous gap, as it is shown in Fig. 2.5.

At middle discharge currents it is favorable another form of gas discharge where two principal regions exist inside a gap, as it is shown in Fig. 2.5. The region near cathode, or the cathode region, with a heightened electric field strength is responsible for reproduction of electrons and ions, and in the other region, the positive column, the electric field strength is small. Such form of the voltage distribution is favorable because it leads to a lower voltage between electrodes compared with Townsend gas discharge. This form of gas discharge is glow gas discharge. Of course, the transition from Townsend to glow discharge may be more complicate because charge distribution in glow discharge may be not uniform not only along an electric field, but also in transverse directions. In this way the optimal number density of electrons and ions is chosen with respect to the total voltage between electrodes. In addition, there are intermediate regions between the cathode region and positive column and

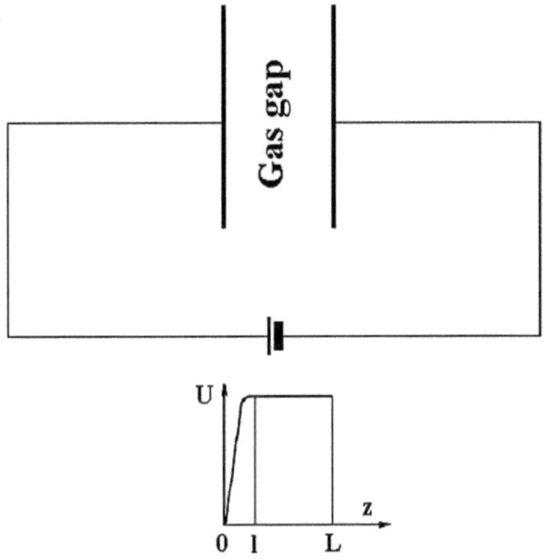

Fig. 2.5 Glow gas discharge in a gaseous gap between two parallel electrodes and the distribution of the electric voltage along this gap

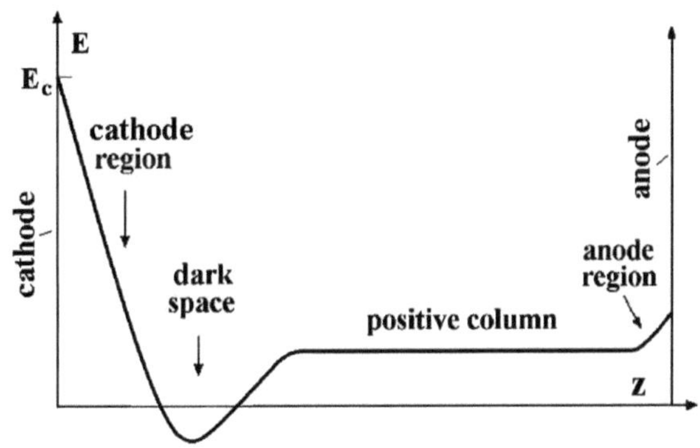

Fig. 2.6 Character of distribution of the electric field in glow gas discharge inside cylinder tube; E_c is the electric field strength at the cathode

near the anode in glow discharge. Figure 2.6 represent regions of glow discharge if it is burnt in a cylinder tube.

2.5 Thermoemission of Electrons from the Cathode

In glow discharge ions bombarded the cathode do not influence the cathode heat balance. At large currents the cathode is heated and electrons formed at the cathode as a result of the thermoemission mechanism. This form of gas discharge, arc, is characterized by a low cathode voltage because of an energy consumed for formation of one electron is close to the electron binding energy in the cathode material and exceeds it by a thermal energy. Note that transition from glow discharge to arc is determined by processes at the cathode, though the positive column of an arc plasma may have another behavior compared to that of the positive column of glow discharge.

Let us find the thermoemission current density from the cathode with use the analogy between bound of an electron in an atom and metal. Introduce the metal working function W that is the minimum binding energy of an electron in a metal and consider the ionization equilibrium for the metal

$$M \leftrightarrow e + M^+, \tag{2.5.1}$$

where m is a metal. Using the analogy of this ionization equilibrium with (2.1.7), we assume that liberation of one electron does not change a metal state, i.e. metal parameters are not varied after release of one electron. The latter gives in formula (2.1.8) $N_i = N_a$ and $g_i = g_a$. This leads to the equilibrium number density of electrons over the metal surface

$$N_e = \left(\frac{m_e T}{2\pi\hbar^2}\right)^{3/2} \exp\left(-\frac{W}{T}\right) \tag{2.5.2}$$

We assume that this equilibrium is established inside a metal for a small time, and then electrons may leave a metal surface. Under such assumptions the electron current density i from a metal surface of a temperature T is equal

$$i = \frac{1}{2} v_T e N_e,$$

where the first factor takes into account that electrons are directed outside the metal surface, $v_T = \sqrt{8T/(\pi m_e)}$ is the electron thermal energy. From this we obtain the Richardson-Dushman formula [100–104] for emission of the electron current density from a hot metal surface

$$i = A_R T^2 \exp\left(-\frac{W}{T}\right), \tag{2.5.3}$$

Table 2.1 Parameters of formula (2.5.3) for electron thermoemission from the metal surface [10, 16, 106] of the polycrystal structure; A_R is expressed in $A/(cm^2K^2)$

Metal	A_R	W(eV)	Metal	A_R	W(eV)	Metal	A_R	W(eV)
Be	300	3.8	*Zr*	330	4.1	Re	720	4.7
Ti	60	3.9	Nb	120	4.2	Os	1100	6.0
Cr	120	3.90	Mo	55	4.2	Ir	120	5.3
Feα	26	4.5	Pd	60	5.0	Pt	33	5.3
Feγ	1.5	4.2	Ag	60	4.6	Au	60	5.4
Co	41	4.4	*Hf*	22	3.6	*Th*	70	3.4
Ni	30	4.6	Ba	60	2.1	U	60	3.3
Cu	120	4.6	Ta	120	4.2	Pa	60	4.9
Y	100	3.3	W	60	4.5	–	–	–

where A_R is the Richardson parameter. If we use the above assumption that an outgoing electron does not interact with other metal electrons, on the basis of the above formulas we have for the Richardson parameter [105]

$$A_R = \frac{em_e}{4\pi^2\hbar^3} = 60A/(\text{cm}^2\text{K}^2) \tag{2.5.4}$$

Table 2.1 gives the values of the Richardson parameter for real metals.

2.6 Parameters of Helium and Argon Atoms and Ions

We below consider gas discharge plasmas on the basis of helium and argon. Hence, information is required about properties of these atoms and processes with their participation, and we represent below this information partially. Helium has two stable isotopes with the occurrence for ^3He of 1.4×10^{-4} % and almost 100 % for the isotope ^4He. The average atomic weight of helium is 4.003 a.u.m. or 6.647×10^{-24} g. The natural occurrence of helium in the Earth crust is 8×10^{-7}. The helium atom in the ground state ^1S contains two 1s-electrons. Lower excited states of the helium atom with the electron shell $1s2s$ are metastable. The excitation energy of the lowest excited state $He(2^3S)$ is 19.820 eV and for another metastable state $He(2^1S)$ it is 20.616 eV. The lowest resonantly excited state of the helium atom 2^1P with the excitation energy 21.218 eV corresponds to the wavelength $\lambda = 58.433$ nm for transition in the ground state; the oscillator strength for this transition is $f = 0.276$, and the radiative lifetime with respect to transition in the ground state is $\tau = 0.56$ ns.

The radiative transition from the lowest resonantly excited state $He(2^1P)$ in the metastable state $He(2^1S)$ proceeds with the wavelength $\lambda = 2058.1$ nm; the oscillator strength for this transition is $f = 0.38$, and the radiative lifetime with respect to transition in the ground state is $\tau = 5$ ns. The next excited state of the helium atom $He(2^3P)$ with the same electron shell $1s2p$ is characterized by the excitation energy of 20.964 eV; the dipole radiative transition in the metastable state $He(2^3S)$ proceeds with the wavelength of $\lambda = 1083.0$ nm. The oscillator strength $f = 0.539$ and the radiative lifetime $\tau = 0.98$ ns relates to this transition. Along with this, the fine splitting of levels, i.e. the energy difference for levels 2^3P_1-2^3P_0 is 0.076 cm^{-1}, and the energy difference for levels 2^3P_0-2^3P_1 is 0.988 cm^{-1}. Spectrum of the helium atom is given in Fig. 2.7, and spectrum for indicated states of the helium atom is represented in Fig. 2.8. The polarizability for the helium atom in the ground state is 1.383 a_o^3.

Stable isotopes of argon ^{40}Ar, ^{39}Ar and ^{37}Ar are characterized by the relative occurrence of 99.6 %, 0.34 % and 0.06 % correspondingly, and the average atomic weight of argon is 39.948 a.u.m. or 6.634×10^{-23} g. The natural occurrence of argon in the Earth crust is 1.2×10^{-4} %.

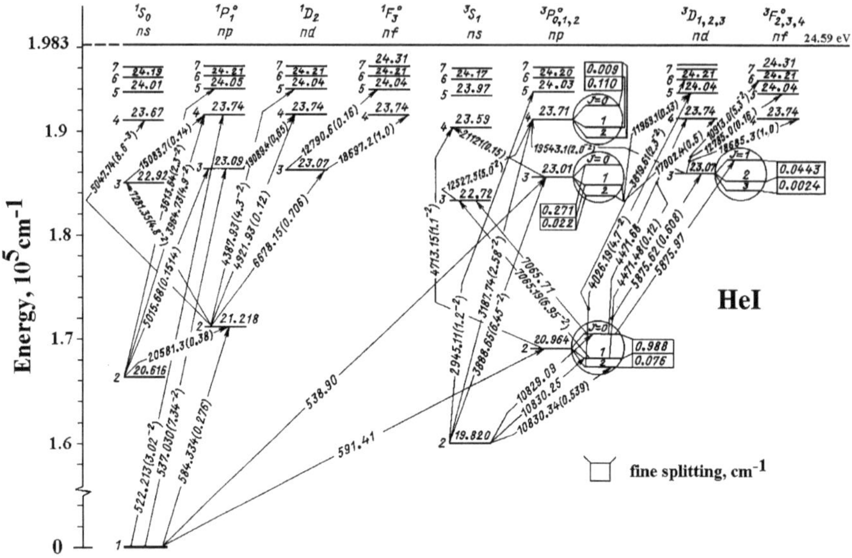

Fig. 2.7 The Grotrian diagram for the helium atom involving lower excited states [107]

Fig. 2.8 Expressed in eV energies of excitation for lower levels of the helium atom and radiative transitions from these states

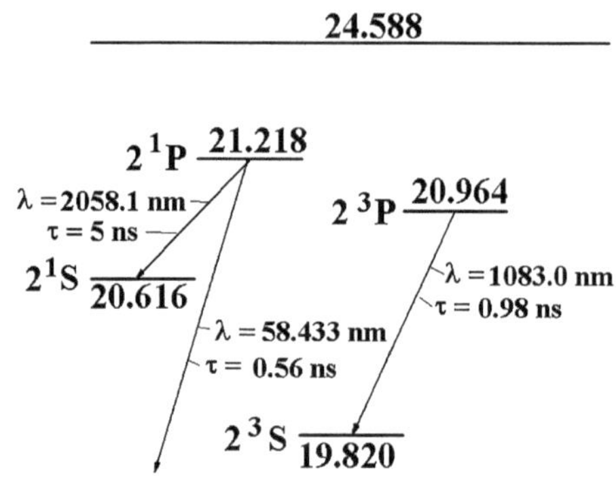

The argon atom in the ground state is characterized by the electron shell $1s^22s^22p^63s^23p^6$, and the ground state of the argon atom is ^1S. Notations for the lowest excited states of argon atom with the electron shell $3p^54s$ are given in Fig. 2.8 according to Pashen notations, and also within the framework of the LS and $j-j$ schemes of momentum summation in the atom. Note that according to notations within the framework of the jj scheme of momentum coupling the total electron momentum of the atom core is given inside the square parentheses, and the orbital momentum of a valence electron is given outside the square parentheses right above, and the multiplicity of the atom state is characterized by a low index outside the square parentheses. Figure 2.9 contains also the energies of levels for first excitations of atom

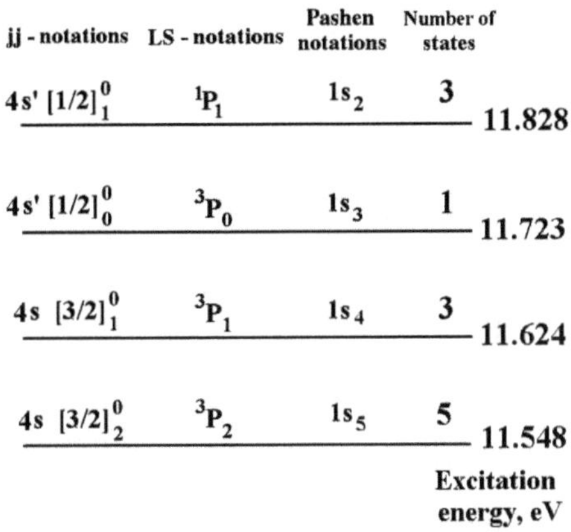

Fig. 2.9 Notations of lowest excited states of the argon atom with the electron shell $3p^5 4s$ and the excitation energies from these states

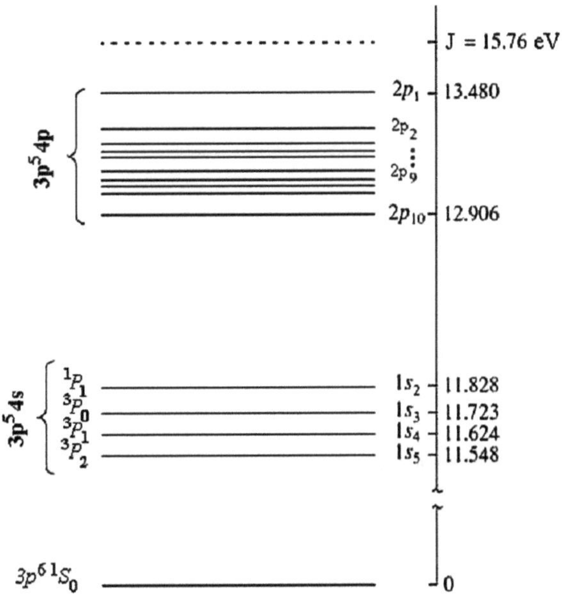

Fig. 2.10 Energies of argon atom levels of groups with the electron shells $3p^5 4s$ and $3p^5 4p$ in eV. Pashen notations of excited states are given *right*, and notations for LS-coupling of the lowest excited group of states are represented *left*

states with the electron shell $3p^5 4s$, and Fig. 2.10 includes the following group of levels with the electron shell $3p^5 4p$.

Note that the second and forth states of the group given in Fig. 2.8 are resonantly excited. The wavelengths for transition from these states in the ground one are $\lambda = 106.66$ nm and $\lambda = 104.82$ nm, the oscillator strengths for transitions in these states from the ground state are $f = 0.05$ and $f = 0.25$ respectively that correspond to $\tau = 10$ ns and $\tau = 2$ ns as the radiative lifetimes of these states with respect to radiative transitions in the ground state. Parameters of the lowest excited states of the

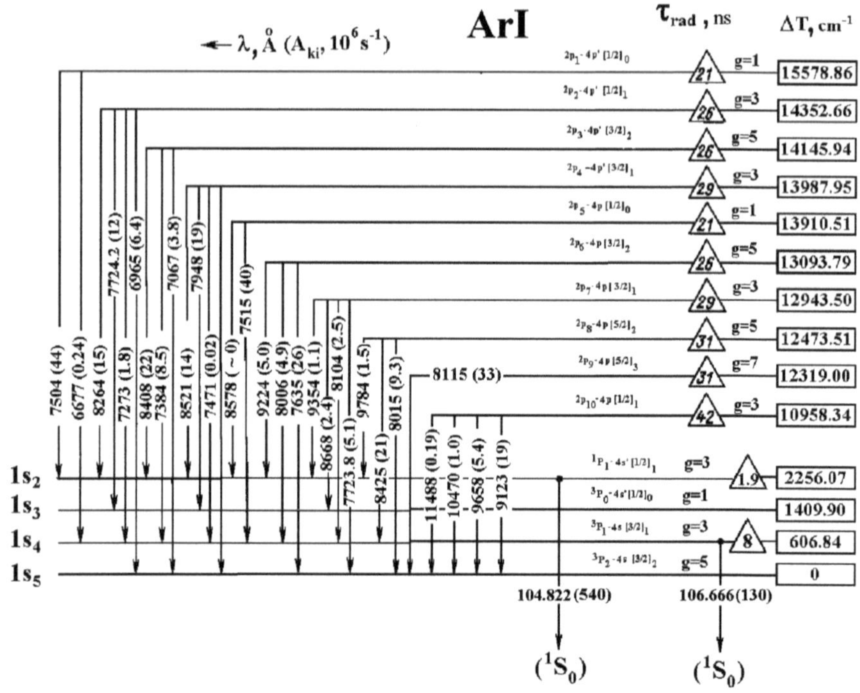

Fig. 2.11 Parameters of argon atom levels of groups with the electron shells $3p^5 4s$ and $3p^5 4p$ and radiative parameters of transitions between these states. Along with Pashen notations for excited states, the notations of LS-coupling and jj-coupling are used. Here ΔT is the excitation energy for a given state in cm^{-1} with respect to the lowest excited state $1s_5$, τ_r is the radiative lifetime of states in ns, the wavelengths of radiative transitions is expressed in Å, and the rate of radiative transitions A_{ik} is represented in $10^6 \mathrm{s}^{-1}$

Table 2.2 Parameters of molecular ions He_2^+ and Ar_2^+ in the ground electron state, where a molecular ion consists of an atom and atomic ion in the ground states

Molecular ion	He_2^+	Ar_2^+
D_e (eV)	2.47	1.23
$\hbar\omega_e$ (cm)$^{-1}$	1698	309
$\hbar\omega_e x_e$ (cm)$^{-1}$	35.3	1.66
B_e (cm)$^{-1}$	7.21	0.143

argon atom with the electron shells $3p^5 4s$ and $3p^5 4p$ are represented in Figs. 2.10 and 2.11 together with parameters of radiative transitions [108, 109] in the latter case. In addition, the polarizability of the argon atom in the ground state is $11.1\, a_o^3$.

The ionization potential of the helium atom is 24.588 eV, and for the helium ion the ionization potential equals to 54.418 eV. The ionization potential for the argon atom and its atoms are equal to 15.760 eV and 27.630 eV correspondingly. Both elements have stable molecular ions, and the electron state of the molecular ion is

$^2\Sigma_g^+$ within the framework of the LS scheme of momentum summation. Table 2.2 contains parameters of molecular ions in this state, so that D_e is the potential well depth for this electron state, $\hbar\omega_e$ is the vibration energy, $\hbar\omega_e x_e$ is the anharmonic constant, B_e is the rotation constant.

Chapter 3
Elementary Processes in Gas Discharge Plasma

Abstract Various elementary processes in a gas discharge plasma are analyzed. Collision electron-atom processes include elastic collisions, atom excitation by electron impact, quenching of excited atoms, atom ionization and recombination of electrons and ions, collision transitions between nearby atom levels, and ion-atom processes result from elastic collisions and resonant charge exchange. Appropriate cross sections and rate constants of the processes under consideration are used for the subsequent analysis of kinetics of a gas discharge plasma.

3.1 Elastic Collisions of Classical Atomic Particles

Being guided by an atomic gas discharge plasma, we below consider elementary processes resulted from collision of atomic particles in a gas discharge plasma. The characteristic of elementary act of collision of two atomic particles (atoms, molecules, ions, electrons) is the cross section of collision. According to definition (for example, [110]), the differential cross section for collision of two atomic particles $d\sigma$ is the ratio of a number of scattered particles per unit time per unit solid angle $d\Omega$ to the flux of incident particles in the center-of-mass frame of reference. The cross section is a basis for parameters of processes resulted from elastic scattering of particles. We demonstrate it on an example of braking of a fast particle in a gas of slow particles due to elastic scattering of these particles. Then the variation of the energy ε of a fast particle per unit time is equal to

$$\frac{d\varepsilon}{dt} = 2\varepsilon \frac{\mu}{M} N v \sigma^*(v) \tag{3.1.1}$$

where the velocity v of a fast test particle exceeds significantly that of gas atoms, $\mu = m_1 m_2/(m_1 + m_2)$ is the reduced mass of two colliding particles (m_1 is the mass of a fast particle, m_2 is the mass of a gas atom), $M = m_1 + m_2$ is the total mass of colliding particles, $\sigma^*(v)$ is the diffusion or transport cross section that is defined as

$$\sigma^*(v) = \int (1 - \cos\vartheta) d\sigma \tag{3.1.2}$$

B.M. Smirnov, *Theory of Gas Discharge Plasma*, Springer Series on Atomic, Optical, and Plasma Physics 84, DOI 10.1007/978-3-319-11065-3_3

The diffusion cross section of particle collision is the most wide used integral cross section of particles.

In the classical case the main contribution to the scattering cross section is determined by large collision momenta. This relates to thermal collisions of atoms and molecules in gases. In particular, let us introduce the gas-kinetic cross section of collision as the cross section in the expression of the diffusion coefficient of atoms in a gas under the assumption that this cross section is independent of the collision velocity. For example, the gas-kinetic cross section is equal to $37\,\text{Å}^2$ for collision of two argon atoms, and is equal approximately $13\,\text{Å}^2$ for collision of two helium atoms [87].

The cross section of elastic collision of two classical particles is connected unambiguously with the interaction potential of these particles. In particular, in the case ion-atom scattering when the interaction potential U is the polarization one [111] $U(R) = -\alpha e^2/(2R^4)$ at large distances R between colliding particles (α is the atom polarizability, e is the ion charge), the diffusion ion-atom cross section is equal to [112–115]

$$\sigma^*(v) = 2.2\pi\sqrt{\frac{\alpha e^2}{\mu v^2}}, \tag{3.1.3}$$

and this exceeds by 10 % the cross section of ion-atom polarization capture [110]. Here v is the relative ion-atom velocity, μ is their reduced mass. This cross section determines the ion mobility in a foreign gas.

In collisions of two classical charged particles with the Coulomb interaction potential between them (electron-electron or electron-ion collisions) the diffusion cross section is diverged due to large impact parameters of collision. Let us determine this cross section assuming the classical character of scattering and taking into account that small scattering angles give the main contribution to the cross section. Transferring to the center-of-mass frame of reference, we reduce the collision problem to scattering of one particle of the reduced mass μ on a force center, and the interaction potential between them at a distance R between colliding particles is e^2/R, where e is a charge of each colliding particle. Because of scattering on a small angles, i.e. particles are moving along straightforward trajectories, the momentum variation $\Delta\mathbf{p}$ for a reduced particle equals to

$$\Delta\mathbf{p} = \int_{-\infty}^{\infty} \mathbf{F}dt = \frac{2e^2\mathbf{n}}{\rho g}, \tag{3.1.4}$$

where $\mathbf{F}$ is the Coulomb force acted on each particle in the course of their collision, ρ is the impact parameter of collision, $\mathbf{n}$ is the unit vector directed along the impact parameter of collision, g is the relative velocity of colliding particles. Since the scattering angle in the center-of-mass frame of reference is

$$\vartheta = \frac{\Delta p}{p} = \frac{2e^2}{\rho \mu g^2},$$

we have from this for the differential cross section of particle scattering on a small angle [110]

$$d\sigma = 2\pi \rho d\rho = \frac{\pi e^4 d\Delta\varepsilon}{\varepsilon (\Delta\varepsilon)^2} \tag{3.1.5}$$

Here the energy variation in the center-of-mass frame of reference is equal to

$$\Delta\varepsilon = \frac{\Delta p^2}{2m} = \frac{2e^4}{\rho^2 \mu g^2} = \frac{e^4}{\rho^2 \varepsilon}$$

This gives for the diffusion cross section of scattering of two charged particles

$$\sigma^* = \int_0^\infty (1 - \cos\vartheta) \times 2\pi\rho d\rho = \int_0^\infty \vartheta^2 \pi \rho d\rho = \frac{4\pi e^4}{\mu^2 g^4} \int \frac{d\rho}{\rho}$$

As is seen, this integral is diverged at large and small scattering angles. One can eliminate the divergence t large scattering angles by refusing from the approach of weak interaction between colliding particles. The divergence at small angles is of principle and is determined by properties of a matter where the scattering takes place. We below assume the scattering to proceed in a plasma where interaction between charged particles is screened at distances of the order of the Debye-Hückel radius.

Thus we have formally for the diffusion cross section of elastic scattering of two charged particles

$$\sigma^* = \frac{4\pi e^4}{\mu^2 g^4} \ln \frac{\rho_{max}}{\rho_{min}},$$

Here the minimum impact parameter of collision ρ_{min} follows from the condition that the interaction energy of colliding particles is comparable to their kinetic energy $\rho_{min} \sim e^2/(\mu g^2)$, i.e. the assumption of scattering on small angles is violated. The maximum impact parameter of collision in an ideal plasma is of the order of the Debye-Hückel radius $\rho_{max} \sim r_D$. From this we obtain for the diffusion cross section of elastic scattering of two charged particles [116]

$$\sigma^* = \frac{4\pi e^4}{\mu^2 g^4} \ln\Lambda, \;\; \ln\Lambda = \ln \frac{e^2}{r_D \mu g^2} \tag{3.1.6}$$

where $\ln\Lambda$ is the Coulomb logarithm.

From this it follows that the diffusion cross sections of electron scattering on electrons and ions are different because of a different reduced mass of colliding particles. In particular, we have for the diffusion cross section of scattering of a fast electron with a velocity v (or the energy $\varepsilon = m_e v^2/2$) on an ion because the reduced mass is $\mu = m_e$, where m_e is the electron mass

$$\sigma_{ei}^* = \frac{\pi e^4}{\varepsilon^2} \ln \Lambda, \tag{3.1.7}$$

If a fast electron of energy ε collides with a slow electron, the diffusion cross section of elastic scattering equals to

$$\sigma_{ee}^* = \frac{4\pi e^4}{\varepsilon^2} \ln \Lambda, \tag{3.1.8}$$

Let us evaluate a typical value of the Coulomb logarithm (3.1.6) taking it for a gas discharge plasma of glow discharge for which a typical number density of charged particles is $N_e \sim 10^{12}\,\text{cm}^{-3}$, the electron temperature (an average electron energy) is $T_e \sim 1\,\text{eV}$, and the gas temperature $T = 400\,\text{K}$ coincides with the ion temperature. The Debye-Hückel radius of this plasma is determined by the ion temperature and is estimate as $r_D \sim 10^{-4}\,\text{cm}$, a distance of a strong interaction between charged particles is $\rho_{min} \sim e^2/T_e \sim 10\,\text{Å}$, and this gives for the Coulomb logarithm $\ln \Lambda = 7$. As is seen, in this example the values of ρ_{max} and ρ_{min} differ by three orders of magnitude that justifies the assumption used. Since the Coulomb logarithm value depends weakly on plasma parameters, we will use below the value $\ln \Lambda = 7$ in estimations.

3.2 Elastic Electron-Atom Scattering in Slow Collisions

The analysis of electron-atom collisions requires a quantum description, and the Scrödinger equation for the total electron-atom system is separated in spherical coordinates in the simplest case of a spherically symmetric effective electron-atom interaction potential. This means that electron-atom scattering proceeds independently for different electron momenta l, and electron-atom scattering is described within the framework of the phase theory [117], where the scattering amplitude $f(\vartheta)$ (ϑ is the scattering angle), and the total $\sigma_t = \int d\sigma$ cross section ($d\sigma$ is the differential scattering cross section) as well as the diffusion cross section $\sigma_t = \int(1-\cos\vartheta)d\sigma$ are expressed through partial scattering phases δ_l [111, 117, 118]. In the limit of low energies, only the zero-th scattering phase δ_0 is responsible for the scattering process that means that scattering of s-electron is determined the collision cross section. In this limit we have $\delta_0 = -Lq$, where q is the electron wave vector, and L is the electron scattering length. The cross sections of electron-atom elastic collision are equal at low electron energies

$$\sigma^*(0) = \sigma_t(0) = 4\pi L^2 \tag{3.2.1}$$

and the the electron scattering length is $L = 1.2a_o$ for $e - He$ scattering and $L = -1.6a_o$ for $e - Ar$ scattering [119, 120], where $a_o = \hbar^2/(m_e e^2)$ is the Bohr radius. We note that determination of the electron scattering length L on the basis of the Schrödinger equation is not reliable even with using contemporary computer methods. Therefore information about the scattering lengths follow from direct and indirect experiments.

Since electron-atom scattering is isotropic at zero energy, the electron scattering length is connected with the wave function Ψ of the scattered electron by the relation

$$\frac{d \ln \Psi}{dr} |_{r=0} = -\frac{1}{L}, \tag{3.2.2}$$

where r is a distance of the scattering electron from the atom center. Note that the scattering length is determined by electron-atom interaction inside the atom, where a one-electron approximation is not correct, i.e. the wave function of a scattering electron is entangled with the wave functions of atomic electrons, and a resultant exchange interaction has a complex form. Hence, we consider the scattering length to be a parameter which results from a combination of a short-range interaction including an exchange between scattered and atomic electrons, and a long-range interaction between them. Transferring to one-electron interaction, one can express the short-range electron-atom interaction potential through the Fermi formula [121–123]

$$U_{sh}(r) = -2\pi L \frac{\hbar^2}{m_e} \delta(\mathbf{r}) \tag{3.2.3}$$

where r is the electron coordinate. Along with a short-range electron-atom interaction, a long-range interaction may give a contribution to scattering parameters. In contrast to a short-range interaction, at low collision energies the contribution from a long-range interaction is determined by an electron region far from the atom where coordinates of scattered and atomic electrons may be separated.

Consideration a short-range electron-atom interaction in the form of the Fermi formula that includes both exchange and electrostatic interaction at small distances between the electron and atom, allows one to separate the short-range and long range electron-atom interactions, so that the effective electron-atom interaction potential $U(\mathbf{r})$ takes the form

$$U(\mathbf{r}) = U_{sh}(\mathbf{r}) + U_l(\mathbf{r}), \tag{3.2.4}$$

where $U_l(\mathbf{r})$ is the long-range part of the electron-atom interaction potential, and we use that the short-range and long-range parts are determined by different electron space regions. Note that a short-range interaction is enough strong and hence dependence of scattering parameters on the electron energy is determined by a long-range electron-atom interaction. Basing on this fact, we below find this dependence.

Assuming the long-range electron-atom interaction potential to spherically symmetric, one can construct by analogy with the Born approximation [111] the perturbation theory for the electron-atom scattering amplitude that includes both the short-range and the long-range interactions and has the form [124]

$$f(\vartheta) = -L + \frac{m_e}{2\pi\hbar^2}\int [1 - \exp(-i\,\mathbf{Kr})]U_l(r)d\mathbf{r}, \tag{3.2.5}$$

where $K = \left|\mathbf{q}\text{-}\mathbf{q}'\right| = 2q\sin(\vartheta/2)$ is the variation of the electron wave vector as a result of scattering, so that $\mathbf{q}$ and $\mathbf{q}'$ are the electron wave vectors before and after collision. Let us apply this formula to electron-atom scattering for the polarization long-range interaction potential $U(r) = -\alpha e^2/(2r^4)$ which is realized at large electron-atom distances r (α is the atom polarizability). We have

$$f(\vartheta) = -L - \frac{\pi\alpha}{4a_o}K = -L - \frac{\pi\alpha q}{2a_o}\sin\frac{\vartheta}{2}, \tag{3.2.6}$$

This is an equivalent to expansion of the scattering phases over a small parameter $x = -\pi\alpha q/(2La_o)$ at low collision energies [125, 126]. As a matter, this formula gives an expansion of the scattering amplitude on a small parameter that is proportional a low collision energy. In particular, we have from this for the diffusion cross section of electron-atom scattering at low electron energies in this approximation [127]

$$\sigma^*_{ea} = 4\pi\left(L^2 + \frac{4}{5}\pi\frac{\alpha qL}{a_o} + \frac{\pi^2}{6}\frac{\alpha^2 q^2}{a_o^2}\right) = 4\pi L^2\left(1 - \frac{8}{5}x + \frac{2}{3}x^2\right), \quad x = -\frac{\pi\alpha q}{2La_o} \tag{3.2.7}$$

The important conclusion resulting from these formulas consists in a sharp minimum for the cross sections of electron-atom scattering at small collision energies if the scattering length L is negative. This is called the Ramsauer effect [128, 129] which firstly was observed experimentally. A simple interpretation of a sharp minimum in the cross sections is that the zero-th phase δ_0 in electron-atom scattering becomes zero at low electron energies where the contribution of other phases to the cross section is relatively small. The Ramsauer effect is observed in elastic scattering of electrons on argon, krypton and xenon atoms where the electron scattering length is negative. Note that one can expect the Ramsauer effect for atoms with completed electron shell, because in other cases electron-atom scattering for different quantum numbers of the total electron-atom system proceeds independently, and the observed cross section is the combination of the cross sections for different quantum numbers of the total system. Next, as it follows from formula (3.2.7), the diffusion cross section σ^* in this approximation has the minimum $\sigma_{min} = 4\pi L^2/25$ at the electron wave vector $q_{\min} = -12La_o/(5\pi\alpha)$ ($x = 6/5$). Let us apply these formulas to the case of electron scattering on the argon atom where $L = -1.7a_o$, $\alpha = 11a_o^3$, and the Ramsauer effect is observed. We obtain on the basis of formula

Table 3.1 The diffusion cross section $\sigma^*(\varepsilon)$ for electron scattering on helium and argon atoms according to [130]

ε, eV	He	Ar	ε eV	He	Ar
0	4.95	10.0	0.8	6.79	1.05
0.001	4.96	8.35	1.0	6.87	1.38
0.002	4.99	7.80	1.5	6.98	2.07
0.005	5.14	6.61	2.0	6.97	2.70
0.01	5.27	5.60	2.5	6.95	3.37
0.02	5.38	4.15	3	6.93	4.10
0.04	5.58	2.50	4	6.8	6.00
0.06	5.63	1.60	5	6.6	7.60
0.08	5.71	1.00	6	6.3	9.3
0.10	5.80	0.59	7	5.9	11.0
0.15	6.00	0.23	8	5.5	14.0
0.20	6.20	0.10	10	5.0	14.6
0.25	6.26	0.091	15	3.4	14.1
0.30	6.32	0.15	20	2.69	9.5
0.35	6.38	0.24	30	1.60	6.0
0.40	6.44	0.33	40	1.00	4.7
0.50	6.55	0.51	50	0.70	3.5
0.60	6.66	0.68	75	0.36	2.3
0.70	6.74	0.86	100	0.22	1.7

The electron energy ε is expressed in eV, and the diffusion cross section is given in Å^2

(3.2.7) $q_{min} = 0.12/a_o$, $\varepsilon_{min} = 0.19\,\text{eV}$, $\sigma^*_{min} = 0.40\,\text{Å}$, $\sigma^*(0) = 10\,\text{Å}$. Thus, within the framework of this approximation, the scattering cross section drops by the order of magnitude at low electron energies that it is of importance for processes in gases or plasmas involving electrons.

We give in Table 3.1 the diffusion cross sections of electron scattering on helium and argon atoms [130] since helium and argon are the objects of this book. These cross sections are obtained in [130] from treatment of drift electron parameters in helium and argon. Figures 3.1 and 3.2 contain the diffusion cross sections $\sigma^*(\varepsilon)$ for electron scattering on helium and argon atoms which are the object of this consideration. We give also in Fig. 3.1 the following approximation for the electron-atom diffusion cross section in the helium case which correspond to a sum of experimental data [130] with the accuracy of 20–30 %

$$\sigma^*_{ea}(\varepsilon) = (6 \pm 1)\,\text{Å}^2, \ \varepsilon < 10\,\text{eV};$$
$$\sigma^*_{ea}(\varepsilon) \approx \frac{A}{\varepsilon}, \ A \approx 60\,\text{eV} \cdot \text{Å}^2, \ 10\,\text{eV} < \varepsilon < 40\,\text{eV} \tag{3.2.8}$$

In addition, Figs. 3.3 and 3.4 give the rate constants $k_{el} = v\sigma^*(\varepsilon)$ for elastic electron scattering on helium and argon atoms.

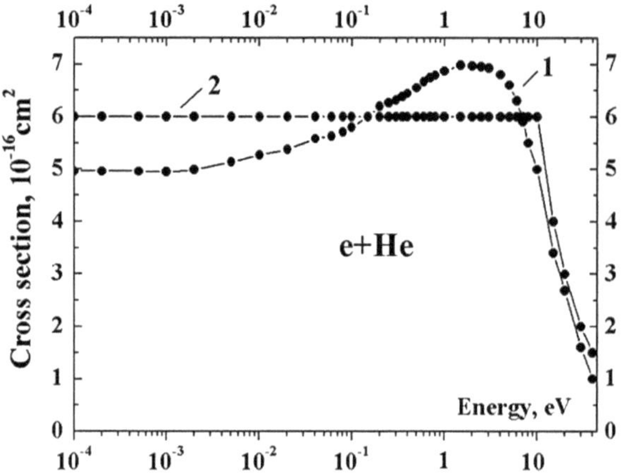

Fig. 3.1 The diffusion cross section of electron scattering on the helium atom. Stars correspond to results of various measurements [130], straightforward lines respect to approximation according to formula (3.2.8)

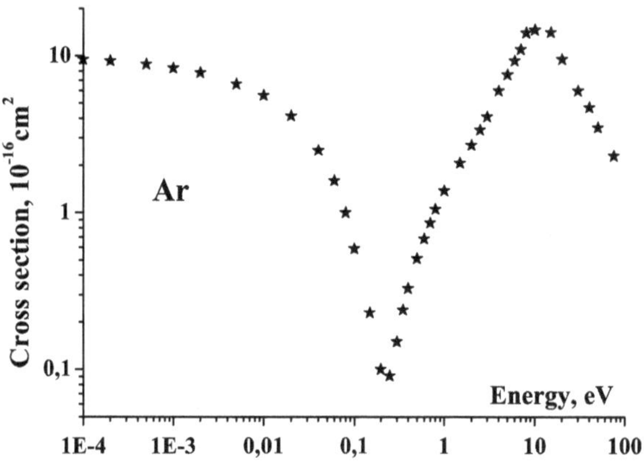

Fig. 3.2 The diffusion cross section of electron collision with the argon atom [130]

Comparing the behavior of the diffusion cross section of electron-atom scattering that follows from formula (3.2.7) and from the data of Table 3.1, we find that formula (3.2.7) transfers the character of this behavior at low electron energies, but the value of the minimum cross section for electron scattering on the argon atom differs approximately in 4 times compared to Table 3.1 data. Because the expansion of the cross section over a small parameter is valid at low electron energy, this difference leads to the contradiction. One can expect that this difference is that the polarization interaction potential is used at all the electron-atom distances, whereas this holds true at large distances. To escape this, we will follow to [131, 132] and will use the

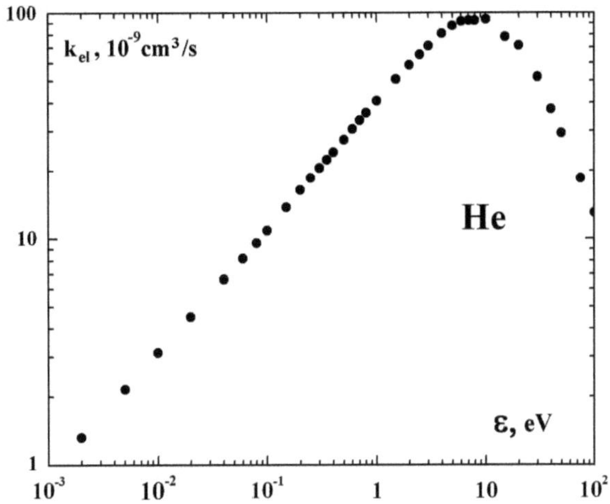

Fig. 3.3 The rate constant $k_{el} = v\sigma^*$ of electron elastic collision with the helium atom calculated on the basis of the Table 3.1 data

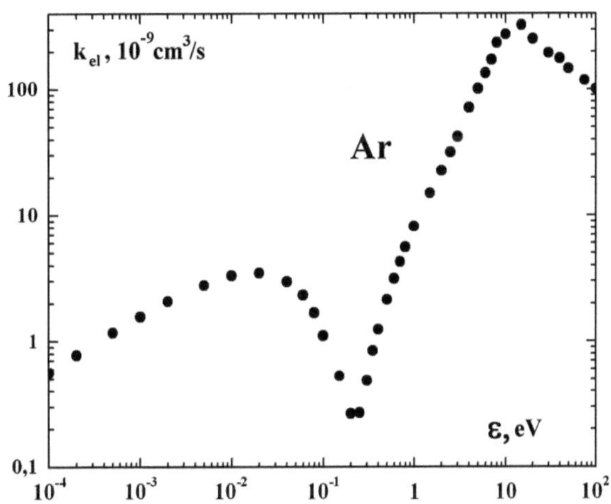

Fig. 3.4 The rate constant $k_{el} = v\sigma^*$ of electron elastic collision with the argon atom calculated on the basis of the Table 3.1 data

long-range interaction potential in the form

$$U(\mathbf{r}) = 2\pi L \frac{\hbar^2}{m_e}\delta(\mathbf{r}) - \frac{\alpha e^2}{2(r^2 + r_o^2)^2}, \qquad (3.2.9)$$

that is transformed into the polarization interaction potential at large electron-atom distances r and contains an additional parameter r_o compared to the polarization interaction potential. For this long-range interaction potential, we obtain for the scattering amplitude instead of formula (3.2.7) [132]

$$f(\vartheta) = -L - \frac{\pi\alpha}{4a_o r_o}\left[1 - \exp\left(-2r_o q \sin\frac{\vartheta}{2}\right)\right] \qquad (3.2.10)$$

Table 3.2 The parameters of the Ramsauer minimum for electron scattering on an argon atom [133]

r_o/a_o	$\varepsilon_{\min}$, eV	$\sigma_{\min}/4\pi L^2$	$x_{\min}$
0	0.35	0.04	1.20
0.1	0.36	0.039	1.22
0.3	0.39	0.038	1.26
0.7	0.44	0.035	1.34
1.0	0.50	0.032	1.42
1.5	0.61	0.028	1.58
2.0	0.79	0.023	1.79
2.5	1.11	0.017	2.12
3.0	1.83	0.011	2.73

From this we have for the reduced diffusion cross section

$$\frac{\sigma^*}{4\pi L^2} = \frac{1}{4(r_o q)^4} \int_0^{2r_o q} z^3 dz \left[1 - y\left(1 - e^{-z}\right)\right]^2, \quad y = -\frac{\pi\alpha}{4a_o r_o L} = \frac{x}{2r_o q}, \tag{3.2.11}$$

and $z = 2r_o q \sin(\vartheta/2)$. Introducing the electron wave vector $q_{\min}$ corresponding to the cross section minimum and a new variable $t = 2r_o q$, we find from the minimum condition $d\sigma^*/dt = 0$ the relation between the parameters x and t [133]

$$y = 2\frac{\frac{t^4}{4}\left(1 - e^{-t}\right) - \int_0^t z^3\left(1 - e^{-z}\right) dz}{\frac{t^4}{4}\left(1 - e^{-t}\right)^2 - \int_0^t z^3\left(1 - e^{-z}\right)^2 dz} \tag{3.2.12}$$

In particular, in the limit $r_o \to 0$ we have from this $t \to 0$, that gives $x = yt = 6/5$ and $\sigma^*/(4\pi L^2) = 1/25$ in accordance with formula (3.2.6) and (3.2.7) for the polarization interaction potential. The results for finite values of r_o are given in Table 3.2 where the parameters of the cross section minimum are given for electron scattering on a xenon atom. As it follows from this Table, an r_o increase leads to a decrease the minimum cross section and an increase of the electron energy $\varepsilon_{\min} = \hbar^2 q_{\min}^2/(2m_e)$ at which this minimum is observed.

3.3 Inelastic Electron Collisions with Atoms

In the case of inelastic collision processes, the rate constant $k = v\sigma(v)$ of an inelastic process is a convenient characteristic of this process along with the cross section $\sigma(v)$ of the transition. This is defined as

$$k_{ij}(v) = v\sigma_{ij}(v), \tag{3.3.1}$$

where v is the collision velocity, σ_{ij} is the cross section of transition between i and j states as a result of collision. If colliding particles are distributed over velocities in a plasma, in accordance with the velocity distribution function $f(v)$ it is convenient to use the average rate constants of collision transitions defined according to the relation

$$\overline{k_{ij}}(v) = \int f(v) k_{ij}(v) d\mathbf{v}, \tag{3.3.2}$$

where the distribution function is normalized here to unity $\int f(v) d\mathbf{v} = 1$. We consider below evolution of atom states if transitions between atom electron states result from collisions with electrons in a plasma. Then the balance equation for the number density of atoms N_i in a state i takes the following form if transitions result from collisions with electrons of the number density N_e

$$\frac{dN_i}{dt} = -N_i N_e \sum_j \overline{k_{ij}} + N_e \sum_j N_j \overline{k_{ij}} \tag{3.3.3}$$

Let us analyze peculiarities of inelastic electron-atom collisions. According to the principle of detailed balance, the partial rates of direct and inverse transitions are equal for an equilibrium electron distribution over energies. For the Maxwell energy distribution of electrons this principle gives

$$\overline{k_{ij}} = \frac{g_j}{g_i} \overline{k_{ji}} \times \exp\left(-\frac{\varepsilon_{ij}}{T_e}\right), \tag{3.3.4}$$

where for definiteness we consider a level i to lie below a level j, and ε_{ij} is the excitation energy; g_i, g_j are the statistical weights for lower and upper transition states.

According to the principle of detailed balance, the cross section of atom excitation $\sigma_{ex}(\varepsilon)$ in collision with an electron of energy ε is connected with the quenching cross section of this excited atom $\sigma_q(\varepsilon - \Delta\varepsilon)$ by electron impact in the following manner [134]

$$\sigma_{ex}(\varepsilon) = \frac{g_*}{g_o} \frac{(\varepsilon - \Delta\varepsilon)}{\varepsilon} \sigma_q(\varepsilon - \Delta\varepsilon) \tag{3.3.5}$$

Here ε is the energy of a fast electron, $\Delta\varepsilon$ is the excitation energy, g_*, g_o are the statistical weights of the atom in the excited and ground states; an argument of the cross section indicates an electron energy at which the cross section is taken.

The threshold law for the excitation cross section of an atom by electron impact has the form [135, 136]

$$\sigma_{ex}(\varepsilon) \sim \sqrt{\varepsilon - \Delta\varepsilon} \tag{3.3.6}$$

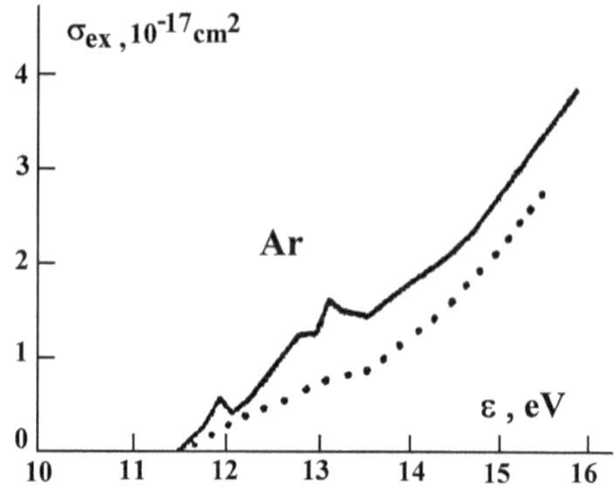

Fig. 3.5 The threshold excitation cross section of an argon atom by electron impact [137]

This simple form of the threshold excitation assumes that only one channel partake in the excitation process. In reality other excitation states may partake in this process if their excitation energy is below the electron energy and also excitation states with a higher excitation energies through resonances in electron-atom interaction. As a result, the excitation cross section has a not smooth dependence on the electron energy, as it follows from the example given in Fig. 3.5 where the excitation cross section of an argon atom by electron impact is represented near the threshold. From this threshold law it follows that the rate constant of quenching of an excited atom by electron impact k_q is independent of the electron energy in the limit of low electron energies. The principle of detailed balance gives for the rate constant k_{ex} of atom excitation by electron impact

$$k_{ex} = k_q \frac{g_*}{g_o} \sqrt{\frac{\varepsilon - \Delta\varepsilon}{\varepsilon}}, \tag{3.3.7}$$

where ε is the energy of an incident electron, g_o, g_* are the atom statistical weights for the ground and excited states.

If atoms are excited by electrons located in a plasma and have the Maxwell distribution over energies, an average of the excitation rate constant (3.3.7) over electron energies gives

$$k_{ex} = k_q \frac{g_*}{g_o} \exp\left(-\frac{\Delta\varepsilon}{T_e}\right), \tag{3.3.8}$$

where T_e is the temperature of electrons which establishes the equilibrium between the atom states as a result of collisions with them. Here we use that the quenching rate constant k_q is independent of an electron energy if the latter is relatively small. Therefore formula (3.3.8) holds true if $T_e \ll \Delta\varepsilon$.

If an atom is found in a resonantly excited state, one can assume that the quenching process is determined by dipole interaction between an incident electron and atom, as it takes place for the radiative transition with the dipole atom interaction with an electromagnetic field. This allows us to connect the rate constant of quenching of a resonantly excited state by electron impact with parameters of the corresponding radiative transition. Let us use this analogy in excitation and quenching of resonantly excited states. We consider first the case where the electron velocity is large compared with a characteristic electron velocity of a bound electron, and the Born approximation [111, 118, 136] holds true. Just at large collision velocities one can construct the perturbation theory [138–140] which allows us to separate interaction of colliding atomic particles from their motion and internal degrees of freedom. The strongest interaction corresponds to transitions between resonantly bound states where the transition between an initial and final atom states are possible as a result of dipole radiation. For excitation of resonantly excited states this perturbation theory leads to the Bethe formula [111, 118, 136, 141] which may be represented in the form

$$\sigma_{0*} = \frac{4\pi}{\varepsilon a_o} |(D_x)_{0*}|^2 \, \Phi\left(\frac{\varepsilon}{\Delta\varepsilon}\right) = \sigma_* \Phi\left(\frac{\varepsilon}{\Delta\varepsilon}\right),$$
$$\sigma_* = \frac{2\pi e^4}{\Delta\varepsilon^2} f_{0*}, \quad \Phi(x) \to \frac{\ln C\sqrt{x}}{x}, \quad x \to \infty \tag{3.3.9}$$

Here indices 0, $*$ refer to certain initial and final states of the excitation process, ε is the electron energy, $\Delta\varepsilon$ is the excitation energy, e is the electron charge, a_o is the Bohr radius, $(D_x)_{0*}$ is the matrix element of the dipole atom moment projection between the transition states, C is a constant, and f_{0*} is the oscillator strength for this transition which being averaged over states of the initial state group 0 and being summarized over states of the upper group $*$, is given by

$$f_{0*} = \frac{2m_e\omega}{3e^2\hbar g_o} |(\mathbf{D})_{0*}|^2 \tag{3.3.10}$$

Here the matrix element square $(\mathbf{D})_{0*}|^2$ is summarized over states of the initial and final group. As a result, we express the excitation cross section of resonantly excited atom states by electron impact through parameters of radiative dipole transitions of this atom. This analogy follows from an analogy in the interaction operators for a fast charged particle with an atom and between an electromagnetic wave and the atom [118, 141–143]. Near the excitation threshold the function $\Phi(x)$ introduced in formula (3.3.9) is given by

$$\Phi(x) = a\sqrt{x-1}, \quad x \to 1, \tag{3.3.11}$$

and $a = 0.130 \pm 0.07$ [38].

3.4 Resonant Atom Transitions in Collisions with Electrons

In evaluation the kinetic parameters of a gas discharge plasma, we need in data for collision processes related for the most part with processes involving electrons.

But numerical calculations of the cross sections of electron-atom processes are not reliable, and hence experimental data are the basis of such information. Nevertheless, the theoretical analysis helps us to use experimental results, and we exhibits it for inelastic electron-atom collisions with transition between resonant levels, i.e. the dipole radiative transition is possible between these levels. In this case at large collision energies the cross section of this transition is given by formula (3.3.9), but low collision energies are of interest for a gas discharge plasma and our task is to continue formula (3.3.9) to low collision energies with using experimental data. We assume in this operation that by analogy with large energies the cross section of a resonant transition is determined by the dipole electron-atom interaction. In this case a general form of the atom excitation cross section in collision with an electron has the form (3.3.9). Then accounting for the threshold dependence (3.3.6) of the excitation cross section [111], we find for the cross section near the threshold for excitation of resonantly excited states [144–146]

$$\sigma_{0*}(\varepsilon) = \sigma_* \times \sqrt{\frac{\varepsilon}{\Delta\varepsilon} - 1}, \qquad (3.4.1)$$

where ε is the electron energy, $\Delta\varepsilon$ is the excitation energy, σ_* is a constant. Just this range of electron energies is of importance for a gas discharge plasma, and we assume here [144] that the dipole interaction remains the main contribution to the electron-atom interaction in slow collisions as well as at large collision energies. Comparison with experimental data allows us to find the numerical coefficient in formula (3.4.1) that is [38, 145]

$$a = 0.130 \pm 0.007 \qquad (3.4.2)$$

We below analyze this result.

At low electron-atom collision energies it is convenient to operate with the rate constant k_q of quenching of an excited state because this quantity is independent of the electron energy ε at small energies. Let us use the expression [147–149] for the average rate of the spontaneous dipole radiative transition $1/\tau_r$ from the group $*$ to the group 0 of atom states

$$\frac{1}{\tau_r} = \frac{4\omega^3 g_o}{3g_* \hbar c^3} |\langle 0|\mathbf{D}|*\rangle|^2 \qquad (3.4.3)$$

Then on the basis of the principle of detailed balance (3.3.7) we have on the basis of formulas (3.3.9) and (3.4.3) for the rate constant of quenching of a resonantly excited state by electron impact [38, 133]

$$k_q = const \frac{g_o f_{0*}}{g_*(\Delta\varepsilon)^{3/2}} = \frac{k_o}{(\Delta\varepsilon)^{7/2}\tau_r} \qquad (3.4.4)$$

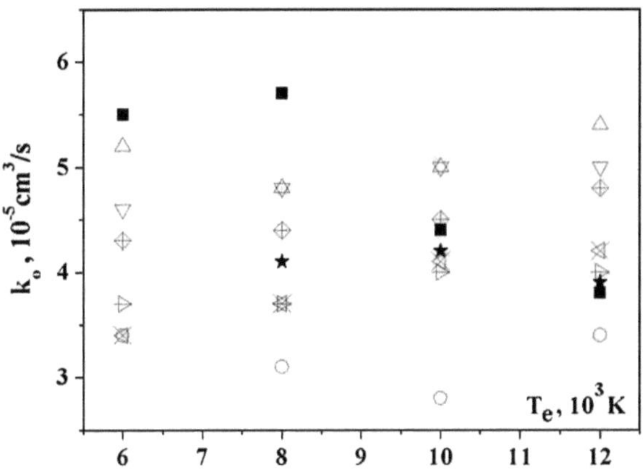

Fig. 3.6 The rate constant k_o obtained from formula (3.4.5) if k_{ex} is the experimental constant for excitation of resonant levels $K(4^2P)$, $Rb(5^2P)$, $Cs(6^2P)$ from the ground state by electron impact, if $\Delta\varepsilon$ is given in formula (3.4.5) in eV, and τ_r is taken in ns

As is seen, this formula contains the average rate of spontaneous emission from levels of the upper state group and is independent of statistical weights of states of these groups.

From the principle of detailed balance (3.3.5) we have for the rate constant of atom excitation by electron impact on the basis of formula (3.4.4)

$$k_{ex} = \frac{k_o g_*}{g_o(\Delta\varepsilon)^{7/2}\tau_r}\sqrt{\frac{\varepsilon - \Delta\varepsilon}{\varepsilon}} \qquad (3.4.5)$$

The value of the rate constant k_o in formulas (3.4.4) and (3.4.5) follows from comparison of the above excitation rate constant with experimental data. Figure 3.6 contains the results of treatment of experimental data [141, 150–152] for excitation of potassium, rubidium and cesium atoms by electron impact from the ground state to the states $K(4^2P)$, $Rb(5^2P)$, $Cs(6^2P)$. A statistical average of the data of Fig. 3.6 gives $k_o = (4.3 \pm 0.7) \times 10^{-5}\,\text{cm}^3/\text{s}$ for s–p electron transition if the atom excitation energy $\Delta\varepsilon$ is expressed in eV in formula (3.4.4), and the radiative lifetime is given in ns. The accuracy of this approximation according to data of Fig. 3.6 is estimated as 20–30 %.

If we deal with processes of atom excitation and quenching by electron impact in a gas discharge plasma with excited states in the initial state, one can remove from the threshold range because of a not large excitation energy for excited state. Hence we return to a general formula (3.3.9) for the atom excitation cross section by electron impact and will be based on experimental data. Figure 3.7 gives the reduced cross section (3.3.9) with using experimental data for atoms with a simple structure of excited state, and Table 3.3 gives the parameters of resonant excitation of these atoms.

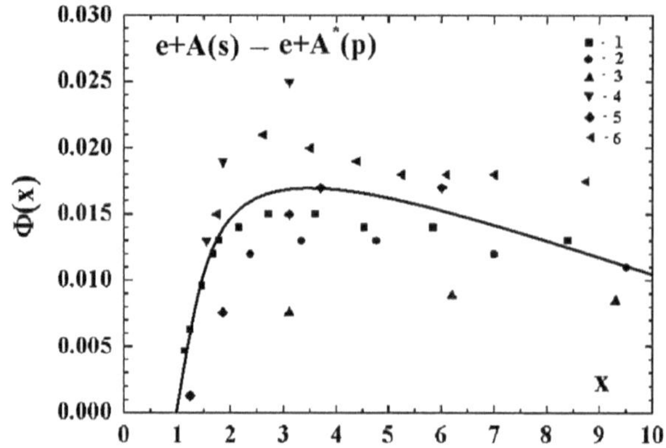

Fig. 3.7 The reduced cross section according to formula (3.3.9) for atom resonant excitation by electron impact from the ground states of lithium, sodium and potassium atoms and from the metastable states $He(2^3S)$ for the helium atoms. Experimental data are taken: *1*—[153] for Li, *2*—[154] for Na, *3*—[155] for K, *4*—[156] for K, *5*—[152] for K, *6*—[157, 158] for $He(2^3S)$

Table 3.3 Parameters of excitation of the lowest resonant states for atoms with a simple structure of excited states

Atom	Transition	$\Delta\varepsilon$, eV	f_{o*}	σ_*, $10^{-14}\,\mathrm{cm}^2$
Li	$2^2S \to 2^2P$	1.848	0.73	2.8
Na	$3^2S \to 3^2P$	2.103	0.96	2.8
K	$4^2S \to 4^2P$	1.614	1.05	5.2
He	$2^3S \to 2^3P$	1.144	0.54	5.4

Formula (3.4.5) near the threshold of excitation may be represented in the form

$$\sigma_{ex} = \frac{\sigma_o g_*}{g_o(\Delta\varepsilon)^4\tau_r}\sqrt{x-1}, \tag{3.4.6}$$

where we have on the basis of the above data $\sigma_o = (7 \pm 1) \times 10^{-13}\,\mathrm{cm}^2$, the atom excitation energy $\Delta\varepsilon$ is expressed in eV in formula (3.4.4), and the radiative lifetime is given in ns, $x = \varepsilon/\Delta\varepsilon$, and formula (3.4.6) holds true near the excitation threshold.

Basing on the position that the theory cannot give reliable cross sections of inelastic electron-atom processes at not large collision energies which are of importance for a gas discharge plasma, and using experimental data, we analyze briefly peculiarities of measurements of these cross sections. Monochromatic electron beams are used for this goal, and because of an energy dispersion in the electron beam, the accuracy of such measurements is restricted near the excitation threshold. Next, for determination of these cross sections it is necessary to know the atom number densities at the initial and final states. But formation of excited atoms takes place as a result of direct and cascade populations of these states, and the latter hampers the determination of the excitation cross sections (for example, [159–163]). Removing the cascade transitions in population of excited states influences the accuracy of results.

It should be noted that the accuracy of measurements of the average rate constants of excitation of atoms by electron impact in a plasma and correspondingly the accuracy the rate constant of excited atom quenching in slow collisions with electrons is better than the accuracy of measurements of the excitation cross sections. In particular, the summarized error in the measured cross sections for excitation of metastable atoms $He(2^3S)$ by electron impact at collision energies compared with the atom excitation energy is estimated as 45 % [157, 164]. Evidently, the cross sections of excitation of excited argon atoms with the electron shell $Ar(3p^54s)$ in states $Ar(3p^54s)$ which are measured in [79, 164, 165] is not lower that the above value because of a large number of transitions.

We note that formula (3.4.4) has not the universal character, but relates to the $s-p$ electron transition. We are based on formula (3.3.9) where the cross section of the rate constant of this transition is proportional to the square $|(D)_{sp}|^2$ of the dipole momentum matrix element between these states. Let us take into consideration the role of level splitting for this transition due to additional interaction. This situation takes the place in reality for light atoms where in the first approximation we have the shell atom structure and in the next approximation each shell gives several levels due to exchange, spin-orbit and other interactions inside the atom. Then if the transition proceeds between some groups of states, and this statistical weight must be taken into account. According to the definition, the rate constant of quenching in this case must be summed over final states and averaged over initial ones. But the rate of a dipole radiative transition is proportional to the square of the above matrix element also and in the same manner the summation takes place over final states and averaging over the initial ones. Hence, formula (3.4.4) remains valid also in the case where levels for s and p electrons are split. In addition, because of the symmetry of the matrix element this formula holds true for $s-p$ electron transitions and $p-s$ electron transitions.

As an example, we consider the transition from the any state of the electron shell $Ar(3p^54p)$ to the state $Ar(3p^54s^3P_2)$, where the upper state is characterized by the statistical weight $g = 36$, while the statistical weight of the lower state is $g = 5$. Then $1/\tau_r$ in formula (3.4.4) is the rate of the radiative transition in any level of the group $Ar(3p^54s^3P_2)$ that is equal $1.45 \times 10^7\,\mathrm{s}^{-1}$. Correspondingly according to formula (3.4.4) the quenching rate constant for this transition is $k_q = 1.2 \times 10^{-7}\,\mathrm{cm}^3/\mathrm{s}$ (the average energy of this transition is $\Delta\varepsilon \approx 1.62\,\mathrm{eV}$. In evaluation the excitation rate constant in accordance with formula (3.4.5) we use $g_* = 36$, $g_o = 5$.

3.5 Transitions Between Neighboring Atom States in Collisions With Electrons

Let us consider one more type of exchange inelastic processes in electron collisions with excited atoms where transitions take place between states with nearby energies and these states belong to the same electron shell. As an example, we first consider the collision process

$$e + He(2^1S) \rightarrow e + He(2^3S) \qquad (3.5.1)$$

that leads to quenching of the metastable state $He(2^1S)$. Both atom states of this process belongs to the electron shell $He(1s2s)$ of the helium atom. The mechanism of this process consists in exchange of an incident and valence electron in electron-atom collision. As a result, a valence electron acquires a new spin direction. Hence one can connect the rate constant of this process with that of the process

$$e \downarrow + He(2^3S) \rightarrow e \uparrow + He(2^3S), \qquad (3.5.2)$$

if an electron energy exceeds remarkably the energy difference between transition states. This mechanism for transitions under consideration results from the exchange interaction of incident and valence electrons, i.e. we ignore weak relativistic interactions in these processes. Let us use the statistical approach according which the atom after an exchange event "forgets" its initial state. Then we have the following relations between the rate constants of corresponding processes

$$\begin{aligned} k[e + He(2^1S)] &\rightarrow e + He(2^3S)] = \tfrac{3}{4}k_{exch}, \\ k[e + He(2^3S)] &\rightarrow e + He(2^1S)] = \tfrac{1}{4}k_{exch}, \\ k[e \downarrow + He(2^3S)] &\rightarrow e \uparrow + He(2^3S)] = \tfrac{3}{4}k_{exch} \end{aligned} \qquad (3.5.3)$$

Here k_{exch} is the rate constant of the process (3.5.2) with exchange of incident and valence electrons if these electrons have different spin directions. According to experimental data [166], the cross section of the process (3.5.2) in thermal collisions is $(160 \pm 40)\,\text{Å}^2$, that corresponds to the rate constant of this process approximately $k_{pol} = (1.7 \pm 0.4) \times 10^{-7}\,\text{cm}^3/\text{s}$. This leads to the rate constant of the exchange process (3.5.1) the value $k_{exch} \approx 2 \times 10^{-7}\,\text{cm}^3/\text{s}$. Direct measurement of the rate constant of the process (3.5.1) according to measurement [167] is $k_{qu} \approx 3.5 \times 10^{-7}\,\text{cm}^3/\text{s}$ that exceeds twice the above value. On the basis of these data we have $k_{exch} = (3 \pm 1) \times 10^{-7}\,\text{cm}^3/\text{s}$.

Another example of electron exchange collisions under consideration relates to an excited argon atom with the electron shell $3p^54s$. The excitation energy for these states lies between 11.624 eV for $^3P_2$ metastable state up to 11.828 eV for the resonantly excited $^1P_1$ state, and we take the energy ε of an incident electron to be $\varepsilon \gg 0.2\,\text{eV}$, so that the difference of excitation energies for these states does not influence on the character of electron-atom collision. Then as well as in the helium case, mixing of these states results from exchange collisions

$$e \downarrow + Ar(3p^54s \uparrow) \rightarrow e \uparrow + Ar(3p^54s \downarrow), \qquad (3.5.4)$$

where vertical arrows indicates spin directions for an incident electron and $4s$-valence electron. The rate constant of mixing of these levels is estimated as [168, 169]

$k_{exch} \sim 1 \times 10^{-7}\,\mathrm{cm^3/s}$, and then the rate constant of transition between states i and j of this group is

$$k_{ij} = \frac{g_j}{12} k_{exch}, \tag{3.5.5}$$

where $g_j = 2J + 1$ is the statistical weight of the final state, so the J is the total electron moment for a final state.

We also give a general expression for the cross section of the exchange process that proceeds according to the scheme

$$e(free) \downarrow + e(bound) \uparrow \rightarrow e(free) \uparrow + e(bound) \downarrow, \tag{3.5.6}$$

where $e(free)$ and $e(bound)$ mean incident and valence electrons respectively, and according to this scheme the exchange interaction between incident and valence electrons exceeds that between valence electron and atom core. Hence during collision one can characterized the states of free and valence electrons by the symmetry of the wave function of these two electrons and separate these states in symmetric (s) and antisymmetric (a) in accordance with the symmetry of the coordinate wave function of two electrons. Correspondingly, the phase theory of electron-atom scattering [111, 117, 118] gives for the cross section σ_{exch} of exchange process (3.5.6) [170]

$$\sigma_{exch} = \frac{4\pi}{q^2} \sum_l (2l+1) \sin^2(\delta_l^s - \delta_l^a), \tag{3.5.7}$$

where q is the wave vector of an incident electron, l is the electron momentum, δ_l^s and δ_l^a are the scattering phases for the symmetric and antisymmetric states of the total system. In particular, in the limit of small collision energies ($q \rightarrow 0$) this cross section is equal to

$$\sigma_{exch} = 4\pi (L_s - L_a)^2, \tag{3.5.8}$$

L_s and L_a are the electron scattering lengths for the symmetric and antisymmetric states of the total electron system correspondingly.

Let us consider from this standpoint the process

$$e + Ar(3p^5 4s^3 P_2) \rightarrow e + Ar(3p^5 4s), \tag{3.5.9}$$

i.e. this process of electron collision with the metastable $^3P_2$ atom state transfers the argon atom into any state of the electron shell $Ar(3p^5 4s)$. The ionization potential of $Ar(^3P_2)$ is 4.21 eV and we will model it by the rubidium atom $Rb(4s)$ in the ground state with the ionization potential 4.18 eV. Both atoms have the valence s-electron and nearby ionization potentials. For the Rb—atom we have [119, 120] $L_s = 2a_o$ and $L_a = -17a_o$. This gives according to formula (3.5.8) $\sigma_{exch} = 1.3 \times 10^{-13}\,\mathrm{cm^2}$ for the cross section σ_{exch} of the exchange process and leads to an estimate $k_{exch} \sim 10^{-6}\,\text{cm}^3/\text{s}$ for the rate constant of this process. Note that because the electron scattering lengths on alkali metal atoms is determined in the most degree by parameters of the autodetachment state ^3P in electron-atom interaction [119, 120], the estimation for the rate constant of the exchange process (3.5.9) is rough.

3.6 Atom Ionization by Electron Impact

Among inelastic processes of electron-atom collisions, the process of atom ionization by electron impact

$$e + A \to 2e + A^+ \tag{3.6.1}$$

is of importance. We analyze this process on the basis of the simplest model of this process—the Thomson model [171]. Within the framework of this model one can ignore interaction of incident and valence electrons with an atomic core in the course of their scattering that is assumed to have the classical character. Atom ionization corresponds to exchange by energy between an incident and valence electrons that exceeds the atom ionization potential J, and the ionization cross section for an atom with one valence electron according to the Thomson model is equal to [171]

$$\sigma_{ion} = \frac{\pi e^4}{\varepsilon}\left(\frac{1}{J} - \frac{1}{\varepsilon}\right) \tag{3.6.2}$$

If several electrons partake in the ionization process, summation over partial cross sections for individual electrons is made.

Analyzing lacks of the Thomson model, we note first the quantum character of electron-atom scattering. But because the classical and quantum cross sections coincide for Coulomb interaction of elastically colliding particles, this lack is not of principle. Second, the scattering cross section depends on the initial velocity of a bound electron [172, 173] that is assumed within the Thomson model to be motionless. Third, at large collision energies the cross section of inelastic collision contains a factor $\ln(\varepsilon/J)$ [174], that is determined by large impact parameters of collision [175], where the ionization probability is small in the quantum case and is zero in the classical case. In principle, the classical theory cannot give the logarithm dependence of the ionization cross section on the collision energy even for ionization of highly excited atoms where the classical description is applicable. This principal distinction between a classical and quantum description was eliminated by Kingston [176–178], who proved that in spite on difference between classical and quantum formulas, they give close values for the ionization cross sections in a range of validity of both approaches. Therefore, accounting for a roughness of the Thomson model (3.6.2), nevertheless it is convenient to use this model for a quantitative analysis of direct atom ionization by electron impact because of the simplicity and principal validity of this model.

Let us modify the Thomson model conserving the classical character of electron motion and considering the bound electron to be free in the course of collision, but accounting for a nonzero velocity of a bound electron. For simplicity we consider scattering on small angles, and the momentum change $\Delta\mathbf{p}$ of a valence electron according to formula (3.1.4) is given by

$$\Delta\mathbf{p} = \frac{2e^2\mathbf{n}}{\rho g},$$

where $\mathbf{n}$ is the unit vector directed along the impact parameter ρ of collision, and g is the relative velocity of colliding electrons. If $\mathbf{u}$ is the velocity of a valence electron, the change $\Delta\varepsilon$ of its energy is equal

$$\Delta\varepsilon = \frac{(m_e\mathbf{u} + \Delta\mathbf{p})^2 - m_e^2u^2}{2m_e}$$

We note that within the framework of the Thomson model we ignore the bond between the valence electron and atomic core during collision, i.e. assume the valence electron to be free. From this we have for the momentum change Δp of the valence (and the incident) electron if we express it through the energy change $\Delta\varepsilon$

$$\Delta p = \sqrt{(m_e\mathbf{uk})^2 + 2m_e\Delta\varepsilon} - m\mathbf{uk}\ ,$$

where $\mathbf{k}$ is the unit vector directed along $\Delta\mathbf{p}$. This gives for ionization cross section $\sigma_{ion} = \int 2\pi\rho d\rho$ which include collision impact parameters ρ at which $\Delta\varepsilon \geq J$. Using the connection between the momentum change $\Delta\mathbf{p}$ and the impact collision parameter ρ, we obtain for the ionization cross section

$$\sigma_{ion} = \int_J^\varepsilon d\sigma = \frac{4\pi e^4}{g^2}\int_J^\varepsilon \left\langle \frac{m_e d\Delta\varepsilon}{(\sqrt{(m_e\mathbf{uk})^2 + 2m_e\Delta\varepsilon} - m\mathbf{uk})^3\sqrt{(m_e\mathbf{uk})^2 + 2m_e\Delta\varepsilon}} \right\rangle,$$

where an average in the angle brackets is made over an angle between vectors $\mathbf{u}$ and $\mathbf{k}$, and over the velocities u of the valence electron. Averaging over this angle and assuming the energy ε of an incident electron to be large compared with the kinetic energy of an incident electron, we obtain

$$\sigma_{ion} = \frac{\pi e^4}{\varepsilon}\int_J^\varepsilon \left\langle \frac{m_e d\Delta\varepsilon}{\Delta\varepsilon^3}(\Delta\varepsilon + \frac{2}{3}m_e u^2 \right\rangle = \frac{\pi e^4}{\varepsilon}\left[\frac{1}{J} - \frac{1}{\varepsilon} + \frac{2\overline{\varepsilon}}{3}\left(\frac{1}{J^2} - \frac{1}{\varepsilon^2}\right)\right], \tag{3.6.3}$$

where $\varepsilon = m_e g^2/2$ is the energy of an incident electron, $\overline{\varepsilon} = \overline{m_e u^2/2}$ is the average kinetic energy of a valence electron, and an average is made over the electron distribution inside the atom. The used condition $\varepsilon \gg J$ allows us to separate parameters of the incident and valence electrons. If we assume the valence electron to be motionless, i.e. $\overline{\varepsilon} = 0$, this formula is converted into the Thomson formula (3.6.2). Assuming the valence electron to be located mostly in the Coulomb field of the atomic core, we obtain $\overline{\varepsilon} = J$ according to the virial theorem. As a result, we obtain [172, 173]

$$\sigma_{ion} = \frac{\pi e^4}{\varepsilon}\left(\frac{5}{3J} - \frac{1}{\varepsilon} - \frac{2J}{3\varepsilon^2}\right) \tag{3.6.4}$$

The Thomson formula (3.6.2) leads to the maximum of the cross section at $\varepsilon = 2J$ with the value $\sigma_{max} = \pi e^4/(4J^2)$, whereas the maximum of the cross section according to formula (3.6.4) is approximately twice compared to that of the Thomson formula, and this maximum occurs at the electron energy $\varepsilon = 1.85J$. Next, the cross section according to the modified Thomson model (3.6.4) is in 5/3 times exceeds that according to the Thomson formula (3.6.2). In addition, at the threshold the Thomson formula gives

$$\sigma_{ion} = \frac{\pi e^4}{J^2}(\varepsilon - J), \quad \varepsilon - J \ll J\ ,$$

while formula (3.6.4) gives near the threshold

$$\sigma_{ion} = \frac{7\pi e^4}{3J^2}(\varepsilon - J), \ \varepsilon - J \ll J$$

Thus, the values of values of the ionization cross sections according to the Thomson formula (3.6.2) and formula (3.6.4) for the modified Thomson model differs up to factor 2.

Our task is to construct a simple and reliable model for the cross section of atom ionization by electron impact. As is seen, the Thomson model and its modification satisfy to these requirements and are able to give the accuracy of the ionization cross section roughly 50 %. One can improve this accuracy if we use experimental results along with the classical approach of the Thomson model and its modifications. Let us construct the cross section of atom ionization by electron impact σ_{ion} assuming the classical character of this process as a result of collision of free and bound electrons. This cross section is composed on the basis of the following parameters: e is the electron charge, m_e is the electron mass, ε is the energy of an incident electron, J is the atom ionization potential. In a general form this cross section constructed from the dimensional consideration has the form

$$\sigma_{ion} = \frac{\pi e^4 n}{J^2} f(\varepsilon/J), \tag{3.6.5}$$

where n is a number of identical valence electrons, and $f(x)$ is an universal function that is equal

$$f(x) = \frac{1}{x} - \frac{1}{x^2}$$

for the Thomson model (3.6.2) and

$$f(x) = \frac{5}{3x} - \frac{1}{x^2} - \frac{2}{3x^3}$$

for the modified Thomson model (3.6.4). We give in Fig. 3.8 the reduced ionization cross section for atomic particles with valence s-electrons [170]. Experimental data are approximated by the function [170, 190]

$$f(x) = \frac{10(x-1)}{\pi x(x+8)} \qquad (3.6.6)$$

with the accuracy about 20 % at electron energies above the cross section maximum and about 40 % at electron energies below the cross section maximum. This dependence is taken from Fig. 3.8. The maximum of the approximated reduced cross section (3.6.6) is equal to $f(x_{max}) = 0.2$ at $x_{max} = 4$ instead of $f(x_{max}) = 0.25$ at $x_{max} = 2$ for the Thomson model. For the modified Thomson model (3.6.4) we have $f(x_{max}) = 0.5$ at $x_{max} = 1.85$. At large electron energies the ratio of the approximated reduced cross section (3.6.6) to that for the Thomson model (3.6.2) is $10/\pi$, whereas this ratio near the ionization threshold is $10/(9\pi)$. Thus, as it follows from the above analysis various modifications of the Thomson model gives the same form of the ionization cross section as a function of the electron energy,

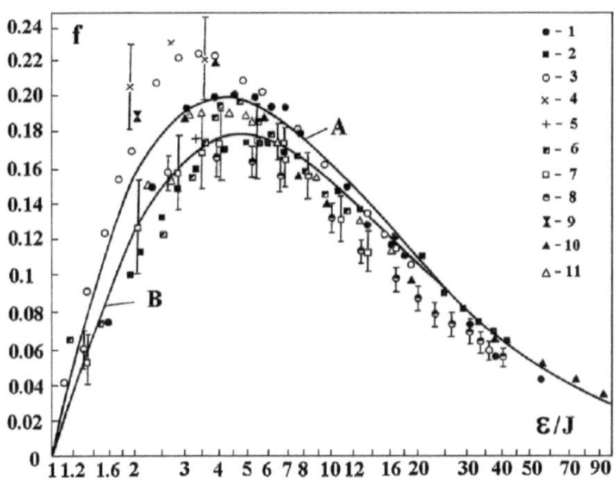

Fig. 3.8 The reduced cross section f for atom ionization by electron impact constructed on the basis of experimental cross section for atoms with valence s-electrons. Experimental data: *1*—H [179], *2*—He [180], *3*—He^+ [181], *4*—$He(2^3S)$ [182], *5*—$He(2^3S)$ [183], *6*—Li^+ [184], *7*—Li^+ [185], *8*—Li^+ [186], *9*—Li [187], *10*—Li [188], *11*—H_2 [189]. Approximation functions: A – $f(x) = \frac{10(x-1)}{x(x+8)}$, B – $f(x) = \frac{10(x-1)}{x+0.5)(x+8)}$

while the values of cross section may differ by factor ~ 2. We below will be guided by the approximated ionization cross section (3.6.6) that is based on experimental data.

On the basis of formula (3.6.6) for the approximated reduced cross section $f(x)$ one can determine the differential cross section $d\sigma(\Delta\varepsilon)$ for collision of incident and valence electrons if an energy exchange is $\Delta\varepsilon$. We have

$$\frac{d\sigma}{d\Delta\varepsilon} = -\frac{d\sigma_{ion}}{dJ}(J = \Delta\varepsilon) = \frac{\pi e^4 n}{\Delta\varepsilon^3}\left[2f(y) + y\frac{df(y)}{dy}\right], \quad (3.6.7)$$

where $y = \varepsilon/\Delta\varepsilon$. From this we obtain for the Thomson model $[f(y) = 1/y - 1/y^2]$

$$\frac{d\sigma}{d\Delta\varepsilon} = \frac{\pi e^4 n}{\varepsilon(\Delta\varepsilon)^2},$$

that coincides with formula (3.1.5). In the case of the modified Thomson model in accordance with formula (3.6.4) we have $f(y) = 5/(3y) - 1/y^2 - 2/(3y^3)$, that gives on the basis of formula (3.6.7)

$$\frac{d\sigma}{d\Delta\varepsilon} = -\frac{d\sigma_{ion}}{dJ}(J = \Delta\varepsilon) = \frac{\pi e^4 n}{3\varepsilon}\left(\frac{5}{\Delta\varepsilon^2} + \frac{2}{\varepsilon^2}\right) \quad (3.6.8)$$

Repeating these operations in the case of the approximated reduced cross section [formula (3.6.6)], we obtain

$$\frac{d\sigma}{d\Delta\varepsilon} = \frac{10e^4 n(\varepsilon^2 + 16\varepsilon\Delta\varepsilon - 8\Delta\varepsilon^2)}{\varepsilon\Delta\varepsilon^2(\varepsilon + 8\Delta\varepsilon)^2} \quad (3.6.9)$$

Let us use the above expressions for the cross sections of a given exchange by energy for colliding electrons in order to determine the average kinetic energy $\overline{\varepsilon}$ of a released electron. We have by definition

$$\overline{\varepsilon} = \overline{\Delta\varepsilon} - J = \int \Delta\varepsilon \frac{d\sigma}{\Delta\varepsilon} d\Delta\varepsilon - J \quad (3.6.10)$$

In the case of the Thomson model this formula along with the above expressions for electron-electron cross sections gives

$$\frac{\overline{\varepsilon}}{J} = \frac{x\ln x}{x-1} - 1 \quad (3.6.11)$$

in the same manner we have in the case of the modified Thomson model

$$\frac{\overline{\varepsilon}}{J} = \frac{5x^2\ln x + x^2 - 1}{(x-1)(5x+2)} - 1 \quad (3.6.12)$$

Similarly we have for the reduced energy of a released electron as a result of atom ionization by electron impact in the case where the ionization cross section is approximated by formulas (3.6.5) and (3.6.6) on the basis of experimental data and formula (3.6.9) for the cross section of exchange by a given energy as a result of collision of two free electrons

$$\frac{\overline{\varepsilon}}{J} = \frac{(x+8)}{(x-1)} \int_1^x \frac{dy}{y} \frac{(y^2+16y-8)}{(y+8)^2} - 1 \qquad (3.6.13)$$

Figure 3.9 gives the dependence of the reduced average exchange energy between a free and valence electron $\Delta\varepsilon/\varepsilon$ in the course of the ionization process as a function of the reduced energy of an incident electron ε/J on the basis of formulas (3.6.11), (3.6.12) and (3.6.13). Note that the average energy of a released electron $\overline{\varepsilon}$ is equal to the binding energy of a valence electron J at the reduced energy $\varepsilon/J = 4.9$ of an incident electron in the case of the Thomson model. This value is equal to 4.5 for the modified Thomson model, and $\varepsilon/J = 10$ if the ionization cross section is constructed on the basis of experimental data and is given by formulas (3.6.5) and (3.6.6).

We also evaluate the ionization rate constant k_{ion} for the approximated formula (3.6.6) with averaging over the Maxwell distribution function of electrons (2.1.2). This gives for the ionization rate constant

$$k_{ion} = \int \sqrt{\frac{2\varepsilon}{\pi m_e}} \frac{2}{\sqrt{\pi} T_e^{3/2}} \sqrt{\varepsilon} \exp\left(-\frac{\varepsilon}{T_e}\right) d\varepsilon \frac{\pi e^4}{J^2} f(x) \qquad (3.6.14)$$

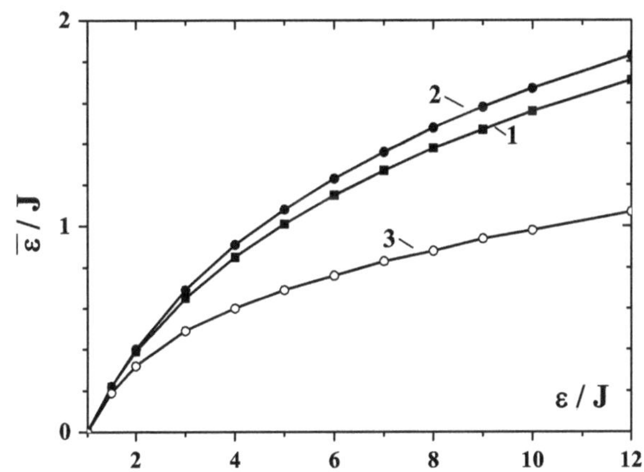

Fig. 3.9 The reduced average energy of a released electron $\overline{\varepsilon}/J$ as a function of the reduced energy ε/J of an incident electron. *1*—the Thomson model—formula (3.6.11), the modified Thomson model—formula (3.6.12), *3*—the classical model of atom ionization based on experimental data—formula (3.6.13)

where $f(x)$ is given by formula (3.6.6), T_e is the electron temperature. Evaluating the integral in the limiting cases $T_e \ll J$ and $T_e \ll J$ and sewing the results for these limiting cases, we obtain the rate constant of atom ionization within the framework of the classical consideration with the reduced cross section (3.6.6) that has the form

$$k_{ion} = 10\sqrt{\frac{8\pi T_e}{m_e}}\frac{e^4}{J^2}\exp(-J/T_e)\times(9+T_e/J)^{-1} \quad (3.6.15)$$

We will use below this expression for the ionization rate constant of an atom by electron impact. Usually for a gas discharge plasma we have $T_e \ll 9J$, and formula (3.6.15) in this limiting case for the Maxwell distribution of electrons takes the form

$$k_{ion}(T_e) = 4.85\times 10^{-6}\frac{\sqrt{T_e}}{J^2[1+T_e/(9J)]}\exp\left(-\frac{J}{T_e}\right), \quad (3.6.16)$$

where the rate constant is expressed in $\mathrm{cm^3/s}$, while the electron temperature T_e and the atom ionization potential J are given in eV.

It is convenient to represent this rate constant in the form

$$k_{ion}(\varepsilon) = k_o F(z),\ k_o = 10\sqrt{\frac{8\pi J}{m_e}}\frac{e^4}{J^2},\ z = T_e/J,\ F(z) = k_o\frac{\sqrt{z}}{(9+z)\exp(1/z)} \quad (3.6.17)$$

The universal function $F(z)$ has the maximum $F(z_{max}) \approx 0.15$, at $z_{max} \approx 10$. In reality the ionization rate constant k_{ion} is of interest at low values z, and the dependence $F(z)$ is represented in Fig. 3.10.

It should be noted that ionization of atoms by electron impact in a gas discharge plasma has an interest for excited atoms mostly. Since helium and argon are objects of our consideration, we give in Table 3.4 the values of the ionization rate constants which relate to excited helium and argon atoms.

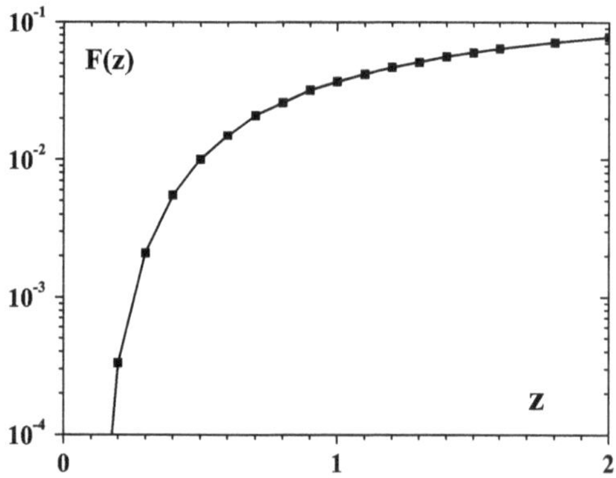

Fig. 3.10 Function F(z) according to formula (3.6.17)

Table 3.4 The reduced rate constant k_o of atom ionization by electron impact in accordance with formula (3.6.16)

Atom	J, eV	k_o, $10^{-6}\,cm^3/s$
$He(2^3S)$	4.77	3.0
$He(2^1S)$	3.97	3.9
$He(2^3P)$	3.62	4.5
$He(2^1P)$	3.37	5.0
$Ar(^3P_2)(1s_5)$	4.21	3.6
$Ar(^1P_1)(1s_2)$	3.93	4.0
$Ar(2p_{10})$	2.85	6.4
$Ar(2p_1)$	2.28	9.0

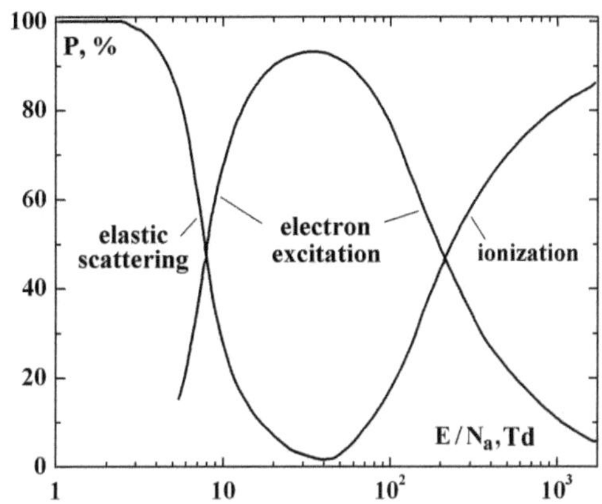

Fig. 3.11 The contributions P to consumption of the electron energy as a result of collisions with argon atoms depending on the electron energy [191]

In considering the energy loss for a fast electron moved in an atomic gas note that the cross section of electron-atom elastic scattering exceeds that for inelastic collisions. But the factor m_e/M for the mass ratio that is included in formula for the rate of energy loss, makes the elastic channel to be weak compared to inelastic channels at energies where inelastic collisions are effective. This is demonstrated in Fig. 3.11 where the contribution of some channels to the electron energy loss is given depending on the electron energy. As is seen, at low energies below the excitation threshold the electron energy loss is determined by elastic electron-atom collisions. An increase of the incident electron energy makes as a main channel of the energy lass excitation of atoms and then their ionization. In particular, fast secondary electrons resulted from cathode bombardment by fast ions consume the energy mostly in atom ionization.

3.7 Recombination of Electrons and Ions in Plasma

Recombination of electrons and ions requires an additional degree of freedom for transfer of an excess energy that releases due to formation of a bound state of an electron in the ion field. In the case of molecular ions the energy excess is consumed on breaking of the chemical bond, and the process of dissociative recombination proceeds according to the scheme

$$e + AB^+ \rightarrow A + B^* \tag{3.7.1}$$

The rate constant of dissociative recombination (3.7.1), that is called the coefficient of dissociative recombination α_{dis} is of the order of an atomic value under optimal conditions of this process. In the helium case positions of the electron terms of molecular helium ion He_2^+ and excited helium molecule He_2^* for states which partake in process (3.7.1) is unfavorable, and hence the coefficient of dissociative recombination depends strongly from vibrational and rotational states of the molecular ion. Averaging over many measurements [190] gives for the coefficient of dissociative recombination with participation of the helium molecular ion He_2^+ the value $\alpha = 5 \times 10^{-9\pm0.3}\,\mathrm{cm^3/s}$ at room temperature. The coefficient of dissociative recombination of electrons and molecular argon ions Ar_2^+ at room temperature is equal to $\alpha = (7 \pm 2) \times 10^{-7}\,\mathrm{cm^3/s}$ as it follows from an average of a sum of experimental results many measure $\alpha_{dis} = 6.9 \times 10^{-7}\,\mathrm{cm^3/s}$ [190].

The coefficient of dissociative recombination depends both on the electron temperature T_e and on a vibration-rotation state. Usually is assumed that the dependence of α on the electron temperature T_e at constant gas temperature $\alpha_{dis} \sim T_e^{1/2}$; if the gas temperatures varies simultaneously with the electron one, this dependence is more sharp $\alpha_{dis} \sim T_e^{3/2}$. Note that these dependencies are rough.

Associative ionization, the inverse process with respect to dissociative recombination, proceeds according to the scheme

$$A + B^* \rightarrow e + AB^+ \tag{3.7.2}$$

and may be responsible for formation of ions in a plasma. Another ionization process, the Penning process [192, 193], takes place in a mixture of gases in the case when the excitation energy of one atom exceeds the ionization potential of another one. This process proceeds according to the scheme

$$A + B^* \rightarrow e + A^+ + B \tag{3.7.3}$$

This process is of importance for gas breakdown when a small admixture of atoms with a small ionization potential varies the breakdown parameters. In particular, the rate constant for the Penning process involving helium metastable atoms $He(2^3S)$ and $He(2^1S)$ and argon atoms is equal to $8 \times 10^{-11}\,\mathrm{cm^3/s}$ and $3 \times 10^{-10}\,\mathrm{cm^3/s}$ at room temperature [194]. In the case of the metastable argon atom $Ar(^3P_2)$ that

Table 3.5 The rate constant k_P of the Penning process (3.7.4) and the energy of a released electron $\Delta\varepsilon$ that corresponds to the maximum of the energy distribution function for released electrons

Colliding atoms	k_P, $10^{-9}\,\mathrm{cm}^3/\mathrm{s}$	$\Delta\varepsilon$, eV
$2He(2^3S)$	2.0 ± 0.2	14.4
$He(2^3S) + He(2^1S)$	4.0 ± 0.5	15.4
$2He(2^1S)$	4.1 ± 0.9	16.2
$2Ar(^3P_2)$	1.2 ± 0.2	7.3

collides with krypton and xenon atoms, and also hydrogen, nitrogen and oxygen molecules the rate constant of the Penning process at room temperature in units cm^3/s is equal to 5×10^{-12}, 2×10^{-10}, 9×10^{-11}, 2×10^{-10} 4×10^{-11} correspondingly. The ratio of the cross section of the Penning process (3.7.4) involving metastable atoms $He(2^3S)$ and $He(2^1S)$ collided with an argon atom equals to 3.1 according to measurement [195].

The Penning process (3.7.3) proceeds also in collisions of two metastable atoms that proceeds according to the scheme

$$2A^* \rightarrow A + A^+ + e \tag{3.7.4}$$

The rate constants of the process (3.7.4) at room temperature are given in Table 3.5 together with the energy values $\Delta\varepsilon$ which correspond to the maximum of the distribution function of released electrons. These data follow from measurements [196–199]. Experiment [200] gives additional information for collisions of two atomic beams including metastable atoms.

3.8 Three Body Processes and Stepwise Ionization of Atoms

Three body processes are responsible for formation of bound state for free colliding particles, and a third particle takes an energy excess that releases as a result of formation of a bound state of initially free particles. Recombination of electrons and ions in a dense plasma results from the three body process according to the scheme

$$2e + A^+ \rightarrow e + A^* \tag{3.8.1}$$

and is a reverse process with respect to the ionization process (3.6.1). Usually the electron temperature T_e in a plasma is relatively small

$$T_e \ll J \tag{3.8.2}$$

Then electron capture on a high level results from three-body collisions, and then an excited electron transfers in lower levels as a result of radiative transitions and collisions with plasma electrons. This is the scheme of Bates and Kingston [201],

Bates et al. [202, 203] for evolution of excitations in processes of ionization or recombination.

Note that in contrast to the above elementary inelastic processes of electron-atom collisions which result from one pair collision, the scheme of Bates and Kingston [201], Bates et al. [202, 203] describes a many-stage process of evolution of atom excitation states involving many excited atom states. Hence three body recombination of electrons and ions (3.8.1), as well the inverse process of stepwise ionization, is many-step evolution on excited states. In considering transitions between excited atom states, the scheme of Bates and Kingston [201], Bates et al. [202, 203] accounts for both collisions of excited atoms with electrons of an equilibrium plasma and radiative transitions involving excited atom states. Some models of this scheme is represented in book [15].

In this context it is of interest the limit of a dense plasma when electron-atom collision processes dominate and radiative processes are not essential. Next, due to the criterion (3.8.2) electron energies are small and the classical description of collision processes holds true. Then the temperature dependence for the rate constant K_{ei} of the three body process (3.8.1) may be determined from dimension consideration that gives [204]

$$K_{ei} = \frac{\alpha}{N_e} = C\frac{e^{10}}{m_e^{1/2}T_e^{9/2}} \tag{3.8.3}$$

The same dependence of the rate constant on the process parameters follows from the Thomson theory [205] for the three-body process that accounts for the process nature [15]. The numerical factor in formula (3.8.3) taken from a sum of numerical calculations is equal to [83, 190]

$$C = 4 \times 10^{\pm 0.2} \tag{3.8.4}$$

Expressing the electron temperature in formula (3.8.3) in eV, one can rewrite this formula on the basis of formula (3.8.4) in the form

$$K_{ei} = \frac{\alpha}{N_e} = \frac{1 \times 10^{\pm 0.2} \times 10^{-26}\mathrm{cm}^6/\mathrm{s}}{T_e^{9/2}} \tag{3.8.5}$$

Along with the recombination process (3.8.1) with formation of an excited atom state, a bound state of an electron and ion, it can be responsible for formation of a bound state of heavy atomic particles. Conversion of atomic ions into molecular ones in gases at a low temperature proceeds according to the scheme

$$A^+ + 2A \rightarrow A_2^+ + A \tag{3.8.6}$$

Subsequently molecular ions can be destructed as a result of dissociative recombination (3.7.1). The rate constant of this process at room temperature according to a sum of measurements in the helium case ($A \equiv He$) equals to $K_{as}(He) = (1.0 \pm 0.2) \times 10^{-31}\,\text{cm}^6/\text{s}$, and in the argon case it is $K_{as}(Ar) = (2.2 \pm 0.7) \times 10^{-31}\,\text{cm}^6/\text{s}$ as a result of an average of the results of measurements [206–221]. The indicated error results from the statistics of indicated measurements [190].

In a dense gas excited atoms are converted in excited molecules in three body collisions according to the scheme

$$A^* + 2A \rightarrow A_2^* + A \tag{3.8.7}$$

The rate constant of this process involving the metastable helium atom $He(2^3S)$ in helium is equal to $2.3 \times 10^{-34}\,\text{cm}^6/\text{s}$ at room temperature, while for lowest excited states of the argon atoms with the electron shell $Ar(3p^54s)$ in argon at room temperature is $1.2 \times 10^{-32}\,\text{cm}^6/\text{s}$.

Stepwise ionization of atoms in a plasma is the inverse process with respect to (3.8.1), and in the course ionization in this process an atom passes through many excited states. This process is of importance for a dense plasma where transitions between excited atom states and destruction of excited atoms in collisions with electrons dominate. Because this process is inverse with respect (3.8.1), one can connect the rate constants of these processes on the basis of the principle of detailed balance that gives for the rate constant of stepwise ionization in the limit of a high electron density and small electron temperatures [190, 222]

$$\kappa_{ion} = 2\frac{g_i}{g_a}\frac{m_e e^{10}}{\hbar^3 T_e^3}\exp\left(-\frac{J}{T_e}\right), \quad T_e \ll J \tag{3.8.8}$$

Here the coefficient C is given by formula (3.8.4), g_i, g_a are the statistical weights of an atom in an initial state and ion in the ground state, m_e is the electron mass, J is the atom ionization potential.

This rate constant describes stepwise ionization at low electron temperature if the ionization process is a sum of processes which proceed through different excited states, and decay of these excited states results from collisions with electrons. Hence the real rate constant of ionization of an atom in a given state is less than that according to formula (3.8.8) and exceeds the rate constant (3.6.16). The ratio of the limiting rate constant is

$$\frac{k_{ion}}{\kappa_{ion}} = 3 \times 2^{\pm 0.2}\frac{g_a}{g_i}\left(\frac{T_e}{J}\right)^2\left(\frac{\hbar^2 T_e}{m_e e^4}\right)^{3/2} \tag{3.8.9}$$

One can see from this that in the limit of low electron temperatures in a dense plasma the stepwise ionization process is more effective than the direct one. This holds true at low electron temperatures until processes of collision of excited atoms with electrons suppress other channels of destruction of excited atoms as radiative processes and collisions of excited atoms with walls. In violation of this, direct ionization of atoms in electron-atom collisions may dominate in a plasma.

3.9 Collision Processes Involving Ions

Processes of ion-atom collisions with a remarkable variation of ion momentum are of importance for transport of ions in a plasma which in turn can influence on displacement of a plasma as a whole in a space. Such processes involving ions are elastic ion-atom collisions and the charge exchange process. Elastic scattering of ions on atoms at not high plasma temperatures is determined by polarization interaction between ion and atom. Then the ion-atom interaction potential $U(R)$ at large distances R between them has the form

$$U(R) = -\frac{\alpha e^2}{2R^4} \tag{3.9.1}$$

where α is the atom polarizability, the ion is assumes to be single charged, i.e. e is the electron charge. The peculiarity of the interaction potential (3.9.1) is the possibility of particle capture that means approach of colliding particles up to $R = 0$. The cross section of capture σ_{cap} for the polarization interaction potential is [110]

$$\sigma_{cap} = 2\pi\sqrt{\frac{\alpha e^2}{\mu g^2}}, \tag{3.9.2}$$

where μ is the reduced mass of colliding ion and atom, g is their relative velocity.

In reality, the interaction potential differs from the polarization one (3.9.1) at distances compared with atoms size and corresponds to repulsion at small ion-atom distances. Therefore the capture cross section (3.9.2) means only a strong approach of colliding ion and atom compared to the impact parameter of collisions. Nevertheless, the diffusion cross section of ion-atom collision for the polarization interaction potential (3.9.1) between them σ^*_{ia} is close to the capture cross section (3.9.2) and is equal to [225] $\sigma^*_{ia} = 1.10\sigma_{cap}$.

The resonant charge exchange in ion collision with own atom proceeds according to the scheme

$$A^+ + \widetilde{A} \rightarrow \widetilde{A^+} + A, \tag{3.9.3}$$

where tilde marks one of colliding particles. Usually at room temperature and higher collision energies the cross section of resonant charge exchange exceeds remarkable the cross section of ion-atom elastic collision, and therefore transport of ions in the parent gas is determined by resonant charge exchange.

The cross section of resonant charge exchange is large because this process results from interference the even and odd states of the system of colliding ion and atom [224]. This nature of interaction causes the transition of a valence electron between atomic cores at large distances between them also for the classical character of electron motion [225]. A tunnel character of electron transition leads to a weak

Table 3.6 The cross sections of resonant charge exchange σ_{res} are expressed in $10^{-16}\,\mathrm{cm}^2$, an argument indicates colliding particles, and the collision energy ε is taken in the laboratory frame of reference [41]

ε, eV	$\sigma_{res}(He^+ + He)$	$\sigma_{res}(Ar^+ + Ar)$
0.01	43	83
0.03	39	77
0.1	35	70
0.3	32	64
1	28	58
3	25	52
10	24	47

logarithm dependence of the cross section of resonant charge exchange σ_{res} on the collision velocity v [226, 227]

$$\sigma_{res}(v) = C \ln^2\left(\frac{v_o}{v}\right), \tag{3.9.4}$$

where C and v_o are constants. This dependence can be represented in the form

$$\sigma_{res}(v) = \frac{\pi}{2}\left[R_o + \frac{1}{\gamma}\ln\left(\frac{v_o}{v}\right)\right]^2, \tag{3.9.5}$$

where πR_o^2 is the cross section at the collision velocity v_o; the ionization potential J and the parameter γ of this formula are connected by the relation $J = \hbar^2\gamma^2/(2m_e)$.

Because we are guided by a gas discharge plasma of helium and argon, we consider the resonant charge exchange in these cases. Table 3.6 contains the cross sections of resonant charge exchange for these cases [228–230]. Then formula (3.9.4) for the cross sections of resonant charge exchange in a given range of collision energies has the form

$$\sigma_{res}(He^+ + He) = (5.5 - 0.28\ln\varepsilon)^2$$

in the helium case and

$$\sigma_{res}(Ar^+ + Ar) = (7.62 - 0.326\ln\varepsilon)^2$$

in the argon case. In these formulas the cross section of resonant charge exchange σ_{res} is expressed in Å^2, and the ion energy ε is given in eV in the laboratory frame of reference (an atom is motionless). An argument indicates particles involving in this process. In addition, Table 3.6 contains the values of the resonant charge exchange cross sections for the helium and argon cases at collisions energies which are of interest for a gas discharge plasma.

Chapter 4
Radiative Processes in Gas Discharge Plasma

Abstract Radiative plasma processes under consideration includes radiative transitions between discrete atom levels, photoionization of atoms and photorecombination of electrons and ions. Propagation of resonant photons in a gas including reabsorption of photons in the course of this process is analyzed, and parameters of this process for helium and argon are represented.

4.1 Radiative Transitions in Atoms

Electrons in a gas discharge plasma are able to excite atoms, and therefore a gas discharge plasma may be used as a basis of lamps in light sources [235–237] and lasers [238–242]. In considering the radiation processes in a gas discharge plasma, we pay attention to two of them—dipole radiative transitions in atoms and photorecombination radiation involving electrons and ions of a gas discharge plasma. We analyze interaction of atomic particles of a gas-discharge plasma with a weak electromagnetic field as it take place in gas discharges, so that the intensity of this electromagnet field is small in comparison with the corresponding atomic value, and therefore typical times of radiative transitions are large compared to typical atomic times; one more small parameter respects to this interaction when valence electrons partake in radiative transitions and their typical velocity is small compared to the light speed c. A small parameter of radiative transitions is the fine structure constant $\alpha = e^2/(\hbar c) = 1/137$, where e is the electron charge, $\hbar$ is the Planck constant, c is the light velocity, and $e^2/\hbar$ is a typical atomic velocity that corresponds to a typical velocity of valence electrons on their orbits. If we characterize a radiative transition between two atomic states by the radiative lifetime of the upper state with respect to the radiative transition τ_r, for strongest transitions this time is $\tau_r \sim \tau_r^o = \alpha^3 \tau_o$, where $\tau_o = 2.42 \times 10^{-17}$ s is the atomic time, and $\tau_r^o = 6.2 \times 10^{-11}$ s.

In addition, we will consider the strongest interaction between the radiation field and atomic system whose interaction operator has the form $\widehat{V} = -\mathbf{ED}$ [151, 153], where $\mathbf{E}$ is the electric field strength of the radiation field, $\mathbf{D}$ is the dipole moment operator of the atomic system. Therefore the strongest radiative transitions are called dipole transitions and connect transition states with non-zero matrix element of

B.M. Smirnov, *Theory of Gas Discharge Plasma*, Springer Series on Atomic, Optical, and Plasma Physics 84, DOI 10.1007/978-3-319-11065-3_4

Table 4.1 Parameters of radiative transitions involving lowest excited states of helium and argon atoms

Radiative transition	$\Delta\varepsilon$ (eV)	λ (nm)	τ_r (ns)
$He(2^1P \to 1^1S)$	21.22	58.433	0.56
$He(2^3P \to 2^3S)$	1.144	1083	98
$He(2^1P \to 2^1S)$	0.602	2058	500
$Ar(3p^54s^3P_1 \to 1^1S)$	11.62	106.67	10
$Ar(3p^54s^1P_1 \to 1^1S)$	11.83	104.82	2

$\Delta\varepsilon$ is the transition energy, λ is the wavelength of an emitting photon, τ_r is the radiation lifetime with respect to this transition [109]

the dipole moment operator. We restrict below by only such radiative transitions. The upper excited state of a radiative transition is named as a resonantly excited state if the dipole radiative transition in the ground state is possible from this state. For example, for the helium atom this state is $He(2^1P)$ and for the resonantly excited argon atoms are $Ar(3^3P_1)$ and $Ar(3^1P_1)$. Table 4.1 lists parameters of radiative transitions involving lowest excited states of helium and argon atoms. Note that the radiative time for these transitions exceeds τ_r^o, because the interaction is shared between many excited states and the transition energy is less than the atomic one.

We also give the expression for the rate of dipole radiation [151–153], i.e. the lifetime τ_r of a resonantly excited state with respect to spontaneous radiative transition into the group of lower transition states according to formula (3.4.3)

$$\frac{1}{\tau_r} = \frac{4\omega^3}{3\hbar c^3}|\langle 0|\mathbf{D}|*\rangle|^2 g_o \tag{4.1.1}$$

Here indices 0, $*$ relate to the group of lower and upper states of the radiative transition, $\mathbf{D}$ is the dipole momentum operator, ω is the transition frequency, g_o is the statistical weight of a lower transition state, c is the light velocity.

Because processes in a helium and argon gas discharge plasma is the subject of the subsequent analysis, we represented in Fig. 2.6 the Grotrian diagram for the helium atom. The Grotrian diagram includes along with positions of excited atoms some parameters of radiative transitions between states, usually dipole radiative transitions. In the argon case we gave in Fig. 2.7 the lower part of this diagram which contains radiative transitions between lower excited states of the argon atom with the electron shells $Ar(3p^54s)$ and $Ar(3p^54p)$. Because these transitions are determined mostly by transitions of a valence electron, radiative transitions from states with the electron shell $Ar(3p^54p)$ in the ground electron state $Ar(3p^6)(^1S)$ are forbidden in the dipole approximation.

4.2 Photoionization and Photorecombination Process

The photoionization process proceeds according to the scheme

$$\hbar\omega + A \to e + A^+, \tag{4.2.1}$$

and the opposite process, the process of photorecombination, is

$$e + A^+ \to \hbar\omega + A \tag{4.2.2}$$

We first derive the connection between the cross sections of processes (4.2.1) and (4.2.2) on the basis of the principle of detailed balance [243]. We have the following

$$g_{rec}\sigma_{rec}j_i = g_{ion}\sigma_{ion}j_p,$$

where σ_i, σ_r are the cross sections of photoionization and photorecombination correspondingly, g_{ion}, g_{rec} are the statistical weights of for these states, j_e, j_p are the fluxes of electrons and photons in these channels. Taking the statistical weights as a number of states per one electron or photon, we have

$$g_{rec} = g_e g_i \frac{4\pi k^2 dk}{(2\pi)^3}, \quad g_{ion} = g_a \cdot 2\frac{4\pi q^2 dq}{(2\pi)^3},$$

where g_e, g_i, g_a are the statistical weights of an electron, ion and atom with respect their electron states, k, q are the wave vectors of a photon and electron, the factor 2 accounts for two photon polarizations, and we consider the ion to be motionless in this consideration. Because one photon is located in this volume V, we have for a photon flux $j_p = c/V$ and an electron flux $j_e = v_e/V$, where c is the light velocity, and the electron velocity is $v_e = \hbar q/m_e$.

Using the conservation energy for these transitions

$$\hbar\omega = J + \frac{\hbar^2 q^2}{2m_e},$$

where J is the atom ionization potential, and the dispersion relation for the photon $\omega = kc$, we have

$$\sigma_{rec} = \frac{2g_a}{g_e g_i}\frac{k^2}{q^2}\sigma_{ion} \tag{4.2.3}$$

We now give the expression for the photoionization cross section in the limiting cases. When an atom A in processes (4.2.1) and (4.2.2) is the hydrogen atom in the ground state [244–246]

$$\sigma_{ion} = \frac{2^9 \pi^2}{3} \frac{e^2}{\hbar c} a_o^2 F(a_o q), \quad F(x) = \frac{\exp\left(-\frac{4}{x} \arctan x\right)}{(1+x^2)^4 [1 - \exp(-\frac{2\pi}{x})]} \tag{4.2.4}$$

In the limit $x \to 0$ we have

$$F(0) = \frac{1}{\exp(4)} \left(\frac{\omega_o}{\omega}\right)^{8/3}$$

where $\hbar\omega_o = Ry = 13.6\,\mathrm{eV}$ is the ionization potential for the hydrogen atom in the ground state. This give in the limit when the photoionization process proceeds near the threshold

$$\sigma_{ion} = \sigma_o \left(\frac{\omega_o}{\omega}\right)^{8/3}, \tag{4.2.5}$$

where the parameter σ_o, the photoionization cross section at the threshold, is equal

$$\sigma_o = \frac{2^9 \pi^2}{3 \exp(4)} \frac{e^2}{\hbar c} a_o^2 = 0.225 a_o^2 = 6.3 \times 10^{-18}\,\mathrm{cm}^2 \tag{4.2.6}$$

The principle of detailed balance (4.2.3) gives on the basis of formula (4.2.5) the following expression for the photorecombination cross section involving a slow electron in the ground state of the hydrogen atom

$$\sigma_{rec} = \frac{2\omega^2}{c^2 q^2} \sigma_{ion} = \frac{2^9 \pi^2}{3 \exp(4)} \left(\frac{e^2}{\hbar c}\right)^3 \frac{\omega_o^{5/3}}{\omega^{2/3}(\omega - \omega_o)} a_o^2 = \sigma_1 \frac{\omega_o^{5/3}}{\omega^{2/3}(\omega - \omega_o)}, \tag{4.2.7}$$

where

$$\sigma_1 = \frac{2^8 \pi^2}{3 \exp(4)} \left(\frac{e^2}{\hbar c}\right)^3 a_o^2 = 0.225 a_o^2 = 1.7 \times 10^{-22}\,\mathrm{cm}^2 \tag{4.2.8}$$

In the case of a highly excited initial atom state when a transferring electron is described due to classical laws, the photoionization cross section is given by the Kramers formula [247]

$$\sigma_{ion} = \frac{\sigma_K}{n^5} \left(\frac{\omega_o}{\omega}\right)^3, \tag{4.2.9}$$

where n is the principal quantum number of a bound electron in the initial state, and averaging is fulfilled over other quantum numbers of this electron; the parameter σ_K in this formula is equal

$$\sigma_K = \frac{64\pi}{3\sqrt{3}}\frac{e^2}{\hbar c}a_o^2 = 7.9 \times 10^{-18}\,\text{cm}^2 \tag{4.2.10}$$

Correspondingly, according to the principle of detailed balance (4.2.3) formula (4.2.9) gives for the photorecombination cross section when the final atom state is a highly excited state with the principal quantum number n

$$\sigma_{rec} = \frac{32\pi}{3\sqrt{3}}\left(\frac{e^2}{\hbar c}\right)^3 \frac{\omega_o^2}{\omega(\omega - \omega_o/n^2)}\frac{a_o^2}{n^3} = \frac{8\pi}{3\sqrt{3}}\left(\frac{e^2}{\hbar c}\right)^3 \frac{e^4}{\hbar\omega\varepsilon n^3}, \tag{4.2.11}$$

where $\hbar\omega_o = m_e e^4/\hbar^3$, $2\omega_o a_o = e^2/\hbar$, n is the principal quantum number of a transferred electron in the final state. Kramers formulas correspond to the limit when both transition states may be described classically [248]. Transitions between excited states respect to this limit, and this follows from Tables for parameters of radiative transitions of an electron that is located in the central field of an atomic core [249].

It is of interest to use the Kramers formula (4.2.9) for photoionization of the hydrogen atom in the ground state when the classical description of a transferring electron is not valid. In this case according to this classical description the threshold photoionization cross section is equal to σ_K of formula (4.2.10) that is 25 % higher than the accurate value (4.2.6) of this quantity. This means that two limiting cases under consideration are able to give correct estimations of radiative parameters for real cases. Note that if the electron shell of an atom contains several valence electrons, the above general peculiarities are conserved [250].

4.3 Broadening of Spectral Lines

Radiation due to a certain radiative transition is monochromatic in scales of emitting photons, i.e. a width of frequencies of emitting photons is small compared with the frequency of this radiative transition. Nevertheless this narrow spectrum is of importance for reabsorption processes, and therefore we analyze now the mechanisms of broadening of spectral lines.

Let us introduce in this manner the distribution function of emitting photons a_ω [152], so that $a_\omega d\omega$ is the probability that the emitting photon frequency lies in a range between ω and $\omega + d\omega$. Because the probability is normalized to one, the frequency distribution function of photons a_ω satisfies to the relation

$$\int a_\omega d\omega = 1, \tag{4.3.1}$$

and in scales of emitting photons the photon distribution function is given by

$$a_\omega = \delta(\omega - \omega_o)\,, \tag{4.3.2}$$

where ω_o is the frequency of an emitting photon. This means that spectral line is narrow, i.e. the line width $\Delta\omega$ is small compared to its frequency

$$\Delta\omega \ll \omega_o$$

We first find the frequency distribution function a_ω due to a finite lifetime of transition states. Note that the amplitude of radiative transitions is expressed through the matrix element between transition states, and we analyze the time dependence of the matrix element. Since the stationary wave function of a state k contains a time factor $\exp(-iE_k t/\hbar)$, where E_k is the energy of this state, the matrix element for transition between stationary states i and k includes the time factor $\exp(-i\omega_o t)$, where $\omega_o = (E_i - E_k)/\hbar$. If we account for a finite lifetime τ_r of an upper state and represent the wave function of this state as $\exp(-iE_i t/\hbar - t/2\tau_r)$, the time dependence of the matrix element for transition between two states has the form $c(t) = (-i\omega_o t - t/2\tau_r)$. Hence, the frequency dependence for a matrix element, the Furie component from the function $c(t)$, is given as $c_\omega \sim |i(\omega - \omega_o) + 1/2\tau_r|$, and therefore the frequency distribution function of emitting photons is

$$a_\omega = |c_\omega|^2 = \frac{\nu}{\pi} \cdot \frac{1}{(\omega - \omega_o)^2 + \nu^2}, \tag{4.3.3}$$

where $\nu = 1/(2\tau_r)$ is the spectral line width. We above account for the normalization of the frequency distribution function according to the condition (4.3.1). This shape of the frequency distribution function is named the Lorenz one.

We now consider another mechanism of broadening of spectral lines due to motion of emitting atoms. Indeed, if an emitting particles moves with respect to a receiver with a velocity v_x and emits a photon of frequency ω_o, according to the Doppler effect it is perceived by the receiver as having the frequency

$$\omega = \omega_o \left(1 + \frac{v_x}{c}\right), \tag{4.3.4}$$

where c is the light velocity. Therefore if radiating atoms are distributed over velocities, their radiation by a motionless receiver will be detected as to be frequency distributed. This distribution is determined by the velocity distribution function $f(v_x)$ of radiating atoms.

If the atom distribution function is normalized by unity ($\int f(v_x)dv_x = 1$), the frequency distribution function follows from the relation

$$a_\omega d\omega = f(v_x)dv_x ,$$

Let us consider a spread case with the Maxwell distribution over velocities and are characterized by the distribution function

$$f(v_x) = C \exp\left(-\frac{mv_x^2}{2T}\right),$$

where T is the atom temperature expressed in energetic units, m is the radiating particle mass, C is the normalized constant. From this we find for the frequency distribution function that is normalized according to the condition (4.3.1)

$$a_\omega = \frac{1}{\omega_o}\left(\frac{mc^2}{2\pi T}\right)^{1/2} \cdot \exp\left[-\frac{mc^2(\omega-\omega_o)^2}{2T\omega_o^2}\right] \quad (4.3.5)$$

Note that the ratio of a typical spectral line width due to Doppler broadening $\Delta\omega_D$ to the photon frequency ω_o, that is the ratio of atom thermal velocity to the light velocity, is relatively small

$$\frac{\Delta\omega_D}{\omega_o} \sim \sqrt{\frac{T}{mc^2}}$$

For example, for helium atoms at room temperature the right-hand side of this relation equals to 2.6×10^{-6}.

Broadening of spectral lines is of importance for a gas discharge plasma where a radiating atom interacts with surrounding atomic particles. This broadening may be divided in two limiting cases. In the first case collision of individual particles of this plasma with a radiating atom proceeds fast, and these collisions are seldom, i.e. at each time moment an interaction of a radiating atom is possible with one surrounding atomic particles only. This mechanism relates to a rare plasma. The other limiting case corresponds to interaction of a radiating atom with many surrounding atomic particles and relates to a dense plasma. The first case of broadening of spectral lines is described by the impact theory, and the second case respects to the quasistatic theory of broadening of spectral lines.

In the case of impact broadening of spectral lines, the phase of the transition matrix element get a change during each interaction of a radiating atom with an incident atom. Then the width of a spectral line is characterized by the rate of collisions of an radiating atom with atomic particles (atoms, electrons, ions) of a plasma and is given by the cross section σ_b that determines broadening of a spectral line as a result of collisions. Describing the frequency distribution function of photons a_ω by the Lorenz formula (4.3.3), we have for the collision rate in this formula $\nu = N_a\langle v\sigma_b\rangle$. The cross section σ_b that characterizes broadening of a spectral line is connected with the total cross section by the relation [152] $\sigma_b = \sigma_t/2$, where $\sigma_t = \int d\sigma$ is the total cross section of collision between a radiating atom and atomic particles, and an average is made over the velocity distribution function of colliding particles.

Note that though perturbation of a radiating atom proceeds in both upper and lower transition states, we reduce the interaction to the upper state only because of a more strong interaction for this state. Next, the total cross section σ_t is based on the phase shift, that in its turn is determined in the classical limit as $\int U(R)dt/\hbar$, so that $U(R)$ is the interaction potential between colliding particles at the distance R

between them. Hence, in the classical limit the total cross section σ_t of collision is estimated as

$$\sigma_t = \pi\rho_t^2, \quad \frac{U(\rho_t)}{\hbar} \sim \frac{v}{\rho_t}, \tag{4.3.6}$$

where v is the relative velocity of colliding particles, and the quantity R_o is called the Weiskopf radius. The criterion of validity of this broadening mechanism corresponds to a small probability of perturbed particle location in a region of a Weiskopf radius size formula that gives

$$N_a\sigma_t^{3/2} \ll 1\,, \tag{4.3.7}$$

In the case of the opposite criterion with respect to (4.3.7) the quasistatic theory of spectral line broadening is realized. This theory assumes perturbed particles to be motionless during times which are responsible for broadening. Assuming the interaction potential of a radiating atom with perturbed atomic particles to be pairwise and isotropic, we find the frequency distribution function on the spectral line wing.

The wing of a spectral line is determined by location of perturbed atomic particles on a small distances from a radiating atom where the probability of location of perturbed particles is small. This gives for the frequency distribution function

$$a_\omega d\omega = w(R)dR = N_a \cdot 4\pi R^2 dR$$

where $w(R)dR$ is the probability of location of a perturbed particle in a distance range from R up to $R + dR$ for the radiating atom. This gives

$$a_\omega = 4\pi R^2 N_a \hbar (dU/dR)^{-1} \tag{4.3.8}$$

where we assume a monotonic dependence $U(R)$. One can estimate the spectral line width for the quasistatic theory of broadening as

$$\Delta\omega \sim \frac{U(N_a^{-1/3})}{\hbar} \tag{4.3.9}$$

We now write the criterion of validity of the quasistatic theory of broadening that is based on the assumption of immobility of perturbed atomic particles during times which are responsible for broadening. We have for a typical time of a particle shift $\tau \sim N^{-1/3}/v$, because $N^{-1/3}$ is the average distance between perturbed particles. Because $\tau \sim (\Delta\omega)^{-1}$, on the basis of formula (4.3.9) we have for this criterion

$$\frac{U(N_a^{-1/3})N_a^{-1/3}}{\hbar v} \gg 1 \tag{4.3.10}$$

Comparing this criterion with formula (4.3.6) of definition of the Weiskopf radius and assuming the monotonic dependence $U(R)$, we find from this the criterion of validity of the quasistatic theory in the form

$$\rho_t N_a^{1/3} \gg 1 \qquad (4.3.11)$$

One can see that this criterion is opposite to (4.3.7) and is based on the assumption that the probability to locate for two and more surrounding particles in a region of a strong interaction with a radiating atom is small. A typical size of this region correspond to $\rho_t \sim \sqrt{\sigma_t}$, and the criterion of the collision broadening of a spectral line for a typical frequency shift is (4.3.7), i.e. the impact theory and quasistatic theory of broadening of spectral lines relate to opposite cases of interaction of a radiating atom with surrounding atomic particles.

We now consider in detail broadening of the spectral line that results from transition from the resonantly excited state in the ground atom state and broadening proceeds due to collision between atoms in the ground and resonantly excited states. The operator of the interaction that governs by broadening of the spectral line in this case has the form

$$U(\mathbf{R}) = \frac{\mathbf{D}_1\mathbf{D}_2 - 3(\mathbf{D}_1\mathbf{n})(\mathbf{D}_2\mathbf{n})}{R^3} \qquad (4.3.12)$$

Here $\mathbf{R}$ is the distance between colliding atoms, $\mathbf{n}$ is the unit vector directed along $\mathbf{R}$, $\mathbf{D}_1, \mathbf{D}_2$ are the dipole moment operator for an indicated atom. Since the matrix element $d = |\langle o|\mathbf{D}|*\rangle|$ is nonzero, the total cross section for collision of these atoms that is responsible for broadening in this case, according to formula (4.3.6) equals to $\sigma_t \sim d^2/\hbar v$, where v is the collision velocity.

In this case of spectral line broadening the frequency distribution of emitting photons has the Lorenz form according to formula (4.3.3), where the line width is given by [152]

$$\nu = \frac{1}{2} N_a \langle v\sigma_t \rangle, \qquad (4.3.13)$$

N_a is the number density of atoms in the ground state, σ_t is the total cross section of collisions of atoms in the ground and resonantly excited state, and averaging is made over collision velocities. In the case under consideration the interaction potential between colliding atoms is $U \sim d^2/R^3$, where d is the matrix element from the operator of the atom dipole moment, R is a distance between colliding atoms. One can estimate a typical cross section for scattering of atoms by large angles in the standard way [208, 251]

$$\sigma \sim R_o^2, \quad \frac{U(R_o)}{\hbar} \sim \frac{v}{R_o},$$

that gives for the cross section of atom collision $\sigma \sim d^2/v$. The numerical coefficient in this expression for the cross section depends on the symmetry of the ground and excited atom states. In particular, if the ground atom state is ^1S, and resonantly excited atom state is ^1P, the total cross section of for collision of atoms in the ground and resonantly excited states is [252]

$$\sigma_t = 4.8\pi \frac{d^2}{\hbar v}, \tag{4.3.14}$$

and the matrix element d from the dipole moment operator taken between radial wave functions is connected with the oscillator strength f of this transition by the relation

$$f = \frac{2m_e \Delta\varepsilon d^2}{e^2\hbar^2}, \tag{4.3.15}$$

where $\Delta\varepsilon$ is the energy of this transition. From this one can find the width of the spectral line for the radiative transition $A(^1S) + \hbar\omega \rightarrow A(^3P)$ [252]

$$\nu = \frac{2.4\pi N_a d^2}{\hbar} \tag{4.3.16}$$

As is seen, the width of the spectral line is independent of atom velocities. In particular, in the case of collision of helium atoms in the ground $He(1^1S)$ and the lowest resonantly excited state $He(^3P)$, where the wavelength of this transition is $\lambda = 58.4\,\text{nm}$, the radiative lifetime of the excited state is $\tau = 0.56\,\text{ns}$, the oscillator strength for transition between these states is $f = 0.276$, and $d^2 = 0.177e^2a_o^2$, we have for the reduced width of the spectral line

$$\frac{\nu}{N_a} = \frac{2.4\pi d^2}{\hbar} = 8.2 \times 10^{-9}\,\text{cm}^3/\text{s} \tag{4.3.17}$$

Let us consider briefly the quasistatic theory of broadening of a spectral line in the case of a resonantly excited radiating atoms where the interaction potential operator has form (4.3.12). There are in this case several states of the quasimolecule consisting of interacting atoms, and these states differ by the momentum projections and the state parity. Averaging over these states, we have the estimation for interaction in this case $U \sim d^2/R^3$, and formula (4.3.8) gives for the frequency distribution function at the line wings

$$a_\omega d\omega \sim \frac{N_a d^2}{\hbar} \frac{d\omega}{|\omega - \omega_o|^2} \tag{4.3.18}$$

As is seen, the character of line broadening is the same as for the Lorenz mechanism of line broadening, but an average over states has another form.

One can see that if broadening of a spectral line is connected with interaction of a radiating atom with surrounding atomic particles, the character of broadening transfers from the Lorenz one, where it is determined by collision processes, to the quasistatic case with interaction of motionless atomic particles, as the number density of atomic particles or the frequency shift are increased. The criterion of the Lorenz broadening of spectral lines requires a fast flight of a perturbed atomic particle and has the form

$$\frac{v}{\rho_b} \gg |\omega - \omega_o|, \tag{4.3.19}$$

where ρ_b is the impact parameter for collision created broadening that follows from the relation for the total cross section $\sigma_t = \pi\rho_b^2/2$ of scattering of a radiating atom and a perturbed atomic particle. In particular, if the broadening of a spectral line is determined by interaction of a resonantly excited radiating atom with a parent atom in the ground state, i.e. if the interaction potential operator has form (4.3.12), the criterion of the Lorenz mechanism of line broadening is given by

$$|\omega - \omega_o|^2 \ll \frac{\hbar v^3}{d^2} \tag{4.3.20}$$

This criterion for the central part of a spectral line $|\omega - \omega_o| \sim \nu$ takes the form

$$N_a \ll 0.1 \left(\frac{\hbar v}{d^2}\right)^{3/2} \tag{4.3.21}$$

One can see that it is opposite to the criterion (4.3.10), if we neglect by numerical coefficients. As we remove from the spectral line center, conditions for the Lorenz mechanism of spectral line broadening become worse, and the criterion of the validity of this broadening mechanism is given by (4.3.20). Note also that in the above consideration we assume $\hbar|\omega - \omega_o| \ll T$ and for room temperature this gives $|\omega - \omega_o| \ll 200\,\mathrm{cm}^{-1}$. This criterion allows us to consider the collision energy of colliding atoms in the ground and excited states to be large compared to the splitting between energies of electron terms which determine the broadening cross section.

4.4 Cross Section and Absorption Coefficient for Resonant Photons

Let us consider the behavior of an atom that is located in an electromagnetic field of frequency ω. This field can cause atom transitions between a lower o and excited $*$ states. Then the rates of transitions between these states (the transition probabilities per unit time) may be represented in the form

$$w(o, n_\omega \to *, n_\omega - 1) = A \cdot n_\omega; \quad w(*, n_\omega \to o, n_\omega + 1) = \frac{1}{\tau} + B \cdot n_\omega, \qquad (4.4.1)$$

where τ is the radiative lifetime of an excited state $*$ with respect to spontaneous radiative transition to a state o, n_ω A and B are the Einstein coefficients [253]. The arguments of the transition rates indicates that atom excitation is accompanied by absorption of one photon of a given frequency, and atom quenching leads to formation of one photon.

The expressions of the Einstein coefficients A and B may be obtained from the balance of these radiative transitions under thermodynamic equilibrium. Indeed, under equilibrium the number of transitions in both direction are equal, i.e.

$$N_o w(o, \overline{n}_\omega \to *, \overline{n}_\omega - 1) = N_* \, w(*, \overline{n}_\omega - 1 \to o, \overline{n}_\omega),$$

where N_o, N_* are the number densities of atom in the lower and upper states of transition, $\overline{n}_\omega$ is the number of photons of a given frequency in one state. Under thermodynamic equilibrium the number density of atoms in a given state is given by Boltzmann formula, and the number of photons in one state of frequency ω is determined by the Planck formula [254, 255], and hence we have

$$N_* = \frac{g_*}{g_o} N_o \exp\left(-\frac{\hbar\omega}{T}\right), \; \overline{n}_\omega = \left[\exp\left(-\frac{\hbar\omega}{T}\right) - 1\right]^{-1},$$

where T is the temperature expressed in energy units, g_o, g_* are the statistical weights of the lower and upper states, and the photon energy $\hbar\omega$ coincides with the energy difference between the two states. Substituting these expressions in the balance equation and equalizing terms which do not contain terms $\exp(-\hbar\omega/T)$ and contain this, we get for the Einstein coefficients

$$A = \frac{g_*}{g_o \tau}, \; B = \frac{1}{\tau} \qquad (4.4.2)$$

Let us derive the expression for the absorption cross section that is according to the cross section definition the ratio of the transition rate $w(o, n_\omega \to *, n_\omega - 1)$ to the photon flux j_ω. The photon flux is $j_\omega = c dN_\omega$, where the number density of photons in a range of photon wave vectors is $dN_\omega = 2n_\omega d\mathbf{k}/(2\pi)^3$, so that $\mathbf{k}$ is the photon wave vector, $d\mathbf{k}/(2\pi)^3$ is the number of states in a range $d\mathbf{k}$ of the wave vector values, and the factor 2 accounts for the two independent polarization states. From this it follows for the photon flux $j_\omega = \omega^2 d\omega/(\pi^2 c^2)$ if we use the dispersion relation $\omega = kc$ for photons. Finally, we find for the absorption cross section σ_ω

$$\sigma_\omega = \frac{w(o, n_\omega \to *, n_\omega - 1)}{j_\omega} = \frac{A n_\omega}{j_\omega} = \frac{\pi^2 c^2}{\omega^2} \cdot \frac{g_*}{g_o} \cdot \frac{a_\omega}{\tau} = \frac{\lambda^2}{4} \cdot \frac{g_*}{g_o} \cdot \frac{a_\omega}{\tau} \qquad (4.4.3)$$

In the same manner one can derive the expression for the stimulated cross section σ'_ω as the ratio of the rate of photon formation under action of the radiation field Bn_ω to the photon flux j_ω. We obtain

$$\sigma'_\omega = \frac{Bn_\omega}{j_\omega} = \frac{\pi^2 c^2}{\omega^2} \cdot \frac{a_\omega}{\tau} \tag{4.4.4}$$

Let us introduce the absorption coefficient k_ω such that the intensity of radiation I_ω of frequency ω that propagates in a gas of atoms in direction x, is given by

$$\frac{dI_\omega}{dx} = -k_\omega I_\omega \,, \tag{4.4.5}$$

One can see that $1/k_\omega$ is the mean free path of photons in a gas of atoms. According to the definition, the absorption coefficient k_ω is expressed through absorption cross section σ_ω and the cross section of stimulated radiation σ'_ω as

$$k_\omega = N_o\sigma_\omega - N_*\sigma'_\omega = N_o\sigma_\omega\left(1 - \frac{N_*}{N_o}\frac{g_o}{g_*}\right) \tag{4.4.6}$$

If the distribution of atom densities is determined by the Boltzmann formula (2.1.5), the expression for the absorption coefficient takes the form

$$k_\omega = N_o\sigma_\omega\left[1 - \exp\left(-\frac{\hbar\omega}{T}\right)\right] \tag{4.4.7}$$

We have for the absorption coefficient (4.4.7) in the spectral line center with using the expression (4.4.3) for the absorption cross section

$$k_o = N_o\sigma_\omega \sim N_o\frac{c^2}{\omega^2}\frac{a_o}{\tau} \tag{4.4.8}$$

Since the frequency distribution function at the cental frequency is $a_o \sim 1/\Delta\omega \sim \hbar/(N_o d^2)$ and the radiative time according to formula (4.1.1) is $1/\tau \sim \omega^3 d^2/(\hbar c^3)$, the absorption coefficient at the spectral line center is estimated as

$$k_o \sim \frac{\omega}{c} \sim \frac{1}{\lambda}, \tag{4.4.9}$$

where λ is the wavelength for radiation. As it follows from this, the mean free path of resonant photons at the center of the resonant spectral line is of the order of the wavelength λ of resonant radiation. This is determined by the identical nature of interaction for atom resonant radiation and for collisions of atoms in the ground and resonantly excited states.

Using this fact, we determine now the absorption coefficient for the line center for the radiative transition

$$A(^1P) \to A(^1S) + \hbar\omega, \qquad (4.4.10)$$

where $A(^1S)$ is the ground atom state, so that the total collision cross section σ_t involving the ground ^1S and resonantly excited atom state ^1P is given by formula (4.3.14) [252]

$$\sigma_t = \frac{4.8\pi d^2}{\hbar v}$$

Here d is the matrix element of the dipole atom moment for transition between these states, and the line width due to collisions is [152] $\nu = N_a v \sigma_t/2$. According to formulas (4.3.3), (4.3.16), (4.4.3), and (4.4.6) we have for the absorption coefficient in the line center

$$k_o = \frac{\pi c^2}{2\omega^2} \frac{1}{v\sigma_t \tau_r} \frac{g_*}{g_o}, \quad \hbar\omega \gg T, \qquad (4.4.11)$$

The radiative lifetime τ_r of an isolated atom in the ground state due to the dipole-dipole interaction is given by [152, 153]

$$\frac{1}{\tau_r} = \frac{4\omega^3}{3\hbar c^3} d^2 g_o, \qquad (4.4.12)$$

where the matrix element of the dipole moment operator d is taken between radial wave functions of the valence electron. We note that introducing the matrix element of the operator of the dipole moment d, we average it over polarization of the electromagnetic wave. If we account for the polarization of an electromagnet wave, we also take into account the selection law for a radiative transition, and as a consequence from this the final state of the transition between states s and p of a valence electron is chosen depending on the wave polarization. This means that for a given wave polarization we have $g_o = g_* = 1$, and after averaging over a wave polarization we obtain

$$k_o = \frac{\omega g_*}{1.8c} = \frac{3.5}{\lambda} \qquad (4.4.13)$$

In particular, for the resonant transition in helium

$$He(2^1P) \to He(1^1S) + \hbar\omega$$

with the wavelength $\lambda = 58.43\,\text{nm}$ this formula gives $k_o = 6.0 \cdot 10^5\,\text{cm}^{-1}$. In the case of the resonant transitions in argon

$$Ar(3p^5 4s)(^1P_1) \rightarrow Ar(3p^6)(^1S_1) + \hbar\omega, \; Ar(3p^5 4s)(^3P_1) \rightarrow Ar(3p^6)(^1S_1) + \hbar\omega$$

with the wavelengths $\lambda = 104.8\,\text{nm}$ and $\lambda = 106.7\,\text{nm}$ we have approximately $k_o = 3.3 \cdot 10^5\,\text{cm}^{-1}$ for these states $Ar(^1P_1)$ and $Ar(^3P_1)$. As is seen, the mean free path of a resonant photon at the line center $\sim 1/k_o$ is small compared with a typical laboratory size.

4.5 Propagation of Resonant Radiation in the Gas

Atomic radiative transitions are of interest for a gas discharge plasma in two relations—for kinetics of excited atoms and when this plasma is a source of radiation. Let us consider kinetics of excited atoms when it is governed by the following processes

$$e + A \longleftrightarrow e + A^* \; ; \; A^* \rightarrow A + \hbar\omega \, , \tag{4.5.1}$$

where A, A^* are the atoms in the ground and excited states, and $\hbar\omega$ denotes a resonance photon. Taking A^* to be a resonantly excited state, we obtain the following balance equation on the basis of the above scheme of processes

$$\frac{dN_*}{dt} = N_e N_o k_{ex} - N_e N_* k_q - \frac{N_*}{\tau}, \tag{4.5.2}$$

where N_o, N_* are the number densities of atoms in the ground and resonantly excited states, N_e is the number density of electrons, k_{ex} is the rate constant of atom excitation by electron impact, k_q is the rate constant of quenching of a resonantly excited atom by electron impact, and τ is a lifetime of the resonantly excited atom with respect to radiation, and this lifetime accounts for the reabsorption process if it takes place, i.e. $\tau \geq \tau_r$, where τ_r is the radiative lifetime of an isolated atom. The solution of this equation under stationary conditions gives

$$N_* = N_o \frac{k_{ex}}{k_q} \left(1 + \frac{1}{N_e k_q \tau}\right)^{-1} \tag{4.5.3}$$

From this formula it follows that radiation does not violate an equilibrium in the plasma if the following criterion is fulfilled

$$N_e k_q \tau \gg 1 \tag{4.5.4}$$

Under the opposite criterion, when electron-atom collisions do not restore the equilibrium number density of excited atoms, it is given by

$$N_* = N_o N_e k_{ex} \tau \tag{4.5.5}$$

Let us introduce the optical thickness u of a plasma for a given photon frequency ω as $u = k_\omega L$, where L is a plasma size. We below consider the case of optically thick plasma for the center of the resonant spectral line $u_o = k_o L \gg 1$ and analyze the character of plasma radiation in this case. We find first the probability $P(L)$ that a resonant photon propagates on a distance L without absorption. This probability is the product of the probability $a_\omega d\omega$ for photon emission at a given frequency ω and the probability $\exp(-k_\omega L)$ of photon surviving, i.e. this quantity is equal to [256, 257]

$$P(L) = \int a_\omega \mathrm{d}\omega \exp(-k_\omega L) \tag{4.5.6}$$

For the Lorentz line shape (4.3.3) we introduce a new variable $s = (\omega - \omega_o)/\nu$ and take the optical thickness of this plasma for the line center as $u = k_o L$. Then we have

$$a_\omega d\omega = \frac{ds}{\pi(1+s^2)}, \quad k_\omega = \frac{k_o}{1+s^2},$$

and formula (4.5.6) gives for the probability of photon propagation on a given distance for a large optical thickness u [256, 257]

$$P(L) = \frac{1}{\sqrt{\pi u}}, \quad u \gg 1 \tag{4.5.7}$$

In the case of the Doppler shape of the spectral line, using a new variable

$$t = u \exp\left[-\frac{mc^2}{2T}\left(\frac{\omega - \omega_o}{\omega_o}\right)^2\right],$$

we obtain for the probability for photon surviving when it passes a distance L

$$P(L) = \frac{1}{\sqrt{\pi u}} \int_0^u e^{-t} dt \left(\ln \frac{u}{t}\right)^{-1} = \frac{1}{\sqrt{\pi u}\sqrt{\ln u + C}}, \quad u \gg 1 \tag{4.5.8}$$

where $C = 0.577$ is the Euler constant.

Let us consider competition of these mechanisms of broadening in order to choose the strongest one among them. In the case of reabsorption of resonant photons we obtain different contribution to the broadening of spectral lines depending on a distance passed by radiation. The ratio of probabilities of surviving of resonant photons after propagation on a distance L is equal according to formulas (4.5.7) and (4.5.8)

$$\frac{P_L(L)}{P_D(L)} = \frac{k_D L \sqrt{\ln(k_D L) + C}}{\sqrt{k_L L}}, \tag{4.5.9}$$

where $P_L(L)$ and $P_D(L)$ are the probability of photon surviving as a result of its propagation on a distance L, k_L and k_D are the absorption coefficients at the line centers for the Lorentz and Doppler broadening mechanisms. In particular, one can find from formula (4.5.9) that if the Doppler width of the spectral line for an individual atom is larger than the Lorenz width, i.e. the Doppler mechanism determines radiation of an optically think plasma, for an optically thick plasma the role of the Lorenz mechanism of broadening increases. Indeed, the ratio (4.5.9) increases with an increasing L, and if at small L it is smaller one, it exceeds one at some L.

We note the assumption used that the rate of radiation of a photon of a given frequency is proportional to the photon distribution function a_ω, i.e. we are based on the principles of statistical physics. Then a given distance is reached without reabsorption by photons whose mean free path is compared to this distance. Because wings of the Doppler spectral line drop sharper than that in the case of the Lorentz spectral line, the probability to propagate on large distances for the Lorentz spectral line is larger than that for the Doppler one.

It should be noted that in the case under consideration, where $k_o L \gg 1$ the character of propagation of resonant photons through a gas differs from the diffusion character of propagation if a particle with a small mean free path moves in a gas. Since in this case photons are absorbed at one frequency and are emitted at other one, they reach the plasma boundary as a result single reabsorption at a central part of the spectral line, and by single reabsorption at the wing of the spectral line. This leads to broadening of the wavelength for radiation that passes through an optically thick gas. One can determine the width of this line such that the mean free path of photons at the boundary frequency is comparable with a plasma dimension, i.e. in the case where an uniform plasma fills an infinite cylinder tube the boundary frequency ω is given by $k_\omega L \sim 1$. Then denoting the line width as Δ_L, we have on the basis of formula (4.3.3) we have for the Lorenz mechanism of broadening

$$\Delta\omega_L \sim \nu\sqrt{k_o L}, \tag{4.5.10}$$

and for the Doppler mechanism of broadening (4.3.5) the width of the frequency distribution function for the yield radiation has the form

$$\Delta\omega_D \approx \frac{\Delta\omega_D}{\sqrt{\pi}} \ln(k_o L)\sqrt{k_o L}, \tag{4.5.11}$$

where $\Delta\omega_D = 1/a(\omega_o) = \omega_o\sqrt{(2\pi T/mc^2)}$ is the width of the spectral line for radiation of a single atom. In the case of competition of these mechanisms of broadening it is necessary to choose such of the larger width.

Note that radiation fluxes at frequencies where a uniform plasma is optically thick have the same order of magnitude, but these fluxes are determined by different plasma volumes. For the line center the radiation flux is created by a region near the surface that restricts this plasma, and a typical depth of this region is $\sim 1/k_o$, whereas the main part of the spectral line corresponds to the total plasma volume.

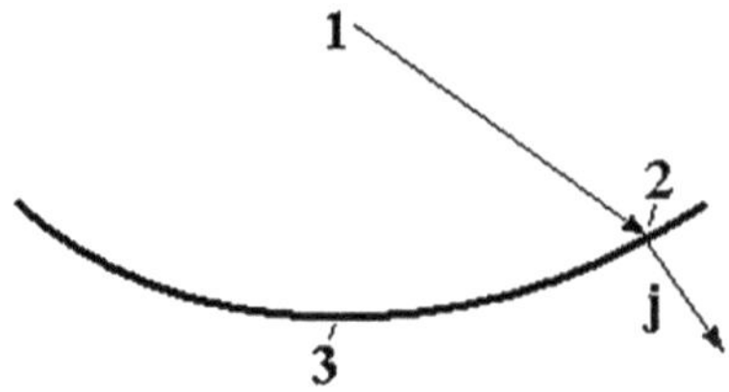

Fig. 4.1 Geometry of formation of the photon flux from a region filled by a uniform plasma. **1**—a point of photon emitting of an individual photon that is directed to a point **2** of the plasma surface and gives a contribution to a total photon flux j through this point. **3** is the plasma boundary

This character of broadening of plasma spectral lines leads to a specific effect that is called a self-reversal of spectral lines and consists in a depression of the spectral line center. This effect relates to a non-uniform plasma when the number density of excited atoms (or the plasma temperature) drops near the plasma boundary. Then the radiation flux for the central part of the spectral line is created by a plasma region near its boundary is lower than that for wings of the spectral line for which the total plasma volume is responsible.

We now determine the flux of resonant radiation through a boundary that separates a uniform plasma from a vacuum. The number density of resonantly excited atoms is N_* the number density of atoms in the ground state is N_a, and resonant photons occur as a result of the radiative transition between these states. Then the radiative flux resulted from the radiative transitions between indicated states is given by

$$j_\omega d\omega = \int \frac{N_*(\mathbf{r})d\mathbf{r}}{\tau} \cdot a_\omega d\omega \cdot \frac{1}{4\pi r^2} \cdot \exp\left(-\int_0^r k_\omega dr'\right) \cdot \cos\theta \qquad (4.5.12)$$

We construct this radiative flux in accordance with notations of Fig. 4.1 as a product of several factors assuming that emitted photons are distributed isotropically. In formula (4.5.12) the first factor is the rate of photon emission in a given volume element, the second factor is the probability of photon generation in a given frequency range, the third factor accounts for that the emission rate relates to the square $4\pi r^2$ where r is the photon way inside a plasma, the fourth factor is the probability that the emitted photon reaches the surface, and the last factor accounts that the photon trajectory intersects the plasma surface with an angle θ. Because of the plasma uniformity, we obtain for the total flux of photons j, introducing the probability $P(r)$ for a photon to propagate on a distance r without absorption

$$j = \int j_\omega d\omega = \frac{N_*}{\tau} \int \frac{P(r)d\mathbf{r}}{4\pi r^2} \cdot \cos\theta \qquad (4.5.13)$$

In integrating formula (4.5.12) over frequencies, we above assume that partial fluxes j_ω have the same direction. It holds true for symmetric volumes filled by a plasma, in particular, if a plasma is located inside a cylinder tube as we consider below.

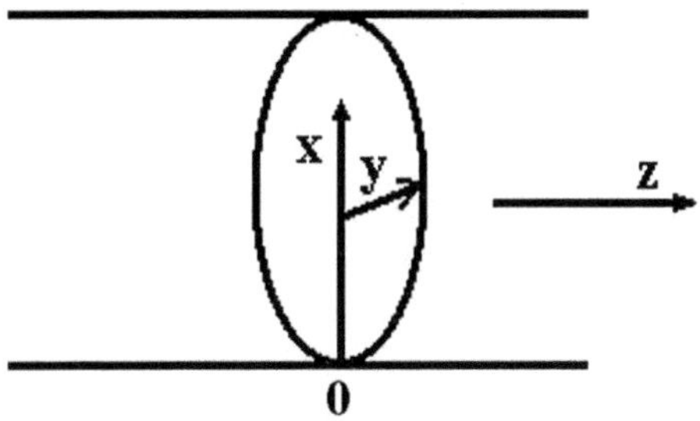

Fig. 4.2 The frame of reference for radiation of a uniform plasma located inside an infinite cylinder tube. **0** is the origin located on the cylinder surface, the axis **x** is directed perpendicular to the plasma surface, the axis **y** lies in the cylinder cross section, and the axis **z** is directed along the cylinder axis

We now evaluate the photon flux emitted by an infinite cylinder tube that is filled by a uniform plasma. Let us use the frame of reference (see Fig. 4.2) with the origin located on the cylinder surface and the axis of this frame of reference passes through this point and is directed perpendicular to the surface, i.e. it passes through the cylinder axis. Taking the flux direction as the axis x, the cylinder axis to be directed along the axis z, we reduce formula (4.5.13) to the form

$$j = \frac{N_*}{4\pi\tau}\int \frac{x\mathrm{d}x\mathrm{d}y\mathrm{d}z}{(x^2+y^2+z^2)^{3/2}}P(r)$$

We below restrict ourselves by the Lorenz shape of the spectral line and optically thick plasma for the center of the spectral line, i.e. $k_o R \gg 1$, where R is the cylinder radius. Then on the basis of formulas (4.5.7) and (4.5.13) we obtain for the photon flux at the cylinder boundary

$$j = j_o J,\ j_o = \frac{N_*\sqrt{R}}{\tau\sqrt{k_o}},\ J = \frac{1}{4\pi^{3/2}\sqrt{R}}\int \frac{x\mathrm{d}x\mathrm{d}y\mathrm{d}z}{(x^2+y^2+z^2)^{7/4}} \tag{4.5.14}$$

Integrating over dz and taking $t = z/\sqrt{x^2+y^2}$, we find

$$\int \frac{x\mathrm{d}x\mathrm{d}y\mathrm{d}z}{(x^2+y^2+z^2)^{7/4}} = \int \frac{x\mathrm{d}x\mathrm{d}y}{(x^2+y^2)^{5/4}}\int_0^\infty \frac{\mathrm{d}t}{(1+t^2)^{7/4}},$$

and

$$\int_0^\infty \frac{\mathrm{d}t}{(1+t^2)^{7/4}} = \frac{1}{2}\int_0^1 \frac{\xi^{1/4}\mathrm{d}\xi}{(1-\xi)^{1/2}} = \frac{1}{2}B\left(\frac{5}{4},\frac{1}{2}\right) = \frac{\sqrt{\pi}\Gamma(5/4)}{\Gamma(7/4)} = 0.875$$

where $\xi = 1/(1+t^2)$. Introducing the polar coordinates $x = \rho\cos\varphi,\ y = \rho\sin\varphi$ in the next integration, we obtain

$$\int \frac{xdxdy}{(x^2+y^2)^{5/4}} = \int_{-\pi/2}^{\pi/2}\int_0^{2R\cos\varphi} \frac{d\rho}{\sqrt{\rho}} = \sqrt{2R}\int_0^1 \frac{d\zeta}{\zeta^{1/2}}(1-\zeta)^{1/4}$$
$$= \sqrt{2\pi R}\frac{\Gamma(5/4)}{\Gamma(7/4)} = 4.95\sqrt{R}$$

Summing up these integrations, we obtain for the flux of resonant photons from the surface of a cylinder uniform plasma [258, 259]

$$j = 0.194\frac{N_*\sqrt{R}}{\tau\sqrt{k_o}} \tag{4.5.15}$$

Let us introduce the effective lifetime of resonantly excited atoms in a cylinder uniform plasma. If this plasma is optically think, the photon flux from its surface equals to $N_*R/(2\tau_r)$. Reducing the photon flux from an optically thick plasma to this expression, we obtain

$$\tau_{ef} = \frac{\tau_r\sqrt{k_oR}}{2\times 0.194} = 2.6\tau_r\sqrt{k_oR} \tag{4.5.16}$$

One can use a simple approximation for the effective radiative lifetime for excited atoms in a plasma

$$\tau_{ef} = \frac{\tau_r}{P(2R)} \tag{4.5.17}$$

where $P(r)$ is the probability for a resonant photon to pass a distance r without absorption in a optically dense gas in accordance with formula (4.5.7). One can find on the basis of this formula the effective radiative lifetime for resonantly excited atoms located in an optically dense gas if the line width is determined by the Doppler effect. By analogy with formulas (4.5.16) and (4.5.17) we have now on the basis of formula (4.5.8)

$$\tau_{ef} = 6.4\tau k_oR\sqrt{\ln(k_oR)+1.3} \tag{4.5.18}$$

4.6 Resonant Radiation in Excited Helium and Argon

We now summarize the results for broadening of resonant spectral lines of transitions in helium and argon

Table 4.2 Parameters of transport of resonant photons in a gas discharge plasma of helium and argon with processes (4.6.1) and parameters of broadening of spectral lines for these transitions in the parent plasma

Initial state $He(2^1P \to 1^1S)$	$He(1^1S)$	$Ar(^1S_0)$	$Ar(^1S_0)$
Final state $He(2^3P \to 2^3S)$	$He(2^1P)$	$Ar(3p^54s^3P_1)$	$Ar(3p^54s^1P_1)$
λ (nm)	58.433	106.67	104.82
$\Delta\varepsilon$ (eV)	21.22	11.62	11.83
ω_o, $10^{16}\mathrm{s}^{-1}$	3.2	1.8	1.8
f_{0*}	0.276	0.051	0.25
τ_r (ns)	0.56	10	2
$d^2/(e^2a_o^2)$	0.177	0.06	0.29
k_o, $(10^5\,\mathrm{cm}^{-1})$	6.0	3.3	3.3
$\tau_{ef}(\rho_o = 1\,\mathrm{cm})$, μs	1.1	15	3.0
$\Delta\omega_D$, cm^{-1}	0.19	0.11	0.11
σ_D, $10^{-13}\,\mathrm{cm}^2$	1.9	2.0	9.4
k_D/p, $10^3\,\mathrm{cm}^{-1}\mathrm{Torr}^{-1}$	4.5	4.8	23
$\Delta\omega_D(0)$, cm^{-1}	0.19	0.10	0.10
$\Delta\omega_L(0)/p$, $10^{-3}\,\mathrm{cm}^{-1}\mathrm{Torr}^{-1}$	3.3	1.1	5.4
$\Delta\omega_{cr}$, cm^{-1}	1.1	15	3.0
$P_L(2\rho_o)$	5.2×10^{-4}	6.9×10^{-4}	7.0×10^{-4}
$P_D(2\rho_o)$	1.4×10^{-5}	2.3×10^{-6}	1.3×10^{-5}

$$He(1^1S) + \hbar\omega \to He(2^1P), \; Ar(^1S_0) + \hbar\omega \to Ar(^{1,3}P_1), \tag{4.6.1}$$

and also to analyze reabsorption of these radiative transitions in a gas discharge plasma of helium and argon. The results are given in Table 4.2, where λ is the wavelength of the corresponding radiative transition, $\Delta\varepsilon$ is the excitation energy for the resonantly excited atom, $\omega_o = \Delta\varepsilon/\hbar$ is the photon frequency, f_{0*} is the oscillator strength for this transition, τ_r is the lifetime of an excited state with respect to the radiative transition, and the matrix element d is given by formula (4.3.15). The absorption coefficient in the line center k_o for the Lorenz mechanism of broadening of spectral lines is given by formula (4.4.13) and is evaluated above. Next, τ_{ef} is the average lifetime of resonantly excited atoms accordance to formula (4.5.16) at the indicate radius ($\rho_o = 1$ cm) of the discharge tube.

For definiteness, we consider a uniform gas discharge plasma with the gas temperature $T = 400$ K and the spectral line width for the Doppler broadening mechanism and an individual atom in accordance with formula (4.3.5) is given by

$$\Delta\omega_D = \omega_o\sqrt{\frac{2\pi T}{mc^2}} \tag{4.6.2}$$

Correspondingly, the cross section of absorption for the spectral line center in the case of the Doppler mechanism of line broadening according to formulas (4.4.3) and (4.3.5) is equal to

$$\sigma_D = \frac{3\lambda^2}{4} \cdot \frac{1}{\omega_o \tau} \cdot \sqrt{\frac{mc^2}{2\pi T}} = 1.2 \times 10^{-12}\,\text{cm}^2, \tag{4.6.3}$$

that leads to the following expression for the absorption coefficient at the line center

$$k_D = N_a \sigma_D = \frac{3N_a\lambda^2}{4} \cdot \frac{1}{\omega_o \tau} \cdot \sqrt{\frac{mc^2}{2\pi T}} \tag{4.6.4}$$

We use these formulas to determine the parameters of broadening of resonant spectral lines for the Doppler broadening mechanism and in the helium and argon cases collect the results in Table 4.2. The argument 0 in Table 4.2 means that the width corresponds to an individual atom.

In the case of the Lorenz broadening of the spectral line the mechanism of broadening results from collision of a resonantly excited atom with a parent atom in the ground state according to the scheme

$$A^* + A \rightarrow A + A^*(A^* + A); \tag{4.6.5}$$

Here the processes of elastic scattering of colliding atoms, excitation transfer and depolarization are entangled. If one colliding atom is found in ^1S state and the state of another atom is ^1P, the total cross section of collision of these atoms is given by formula (4.3.14) [252] and the broadening cross section is one half of the total cross section [152]. The Lorenz width of a spectral line for an individual atom according to formulas (4.3.3) and (4.3.13) is given by

$$\Delta\omega_L(0) = \frac{1}{a(\omega_o)} = \pi\nu = \frac{2.4\pi^2 N_a d^2}{\hbar}$$

Table 4.2 contains the values of this quantity for considering transitions at the gaseous temperature $T = 400$ K that corresponds to the number density of atoms $N_a = 2.4 \times 10^{16}\,\text{cm}^{-3}$ at the pressure 1 Torr.

The absorption coefficient k_o at the center line is determined on the basis of formula (4.4.13) and its values are given in Table 4.2. Note that in the Lorenz case in accordance with formula (4.4.13) the absorption coefficient is independent of the atom number density and gas temperature. Table 4.2 contains also the boundary width of a spectral line $\Delta\omega_{cr}$ for resonant radiation at which the Lorenz and quasistatic broadening mechanisms give an identical contribution to line broadening. The boundary width of a spectral line is determined by formula (4.3.20) and has the form

$$\Delta\omega_{cr} = \frac{v}{\sqrt{\sigma_t}}, \quad v = \sqrt{\frac{8T}{\pi M}}, \tag{4.6.6}$$

where we take the average atom velocity as v, and the total cross section of atom scattering is given by formula (4.3.14).

The probabilities to survive for a resonant photon propagated on a distance L which is large compared to the mean free path of the resonant photon at the line center, i.e. $k_o L \gg 1$ for the Lorenz mechanism of line broadening and $k_D(0)L \gg 1$ for the Doppler mechanism of line broadening, are given by formulas (4.5.7) and (4.5.8) correspondingly for the Lorenz and Doppler mechanisms of line broadening

$$P_L(l) = \frac{1}{\sqrt{\pi k_L l}}, \quad P_D(l) = \frac{1}{k_D l\sqrt{\pi}\sqrt{\ln(k_D l) + C}}, \quad k_L l \gg 1, k_D l \gg 1,$$

where l is a distance of photon propagation, and $C = 0.577$ is the Euler constant. The values $P_L(2l)$ and $P_D(2l)$ are represented in Table 4.2 for a distance $\rho_o = 1\,\text{cm}$ and the pressure $p = 1$ Torr. Under these conditions the width of a spectral line of an individual atom for resonant radiation is determined by the Doppler mechanism of line broadening, while the the width of a band of resonant radiation that leaves a gas discharge plasma of laboratory sizes is determined by Lorenz broadening. This means that the mean free path of a resonant photon in a gas at a spectral line wing is determined by the Lorenz broadening mechanism.

Figure 4.3 gives the dependence on the gas pressure for a passing distance l_o, where the probabilities of photon surviving for Lorenz and Doppler mechanisms of spectral line broadening are equal, i.e. the following relation holds true

$$P_L(l_o) = P_D(l_o) \tag{4.6.7}$$

Figure 4.4 contains the values of the surviving probabilities in accordance with equation (4.6.7) depending on the gas pressure p.

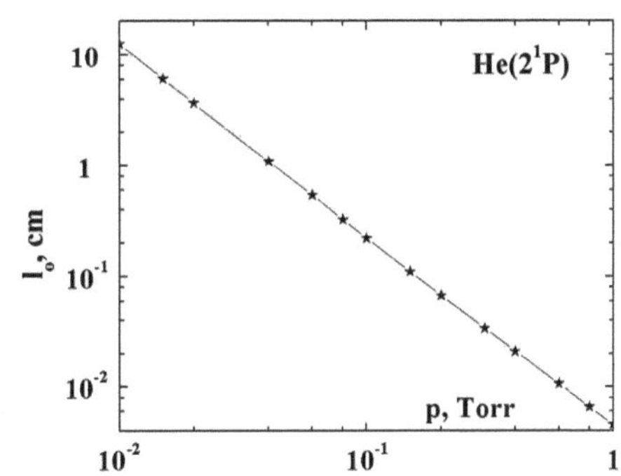

Fig. 4.3 Dependence of the distance l_o of resonant photon propagation on the gas pressure p in accordance with (4.6.7)

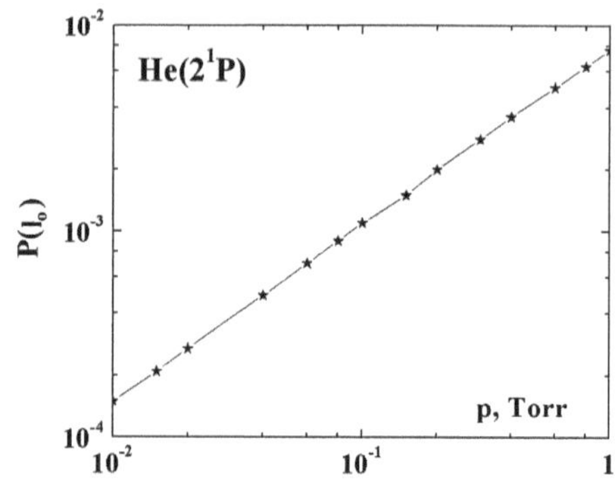

Fig. 4.4 Dependence of the probabilities of photon surviving $P_L(l_o) = P_D(l_o)$ at a distance l_o of resonant photon propagation on the gas pressure p in accordance with (4.6.7)

4.7 Block Model for Atom Levels

The structure of atom levels may be complex due to various types of electron interaction inside the atom. In considering the spectroscopy of atoms in a gas discharge plasma it is necessary to account for positions of each level until the width of a spectral line for radiative transitions involving this level is relatively small. Nevertheless, kinetics of corresponding excited atoms may be identical, and it can be considered on the basis of the model where some nearby levels of the same nature are joined in one level. As a result, atom states are divided in blocks [15], and this block model simplifies the analysis of kinetics of excited atoms. Being guided by an argon gas discharge plasma, we use this model for excited argon atoms.

In considering the population of lower excited states of the argon atoms, we are based on data of Fig. 4.2 and construct form this the block model with joining the states with the electron shell $Ar(3p^5 4p)$ in one state. In reality, this state is characterized by a valence $4p$ electron which is located in the field of the atomic core with the electron shell $3p^5$. Figure 4.5 contains the parameters of this model, where the energy of the lowest excited state $^3P_2$ of the argon atom is used as zero. The excitation energy E of the combined level is taken as

$$E = \frac{\sum_i g_i E_i}{\sum_i g_i}, \quad \sum_i g_i = 36 \tag{4.7.1}$$

and the rates $1/\tau$ of radiative transition in each level of the electron shell $Ar(3p^5 4p)$ is defined according to formula

$$\frac{1}{\tau_r} = \frac{\sum_i \frac{g_i}{\tau_i}}{\sum_i g_i} \tag{4.7.2}$$

Note that if we take the excitation energy of the combined level as an average of excitation energies of $2p_1$ and $2p_{10}$ levels, it is equal $E = 1.65\,\text{eV}$ instead of

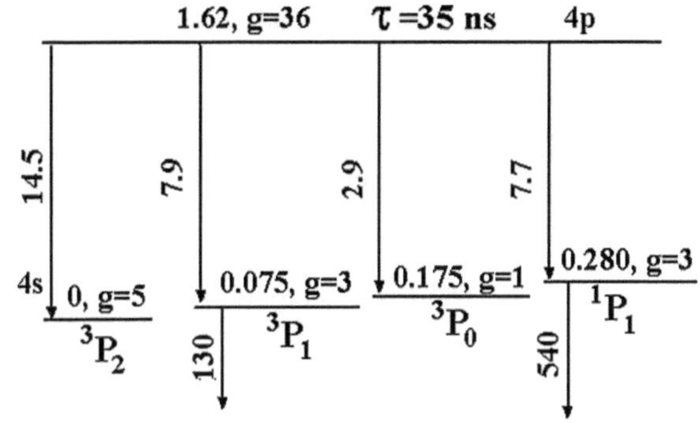

Fig. 4.5 Model for excited states of the argon atom where the levels of the electron shell $Ar(3p^5 4p)$ are joined in one level and the average parameters are used for the combined level. The excitation energy in eV is indicated on a line of each level along with its statistical weight g. The rates of radiative transitions are given in $10^6\,\mathrm{s}^{-1}$ near the corresponding *arrow*. States of the electron shell $3p^5 4s$ are given at the foot of the level line

1.62 eV of Fig. 4.5 with taking into account all the levels. Next, the summarized rate of radiative transitions gives $\tau = 33$ ns in comparison with $\tau = 35$ ns which is obtained on the basis of the radiative lifetimes of $2p$-levels. In addition, the rate of radiative transitions per one state of the electron shell $Ar(3p^5 4\,\mathrm{s})$ in units $10^6\,\mathrm{s}^{-1}$ is equal 2.9, 2.6, 2.9, 2.6 for the states $^3P_2$, $^3P_1$, $^3P_0$ and $^1P_1$ correspondingly (the average value is $(2.8 \pm 0.2) \cdot 10^6\,\mathrm{s}^{-1}$. One can see that introduction of a unit level in the block model leads to an error in atom parameters, and we assume this error to be small.

Let us find the quenching rate constant k_q for atom transition from a state of the electron shell $Ar(3p^5 4\,\mathrm{p})$ to a state with the electron shell $Ar(3p^5 4\,\mathrm{s})$ on the basis of formula (3.4.4) which are based on measurements for alkali metal atoms with transitions from p to s state of a valence electron where transition proceeds from one upper state in all lower states, and the rate of the radiative transition in a group of lower states with the statistical weight g_i is $g_i/(\tau_r g_o)$, where $1/\tau_r$ is the total radiative rate of the radiative transition in all lower states, and g_o is the total statistical weight of states of a given group. Correspondingly, the rate of transition between the lower and upper states that is the transition from one state of the lower state in all upper states is given by

$$k_{ex} = \frac{k_o g_*}{g_o (\Delta\varepsilon)^{7/2} \tau_r} \exp\left(-\frac{\Delta\varepsilon}{T_e}\right) = k_q \frac{g_*}{g_o} \exp\left(-\frac{\Delta\varepsilon}{T_e}\right), \qquad (4.7.3)$$

where we assume the Maxwell energy distribution of electrons, $\Delta\varepsilon$ is the excitation energy, and τ_r is the radiative lifetime of each state of an upper level group. As is seen formulas (4.7.3) and (3.4.5) are identical in the cases where transitions between s and p states proceed in cases of level splitting and in their absence.

From this one can evaluate the average rate constant of quenching of states $Ar(3p^54p)$ by electron impact with transitions in atom states with the electron shell $Ar(3p^54s)$. According to formula (3.4.4) we have for the quenching rate constant k_q with transition between these states is $k_q = 2.3 \cdot 10^{-7}\,\mathrm{cm^3/s}$. The rate of the inverse transition follows from the principle of detailed balance and is given by formula (3.3.8)

$$k_{ex} = k_q \frac{g_*}{g_o} \exp\left(-\frac{\Delta\varepsilon}{T_e}\right)$$

Let us consider the equilibrium between states

$$A_i + e \leftrightarrow A_k + e, \; A_k \to A_i + \hbar\omega \tag{4.7.4}$$

Balance equations for the number density of atoms in ith N_i and kth N_k states lead to the following formal solution in accordance with formula (4.5.3)

$$\frac{N_k}{N_i} = \frac{N_e k_{ex}}{N_e k_q + 1/\tau} = \frac{g_k}{g_i} \exp\left(-\frac{\Delta\varepsilon}{T_e}\right) A_i + e \leftrightarrow A_k + e, \; A_k \to A_i + \hbar\omega, \tag{4.7.5}$$

and $1/\tau$ is the rate of the radiative transition between these states. One can see that this formula coincides with the Boltzmann formula (2.1.5) if the criterion (4.5.4) holds true

$$N_e \gg \frac{1}{k_q \tau}$$

In the limit of a low number density of excited atoms in the lower state τ is the radiative lifetime of an isolated atom. In particular, in the case if the state i is $Ar(3p^54s)$ and the state k is $Ar(3p^54p)$ the criterion (4.5.4) takes the form ($\tau = 35\,\mathrm{ns}$, $k_q = 2.3 \times 10^{-7}\,\mathrm{cm^3/s}$)

$$N_e \gg 10^{14}\,\mathrm{cm^{-3}} \tag{4.7.6}$$

If reabsorption of resonant photons takes place, the above average over states of a given group is not valid. The effective lifetime τ_{ef} of a certain level of a group k is given by formula (4.5.16) with the absorption coefficient $k_o \sim 1/\lambda$ according to formula (4.4.13), and this consideration requires the validity of the criterion

$$N_* \lambda^3 \gg 1,$$

and in the case of the transition between $Ar(3p^54s)$ and $Ar(3p^54p)$ states this criterion $N_* \gg 10^{12}\,\mathrm{cm^{-3}}$ relates to each state of this group. Since the criterion (4.7.6) is not fulfilled under considering conditions, as well as the criterion for the partial number density of excited atoms, the relative population of levels $Ar(3p^54p)$ is lower than that due to the Boltzmann formula (2.1.5).

Part II
Kinetics and Transport Phenomena in Gas Discharge Plasma

Chapter 5
Collision Processes in Kinetics of Gas Discharge Plasma

Abstract Kinetics of gases and plasma is described by the Boltzmann kinetic equation, and the expression for the collision integral of electrons is represented for cases where electron-atom collisions or electron-electron collisions dominate. Depending on the relation between these collision integrals two different regimes of plasma evolution are possible.

5.1 Kinetic Equation for Charged Particles

A gas discharge plasma is a self-consistent state of an ionized gas that is formed and maintained by an external electric field and processes inside it. The electric field creates and supports a plasma and provides passage of the electric current through a gas. An energy from an external electric field is transferred first to electrons and then is transmitted to gas atoms as a result of electron-atom collisions. Ions as a charged component influence on properties of a gas discharge plasma especially by their charge. Thus the properties of the gas discharge plasma result from both interaction of electrons with an external field and various collision processes. Therefore kinetics of atomic particle in gas discharge plasma is of importance for creation of a certain self-consistent state of this system.

In description the kinetics of atomic particles in a gas discharge plasma, we will be based in first turn on the kinetic equation for the distribution function of atomic particles under consideration that has the form [256]

$$\frac{\partial f(\mathbf{v})}{\partial t} + \frac{e\mathbf{E}}{m} \cdot \frac{\partial f(\mathbf{v})}{\partial \mathbf{v}} = I_{col}(f) \qquad (5.1.1)$$

Here $\mathbf{r}$ is a coordinate, $\mathbf{v}$ is the particle velocity, m is the particle mass, $f(\mathbf{v})$ is the velocity distribution function for these particles that is normalized to the number density N of particles ($\int f(\mathbf{v})d\mathbf{v} = N$), and $e\mathbf{E}$ is the force that acts on a charged particle (e is its charge) from an external field. The collision integral $I_{col}(f)$ is of principle in this equation and is responsible for collision of this particle with some particles of a plasma. As a matter, the kinetic equation (5.1.1) is the balance equation for particles of a given type in the velocity space so that transitions of a particle in

B.M. Smirnov, *Theory of Gas Discharge Plasma*, Springer Series on Atomic, Optical, and Plasma Physics 84, DOI 10.1007/978-3-319-11065-3_5

the phase space results both from evolution of an individual particle in external field and from pair collisions with other particles of this gaseous system.

In the case of elastic collisions of these particles with gas atoms the collision integral has the form

$$I_{col}(f) = \int (f\varphi - f'\varphi') \mid \mathbf{v} - \mathbf{v}_a \mid d\sigma d\mathbf{v}' d\mathbf{v}'_a d\mathbf{v}_a, \tag{5.1.2}$$

where $\mathbf{v}$, $\mathbf{v}'$ are the velocities of a test particle before and after collision, $\mathbf{v}_a$, $\mathbf{v}'_a$ are the atom velocities before and after collision, $d\sigma$ is the differential cross section of scattering of a test particle on a gas atom. We use the following notations for the distribution functions of a test particle $f \equiv f(\mathbf{v})$, $f' \equiv f(\mathbf{v}')$ and for the Maxwell distribution function of gas atoms $\varphi \equiv \varphi(\mathbf{v}_a)$, $\varphi' \equiv \varphi(\mathbf{v}'_a)$. This collision integral may be used for the analysis of particle evolution in a gas discharge plasma as a result of elastic collisions with gas atoms. The simplest representation of the collision integral corresponds to "tau-approximation"

$$I_{col}(f) = -\frac{f(\mathbf{v}) - f_o}{\tau}, \tag{5.1.3}$$

where $f_o(\mathbf{v})$ is the equilibrium distribution function, and τ is a relaxation time. In this approximation we consider a gas as a continuous matter, and the parameter τ characterizes the friction force acted on a test atomic particle which moves in a gas.

If inelastic collisions are included in the kinetic equation (5.1.1), the collision transitions between states must be introduced in the right-hand side of equation. In particular, the collision integral for the electron distribution function f has the following form if we take into account the processes of atom excitation in a state $*$ by electron impact as well as quenching of this state

$$I_{col}(f) = -N_a k_{ex}(\varepsilon) f(\mathbf{v}) + N_* k_q(\varepsilon') f(\mathbf{v}'), \tag{5.1.4}$$

where $\varepsilon = m_e v^2/2$ is the electron energy, m_e is the electron mass, $k_{ex}(\varepsilon)$ is the rate constant of atom excitation in collision with an electron of an energy ε, $\Delta\varepsilon$ is the excitation energy ($\varepsilon > \Delta\varepsilon$), $k_q(\varepsilon')$ is the rate constant of atom quenching with an electron of an energy $\varepsilon' = \varepsilon - \Delta\varepsilon = m_e(v')^2/2$; N, N_* are the atom number densities in the ground and excited states correspondingly. In the same manner the collision integral may be represented in the case of other inelastic processes.

We note the property for average parameters of electrons and ions in a gas if the concentration of these atomic particles is small and one can neglect interaction between them. In this regime of low concentration of charged particles the electric field strength E is included in the kinetic equation in the combination E/N_a, where N_a is the number density of gas atoms. Therefore both the distribution function of charged particles and their average parameters depend on the electric field strength through the combination E/N_a. This is demonstrated in Figs. 5.1 and 5.2 for the drift velocity of electrons in helium and argon in the regime of low electron concentration.

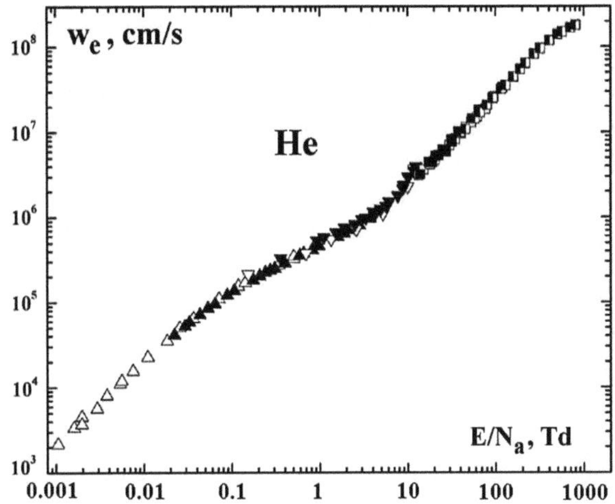

Fig. 5.1 The electron drift velocity w_e in helium as a function of the electric field strength E/N_a in the regime of a low electron number density according to measurements [257–262]

Fig. 5.2 The electron drift velocity w_e in argon as a function of the electric field strength E/N_a in the regime of low electron number densities according to measurements [6, 7, 258, 260, 261, 263–267]

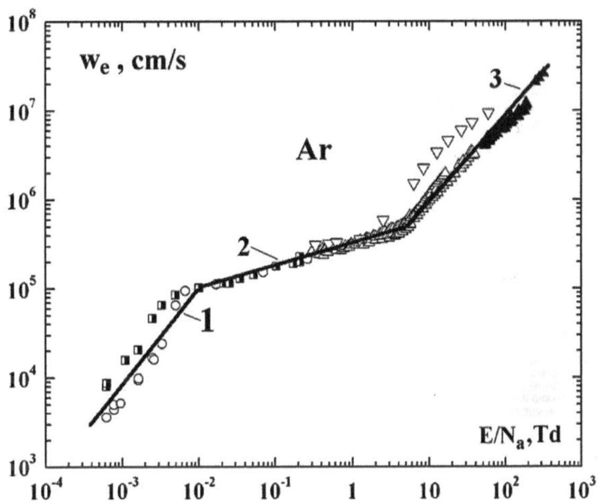

5.2 Integral Relations for an Average Momentum and Energy of a Charged Particle in Gas in Electric Field

We have a specific character of energy transmittance in a gas discharge plasma. An external electric field acts on electrons and does not act on gas atoms. Hence, the equilibrium for gas atoms results from elastic collisions between them, and atoms are characterized by the Maxwell velocity distribution function with a gas temperature T. For electrons and ions such an equilibrium is violated by an action of an external electric field, and the behavior of charged particles in a plasma is described by the kinetic equation that accounts for both collisions of charged particles with gas atoms and the action of external fields on charged particles. One can simplify this description replacing collisions by interaction of charged particles with a continuous matter. This corresponds to a hydrodynamic description of motion of charged particles in a gas. Transferring from the kinetic description to the hydrodynamic one, we really go from the kinetic equation of charged particles to its integral form with additional assumptions. For this reason it is useful to obtain integral relations for parameters of charged particles moved in a gas in an external field directly from the kinetic equation. This operation will be done below [38].

We below derive from this the equation for the average momentum of a charged particle assuming that their elastic collisions with gas atoms are responsible for the energy change for these charged particles. Then we use the kinetic equation (5.1.1) with the collision integral (5.1.2) in this case, multiply the kinetic equation (5.1.1) by the momentum $m\mathbf{v}$ of a charged particle and integrate over velocities $d\mathbf{v}$ of this particle. We assume the space distribution of particles to be uniform and take use the following normalization of the distribution functions

$$\int f(\mathbf{v})d\mathbf{v} = N_i, \quad \int \varphi(\mathbf{v}_a)d\mathbf{v}_a = N_a \tag{5.2.1}$$

Here $\mathbf{v}$, $\mathbf{v}_a$ are the velocities of charged particles and gas atoms, $f(\mathbf{v})$ is the distribution function of test particles, $\varphi(\mathbf{v}_a)$ is the Maxwell distribution function of gas atoms, N_i, N_a are the number densities of charged particles and gas atoms. In the course of integration of the kinetic equation, we use he principle of detailed balance for elastic scattering of charged particles of gas atoms that consists in evolution of the system in the inverse direction as a result of time inverse $t \to -t$ that leads to mutual replacing of velocities of colliding particles before and after collision $\mathbf{v} \leftrightarrow \mathbf{v}'$, $\mathbf{v}_a \leftrightarrow \mathbf{v}'_a$, where the particle velocities after collision are denoted by a superscript $\prime$. This gives

$$\int \mathbf{v} f(\mathbf{v}')\varphi(\mathbf{v}'_a)d\sigma d\mathbf{v}d\mathbf{v}_a = \int \mathbf{v}' f(\mathbf{v})\varphi(\mathbf{v}_a)d\sigma d\mathbf{v}d\mathbf{v}_a$$

This leads to the following relation after integration

$$m\frac{d\mathbf{w}}{dt} = e\mathbf{E} - \frac{m}{N_i}\int (\mathbf{v}' - \mathbf{v})gd\sigma f(\mathbf{v})\varphi(\mathbf{v}_a)d\mathbf{v}d\mathbf{v}_a,$$

where m is the mass of a charged particle, M is the atom mass, $\mathbf{w}$ is the average velocity (drift velocity) of charged particles, $\mathbf{g} = \mathbf{v} - \mathbf{v}_a$ is the relative particle velocity.

The velocities of a charged particle are expressed through the relative velocity of colliding particles $\mathbf{g}$ and the velocity of their center of mass $\mathbf{V}$ through relations

$$\mathbf{v} = \mathbf{V} + \frac{M\mathbf{g}}{m+M}, \quad \mathbf{v}' = \mathbf{V} + \frac{M\mathbf{g}'}{m+M}, \quad m(\mathbf{v} - \mathbf{v}') = \mu(\mathbf{g} - \mathbf{g}')$$

where $\mu = mM/(m + M)$ is the reduced mass of colliding charged and gaseous particle. Let us represent the relative velocity of particles in the form $\mathbf{g}' = \mathbf{g}\cos\vartheta + \mathbf{k}g\sin\vartheta$, where ϑ is the scattering angle, and the unit vector $\mathbf{k}$ directs perpendicular to the relative velocity $\mathbf{g}$. Because of the random character of scattering, the second term gives zero after integration over scattering angles, and

$$\int (\mathbf{g} - \mathbf{g}')d\sigma = \mathbf{g}\sigma^*(g),$$

where $\sigma^*(g) = \int(1-\cos\vartheta)d\sigma$ is the diffusion cross section for collision of a charged particle with a gas atom. As a result, we have the following integral relation

$$m\frac{d\mathbf{w}}{dt} = e\mathbf{E} - \frac{\mu}{N_i}\int \mathbf{g}g\sigma^*(g)f(\mathbf{v})\varphi(\mathbf{v}_a)d\mathbf{v}d\mathbf{v}_a \tag{5.2.2}$$

As a matter, this is the motion equation for a charged particle, where the first term in the right hand side of the equation is a force from an external electric field, and the second term is a friction force due to collisions of a charged particle with gas atoms. Just this term is responsible for energy transfer from the field to a gas through interaction with charged particles. If the rate of elastic collisions of a charged particles with gas atoms $\nu = N_a g\sigma^*(g)$ is independent of the collision velocity, the relation (5.2.2) takes a simple form

$$m\frac{d\mathbf{w}}{dt} = e\mathbf{E} - \mu\nu\mathbf{w}, \tag{5.2.3}$$

that is the motion equation for charged particles.

We now derive the integral relation for the average energy of a charge particle with use the above method. Indeed, multiplying the kinetic equation (5.1.1) by the energy of a charged particle $mv^2/2$ and integrating over particle velocities $d\mathbf{v}$, we obtain the following integral equation with use of the principle of detailed balance

$$\frac{d\overline{\varepsilon}}{dt} = e\mathbf{E}\mathbf{w} - \frac{m}{2N_i}\int [(v')^2 - v^2]gd\sigma f(\mathbf{v})\varphi(\mathbf{v}_a)d\mathbf{v}d\mathbf{v}_a\ ,$$

where $\overline{\varepsilon}$ is the average energy of charged particles. Transferring to the relative velocity of a colliding charged particle and gas atom $\mathbf{g}$ and to velocity of their center of mass $\mathbf{V}$, we have

$$\frac{m}{2}[(v')^2 - v^2] = \mu\mathbf{V}(\mathbf{g} - \mathbf{g}')\ ,$$

because the velocity of the center of masses is conserved as a result of collision, and the relative velocity change a direction only, i.e. $(g')^2 = g^2$. This gives

$$\frac{d\overline{\varepsilon}}{dt} = e\mathbf{E}\mathbf{w} - \frac{\mu}{N_i}\int \mathbf{V}(\mathbf{g} - \mathbf{g}')gd\sigma(g)f(\mathbf{v})\varphi(\mathbf{v}_a)d\mathbf{v}d\mathbf{v}_a$$

Integrating over the scattering angle and averaging over a polar scattering angle in the plane that is perpendicular to the relative velocity of colliding particles, we have as early

$$(\mathbf{g} - \mathbf{g}')d\sigma(g) = \mathbf{g}\sigma^*(g)$$

where $\sigma^*(g)$ is the diffusion cross section of elastic collision between a charged particle and gas atom. As a result, we obtain

$$\frac{d\overline{\varepsilon}}{dt} = e\mathbf{E}\mathbf{w} - \frac{\mu}{N_i}\int(\mathbf{V}\mathbf{g})g\sigma^*(g)f(\mathbf{v})\varphi(\mathbf{v}_a)d\mathbf{v}d\mathbf{v}_a \tag{5.2.4}$$

This is the balance equation for the energy of charged particles. The first term of the right hand side describes energy transfer from an external field to charged particles, and the second term is responsible for energy transfer from charged particles to gas atoms.

The relation (5.2.4) is simplified if the collision rate for a charged particle with gas atoms $\nu = N_a g\sigma^*(g)$ is independent of the collision velocity. Then in a motionless gas the integral relation (5.2.4) takes the form

$$\frac{d\overline{\varepsilon}}{dt} = e\mathbf{E}\mathbf{w} - \mu\nu\langle\mathbf{V}\mathbf{g}\rangle,$$

where an average is made over the distribution functions of charged particles and atoms. Returning to velocities of charged particles and gas atoms, we obtain

$$\langle\mathbf{V}\mathbf{g}\rangle = \frac{m}{m+M}\langle v^2\rangle - \frac{M}{m+M}\langle v_a^2\rangle$$

The average energy of a charged particle equals to $\overline{\varepsilon} = m\langle v^2\rangle/2$, and an average atom kinetic energy is $3T/2 = \langle Mv_a^2/2\rangle$, where T is the gas temperature expressed in energy units. On the basis of the expression $w = eE/\mu\nu$ for the drift velocity of a charged particle, we reduce the equation to the form

$$\frac{d\overline{\varepsilon}}{dt} = \mu w^2\nu - \overline{\varepsilon}\frac{2\mu\nu}{m+M} + T\frac{3\mu\nu}{m+M} \tag{5.2.5}$$

In particular, this gives in the stationary case for the average energy of charged particles

$$\overline{\varepsilon} = \frac{(m+M)w^2}{2} + \frac{3}{2}T \tag{5.2.6}$$

The relations obtained describe the behavior of average parameters for charged particles moving in a gas in an external electric field.

5.3 Integral of Electron-Atom Collisions

Kinetics of electrons in an atomic gas located in an external field is governed by a small parameter m_e/M in an energy range where elastic electron-atom collisions dominate. In particular, colliding electron and atom can be changed by $\sim m_e/M$ part of the electron energy, and hence the velocity distribution function of electrons in a gas is close to a spherically symmetric one [268, 269]. This fact allows us to represent the electron distribution function $f(\mathbf{v})$ as an expansion over spherical harmonics [270–276]

$$f(\mathbf{v}) = f_0(v) + v_z f_1(v), \tag{5.3.1}$$

where the electric field strength $\mathbf{E}$ directs along the axis z. Then the kinetic equation (5.3.1) for the electron distribution function $f(\mathbf{v})$ takes the form

$$\frac{e\mathbf{E}}{m_e} \cdot \frac{\partial f}{\partial \mathbf{v}} = I_{ee}(f) + I_{ea}(f), \tag{5.3.2}$$

where $I_{ee}(f)$, $I_{ea}(f)$ are the integrals of electron-atom and electron-electron collisions correspondingly. Integrating the kinetic equation (5.3.2) over the angle θ between the electron velocity and the electric field with accounting for the expression (5.3.1) for the electron distribution function, and also integrating the equation (5.3.2) after its multiplication by $\cos\theta$, we obtain from (5.3.2) the following equation for spherical harmonics [272–275]

$$a\frac{v_z df_0}{v dv} = I_{ea}(v_z f_1), \quad \frac{a}{3v^2}\frac{d}{dv}\left(v^3 f_1\right) = I_{ea}(f_0), \tag{5.3.3}$$

where $a = eE/m_e$ and we ignore electron-electron collisions ($I_{ee}(f) = 0$), i.e. this equation holds true for low number densities of electrons. The character of kinetics of electrons in gases is similar to that in semiconductors where interaction between electrons is negligible, and therefore the theory of electron kinetics in semiconductors [277–281] was the basis for the theory of electron kinetics in gases.

In determination the collision integral from the nonsymmetric part of the distribution function $v_z f_1(v)$, we take into account that because of a small parameter m_e/M the relative collision velocity $g = |\mathbf{v} - \mathbf{v}_a|$ is not changes as a result of electron-atom collision that gives $v = v'$ and $\mathbf{v}_a = \mathbf{v}'_a$. This gives [134, 275]

$$I_{ea}(v_z f_1) = -\int \left(f_a(v_a)dv_a(\mathbf{v} - \mathbf{v}')_z\, f_1(v)v d\sigma = -\nu_{ea}(v)v_z f_1(v), \tag{5.3.4}$$

where $\nu_{ea}(v) = N_a v\sigma^*(v)$, N_a is the number density of gas atoms, $\sigma^*(v) = \int(1 - \cos\vartheta)d\sigma$ is the diffusion electron-atom scattering, ϑ is the scattering angle, $d\sigma$ is the differential cross section for electron-atom scattering.

The collision integral from the spherically symmetric part of the distribution function f_0 that results from elastic electron-atom collision has the form of the electron diffusion flux in the energy space because the electron energy change in one elastic collision with an atom is relatively small. Hence, a general form of this collision integral is

$$I_{ea}(f_0) = -\frac{\partial j(\varepsilon)}{\rho(\varepsilon)\partial\varepsilon}$$

Here $\rho(\varepsilon)\sim(\varepsilon)^{1/2}$ is the state density in the energy space, and it is convenient to use the following normalization of the distribution function

$$\int_0^\infty f_o\,\varepsilon^{1/2}d\varepsilon = 1$$

The electron flux in the energy space $j(\varepsilon)$ is the right hand side of the Fokker-Planck equation [282, 283] and consists of the hydrodynamic and diffusion fluxes in the energy space [134, 276]

$$j(\varepsilon) = \rho(\varepsilon)A(\varepsilon)f_o(\varepsilon) - \frac{\partial[B(\varepsilon)\rho(\varepsilon)f_o(\varepsilon)]}{\partial\varepsilon} \tag{5.3.5}$$

The connection between the quantities $A(\varepsilon)$ and $B(\varepsilon)$ follows from the condition, that the flux is zero if the electron distribution function is the Maxwell one and the electron temperature coincides with the gas one. This gives

$$I_{ea}(f_0) = \frac{1}{\rho(\varepsilon)}\frac{\partial}{\partial\varepsilon}\left[\rho(\varepsilon)B(\varepsilon)\left(\frac{\partial f_0}{\partial\varepsilon} + \frac{f_0}{T}\right)\right], \tag{5.3.6}$$

and the quantity $B(\varepsilon)$ is an analog of the diffusion coefficient in the energy space. Taking into account elastic electron-atomic collisions only, we have for this quantity according to its definition

$$B(\varepsilon) = \frac{1}{2}\left\langle\int(\varepsilon-\varepsilon')^2 N_a v d\sigma(\varepsilon\to\varepsilon')\right\rangle, \tag{5.3.7}$$

where the angle brackets mean an average over the atom distribution, and $d\sigma$ is the differential cross section of electron elastic scattering on an atom.

We below takes into account that the relative electron-atom velocity changes the direction as a result of their elastic scattering, but the velocity value is conserved. Because of a large atom mass, its momentum and velocity vary weakly as a result of one collision act. Therefore we have

$$|\mathbf{v}-\mathbf{v}_a| = |\mathbf{v}'-\mathbf{v}_a|,$$

where $\mathbf{v}$, $\mathbf{v}'$ are the electron velocities below and after elastic collision with an atom of a velocity $\mathbf{v}_a$. This gives for change of the electron energy after collision

$$\varepsilon - \varepsilon' = \frac{m_e}{2}[v^2 - (v')^2] = m_e \mathbf{v}_a(\mathbf{v} - \mathbf{v}')$$

where m_e is the electron mass. We next substitute this in the expression for $B(\varepsilon)$ and take into account the definition for the average atom kinetic energy $\langle M v_a^2/2 \rangle = 3T/2$, where M is atom mass, T is the gas temperature. Use also use $|\mathbf{v} - \mathbf{v}'| = 2v \sin(\vartheta/2)$, where ϑ is the electron scattering angle. As a result we obtain the following formula for the diffusion coefficient of an electron in the energy space

$$B(\varepsilon) = \frac{m_e^2}{2} \left\langle \frac{v_a^2}{3} \right\rangle \int (\mathbf{v} - \mathbf{v}')^2 N_a v d\sigma = \frac{m_e}{M} T m_e v^2 N_a v \sigma^*(v), \qquad (5.3.8)$$

where $\sigma^*(v) = \int (1 - \cos\vartheta) d\sigma$ is the diffusion cross section for electron-atom scattering. Substituting this expression in formula (5.3.6) and transferring in a space of electron velocities, we obtain for the collision integral

$$I_{ea}(f_0) = \frac{m_e}{M} \frac{\partial}{v^2 \partial v} \left[v^3 \nu_{ea} \left(f_0 + \frac{T \partial f_0}{m_e v \partial v} \right) \right], \qquad (5.3.9)$$

where $\nu_{ea} = N_a v \sigma^*(v)$ is the rate of elastic electron-atom collisions.

5.4 Landau Collision Integral

We now consider the collision integral resulted from electron-electron collisions. This collision integral for the velocity distribution function of electrons, the Landau collision integral, is determined by elastic collision of electrons with a relatively small energy change. Correspondingly, evolution of electrons in the energy space has a diffusion character, and the collision integral can be represented in the form [116]

$$I_{ee}(f) = -\frac{\partial j_i}{\partial v_i}, \qquad (5.4.1)$$

where i, k are vector components (x, y, z) in the electron velocity space. But in contrast to the above case for the collision integral with electron scattering on atoms, now the collision integral is bilinear function with respect to the distribution function of electrons. Following to the Landau derivation [116], we use a general expression for the collision integral with accounting for elastic scattering of particles

$$I_{ee}(f_1) = \int [f(\mathbf{v}_1') f(\mathbf{v}_2' - (v_1) f(\mathbf{v}_2)] W(\mathbf{v}_1, \mathbf{v}_2 \to \mathbf{v}_1', \mathbf{v}_2') d\mathbf{v}_1' d\mathbf{v}_2' d\mathbf{v}_2, \qquad (5.4.2)$$

where $W(\mathbf{v}_1, \mathbf{v}_2 \to \mathbf{v}_1', \mathbf{v}_2')$ is the probability of transition between indicated electron velocities per unit time and unit number density, and the distribution function of electrons is normalized to the electron number density. The principle of detailed balance with respect to elastic scattering of electrons is used in derivation of expression (5.4.2)

$$W(\mathbf{v}_1, \mathbf{v}_2 \to \mathbf{v}_1', \mathbf{v}_2') = W(\mathbf{v}_1', \mathbf{v}_2' \to \mathbf{v}_1, \mathbf{v}_2)$$

This form of the collision integral allows us to use in full measure the problem symmetry. For this goal we represent the specific transition probability as a result of elastic collision of two electrons as

$$W(\mathbf{v}_1, \mathbf{v}_2 \to \mathbf{v}_1', \mathbf{v}_2') \equiv W\left(\frac{\mathbf{v}_1 + \mathbf{v}_1'}{2}, \frac{\mathbf{v}_2 + \mathbf{v}_2'}{2}, \Delta\mathbf{v}\right)$$
$$\equiv W\left(\mathbf{v}_1 + \frac{\Delta\mathbf{v}}{2}, \mathbf{v}_2 - \frac{\Delta\mathbf{v}}{2}, \Delta\mathbf{v}\right),$$

where $\Delta\mathbf{v}$ is an electron velocity change resulted from elastic collision with another electron. According to the principle of detailed balance W is an even function of $\Delta\mathbf{v}$, that is $W(\Delta\mathbf{v}) = W(-\Delta\mathbf{v})$. This form of the probability accounts for conservation of the total electron momentum or the total electron velocity in collision. This decreases a number of integration up $d\Delta\mathbf{v}d\mathbf{v}_2$ instead of $d\,\mathbf{v}_1' d\mathbf{v}_2' d\mathbf{v}_2$.

Expanding the collision integral over a small parameter $\Delta\mathbf{v}$, we obtain in the first approximation over this small parameter

$$I_{ee}(f) = -\int \left[f(\mathbf{v}_2)\frac{\partial f(\mathbf{v}_1)}{\partial \mathbf{v}_1} - f(\mathbf{v}_1)\frac{\partial f(\mathbf{v}_2)}{\partial \mathbf{v}_2}\right] \Delta\mathbf{v} W d(\Delta\mathbf{v}) d\mathbf{v}_2$$

Since the function W is even with respect to $\Delta\mathbf{v}$, this integral is zero.

In the second approximation over a small parameter $\Delta\mathbf{v}$ we have

$$I_{ee}(f) = -\int d\Delta\mathbf{v}d\mathbf{v}_2 W \cdot \left(\frac{1}{2}\Delta_i\Delta_k \frac{\partial^2 f_1}{\partial v_{1i}\partial v_{1k}} f_2 - \Delta_i\Delta_k \frac{\partial f_1}{\partial v_{1i}}\frac{\partial f_2}{\partial v_{2k}} + \frac{1}{2}\Delta_i\Delta_k f_1 \frac{\partial^2 f_2}{\partial v_{2i}\partial v_{2k}}\right)$$
$$-\int \Delta\mathbf{v}d\mathbf{v}_2 \frac{1}{2}\Delta_i \left(\frac{\partial W}{\partial v_{1i}} - \frac{\partial W}{\partial v_{2i}}\right)\Delta_k \left(\frac{\partial f_1}{\partial v_{1k}} f_2 - f_1 \frac{\partial f_2}{\partial v_{2k}}\right),$$

where we use notations $f_1 \equiv f(\mathbf{v}_1)$, $f_2 \equiv f(\mathbf{v}_2)$, $\Delta_i \equiv \Delta v_i$, indices i, k corresponds to components x, y, z, and summation is made over twice repeated indices.

Some terms of this expression may be determined from integration over parts. We have

$$\frac{1}{2}\int d\Delta\mathbf{v}d\mathbf{v}_2 W \cdot \Delta_i\Delta_k \frac{\partial f_1}{\partial v_{1i}}\frac{\partial f_2}{\partial v_{2k}} + \frac{1}{2}\int d\Delta\mathbf{v}d\mathbf{v}_2 \Delta_i\Delta_k \frac{\partial W}{\partial v_{2i}}\frac{\partial f_1}{\partial v_{1k}} f_2$$

$$= \frac{1}{2}\int d\Delta\mathbf{v}d\mathbf{v}_2\Delta_i\Delta_k\frac{\partial f_1}{\partial v_{1i}}\frac{\partial}{\partial v_{2k}}(Wf_2) = 0,$$

$$\frac{1}{2}\int d\Delta\mathbf{v}d\mathbf{v}_2 W\cdot\Delta_i\Delta_k f_1\frac{\partial^2 f_2}{\partial v_{2i}\partial v_{2k}} + \frac{1}{2}\int d\Delta\mathbf{v}d\mathbf{v}_2\Delta_i\Delta_k\frac{\partial W}{\partial v_{2i}}f_1\frac{\partial f_2}{\partial v_{2k}}$$
$$= \frac{1}{2}\int d\Delta\mathbf{v}d\mathbf{v}_2\cdot\Delta_i\Delta_k f_1\frac{\partial}{\partial v_{2i}}\left(W\frac{\partial f_2}{\partial v_{2k}}\right) = 0,$$

because the velocity distribution function of electrons and its derivation equal to zero at infinite electron velocities. From relations obtained we have [116]

$$I_{ee}(f_1) = -\frac{1}{2}\int d\Delta\mathbf{v}d\mathbf{v}_2\Delta_i\Delta_k(W\frac{\partial^2 f_1}{\partial v_{1i}\partial v_{1k}}f_2 - W\frac{\partial f_1}{\partial v_{1i}}\frac{\partial f_2}{\partial v_{2k}}$$
$$+\frac{\partial W}{\partial v_{1i}}\frac{\partial f_1}{\partial v_{1k}}f_2 - \frac{\partial W}{\partial v_{2i}}f_1\frac{\partial f_2}{\partial v_{2k}}$$

Representing the collision integral as the flux divergence in a space of electron velocities (5.4.1), we have for such a flux on the basis of the above expression of the collision integral

$$j_i = \int d\mathbf{v}_2\left(f_1\frac{\partial f_2}{\partial v_{2k}} - \frac{\partial f_1}{\partial v_{1k}}f_2\right)D_{ik}, \quad D_{ik} = \frac{1}{2}\int\Delta_i\Delta_k W d\Delta\mathbf{v} \tag{5.4.3}$$

where we use the notation $\Delta = \Delta\mathbf{v}$.

In evaluating the electron diffusion coefficient D_{ik} in the velocity space, we use the relation

$$Wd\Delta\mathbf{v} = gd\sigma,$$

that follows from the definition of the differential cross section of elastic scattering $d\sigma$, and g is the relative velocity of colliding electrons. We have for a change of the electron velocity $\Delta \equiv \Delta\mathbf{v}$ at scattering on a small angle at the impact parameters of collision ρ

$$\Delta = \frac{2e^2}{\rho g m_e}, \tag{5.4.4}$$

and a velocity change for a test electron $\Delta\mathbf{v}$ directs along the impact parameter of collision, and in components this relation has the form

$$\Delta_i = \frac{2e^2\rho_i}{\rho^2 g m_e}$$

Since the classical differential cross section of elastic scattering is $d\sigma = 2\pi\rho d\rho$, we have for the diffusion tensor in the velocity space [116]

$$D_{ik} = \frac{1}{2}\int \Delta_i \Delta_k W d\Delta\mathbf{v} = \frac{1}{2}\int \Delta_i \Delta_i g d\sigma = \frac{2e^4}{m_e^2 g}\int \frac{\rho_i \rho_k}{\rho^4} d\sigma = \frac{4\pi e^4}{m_e^2 g} n_i n_k \ln \Lambda \tag{5.4.5}$$

Here n_i, n_k are components of the unit vector directed along the impact parameter of collision, and the Coulomb logarithm for collision of two charged particles in an ideal plasma is equal to [116]

$$\ln \Lambda = \int\limits_{\rho_<}^{\rho_>} \frac{d\rho}{\rho} = \ln \frac{r_D \varepsilon}{e^2}, \tag{5.4.6}$$

where r_D Is the Debye-Hückel radius.

Use the frame of reference where the vector of the relative velocity of colliding electrons $\mathbf{g}$ directs along the axis x, and scattering proceeds in the plane xy, i.e. the impact parameter vector directs along the axis y. Then only the component Δ_y is not zero, and correspondingly only the tensor component D_{yy} is not zero and equals to

$$D_{yy} = \frac{2e^4}{m_e g}\int \frac{1}{\rho^2} 2\pi\rho d\rho = \frac{4\pi e^4 \ln \Lambda}{m_e g} \tag{5.4.7}$$

If the relative velocity directs along the axis x, and axes y and z are directed in an arbitrary method with respect to the impact parameter of collision, nonzero components of the diffusion tensor are equal to

$$D_{yy} = D_{zz} = \frac{2\pi e^4 \ln \Lambda}{m_e g},$$

and these components are averaged over the angle between the impact parameter of collision and over the axes. Constructing the diffusion tensor in an arbitrary frame of reference, we use the symmetry of this tensor, so that this can be assembled from tensors δ_{ik} and $g_i g_k$. Taking the diffusion tensor as a combination of these tensors, we have [116]

$$D_{ik} = \frac{2\pi e^4 \ln \Lambda}{m_e g^3}(g^2 \delta_{ik} - g_i g_k) \tag{5.4.8}$$

Thus on the basis of relations (5.4.1), (5.4.3) and (5.4.8) we have the following expression for the collision Landau integral that includes elastic scattering of electrons in an ideal plasma

$$I_{ee}(f) = -\frac{\partial}{\partial v_i}\left[\int d\mathbf{v}_2 \left(f_1 \frac{\partial f_2}{\partial v_{2k}} - \frac{\partial f_1}{\partial v_{1k}} f_2\right) D_{ik}\right], \tag{5.4.9}$$

where summation is taken over twice repeated indices, and the tensor D_{ik} is given by formula (5.4.8).

5.5 Kinetics of Fast Electrons in Plasma

In considering the character of test electrons in a plasma, we focus on the limiting case, where the velocity of motion of a test electron is large compared to a thermal electron velocity in a plasma and braking of a fast electron results from collisions with thermal electron. Variation of the energy of a fast electron in collisions with thermal electrons is given by

$$\frac{d\varepsilon}{dt} = -A(\varepsilon), \tag{5.5.1}$$

where ε is a current energy of a fast electron, and the braking coefficient according to definition (5.3.5) is equal to

$$A(\varepsilon) = \left\langle \int (\varepsilon - \varepsilon')N_e v d\sigma(\varepsilon \to \varepsilon') \right\rangle = \int \frac{(\Delta p)^2}{2m_e} N_e v 2\pi \rho d\rho = \frac{4\pi e^4 N_e \ln \Lambda}{m_e v} \tag{5.5.2}$$

Taking according to formula (5.4.4) $\Delta p = 2e^2/(\rho v)$, we obtain for the rate of variation of the fast electron energy (5.5.1) as a result of collisions with plasma electrons [40]

$$\frac{d\varepsilon}{dt} = -A(\varepsilon) = -\frac{4\pi e^4 N_e \ln \Lambda}{m_e v}, \quad \varepsilon \gg T_e \tag{5.5.3}$$

One can find also the diffusion coefficient $B(\varepsilon)$ of fast electrons in an energy space ε from the general relation [134] $A = B/T$, where a fast electron interacts with an equilibrium gas system of a temperature T. This gives

$$B(\varepsilon) = \left\langle \int (\varepsilon - \varepsilon')^2 N_e v d\sigma(\varepsilon \to \varepsilon') \right\rangle = \int \left[\frac{(\Delta p)^2}{2m_e} \right]^2 N_e v 2\pi \rho d\rho = \frac{4\pi e^4 T_e N_e \ln \Lambda}{m_e v}, \tag{5.5.4}$$

where T_e is the temperature of thermal electrons. It should be noted that the expression (5.5.4) may be obtained from formula (5.3.8), if we replace in this formula the atom mass M by the electron mass m_e, and the gas temperature T by the temperature T_e of thermal electrons. Indeed, in both cases the system of particles is divided in two subsystems ¡ and the energy of test particles varies weakly in single collisions with particles of a thermal system. Therefore, a small parameter $1/\ln \Lambda$ is of principle for the behavior of fast electrons under consideration. Correspondingly, the diffusion cross section in formula (5.5.4) is given by formula (3.1.7) if an energy of one electron exceeds significantly an energy of other collided electron.

$$\sigma^* = \frac{\pi e^4}{\varepsilon^2} \ln \Lambda = \frac{4\pi e^4}{m_e^2 v^4} \ln \Lambda,$$

since ε, the energy of fast electron, is large compared to a thermal electron energy T_e. This also leads to the following expression for the collision integral [40]

$$I_{ee}(f) = \frac{\partial}{v\partial\varepsilon}\left[vB(\varepsilon)\left(\frac{\partial f}{\partial\varepsilon} + \frac{f}{T_e}\right)\right], \quad \varepsilon \gg T_e \tag{5.5.5}$$

where $f = f(\varepsilon)$ is the distribution function of fast electrons, T_e is the electron temperature for thermal electrons, and the diffusion coefficient of fast electrons in an energy space $B(\varepsilon)$ is given by formula (5.5.4). Using this expression, we take into account that on the one hand, evolution of a fast electron in the energy space has a diffusion character (an exchange by energy in one collision is relatively small), and in the other hand, fast electrons are separated from thermal ones, and hence the collision integral is linear with respect to the distribution function of fast electrons. Assuming a number of fast electrons to be relatively small, we can ignore by the influence of fast electrons on the Maxwell distribution function of thermal electrons.

One can use the above expressions for collision integrals in the analysis of electron evolution in a plasma. Let us analyze the character of electron relaxation in a plasma located in an electric field, if an electric field switches off or on. On the basis of the equations for the average electron velocity in an electric field (drift velocity) and average electron energy, we obtain two relaxation times of electrons, for the momentum τ_P and the energy τ_ε, which are defined as

$$\frac{dw_e}{dt} = eE - \frac{w_e}{\tau_P}; \quad \frac{d\overline{\varepsilon}}{dt} = eEw_e - \frac{\overline{\varepsilon}}{\tau_\varepsilon} \tag{5.5.6}$$

Comparing this equation with (5.2.3) for electrons, we obtain for a typical relaxation time of the electron momentum $\tau_P \sim 1/\nu_{ea}$, where ν_{ea} is the effective rate of electron collisions with atoms. In the same manner on the basis of (5.2.4) we obtain for a typical time of electron energy variation τ_ε as a result of elastic electron-atom collisions $\tau_\varepsilon \sim M/(m_e\nu_{ea})$. Thus, the ratio of typical relaxation times of the electron energy and momentum variation is of the order of M/m_e.

Let us consider the relaxation process on the basis of the kinetic equation for the electron distribution function that has the form

$$\frac{\partial f}{\partial t} + \frac{e\mathbf{E}}{m_e}\cdot\frac{\partial f}{\partial \mathbf{v}} = I_{ee}(f) + I_{ea}(f)$$

Multiplying this equation by the electron kinetic energy and integrating over electron velocities, we obtain the relaxation equation for the average electron energy $\overline{\varepsilon} = 3T_e/2$ [133, 170]

$$\frac{d\overline{\varepsilon}}{dt} = eEw_e - \frac{m_e^2}{M}\cdot\left(1 - \frac{T}{T_e}\right) < v^2\nu_{ea} > \tag{5.5.7}$$

We used above the relation

$$\int \frac{m_e(v^2 - v'^2)}{2} I_{ee}(f) d\mathbf{v} = 0$$

This means that exchange by energy between electrons does not change the total energy of the electron subsystem.

Let us introduce the rate constants k_P and k_ε for relaxation of the electron momentum and electron energy in accordance with (5.5.6), so that

$$k_P \equiv \frac{1}{N_a \tau_P} = \frac{eE}{N_a w_e}; \; k_\varepsilon \equiv \frac{1}{N_a \tau_\varepsilon} = \frac{eEw_e}{N_a \overline{\varepsilon}}, \tag{5.5.8}$$

and in accordance with formulas (5.5.6) the balance equation for the average electron momentum and energy have the form

$$\frac{dw_e}{dt} = -\frac{w_e}{N_a k_P}; \; \frac{d\overline{\varepsilon}}{dt} = -\frac{\overline{\varepsilon}}{N_a k_\varepsilon} \tag{5.5.9}$$

We now generalize (5.5.7) for evolution of the average energy $\overline{\varepsilon}$, if excitation of atoms by electron impact is of importance. We then have the following kinetic equation

$$\frac{\partial f}{\partial t} + \frac{e\mathbf{E}}{m_e} \cdot \frac{\partial f}{\partial \mathbf{v}} = I_{ee}(f) + I_{ea}(f) - N_a k_{ex} f, \tag{5.5.10}$$

where N_a is the number density of atoms, and k_{ex} is the rate constant of atom excitation by electron impact. We above neglect by the quenching process, i.e. assume that excited atoms are destructed in other channels and the number density of excited atoms is small compared to its equilibrium value. Multiplying this kinetic equation by the electron energy and integrating over electron velocities, we obtain an equation for the average electron energy that is a generalization of (5.5.7)

$$\frac{d\overline{\varepsilon}}{dt} = eEw_e - \frac{m_e^2}{M} \cdot \left(1 - \frac{T}{T_e}\right) < v^2 \nu_{ea} > - \frac{\Delta\varepsilon}{N_e} \frac{dN_*}{dt} \tag{5.5.11}$$

Here the rate of formation of excited atoms is given by

$$\frac{dN_*}{dt} = N_a \int k_{ex} f d\mathbf{v}, \tag{5.5.12}$$

and we assume that the main contribution to this rate follows from a not large electron energies $\varepsilon - \Delta\varepsilon \ll \varepsilon$. From this it follows that an electron loses its energy as a result of atom excitation if

$$k_{ex} \gg \frac{m_e}{M} k_{el}, \tag{5.5.13}$$

where k_{ex} and k_{el} are typical rate constants for inelastic and elastic electron-atom collisions. One can represent this criterion in another manner if we require in formula (5.5.11) a small contribution to the energetic balance from elastic electron-atom scattering. Let us introduce the parameter

$$\beta = \frac{m_e^2 < v^2 \nu_{ea} >}{MeEw_e}, \qquad (5.5.14)$$

in the limit $T_e \gg T$, and this parameter characterizes a contribution of elastic electron-atom scattering to the total energetics. In particular, if the diffusion cross section of elastic electron-atom scattering σ_{ea}^* is independent of the electron velocity. In this case we have

$$\beta \approx 12 \frac{m_e}{M} \left(\frac{T_e}{eE\lambda} \right)^2, \quad \lambda = (N_a \sigma_{ea}^*)^{-1} \qquad (5.5.15)$$

Figure 5.3 gives the boundary ($\beta = 1$) in the helium case, where we take $\sigma_{ea}^* = 6\text{Å}^2$. In the helium case we have

$$\beta = 0.1 \frac{T_e^2}{x^2},$$

where the electron temperature is expressed in eV, and the reduced electric field strength $x = E/N_a$ is given in Td. The criterion (5.5.13) is fulfilled above the curve 1 ($\beta = 1$) of Fig. 5.3), and this curve gives the connection of the reduced electric field strength and electron temperature if the electric energy introduced in a helium gas discharge plasma is consumed on electron-atom elastic scattering.

We now consider the process of atom excitation from another standpoint. Assuming the criterion (5.5.13) to be fulfilled, we represent the rate of atom excitation by electron impact as the rate of excess of the atom excitation energy, that is

$$\frac{dN_*}{dt} = - \int_{v_o}^{\infty} I_{ee} d\mathbf{v}, \qquad (5.5.16)$$

where $v_o = \sqrt{2\Delta\varepsilon/m_e}$, and $\Delta\varepsilon$ is the atom excitation energy. Taking the boundary condition $f(\Delta\varepsilon) = 0$ under the above assumptions, we obtain the energy distribution function of electrons in the form [288]

$$f_0(v) = N_e \left(\frac{m_e}{2\pi T_e} \right)^{3/2} \left[\exp\left(-\frac{\varepsilon}{T_e} \right) - \exp\left(-\frac{\Delta\varepsilon}{T_e} \right) \right], \qquad (5.5.17)$$

that is normalized by the relation $\int f d\mathbf{v} = N_e$ and is a generalization of the Maxwell distribution function (2.1.2). From this it follows on the basis of expressions (5.5.4) and (5.5.5)

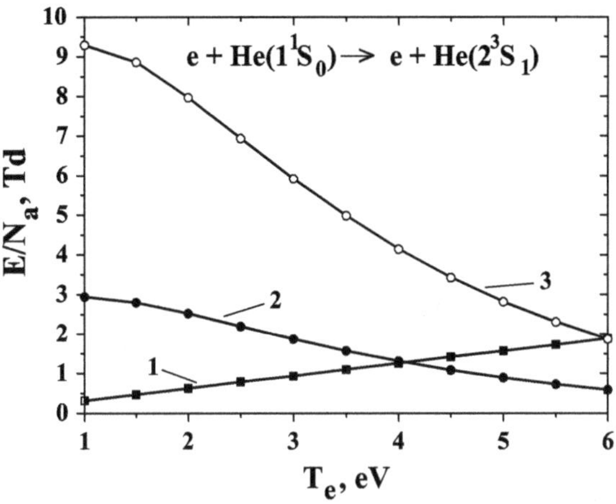

Fig. 5.3 Reduced electric field strength as a function of the electron temperature in the process of excitation of a helium atom in the metastable state $He(2^3S_1)$. *Curve 1* satisfies to the relation $\beta = 1$ for a helium gas discharge plasma, *curves 2* and *3* are determined by formula (5.5.22) for the electron concentration $c_e = 10^{-6}$ and $c_e = 10^{-5}$ correspondingly. Formula (5.5.22) holds true if *curves 2* and *3* are situated above *curve 1*

$$\frac{dN_*}{dt} = \frac{2(\Delta\varepsilon)^{1/2}B(\Delta\varepsilon)N_e}{\sqrt{\pi}T_e^{5/2}}\exp\left(-\frac{\Delta\varepsilon}{T_e}\right) = \frac{4\sqrt{2\pi}e^4N_e^2\ln\Lambda}{m_e^{1/2}T_e^{3/2}}\exp\left(-\frac{\Delta\varepsilon}{T_e}\right) \tag{5.5.18}$$

We now assume that the criterion (5.5.13) holds true, and the balance equation (5.5.11) for an average electron energy has the form

$$\frac{d\overline{\varepsilon}}{dt} = eEw_e - \frac{\Delta\varepsilon}{N_e}\frac{dN_*}{dt} = eEw_e - \frac{4\sqrt{2\pi m_e}\Delta\varepsilon^{3/2}N_e\ln\Lambda}{T_e^2}\exp\left(-\frac{\Delta\varepsilon}{T_e}\right) \tag{5.5.19}$$

Under stationary conditions this gives the relation between the reduced electric field strength E/N_a and the electron temperature T_e at a given concentration $c_e = N_e/N_a$ of electrons, that is

$$\left(\frac{eE}{N_a}\right)^2 = c_e * \frac{12\sqrt{2\pi}e^4\Delta\varepsilon\ln\Lambda}{\langle v/\sigma_{ea}^*\rangle m_e^{1/2}T_e^{3/2}}\exp\left(-\frac{\Delta\varepsilon}{T_e}\right), \tag{5.5.20}$$

where σ_{ea}^* is the diffusion cross section of electron-atom cross section. In the case where the cross section is independent of an electron velocity $\sigma_{ea}^* = const$, as it takes place in the helium case, formula (5.5.20) gives

$$\left(\frac{eE}{N_a}\right)^2 = c_e * 6\pi e^4\sigma_{ea}^*\ln\Lambda * ye^{-y},\ y = \Delta\varepsilon/T_e \tag{5.5.21}$$

In particular, taking $\ln \Lambda = 10$ and in the helium case $\sigma_{ea}^* = 6\,\text{Å}^2$, one can represent formula (5.5.21) in the form

$$\left(\frac{eE}{N_a}\right)^2 = Ac_e * ye^{-y}, \quad A = 6\pi e^4 \sigma_{ea}^* \ln \Lambda \qquad (5.5.22)$$

Figure 5.3 gives the dependence of the electron temperature on the electric field strength according to formula (5.5.22) where it is determined by the balance of an electron energy due to helium atom excitation in the state $He(2^3S)$. It should be noted that formula (5.5.22) is valid at $\beta \ll 1$, that requires for curves 2 and 3 to be located above curve 1.

5.6 Electron Regimes in Gas Discharge Plasma

We above analyze general relations and equations for some parameters of a gas discharge plasmas under various conditions. In reality the results of the theory of gas discharge plasma consist in determination of numerical plasma parameters. The numerical results allows us, on one hand, to understand real parameters of a gas discharge plasma and require, on the other hand, some parameters of elementary processes and kinetics of this plasma, and determination of this parameters is a certain problem.

We below focus on the helium and argon gas discharge plasma. The cross section of elastic electron-atom scattering as an important elementary process for gas discharge plasmas is different for these cases, and hence consideration of these examples gives the possibilities a variety in properties of gas discharge plasmas. From another standpoint, parameters of elementary processes and kinetics used are ambiguous, so that the evaluation of certain plasma parameters gives some experience in this way.

We first consider some plasma properties where this gas discharge plasma may be considered as uniform. We have two different regimes of electron equilibrium in a gas discharge plasma depending on the character of the electron energy change in a plasma. Namely, at high electron concentrations the energy distribution function of electrons is established as a result of collisions between electrons and has the Maxwell form (2.1.2). At low electron concentrations electron-atom collisions determine the way to establish the electron distribution function.

In both regimes the electron distribution function results from change of the electron energy in collisions with atoms or electrons by small portions, and the boundary number density of electrons N_e^* between two regimes under consideration results from equality of the diffusion coefficients $d\langle(\Delta\varepsilon)^2\rangle/dt = B(\varepsilon)$ for electrons in a space of electron energies due to electron-electron and electron-atom collisions

$$B_{ea}(\varepsilon) = B_{ee}(\varepsilon), \qquad (5.6.1)$$

where ε is the electron energy.

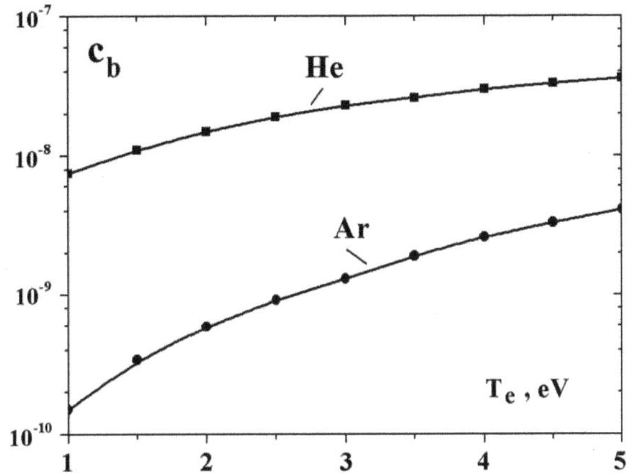

Fig. 5.4 The boundary electron concentration between the regimes of low and high number densities of electrons for helium and argon gas discharge plasmas

Using formulas (5.3.8) and (5.5.4) for the electron diffusion coefficient in the electron energy space for the regimes under consideration, we obtain for the boundary concentration of electrons $c_b^* = (N_e/N_a)_b$ in accordance with formula (5.6.1)

$$c_b^*(\varepsilon) = \frac{1}{\pi}\frac{m_e}{M}\frac{T}{T_e}\frac{\varepsilon^2\sigma_{ea}^*(\varepsilon)}{e^4 \ln \Lambda} \tag{5.6.2}$$

For definiteness, we take the gas temperature $T = 400\,\mathrm{K}$ and a typical value for the Coulomb logarithm $\ln \Lambda = 10$, and the electron temperature $T_e = 1\,\mathrm{eV}$. The boundary electron concentration between two regimes c_b^* according to formula (5.6.2) is represented in Fig. 5.4.

Let us use formula (5.6.2) at the excitation threshold. In the helium case the diffusion cross section of elastic electron-atom scattering is $\sigma_{ea}^*(\Delta\varepsilon) = 2.7\,\text{Å}^2$ at the atom excitation energy $\varepsilon = \Delta\varepsilon = 19.8\,\mathrm{eV}$, that gives at $T_e = 1\,\mathrm{eV}$ the boundary electron number density $c_b^*(\Delta\varepsilon)$ according to formula (5.6.2) $c_b^*(\Delta\varepsilon) = 7.0 \cdot 10^{-7}$ for a helium gas discharge plasma. In the argon case we have $\Delta\varepsilon = 11.7\,\mathrm{eV}$ and $\sigma_{ea}^*(\Delta\varepsilon) = 14.3\,\text{Å}^2$. This gives for the boundary electron number density $c_b^*(\Delta\varepsilon) = 1.4 \cdot 10^{-7}$ at the atom excitation energy for an argon gas discharge plasma.

From the data of Fig. 5.4 in helium and argon plasmas of glow discharge with the number density of atoms $N_a = (10^{17}\text{–}10^{18}\,\mathrm{cm}^{-3})$ and the number density of electrons it follows that the regime of a low number density of electrons $N_e = (10^{10}\text{–}10^{12}\,\mathrm{cm}^{-3})$ that corresponds to the electron concentration $c_e = N_e/N_a \sim 10^{-8}\text{–}10^{-5}$, and a typical electron energy $\overline{\varepsilon} = 2\text{–}4\,\mathrm{eV}$ the regime of a high electron number density competes with the regime of a low electron number density. For an arc plasma of helium and argon the regime of a high electron number density is realized, while in a plasma of Townsend discharge the regime of a low electron number density takes place.

Note that the criterion $B_{ea}(\varepsilon) \ll B_{ee}(\varepsilon)$ means that evolution of the electron energy results from electron-electron collisions. We consider below the case of a weakly ionized gas discharge plasma where the electron momentum is established as a result of electron-atom collisions that corresponds to the criterion

$$\nu_{ea} \gg \nu_{ee} \tag{5.6.3}$$

This criterion takes the form

$$c_e = \frac{N_e}{N_a} \ll \frac{x^2}{4\pi e^2 \sigma^*_{ea} \ln \Lambda}, \tag{5.6.4}$$

where $x = E/N_a$ is the reduced electric field strength. Taking the electron energy ε to be equal to the atom excitation energy $\varepsilon = \Delta\varepsilon$, we rewrite the above criterion

$$c_e \ll \frac{\sigma_{ea}(v_o)\Delta\varepsilon^2}{5\pi e^4 \ln \Lambda} \tag{5.6.5}$$

In particular, in the case of excitation of the metastable state 2^3S of the helium atom this formula gives

$$c_e \ll 0.05 \tag{5.6.6}$$

Chapter 6
Kinetic Processes in Gas Discharge Plasma

Abstract If electrons are located in a gas in an external electric field, one can divide the range of electric field strengths in three ranges, where the electric field influences weakly of the electron energy distribution function (EEDF) in the first range, elastic electron-atom collisions determine the EEDF in the second range, and inelastic electron-atom collisions are of importance for EEDF as well as excited atoms. Atom excitation in a plasma by electron impact is a self-consistent process, because the excitation process acts on the EEDF at energies near the excitation threshold, and the EEDF acts in turn on the excitation rate. Expressions for the rate constants of atom excitation by electron impact are represented under various conditions.

6.1 Distribution Function of Electrons Located in Rareness Gas in Electric Field

Collision processes electrons and atoms in a gas discharge plasma lead to a change of the electron energy as a result of many elastic and inelastic collisions. Because evolution of an electron energy proceeds in many stages, the analysis of this evolution follows from kinetics of electrons. In addition, transport phenomena involving electrons and ions in a gas discharge plasma result from many collisions of charged particles with atoms. Hence below we consider kinetics of electrons and ions in gas discharge plasma.

If the number density of electrons is relatively small, collisions between electrons are not important for establishment of equilibrium of electrons in a gas. In this case when an electron moves in a gas in an electric field of a strength E, its average parameters depend on the ratio E/N_a, where N_a is the number density of atoms, as it follows from the kinetic equation for the electron distribution function. A unit of measurement of the reduced strength E/N_a of the electric field is Townsend (Td) [98] ($1\,\mathrm{Td} = 1 \times 10^{-17}\,\mathrm{V \cdot cm^2}$). Figures 5.1 and 5.2 contain measured values of the drift velocities of electrons, i.e.averaged directed electron velocity, as a function of the reduced electric field strength E/N_a. Here the experimental values of the electron drift velocity [257–262] are used for helium and [6, 7, 258, 260, 261, 263–267] for argon.

B.M. Smirnov, *Theory of Gas Discharge Plasma*, Springer Series on Atomic, Optical, and Plasma Physics 84, DOI 10.1007/978-3-319-11065-3_6

Let us determine the distribution function of electrons in a gas in a constant electric field. The peculiarity of the behavior of electron kinetics in a gas is determined by the character of electron-atom collisions where the electron momentum varies strongly in a single collision, whereas the energy variation is small. This allows one to represent the electron distribution function over spherical harmonics [270–274] that corresponds to expansion of the kinetic equation for electrons over a small parameter. This small parameter allows us to expand the velocity distribution function of electrons over spherical harmonics according to (5.3.2). Let us start from the kinetic equation (5.1.1) for the distribution function $f(\mathbf{v})$ of electrons moved in a gas in a stationary electric field that has the following form

$$\frac{e\mathbf{E}}{m_e}\cdot\frac{\partial f}{\partial \mathbf{v}} = I_{ee}(f) + I_{ea}(f), \tag{6.1.1}$$

where $I_{ee}(f)$, $I_{ea}(f)$ are collision integrals for electron-electron and electron-atom elastic collisions. From this there are two regimes for electron drift in an electric field depending on the electron number density. We consider first the regime of a low number density of electrons N_e according to the criterion

$$N_e \ll \frac{m_e}{M}\frac{\sigma_{ea}}{\sigma_{ee}}N_a, \tag{6.1.2}$$

where M is the atom mass, N_a is the number density of atoms. In this regime one can ignore the term $I_{ee}(f)$ in the kinetic equation (6.1.1), and this limiting case corresponds to drift of a single electron in a gas.

Substitution of the expressions (5.3.4) and (5.3.9) for the collision integrals in the set (5.3.3) of equations leads to the following form of this set

$$a\frac{df_0}{dv} = -\nu v f_1, \quad \frac{a}{3}f_1 = \frac{m_e}{M}T\nu\left(\frac{df_0}{m_e v dv} + \frac{f_0}{T}\right) \tag{6.1.3}$$

This set of equations may be reduced to the following equation for the spherically symmetric part of the distribution function

$$\left(T + \frac{Ma^2}{3\nu^2}\right)\frac{df_0}{d\varepsilon} + f_0 = 0 \tag{6.1.4}$$

Solution of this equation has the form

$$f_0 = C\exp\left(-\int_0^v \frac{m_e v dv}{T + \frac{Ma^2}{3\nu^2}}\right), \tag{6.1.5}$$

where C is the normalization constant, and this solution holds true if the electron distribution function is determined by elastic electron-atom collisions. From this we have for the drift electron velocity

$$w_e = \int v_x^2 f_1 d\mathbf{v} = \frac{eE}{3m_e} \int f_0 d\mathbf{v} \frac{m_e v^2}{T + \frac{Ma^2}{3v^2}}, \tag{6.1.6}$$

and a general form of the expression for the electron drift velocity in this case is [275]

$$w_e = \int v_x^2 f_1 d\mathbf{v} = \frac{eE}{3m_e} \left\langle \frac{1}{v^2} \frac{d}{dv} \left(\frac{v^3}{\nu} \right) \right\rangle, \tag{6.1.7}$$

where an average is made with using the electron distribution function (6.1.5). In particular, in the case if the rate of electron-atom scattering ν is independent of the electron velocity, formula (6.1.7) gives

$$w_e = \frac{eE}{m_e \nu} \tag{6.1.8}$$

Let us consider the Druyvesteyn case [269, 285, 286], if the diffusion cross section of electron-atom scattering $\sigma_{ea}^*(v)$ is independent of the electron energy, and the average electron energy is large compared to a thermal energy of atoms. Then the distribution electron function normalized to the electron number density N_e ($\int 4\pi v^2 f_o(v) dv = N_e$) according to formula (6.1.5) has the form

$$f_0(\varepsilon) = \frac{N_e}{\pi \Gamma(3/4)} \left(\frac{m_e}{2\varepsilon_o} \right)^{3/2} \exp\left(-\frac{\varepsilon^2}{\varepsilon_o^2} \right), \quad \varepsilon_o = eE\lambda \sqrt{\frac{M}{3m_e}}, \tag{6.1.9}$$

and $\varepsilon_o \gg T$. From this on the basis of formula (6.1.7) we have for the electron drift velocity in this case ($\lambda(v) = 1/(N_a \sigma_{ea}^*) = const$)

$$w_e = \frac{2eE\lambda}{3m_e} \cdot \left\langle \frac{1}{v} \right\rangle = \frac{\sqrt{2\pi}}{3^{3/4}\Gamma(3/4)} \left(\frac{m_e}{M} \right)^{1/4} \left(\frac{eE\lambda}{m_e} \right)^{1/2} = 0.90 \left(\frac{m_e}{M} \right)^{1/4} \left(\frac{eE\lambda}{m_e} \right)^{1/2} \tag{6.1.10}$$

Correspondingly, the average electron energy is equal in this case

$$\overline{\varepsilon} = \varepsilon_o \frac{\Gamma(5/4)}{\Gamma(3/4)} = 0.74\varepsilon_o \tag{6.1.11}$$

One can generalize this formula for a case of a monotonic dependence for the diffusion cross section of electron-atom scattering $\sigma_{ea}^*(v)$ on the electron velocity v. Let us use for the rate of elastic electron-atom scattering ν the dependence on the electron velocity v in the form $\nu(v) = bv^k$, i.e. the diffusion cross section of electron-atom scattering is characterized by the dependence $\sigma_{ea}^*(v) \sim v^{k-1}$. Then according to formula (6.1.21) the symmetric part of the energy distribution function of electrons is given by

$$f_0(\varepsilon) = C \exp\left[-\frac{3m_e b^2 v^{2+2k}}{(2+2k)Ma^2}\right], \qquad (6.1.12)$$

where v is the electron velocity, m_e is the electron mass, M is the atom mass, $a = eE/m_e$, E is the electric field strength. The normalized coefficient C corresponds to the following normalization of the distribution function

$$\int_0^\infty 4\pi v^2 f_0 dv = N_e$$

and is equal

$$C = \frac{N_e(2k+2)}{4\pi\Gamma(\frac{3}{2k+2})} \cdot \left[\frac{3m_e b^2}{(2k+2)Ma^2}\right]^{3/(2k+2)}$$

From this we obtain on the basis of formula (6.1.7) for the electron drift velocity

$$w_e = \frac{eE(3-k)}{3m_e b} \cdot \int_0^\infty 4\pi v^{2-k} f_0 dv = \frac{eE(3-k)}{3m_e b}\left[\frac{(2k+2)Ma^2}{3m_e b^2}\right]^{-k/(2k+2)} \cdot \left[\frac{\Gamma(\frac{3-k}{2k+2})}{\Gamma(\frac{3}{2k+2})}\right] \qquad (6.1.13)$$

This formula is converted in formula (6.1.18) in the case $k = 0$ since $b = \nu$ and in formula (6.1.10) in the case $k = 1$, where $b = 1/\lambda$. On the basis of the distribution function (6.6.5) one can determine the average electron energy

$$\overline{\varepsilon} = \frac{m_e}{2} \cdot \left[\frac{(2k+2)Ma^2}{3m_e b^2}\right]^{1/(1+k)} \cdot \frac{\Gamma(\frac{5}{2k+1})}{\Gamma(\frac{3}{2k+1})} \qquad (6.1.14)$$

In the case $k = 0$ we have from this formula $\overline{\varepsilon} = Mw^2/2$.

In a range of electron energies where inelastic processes do not influence on the energy distribution function, one can use the expression (6.1.5) for the distribution function of electrons that may be represented in the form

$$f_0 = C\exp(-\chi), \quad \chi = \int_0^\varepsilon \frac{d\varepsilon'}{T + x^2 F(\varepsilon')}, \quad \varepsilon < \Delta\varepsilon \qquad (6.1.15)$$

where the dependence on the electric field strength is separated. Here $x = eE/N_a$ is the reduced electric field strength, ε is the electron energy, C is the normalization constant, and the function $F(\varepsilon)$ is given by formula

$$F(\varepsilon) = \frac{M}{6m_e\varepsilon[\sigma^*(\varepsilon)]^2} \qquad (6.1.16)$$

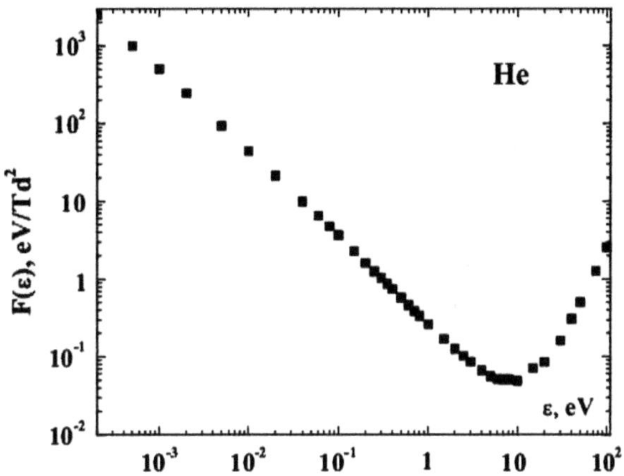

Fig. 6.1 The function $F(\varepsilon)$ for helium according to formula (6.1.16) and the data of Table 3.1 for the diffusion cross section

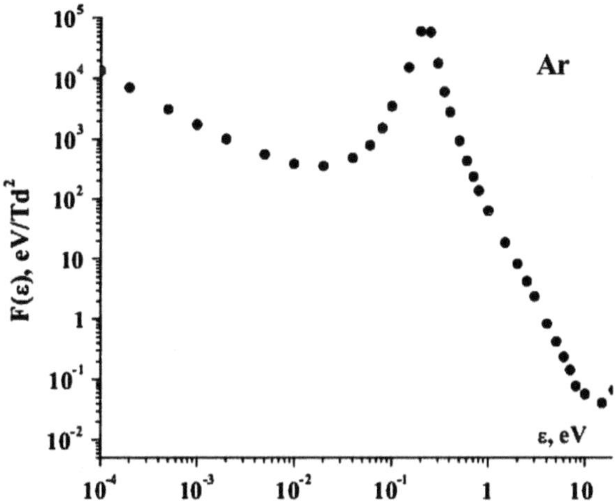

Fig. 6.2 The function $F(\varepsilon)$ for argon according to formula (6.1.16) and the data of Table 3.1 for the diffusion cross section

Figs. 6.1 and 6.2 give the dependence on the electron energy ε for the function $F(\varepsilon)$ if an electron is located in helium and argon. One can see one sharp minimum of this function for helium and two sharp minima in the argon case. Note that this function must be used for electron energies below the excitation energy for atoms where inelastic processes do not influence on the distribution function.

As it follows from formula (6.1.15), the function $T + x^2F(\varepsilon)$ is the effective temperature for electrons. If the effective electron temperature is large at a given electron energy, this energy gives a small contribution to an increase of the value χ given by formula (6.1.15). Let us consider the Druyvesteyn case [269, 285, 286],

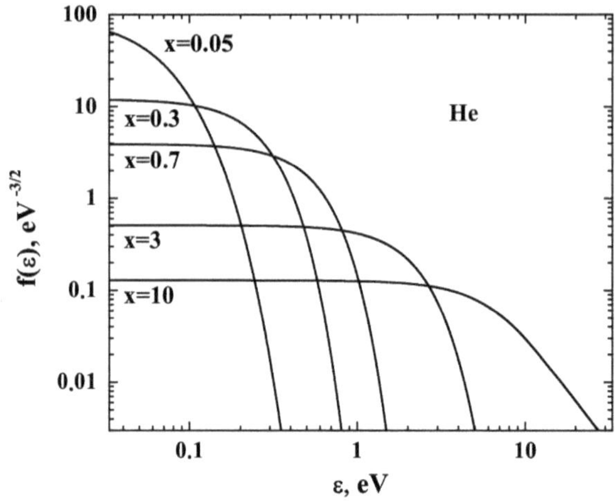

Fig. 6.3 Calculated on the basis of formula (6.1.15) the energy distribution function $f(\varepsilon)$ for electrons located in helium in a constant electric field at some electric field strengths

where the diffusion cross section of electron-atom scattering is independent of the electron energy. In this case the effective electron temperature T_{ef} is

$$T_{ef} = T + \frac{M}{6m_e}\frac{(eE\lambda)^2}{\varepsilon} \tag{6.1.17}$$

The electron distribution functions are given in Figs. 6.3 and 6.4 in the helium and argon cases. If the effective temperature exceeds the electron energy, the distribution

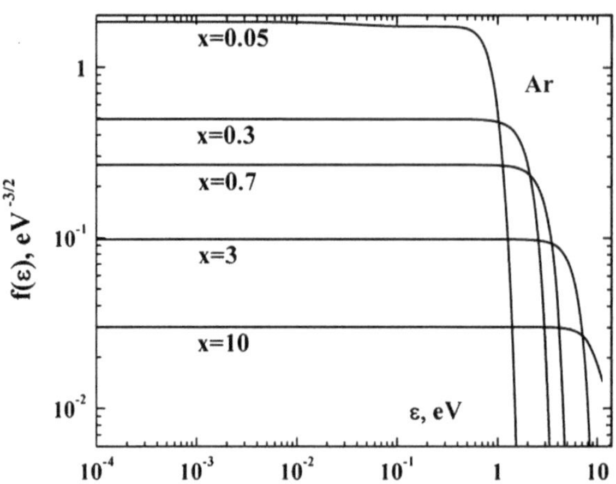

Fig. 6.4 Calculated on the basis of formula (6.1.15) the energy distribution function $f(\varepsilon)$ for electrons located in helium in a constant electric field at some electric field strengths

function at such energies has the form of a plateau as it is seen at some energies of Figs. 6.3 and 6.4. Note that formula (6.1.15) and the distribution functions of Figs. 6.3 and 6.4 are valid below the excitation energies for atoms where inelastic processes in electron-atom collisions do not act on the electron distribution function.

Note that the distribution function (6.1.5) corresponds to the electron energy range where elastic collisions dominate, and the change of the electron energy is determined by elastic electron-atom collisions. Above the excitation threshold this may be violated if

$$\nu_{ex} \gg \frac{m_e}{M}\nu, \tag{6.1.18}$$

where ν_{ex} is the rate of atom excitation by electron impact. Under this criterion, if the electron energy ε attains the atom excitation energy $\Delta\varepsilon$, it excites an atom and loses its energy. Therefore this corresponds to the boundary condition

$$f_0(\Delta\varepsilon) = 0 \tag{6.1.19}$$

We assume that this boundary condition acts only near the excitation energy, i.e. atom excitation corresponds to the distribution function tail only. Then we obtain for the distribution function

$$f_0(\varepsilon) = C\exp\left(-\int\limits_v^{v_o} \frac{m_e v' dv'}{T + \frac{Ma^2}{3\nu^2}}\right) \tag{6.1.20}$$

where $v_o = \sqrt{2\Delta\varepsilon/m_e}$ is the electron velocity at the excitation threshold, and the electron energy is below the excitation threshold $\varepsilon < \Delta\varepsilon$, whereas $f_0(\varepsilon \geq \Delta\varepsilon) = 0$. Note that far from the excitation threshold the distribution function (6.1.20) is converted in (6.1.5). Next, the normalization constant in expression (6.1.20) is equal

$$C = N_e\left[4\pi\int\limits_0^{v_o} v^2 dv \exp\left(\int\limits_v^{v_o} \frac{m_e v' dv'}{T + \frac{Ma^2}{3\nu^2}}\right)\right]^{-1}, \tag{6.1.21}$$

where N_e is the electron number density. In particular, in the case $\nu(v) = const$ this formula gives

$$C = N_e\left(\frac{m_e}{2\pi T_{ef}}\right)^{3/2}, \quad T_{ef} = T + \frac{Ma^2}{3\nu^2} \tag{6.1.22}$$

In the case under consideration we assume the excitation energy $\Delta\varepsilon$ to be found on the tail of the electron distribution function. Then the electron distribution function has the following structure

$$f_0(\varepsilon) = \varphi(\varepsilon) - \varphi(\Delta\varepsilon) \qquad (6.1.23)$$

Here $\varphi(\varepsilon)$ is the distribution function in neglection the inelastic electron-atom processes. In addition, the transition range of energies where excitation (i.e. the second term in formula (6.1.23)) becomes significant, is narrow that corresponds to the criterion

$$\frac{d\ln\varphi}{d\varepsilon}_{|\varepsilon=\Delta\varepsilon} \gg \frac{1}{\overline{\varepsilon}} \qquad (6.1.24)$$

6.2 Regime of High Electron Density in Electron Kinetics

The regime of a high electron density in electron kinetics corresponds to the criterion for the electron number density N_e that is inverse with respect to (6.1.2)

$$N_e \gg \frac{m_e}{M}\frac{\sigma_{ea}}{\sigma_{ee}}N_a, \qquad (6.2.1)$$

Then instead of (6.1.4) the equation set for the electron distribution function takes the form

$$a\frac{df_0}{dv} = -\nu_{ea}vf_1, \quad \frac{a}{3v^2}\frac{d}{dv}\left(v^3 f_1\right) = I_{ea}(f_0) + I_{ee}(f_0), \qquad (6.2.2)$$

Expansion of the second equation of this equation set over a small parameter (6.2.1) gives in the first approximation

$$I_{ee}(f_0) = 0 \qquad (6.2.3)$$

The solution of this equation is the Maxwell distribution function (2.1.2)

$$f_0(v) = N_e\left(\frac{m_e}{2\pi T_e}\right)^{3/2}\exp\left(-\frac{m_e v^2}{2T_e}\right), \qquad (6.2.4)$$

where T_e is the electron temperature. In this regime an ionized gas consists of two subsystems, an atomic and electron gas, and an equilibrium is established in each subsystem independently. The first equation of the set (6.2.2) gives for the drift electron velocity according to its definition

$$w_e = \int v_x f_1 d\mathbf{v} = \frac{eE}{3T_e}\left\langle\frac{v^2}{\nu_{ea}}\right\rangle, \qquad (6.2.5)$$

where an average is made on the basis of the Maxwell distribution function (6.2.4). Figures 6.5 and 6.6 contain the dependencies of the electron drift velocity on the electron temperature in helium and argon correspondingly.

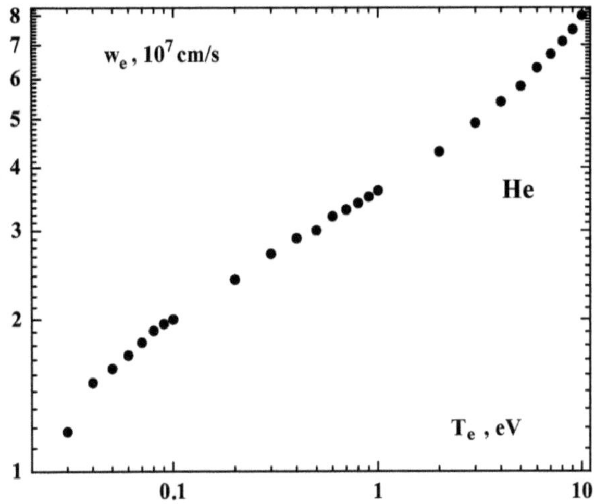

Fig. 6.5 The electron drift velocity w_e in helium as a function of the electron temperature T_e calculated on the basis of formula (6.2.5) with using the diffusion cross sections from Table 3.1

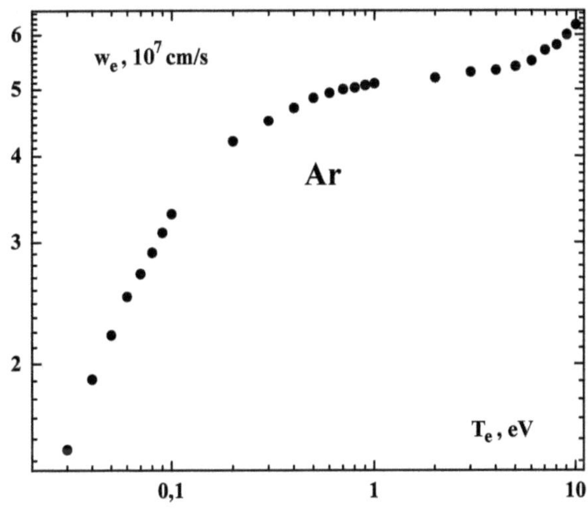

Fig. 6.6 The electron drift velocity w_e in argon as a function of the electron temperature T_e calculated on the basis of formula (6.2.5) with using the diffusion cross sections from Table 3.1

Thus the electron drift velocity is expressed through the cross section of elastic electron-atom collision as well as in the regime of a low electron number density. In addition, because of thermodynamic equilibrium for the electron component, the electron diffusion coefficient D_e and its mobility $K_e = w_e/E$ are connected through the Einstein relation [92, 93]

$$D_e = \frac{K_e T_e}{e} = \frac{w_e T_e}{eE}$$

For determination the difference of the electron T_e and gas temperature T, when the gas is located in a constant electric field, we multiply the kinetic equation (6.1.1) by the electron kinetic energy $m_e v^2/2$ and integrate this over the electron velocities. This gives in the first approximation

$$\int \frac{m_e v^2}{2} I_{ee} d\mathbf{v} = 0,$$

since electron-electron collisions conserve the total electron energy. Next approximation leads to the following balance equation

$$eEw_e = \int \frac{m_e v^2}{2} I_{ea} d\mathbf{v} = \left(1 - \frac{T}{T_e}\right) \frac{m_e^2}{M} \langle v^2 \nu_{ea} \rangle \qquad (6.2.6)$$

The left hand side of this equation is the power that is transferred from the electric field to electrons, and the right hand side corresponds to the power that is transferred from electrons to atoms in their collisions. Using formula (6.2.5) for the electron drift velocity, we obtain for the difference of electron and gas temperatures [83, 133, 170]

$$T_e - T = \frac{Ma^2}{3} \frac{\langle v^2/\nu_{ea} \rangle}{\langle v^2 \nu_{ea} \rangle} \qquad (6.2.7)$$

The dependence of the electron temperature on the electric field strength is given in Fig. 6.7 in helium and argon. Replacing the electron temperature in expression (6.2.5) for the electron drift velocity on the basis of (6.2.7), one can determine the drift velocity dependence on the electric field strength in the regime of high electron number densities. Of course, this dependence may differ from that for the regime of low number densities of electrons.

In considering high electron temperatures $T_e \gg T$, rewrite formula (6.2.7) in the form

$$T_e = \frac{Me^2 x^2}{3m_e^2} \frac{\langle v/\sigma_{ea}^* \rangle}{\langle v^3 \sigma_{ea}^* \rangle}, \qquad (6.2.8)$$

where an average in brackets is made with the Maxwell distribution function (6.2.4). On the basis of formulas (6.2.5) and (6.2.8) we represent the expressions for the electron drift velocity w_e and the electron temperature T_e for simple dependencies $\nu_{ea}(v)$. In the case $\nu(v_{ea}) = const$ these formulas give

$$w_e = \frac{eE}{m_e \nu_{ea}}, \quad T_e - T = \frac{Ma^2}{3\nu_{ea}^2} = \frac{M w_e^2}{3} \qquad (6.2.9)$$

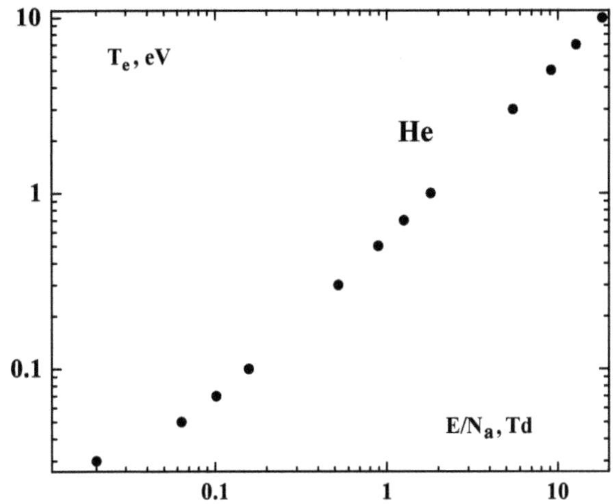

Fig. 6.7 The dependence of the electron temperature T_e on the reduced electric field strength E/N_a in helium. This dependence is calculated on the basis of formula (6.2.7) with using the distribution function of electrons according to formula (6.2.4) and cross sections from Table 3.1

In the case $\lambda = v/\nu_{ea}(v) = const$ we obtain on the basis of formulas (6.2.5) and (6.2.8)

$$w_e = \frac{\sqrt{8}}{3\sqrt{\pi}}\frac{eE\lambda}{\sqrt{m_e T_e}} = 0.532\frac{eE\lambda}{\sqrt{m_e T_e}}, \quad T_e - T = \frac{M}{m_e}\frac{(eE\lambda)^2}{12T_e} = 0.294 M w_e^2 \tag{6.2.10}$$

In particular, if the electron temperature is enough high $T_e \gg T$, we have from formula (6.2.10)

$$T_e = \sqrt{\frac{M}{12m_e}} eE\lambda, \quad w_e = 0.99\left(\frac{m_e}{M}\right)^{1/4}\sqrt{\frac{eE\lambda}{m_e}} \tag{6.2.11}$$

The second formula gives the same dependence of the electron drift velocity on the electric field strength as that in the limit of low electron number densities according to formula (6.1.10), but the factor in this formula exceeds that in formula (6.1.10) by 10 %. In particular, using in the helium case that $\sigma^*_{ea}(\varepsilon) \approx 6\text{Å}^2$ at $\varepsilon < 10\,\text{eV}$, we obtain on the basis of first formula (6.2.11)

$$T_e = 0.41x, \quad T_e \gg T, \tag{6.2.12}$$

where T_e is expressed in eV, the reduced electric field strength $x = E/N_a$ is given in Td. Formula (6.2.12) holds true at low fields.

In the helium case for a wider energy range we use formula (3.2.8) for the diffusion cross section of electron-atom scattering

$$\sigma^*_{ea}(\varepsilon) = (6 \pm 1)\text{Å}^2, \quad \varepsilon < 10\,\text{eV}; \; \sigma^*_{ea}(\varepsilon) \approx \frac{A}{\varepsilon}, \quad A \approx 60\,\text{eV} \cdot \text{Å}^2, \quad 10\,\text{eV} < \varepsilon < 20\,\text{eV},$$

and the accuracy of this cross section is about 20 %. Taking the energy distribution function in the form [288]

$$f_0(\varepsilon) = N_e \left(\frac{m_e}{2\pi T_e}\right)^{3/2} \left[\exp\left(-\frac{\varepsilon}{T_e}\right) - \exp\left(-\frac{\Delta\varepsilon}{T_e}\right)\right], \quad \Delta\varepsilon \gg T_e \qquad (6.2.13)$$

instead of (6.2.4), we obtain (6.2.8) in the form

$$x^2 = \frac{300}{z^2} \frac{2 - e^{-z}(z+2)}{1 + e^{-z}(1 + 2/z)}, \qquad (6.2.14)$$

where $z = \varepsilon_c/T_e = 10\,\text{eV}/T_e$. In the limit of low electric field strengths $z \gg 1$ this formula is transformed into (6.2.12). Figure 6.8 gives these dependencies in the helium case given by formulas (6.2.12) and (6.2.14). Figure 6.9 represents the dependence of the electron temperature on the reduced electric field strength for an argon plasma in accordance with formula (6.2.11) on the basis of data of Table 3.1 for the cross section of electron scattering on the argon atom.

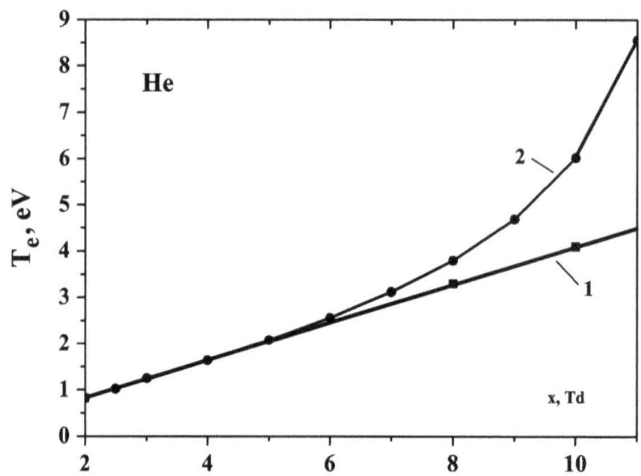

Fig. 6.8 Electron temperature T_e in ionized helium located in the electric field as a function of the reduced electric field strength x. The dependence 1 corresponds to the cross section (3.2.8) for electron scattering on the helium atom, the dependence 2 relates to $\sigma^*_{ea} = 6$ Å^2

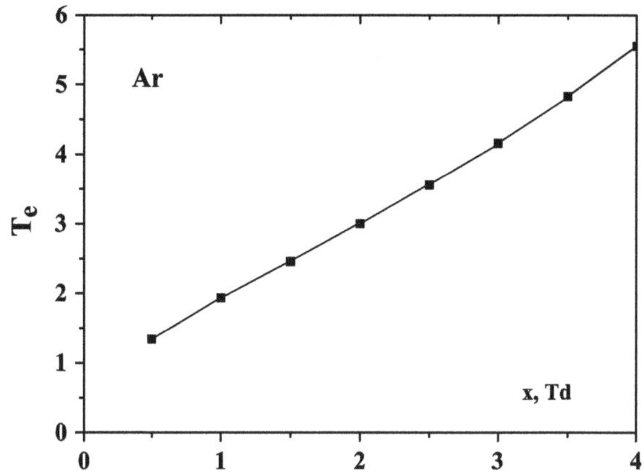

Fig. 6.9 The electron temperature T_e as a function of the reduced electric field strength $x = E/N_a$ for ionized argon located in the electric field

6.3 Atom Excitation as Electron Diffusion in Energy Space

The electron distribution function $f_0(\varepsilon)$ determines the rates of atom excitation and ionization in a plasma. If these processes do not influence on the electron distribution function and the electron distribution function has the Maxwell form, the rate constant of atom excitation is given by formula (3.3.8), and the rate constant of atom ionization by electron impact is given by formula (3.6.15) if the ionization cross section is taken on the basis of experimental data which are approximated by formulas (3.6.5) and (3.6.6) related to a release of s-electron. Let us determine also the rate constants of excitation and ionization of atoms in a gas discharge plasma for the Druyvesteyn distribution function (6.1.9) of electrons if the diffusion cross section of electron-atom elastic scattering is independent of the collision velocity. We have for the rate constant of atom excitation on the basis of formula (3.3.7)

$$k_{ex} = \frac{g_*}{g_o} k_q \int_{\Delta\varepsilon}^{\infty} \sqrt{\frac{\varepsilon - \Delta\varepsilon}{\varepsilon}} f_0(\varepsilon) d\varepsilon / N_e,$$

where the distribution function is normalized in a usual way ($\int f_0(\varepsilon)d\varepsilon = N_e$). Using in the above formula the Druyvesteyn distribution function (6.1.9) of electrons, we find in the limit of low electric field strengths

$$k_{ex} = \frac{\sqrt{\pi}}{2\sqrt{2}\Gamma(3/4)} \frac{g_*}{g_o} k_q \left(\frac{\varepsilon_o^2}{2\Delta\varepsilon}\right)^{3/2} \exp\left(-\frac{\varepsilon_o^2}{\Delta\varepsilon^2}\right), \quad \varepsilon_o \ll \Delta\varepsilon \qquad (6.3.1)$$

In reality this formula is valid only for the limiting case $\varepsilon_o \ll \Delta\varepsilon$ where the assumption holds true that the rate constant of atom quenching k_q by electron impact does not depend on the collision energy. Nevertheless, one can spread it on a wide range of the parameter $\varepsilon_o/\Delta\varepsilon$. Then evaluating the above integral at $\varepsilon_o \gg \Delta\varepsilon$ and sewing the obtained result with formula (6.3.1), we obtain for the rate constant of atom excitation

$$k_{ex} = \frac{g_*}{g_o} k_q \exp\left(-\frac{\varepsilon_o^2}{\Delta\varepsilon^2}\right)\left[1 + 1.95\left(\frac{\Delta\varepsilon}{\varepsilon_o}\right)^{3/2}\right]^{-1} \tag{6.3.2}$$

The above results correspond to the assumption that the process of atom excitation does not influence on the energy distribution function of electrons near the excitation threshold that corresponds to a tail of the distribution function. This holds true at large electric field strengths and depends on the atom excitation rate constant by electron impact. We now consider another limiting case if the excitation process leads to a sharp variation of the energy distribution function of electrons with an increasing electron energy. In analyzing the case of a low electron number density, we note that usually the inelastic cross section of electron-atom collisions is less than the cross section of elastic electron-atom collisions, and therefore evolution of the electron momentum is determined by elastic collisions with atoms. But the energy change in elastic electron-atom collision is of the order of m_e/M portion of the kinetic electron energy. Therefore the inelastic channel of the energy loss may dominate above the excitation threshold. Hence, this case corresponds to the following criterion at electron energies above the atom excitation energy $\Delta\varepsilon$ between the rates of elastic ν_{ea} and inelastic ν_{ex} electron-atom collisions

$$\nu_{ea} \gg \nu_{ex} \gg \nu_{ea}\frac{m_e}{M}\frac{\Delta\varepsilon}{\overline{\varepsilon}}, \tag{6.3.3}$$

In the regime under consideration the electron distribution function at energies below the excitation threshold is determined by elastic electron-atom and electron-electron collisions, while above the threshold the inelastic channel of energy loss dominates. Then the rate of atom excitation in an ionized gas is given by the rate of loss of fast electrons

$$\frac{dN_*}{dt} = -\frac{dN_e}{dt} = -\int 4\pi v^2 dv \frac{\partial f_0}{\partial t}, \tag{6.3.4}$$

where N_* is the number density of excited atoms, and we take into account that the loss of fast electrons is determined by atom excitation. The electron distribution f_0 is normalized by the relation

$$\int_0^\infty f_0(v) \cdot 4\pi v^2 dv = N_e \tag{6.3.5}$$

On the basis of a non-stationary kinetic equation for electrons, we have the following equation set instead of (6.1.4)

$$\frac{\partial f_0}{\partial t}+\frac{a}{3v^2}\frac{\partial(v^3 f_1)}{\partial v}=I_{ea}(f_0),\quad \frac{\partial f_1}{\partial t}+a\frac{\partial f_0}{\partial v}=-\nu_{ea}vf_1$$

A non-stationary is of importance only near the excitation threshold because the rate of atom excitation is small compared to that of elastic electron-atom collisions. From this we have

$$\frac{\partial f_0}{\partial t}=I_{ea}(f_0)+\frac{a^2}{3v^2}\frac{\partial}{\partial v}\left(\frac{v^2}{\nu_{ea}}\frac{\partial f_0}{\partial v}\right) \tag{6.3.6}$$

Using the expression (5.3.9) for the collision integral, one can neglect it because a thermal atom energy T is small compared to a typical electron energy, and the mass ratio m_e/M is small. Integrating over electron velocities, we obtain for $\varepsilon \gg T$

$$\frac{dN_*}{dt}=\frac{4\pi a^2 m_e v_o^3}{3\nu_{ea}}\left|\frac{df_0}{d\varepsilon}\right|_{\varepsilon=\Delta\varepsilon}, \tag{6.3.7}$$

where $\varepsilon = m_e v^2/2$ is the electron energy, $\Delta\varepsilon = m_e v_o^2/2$ is the energy of atom excitation, $\varphi(\varepsilon)$ is the energy distribution function of electrons in ignoring the excitation in accordance with formula (6.1.23), so that $f_0 = \varphi_0$ far from the atom excitation threshold, and we assume that the excitation energy corresponds to the tail of the distribution function. This leads to the rate constant $k_<$ of atom excitation if this process is determined by the electron flux in a space of electron energies below the atom excitation threshold

$$k_< = \frac{1}{N_a N_e}\frac{dN_*}{dt}=\frac{4\pi v_o^3}{3m_e k_o}\left(\frac{eE}{N_a}\right)^2\left|\frac{df_0}{N_e d\varepsilon}\right|_{\varepsilon=\Delta\varepsilon}, \tag{6.3.8}$$

where $k_o = \nu_{ea}(v_o)/N_a$, and the electron distribution functions are normalized in accordance with relation (6.3.5).

We now use formula (6.3.8) for the excitation rate constant for simple expressions of the electron distribution function. Taking the Maxwell distribution function (6.2.4) of electrons, we obtain on the basis of formula (6.3.8)

$$k_< = \frac{4}{3\sqrt{\pi}m_e k_o T_e}\left(\frac{eE}{N_a}\right)^2\exp\left(-\frac{\Delta\varepsilon}{T_e}\right) \tag{6.3.9}$$

In the case of the Druyvesteyn distribution function of electrons (6.1.9) formula (6.3.8) gives

$$k_< = C\frac{1}{m_e k_o \varepsilon_o}\left(\frac{eE}{N_a}\right)^2\left(\frac{\Delta\varepsilon}{\varepsilon_o}\right)^{3/2}\exp\left(-\frac{\Delta\varepsilon^2}{\varepsilon_o^2}\right)^2, \quad \varepsilon_o = eE\lambda\sqrt{\frac{M}{3m_e}}, \qquad (6.3.10)$$

where $C = 2/[3\Gamma(3/4)] = 0.817$.

Let us formulate in a general form the character of electron diffusion in a space of its energy as an element of atom excitation in a gas discharge plasma. Atom excitation in a gas discharge plasma is determined by electrons located at the tail of the energy distribution function, and we evaluate the part of the excitation rate constant of atoms by electron impact which is determined by formation of fast electrons, and subsequently these electrons excite atoms. One can represent the non-stationary kinetic equation for the distribution function $f_0(v)$ of electrons in a space of electron velocities as the continuity equation

$$\frac{\partial f_0}{\partial t} + \frac{1}{v}\frac{\partial}{\partial \varepsilon}(vJ) = 0, \qquad (6.3.11)$$

where the electron flux in a velocity space results from electron interaction with an environment. We above consider the case where formation of fast electrons is determined by the action of an electric field, and the corresponding flux in a space of electron energies has the diffusion character $J = -B\partial f_0/\partial\varepsilon$. We exclude from the consideration the hydrodynamic part of the flux, because it is proportional to the distribution function at the excitation boundary which is equal to zero. Hence, we have from this for the rate of formation of excited atoms

$$\frac{dN_*}{dt} = \int_{\varepsilon_o}^{\infty} 4\pi v^2 dv \frac{\partial f_0}{\partial t} = \frac{4\pi v_o}{m_e} B(\Delta\varepsilon)\left|\frac{df_0}{d\varepsilon}\right|_{\varepsilon=\Delta\varepsilon}, \qquad (6.3.12)$$

and in the case under consideration the electron diffusion coefficient in an energy space due to an electric field is equal in accordance with formula (6.3.8)

$$B_E(\varepsilon) = \frac{(eE)^2 v^2}{3\nu_{ea}} \qquad (6.3.13)$$

In addition, one can use in formula (6.3.12) the unperturbed distribution function $\varphi_0(\varepsilon)$ that coincides with the distribution function $f_0(\varepsilon)$ at lower energies far from the excitation boundary. One can use two more mechanisms of electron diffusion in an energy space due to collisions with atoms and electrons, and the corresponding diffusion coefficients B_{ea} and B_{ee} are given by formulas (5.3.8) and (5.5.4) in the form

$$B_{ea}(\varepsilon) = \frac{m_e}{M} T m_e v^2 \nu_{ea}, \quad B_{ee}(\varepsilon) = \frac{4\pi e^4 T_e N_e \ln\Lambda}{m_e v}$$

Correspondingly, in a general case formula (6.3.12) takes the form

$$\frac{dN_*}{dt} = \frac{4\pi v_o}{m_e}(B_E + B_{ea} + B_{ee})\left|\frac{df_0}{d\varepsilon}\right|_{\varepsilon=\Delta\varepsilon}, \qquad (6.3.14)$$

In particular, in the case the Maxwell distribution of electrons formula (6.3.14) gives for a corresponding mechanism of electron diffusion

$$\frac{dN_*}{dt} = \frac{2}{\sqrt{\pi}}\left(\frac{\Delta\varepsilon}{T_e}\right)^{1/2}\frac{N_e B(\Delta\varepsilon)}{T_e^2}\exp\left(-\frac{\Delta\varepsilon}{T_e}\right), \quad k_< = \frac{2}{\sqrt{\pi}}\left(\frac{\Delta\varepsilon}{T_e}\right)^{1/2}\frac{B(\Delta\varepsilon)}{N_a T_e^2}\exp\left(-\frac{\Delta\varepsilon}{T_e}\right) \qquad (6.3.15)$$

From the above analysis it follows that the contribution of each mechanism of electron diffusion in a space of its energies is determined by the diffusion coefficient $B(\Delta\varepsilon)$ for the corresponding mechanism at the electron energy near the atom excitation energy $\Delta\varepsilon$. We compare in Fig. 6.10 the diffusion coefficients in a space of electron energies B_E and B_{ee} where the energy variation proceeds under the action of an electric field and as a result of electron-electron collisions. We have for the ratio of these diffusion coefficients taking into account that B_E is proportional to the number density of atoms N_a, while B_{ee} is proportional to the electron number density N_e

$$\frac{B_{ee}}{B_E} = c_e\zeta, \quad \zeta = \frac{6\pi e^4 \ln\Lambda\sigma_{ea}^*}{\varepsilon(eE/N_a)^2}, \qquad (6.3.16)$$

where $c_e = N_e/N_a$ is the electron concentration, and v_o is the electron velocity at the atom excitation energy. This gives the rate constant of excitation due to electron kinetics in a space of electron energies in the regime of high electron number densities in

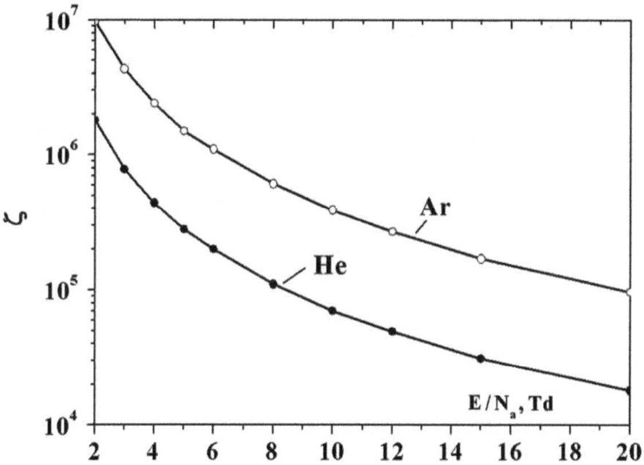

Fig. 6.10 The ratio of electron diffusion coefficients in a space of reduced electric field strengths in accordance with formula (6.3.16) if the Coulomb logarithm is $\ln\Lambda = 10$

accordance with formula (5.5.18) if we account for simultaneously electron diffusion in the electron energy space due to an electric field [287]

$$k_{<} = \frac{1}{N_a N_e}\frac{dN_*}{dt} = \frac{4\sqrt{2\pi} e^4 ln\Lambda}{3} \frac{}{m_e^{1/2} T_e^{5/2}} \exp\left(-\frac{\varepsilon_o}{T_e}\right)\left(c_e + \frac{x^2}{4\pi e^2 \sigma_{ea}(v_o)\ln\Lambda}\right) \tag{6.3.17}$$

The excitation rate constant by electron impact in accordance with formula (6.5.7) is given by

$$k_{>} = \frac{4.6 g_* k_q \Delta\varepsilon^{3/2}}{\pi^{3/2} g_o T_e^{3/2} \kappa^{1.2}} \exp\left(-\frac{\varepsilon_o}{T_e}\right), \tag{6.3.18}$$

and these expressions are valid if the criterion (6.5.4) holds true.

6.4 Efficiency of Atom Excitation by Electron Impact

We also analyze the above results from the standpoint of the energy expense in a gas discharge plasma on excitation of atoms. Let us assume the rate of electron-atom collisions ν_{ea} to be independent of the electron energy. Under the above assumption, if fast electrons expense its energy on atom excitation, we have for the field energy consumed on atom excitation per unit volume and per unit time in the regime of high electron number densities as

$$\Delta\varepsilon \frac{dN_*}{dt} = N_a \frac{4}{3\sqrt{\pi}} \left(\frac{\Delta\varepsilon}{T_e}\right)^{5/2} \frac{N_e \cdot (eE)^2}{m_e \nu_{ea}} \exp\left(-\frac{\Delta\varepsilon}{T_e}\right) \tag{6.4.1}$$

One can find the portion γ of the energy injected in a gas discharge plasma from an electric field that is consumed on atom excitation as the ratio of the power $\Delta\varepsilon dN_*/dt$ for atom excitation to the power $iE = E \cdot eN_e w_e$ transferred to a gas discharge plasma from an external field. We have for this ratio if the rate of electron-atom collisions ν_{ea} does not depend on the electron velocity

$$\gamma = \Delta\varepsilon \frac{dN_*}{dt} / (N_e e E w_e) = \frac{4}{3\sqrt{\pi}} \left(\frac{\Delta\varepsilon}{T_e}\right)^{5/2} \exp\left(-\frac{\Delta\varepsilon}{T_e}\right), \tag{6.4.2}$$

where we use formula (6.2.9) for the electron drift velocity. This dependence is represented in Fig. 6.12. In the same manner one can construct the efficiency of atom excitation by electron impact in a gas discharge plasma in the case where the cross section of elastic electron-atom collisions σ_{ea}^* and respectively the mean free path of electrons $\lambda = 1/(N_a \sigma_{ea}^*)$ are independent of the electron velocity. Then the electron drift velocity is determined by formula (6.2.10), and the power portion consumed on atom excitation is given by

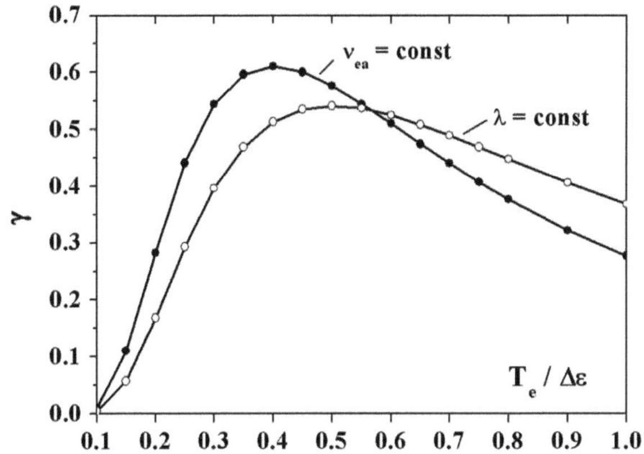

Fig. 6.11 Dependence on the specific electron temperature for the ratio of the specific powers consumed on atom excitation by electron impact in a gas discharge plasma and as a result of elastic electron-atom collisions in the cases where the rate of elastic electron-atom collisions and the mean free path of electrons in a gas discharge plasma are independent of the electron energy in accordance with formulas (6.4.2) and (6.4.3)

$$\gamma = \Delta\varepsilon \frac{dN_*}{dt} / (N_e e E w_e) \approx \left(\frac{\Delta\varepsilon}{T_e}\right)^3 \exp\left(-\frac{\Delta\varepsilon}{T_e}\right), \tag{6.4.3}$$

The ratio of powers resulted from inelastic and elastic electron-atom collisions is given in Fig. 6.11 under corresponding conditions.

Let us define the efficiency ξ of atom excitation in an ionized gas as the ratio of an energy that is consumed on atom excitation to the total electric energy injected in a gas. This value is given by

$$\xi = \frac{\Delta\varepsilon \frac{dN_*}{dt}}{N_e e E w_e + \Delta\varepsilon \frac{dN_*}{dt}} = \frac{\gamma}{1+\gamma}, \tag{6.4.4}$$

where w_e is the electron drift velocity, and the parameter γ is the ratio of inelastic and elastic energy losses and is defined by formula (6.4.2). In the regime of a high electron number density or in the regime of a low number density of electrons in the case $\nu(v) = const$ this parameter is given by formula (6.4.2). One can see the maximum of the function (6.4.4) at $T_e = 0.4\Delta\varepsilon$, that corresponds to $\gamma = 0.61$ and $\xi = 0.38$. This shows that the efficiency of atom excitation may be essential. Figure 6.12 [133] contains the dependence of the excitation efficiency ξ on the reduced average electron energy $\overline{\varepsilon}/\Delta\varepsilon$, where $\overline{\varepsilon} = 3T_e/2$ for the regime of a high electron number density if an ionized gas is located in an external electric field.

Let us determine the efficiency of atom excitation in the regime of a low electron number density for the Druyvesteyn case where the cross section of electron-atom scattering is independent of the electron velocity. Based on the criterion (6.3.3), we assume in this consideration that in the course of electron-atom collisions the electron

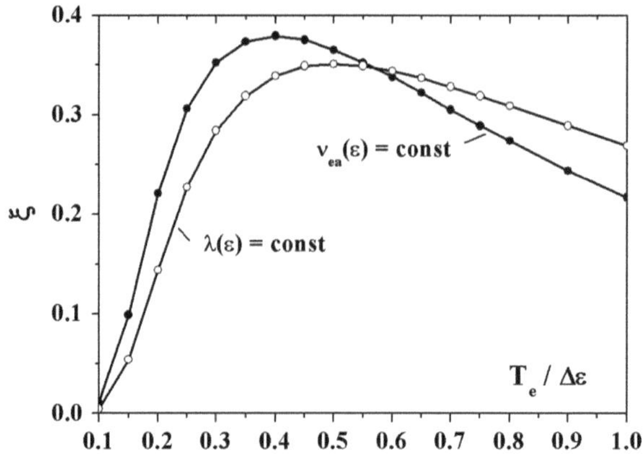

Fig. 6.12 Efficiency of atom excitation by electron impact in a gas discharge plasma depending on the specific electron temperature in the cases where the rate of electron-atom collisions is independent of the electron velocity as well as the mean free path of electrons in a gas discharge plasma

energy reaches the threshold of atom excitation and slightly exceeds this value. Then the electron energy is consumed on atom excitation and an electron obtains almost zero energy. We assume that this excitation energy does not return to an electron (for example, this energy is taken by a radiating photon). Under these conditions we compare the power consumed on atom excitation with that transferred to atoms in the course of electron-atom collisions. The energy distribution function of electrons has the form (6.1.9) far from excitation threshold for the Druyvesteyn case where $\lambda(v) = const$. Near the excitation threshold in accordance with formula (6.1.20) we have ($\Delta\varepsilon \gg T$)

$$f_0(\varepsilon) = \frac{N_e}{\pi\Gamma(3/4)}\left(\frac{m_e}{2\varepsilon_o}\right)^{3/2}\left[\exp\left(-\frac{\varepsilon^2}{\varepsilon_o^2}\right)-\left(-\frac{\Delta\varepsilon^2}{\varepsilon_o^2}\right)\right], \quad \varepsilon_o = eE\lambda\sqrt{\frac{M}{3m_e}}, \ \Delta\varepsilon \gg \varepsilon_o \tag{6.4.5}$$

This gives for the excitation rate on the basis of formula (6.3.7)

$$\frac{dN_*}{dt} = \frac{4\sqrt{2}}{\Gamma\left(\frac{3}{4}\right)}\frac{m_e}{M}\frac{\Delta\varepsilon^2}{\varepsilon_o^{3/2}m_e^{1/2}}\frac{N_e}{\lambda}\exp\left(-\frac{\Delta\varepsilon^2}{\varepsilon_o^2}\right), \tag{6.4.6}$$

where $\lambda = 1/(N_a\sigma_{ea}^*)$, $\Delta\varepsilon$ is the excitation energy, and ε_o is a typical energy $\varepsilon_o = eE\lambda\sqrt{M/(3m_e)}$. From this we obtain for the parameter γ

$$\gamma = \frac{\Delta\varepsilon\frac{dN_*}{dt}}{N_eeEw_e} = \frac{4}{\sqrt{\pi}}\frac{\Delta\varepsilon^3}{\varepsilon_o^3}\exp\left(-\frac{\Delta\varepsilon^2}{\varepsilon_o^2}\right), \tag{6.4.7}$$

where we use formula (6.1.13) for the electron drift velocity in the Druyvesteyn case. This value has a maximum at $\varepsilon_o^2/\Delta\varepsilon^2 = 1.5$ or $\overline{\varepsilon}/\Delta\varepsilon = 0.604$. At this value of the average reduced electron energy we have $\gamma_{max} = 0.925$ and the efficiency of atom excitation is $\xi_{max} = 0.48$. As is seen, in both regimes, the regime of a high electron number density where the efficiency parameter γ is given by formula (6.4.2) and in the Druyvesteyn case of the regime of a low electron number density, the powers transferred from an electron to buffer gas atoms and consumed on atom excitation, are comparable under optimal conditions.

In considering the regime of a high electron number density, we use formula (6.3.7) for this regime

$$\frac{dN_*}{dt} = \frac{4\pi}{3}\frac{m_e a^2 v_o^3}{\nu_{ea}}\frac{\partial f_0}{\partial \varepsilon} \tag{6.4.8}$$

On the basis of the Maxwell distribution function for electrons we obtain for the excitation rate constant

$$k_< = \frac{4}{3\sqrt{\pi}}\left(\frac{\Delta\varepsilon}{T_e}\right)^{3/2}\frac{m_e}{k_{ea}^*(v_o)T_e}\left(\frac{eE}{m_e N_a}\right)^2 \exp\left(-\frac{\Delta\varepsilon}{T_e}\right) \tag{6.4.9}$$

This formula coincides with formula (6.4.2) if we introduce in this formula the electron temperature as $T_e = Ma^2/(3\nu_{ea}^2)$. As it follows from the above analysis, the rate constant $k_<$ is determined by the character of variation of the electron momentum in the course of electron evolution in a space of electron energies.

Let us formulate the problem under consideration. Atom excitation in a gas discharge plasma is determined by electrons located at the tail of the energy distribution function, and we evaluate the part of the excitation rate constant of atoms by electron impact which is determined by formation of fast electrons. For this channel of atom excitation we assume that an electron with an energy above the excitation threshold expends its energy on atom excitation. In this case in accordance with formula (6.3.4) the rate of atom excitation by electron impact is determined by the electron flux in an energy space through the surface $\varepsilon = \Delta\varepsilon$. In this consideration the rate of atom excitation is given by

$$k_< = \frac{4\pi v_o B_E}{m_e}\frac{\partial\varphi_0}{N_e\partial\varepsilon}(\Delta\varepsilon), \quad B_E = \frac{m_e a^2 v_o^2}{3\nu_{ea}}, \tag{6.4.10}$$

where φ_0 is unperturbed distribution function of electrons, B_E is the diffusion coefficient of electrons in an energy space which is determined by electron drift under the action of an external electric field with braking of this motion in collisions with atoms. We have the two other mechanisms of electron diffusion in an energy space due to collisions with atoms and electrons, and the corresponding diffusion coefficients B_{ea} and B_{ee} are given by formulas (5.3.8) and (5.5.4) in the form

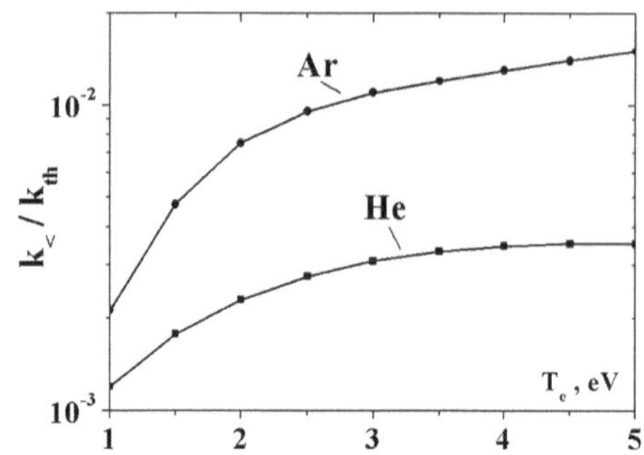

Fig. 6.13 Ratio of the rate constants of atom excitation by electron impact in a gas discharge of a high electron concentration as a function of the electron temperature T_e

$$B_{ea}(\varepsilon) = \frac{m_e}{M} T m_e v^2 \nu_{ea}, \quad B_{ee}(\varepsilon) = \frac{4\pi e^4 T_e}{m_e v} N_e \ln \Lambda,$$

and (6.4.10) takes the form

$$k_< = \frac{4\pi v_o (B_E + B_{ea} + B_{ee})}{m_e N_a} \frac{\partial \varphi_0(\Delta\varepsilon)}{N_e \partial \varepsilon} \tag{6.4.11}$$

In particular, in the case the Maxwell distribution of electrons formula (6.4.11) gives for a corresponding mechanism of electron diffusion this leads to formula (6.3.15). One can compare the rate constant of atom excitation by electron impact $k_<$ according to formula (6.4.11) if it is determined by an electron flux in an energy space at energies below the atom excitation energy, with the excitation rate constant k_{th} according to formula (3.3.8) for thermodynamic equilibrium between excited and nonexcited atoms that is supported by collisions with electrons. The ratio of these quantities is given in Fig. 6.13. As is seen, violation of thermodynamic equilibrium due to excitation process leas to a decrease of this ratio.

6.5 Electron Distribution Function Above the Atom Excitation Threshold

Above we used the model for atom excitation in a gas discharge plasma where the elastic channel for electron-atom collisions acts at the electron energies below the atom excitation threshold, whereas above this electron energy the inelastic channel of electron-atom collisions dominates. Then atom excitation is determined by electrons which energy intersects the threshold of atom excitation, and this allows one to determine the rate of atom excitation in a gas discharge plasma. Now remaining in the regime of a low electron number densities, we find the structure of the electron distribution function accurately, if the criterion (6.3.3) holds true.

Let us account for atom excitation by electron impact in a gas discharge plasma by inserting in the second equation of the set (6.1.4) the excitation channel, that gives

$$-\frac{a}{3v^2}\frac{d}{dv}(v^3 f_1) = I_{ea}(f_0) - \nu_{ex} f_0$$

Using the relation (5.3.9) for the collision integral in this case and assuming the electron energy to be large $\varepsilon \gg T$, we obtain the following equation for the symmetric part of the electron distribution function

$$\frac{a^2}{3v^2}\frac{d}{dv}\left(\frac{v^2}{\nu_{ea}}\frac{df_0}{dv}\right) + \frac{m_e}{M}\frac{1}{v^2}\frac{d}{dv}(v^3 \nu_{ea} f_0) - \nu_{ex} f_0 = 0 \qquad (6.5.1)$$

Accounting for a sharp change of the electron distribution function above the excitation threshold due to the criterion (6.3.3), we solve (6.5.1) in the quasiclassical approach representing the distribution function in the form $f_0 = A\exp(S)$ and assuming $(S')^2 \gg S''$. Ignoring the second term in (6.5.1), i.e., neglecting by the energy loss due to elastic electron-atom collisions, and solving this equation, we have $S' = -\sqrt{3\nu_{ex}\nu_{ea}}/a$, where $a = eE/m_e$ and $\varepsilon \gg \overline{\varepsilon}$. This gives [83]

$$f_0(v) = f(v_o)\exp(-S) = f(v_o)\exp\left(-\int_{v_o}^{v}\frac{dv}{a}\sqrt{3\nu_{ex}\nu_{ea}}\right), \qquad (6.5.2)$$

where the distribution function $\varphi(v)$ is determined by elastic electron-atom collisions only and is introduced in formula (6.1.22), the threshold electron velocity is $v_o = \sqrt{2\Delta\varepsilon/m_e}$. Using the threshold law (3.3.6) for the excitation cross section $\sigma_{ex} \sim \sqrt{\varepsilon - \Delta\varepsilon}$ and expressing according to formula (3.3.7) the atom excitation rate constant k_{ex} through the quenching rate constant k_q that is independent of $\varepsilon - \Delta\varepsilon$ near the threshold, we obtain at electron energies near the threshold

$$S = \kappa\left(\frac{\varepsilon - \Delta\varepsilon}{\Delta\varepsilon}\right)^{5/4}, \quad \kappa = \frac{\alpha}{x}, \quad \alpha = 0.4 v_o k_{ef} m_e/e, \quad \nu_{ef} = \sqrt{3\frac{g_*}{g_o}\nu_q \nu_o} = N_a k_{ef} \qquad (6.5.3)$$

Here $\nu_o = N_a k_{el}(v_o) = N_a v_o \sigma_{el}(v_o)$ is the rate of elastic electron-atom collisions near the excitation threshold, $x = E/N_a$ is the reduced electric field strength, the quenching rate is $\nu_q = N_a k_q$, where g_o, g_* are the statistical weights for the ground and excited atom states. We add to this the criterion of the validity of the quasiclassical exponent form $(S')^2 \gg S''$. We also represent in Table 6.1 the parameters of formula (6.5.3) for the lowest excited states of helium and argon atoms.

In this consideration we assume the electron distribution function (6.5.2) to be decreased sharply near the excitation threshold. This means that for the electron energy $S(\varepsilon_*) \sim 1$ we have $\varepsilon_* - \Delta\varepsilon \ll \Delta\varepsilon$, that is fulfilled at low electric field strengths

$$\kappa \gg 1 \qquad (6.5.4)$$

Table 6.1 Parameters of excitation of helium and argon atoms in a gas discharge plasma

Atom	$He(2^3S)$	$Ar(^3P_2)$	$Ar(^3P_1)$	$Ar(^3P_0)$	$Ar(^1P_0)$
$\Delta\varepsilon$ (eV)	19.82	11.55	11.62	11.72	11.83
v_o (10^8 cm/s)	2.64	2.02	2.02	2.03	2.04
g_*	3	5	3	1	3
$k_o = k_{el}(v_o)$ (10^{-8} cm^3/s)	7.1	29	29	30	30
k_q (10^{-9}cm^3/ s)	3	0.4	0.82	0.4	3.9
k_{ef} (10^{-8} cm^3/s)	4.4	4.2	4.2	1.9	10
α (Td)	263	191	213	86	471

This criterion means that the electron distribution function is distorted above the excitation threshold due to atom excitation. Another limiting case is given by formula (6.1.5) if the rate constant of elastic electron-atom collisions is independent of the collision velocity.

It should be added to criterion (6.4.5) that the atom excitation energy $\Delta\varepsilon$ corresponds to the tail of the electron distribution function (6.1.20), and the regime of low electron number densities may be represented in the form

$$a \ll v_o \nu_o \sqrt{\frac{m_e}{M}}$$

or

$$\kappa \gg \frac{\nu_{ex}}{\nu_o}\sqrt{\frac{m_e}{M}} \tag{6.5.5}$$

One can see that the criteria (6.5.4) and (6.5.5) have an identical structure, and one can expect the criterion (6.5.5) to be more strong.

One can determine the rate of direct atom excitation by electron impact on the basis of the distribution function (6.5.2) under assumption that the main contribution to atom excitation gives a narrow range of electron energies in accordance with criteria (6.5.4) and (6.5.5). Using formula (3.3.7) for the rate constant of excitation, we have for the rate of direct excitation

$$\frac{dN_*}{dt} = N_a \int k_{ex} f_0(v) \cdot 4\pi v^2 dv = \frac{4\pi v_o}{m_e} \frac{g_*}{g_o} \nu_q \int\limits_{\Delta\varepsilon}^{\infty} \sqrt{\frac{\varepsilon - \Delta\varepsilon}{\Delta\varepsilon}} f_0(v) e^{-S} d\varepsilon = N_a N_e \frac{f_0(v_o)}{\varphi_0(v_o)} F, \tag{6.5.6}$$

where $\nu_q = N_a k_q$, and κ is given by formula (6.5.3). One can introduce the rate constant of atom excitation for a given electron distribution over velocities

$$k_> = \frac{1}{N_a N_e} \frac{dN_*}{dt}, \tag{6.5.7}$$

and we have

$$k_{>} = k_{th}\frac{f_0(v_o)}{\varphi_0(v_o)}, \; k_{th} = k_q\frac{g_*}{g_o}\cdot\exp\left(-\frac{\Delta\varepsilon}{T_e}\right), \quad F = \frac{2}{\sqrt{\pi}T_e^{3/2}}\int\limits_0^\infty\sqrt{\varepsilon'}\exp\left[-\frac{\varepsilon'}{T_e} - \kappa\left(\frac{\varepsilon'}{\Delta\varepsilon}\right)^{5/4}\right] \tag{6.5.8}$$

Evaluating the parameter F in the limiting cases where one can remain in the exponent only one term and sewing the limiting cases, we obtain finally

$$k_{>} = k_{th}\frac{f_0(v_o)}{\varphi_0(v_o)}F, \quad F = \left[1 + \kappa^{1.2}\left(\frac{T_e}{\Delta\varepsilon}\right)^{5/4}\right]^{-1} \tag{6.5.9}$$

In analyzing the limiting cases, we account for $f_0 \sim \exp(-\varepsilon/T_e)$, and if in a range $\varepsilon - \Delta\varepsilon \sim T_e$ we have $S = 0$, this limit gives

$$k_{>} = k_{th}\frac{f_0(v_o)}{\varphi_0(v_o)}, \quad k_{th} = k_q\frac{g_*}{g_o}\cdot\exp\left(-\frac{\Delta\varepsilon}{T_e}\right),$$

and the excitation rate constant k_{th} corresponds to the thermodynamic equilibrium. In the other limiting case where $S(v)$ varies sharply near the atom excitation, we obtain

$$k_{>} = k_{th}\frac{f_0(v_o)}{\varphi_0(v_o)}, \quad k_{th} = k_q\frac{g_*}{g_o}\cdot\exp\left(-\frac{\Delta\varepsilon}{T_e}\right),$$

where $f_o = f_0(v_o)$, and v_o is the electron velocity at the excitation threshold, $\nu_o = \nu(v_o)$ is the rate of electron-atom elastic scattering at this energy. Note that formula (6.5.7) for the excitation rate is valid both for regimes of low and high electron number densities. Substituting in this formula the Maxwell distribution function of electrons far from the excitation threshold for the regime of a high electron number density, we obtain for the atom excitation rate by electron impact in the limit if $k_, \ll k_{>}$ and hence $f_o = \varphi_o$

$$\frac{dN_*}{dt} = 0.83 N_e N_a\frac{g_*}{g_o}k_q\left(\frac{\Delta\varepsilon}{T_e}\right)^{3/2}\exp\left(-\frac{\Delta\varepsilon}{T_e}\right)\kappa^{-1.2} \tag{6.5.10}$$

Comparing this expression for direct atom excitation by electron impact in an ionized gas with that due to diffusion motion of electron in an energy space according to formula (6.3.7), we find for the ratio of these rates

$$\zeta\,\frac{f_o}{\varphi_o} = \frac{0.37}{\kappa^{1.2}}\frac{g_*}{g_o}\frac{M}{m_e}\frac{\nu_q}{\nu_o}\frac{f_o}{\varphi_o} \tag{6.5.11}$$

As is seen, both mechanisms of atom excitation may be realized depending on the electric field strength E (or the parameter a) under conditions where criteria (6.5.4) and (6.5.5) are fulfilled. Formula (6.5.11) allows us to find the electric field strength E_* at which both mechanisms gives the same contribution to the rate of atom excitation. This corresponds to $\zeta = 1$ and gives

$$\frac{eE_*}{N_a} = 0.175 m_e v_o k_{ef} \left(\frac{g_o}{g_*} \frac{m_e}{M} \frac{k_q}{k_o} \right)^{5/6}, \qquad (6.5.12)$$

where $k_{ef} = \nu_{ef}/N_a$, $k_q = \nu_q/N_a$, $k_o = \nu_o/N_a$. Note that because of the identical distribution functions of electrons $\varphi(v_o)$ in the expressions for the excitation rates in the diffusion (6.3.7) and direct (6.5.6) mechanisms of atom excitation by electron impact, formulas (6.5.11) and (6.5.12) are valid both for regimes of a low electron number densities and high electron number densities.

Thus, on the basis of formula (6.3.8) we evaluate the rate $k_<$ of travelling of electrons in a space of electron energies to the excitation threshold and also express the atom excitation rate $k_>$ through the electron distribution function at the excitation threshold according to formula (6.5.7). One can introduce also the rate $k'_<$ of electron travelling in a space of electron energies that describes reflection of electrons form the excitation threshold in a space of electron energies. The corresponding fluxes are given in Fig. 6.14, and from the equality of total fluxes $j \le j_> + j'_<$ one can find both the effective excitation rate constant k_{ex} and the electron distribution function f_o at the excitation threshold if φ_o is the distribution function of electrons far before the excitation threshold [287]

$$k_{ex} = \frac{k_< k_>}{k_< + k_>} = 4.6 \frac{g_*}{g_o} \frac{k_q v_o^3 \varphi_o / N_e}{(1+\zeta)\kappa^{1.2}}, \quad f_o = \frac{k_<}{k_< + k_>} \varphi_o \qquad (6.5.13)$$

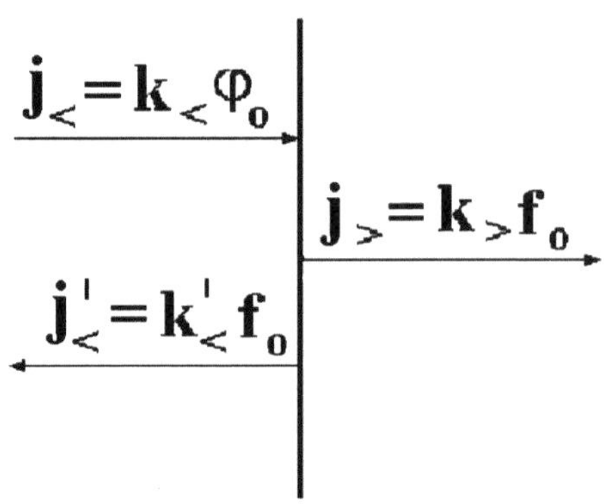

Fig. 6.14 The character of electron fluxes near the atom excitation threshold. The flux of electrons towards the excitation boundary $k_<$ is proportional to the derivation of the electron distribution function at the the excitation threshold and hence is expressed through the distribution function φ_o in ignoring the absorption process for fast electrons. The reflected electron flux $k'_<$ is proportional to the electron distribution function at the absorption boundary [287]

This formula shows that the mechanisms of electron travelling to the excitation rate and the excitation rate above the excitation threshold do not compete, but they establish the distribution f_o at the excitation threshold. In this formula the rate constant $k_<$ of electron motion to the excitation boundary in a space of electron energies corresponds to the diffusion character of this motion where electrons do not reflect from the excitation boundary. In other words, the quantity $k_<$ in formula (6.5.13) relates to the boundary condition (6.1.19), where the distribution function at the excitation boundary f_o is zero. On contrary, the rate constant of atom excitation $k_>$ by fast electrons in formula (6.5.13) is regarded to the case where the electron flux toward the excitation boundary in a space of electron energies is equal to the reflected flux, i.e. the electron distribution function at the excitation boundary f_o is equal to its non-perturbed value φ_o.

Let us sew the expressions (6.1.20) and (6.5.2) for the electron distribution function above the excitation threshold. Let us represent the electron distribution function in this energy range in the quasiclassical form as [83]

$$f_0(\varepsilon) = f_0(\Delta\varepsilon)\exp(-S_1 - S_2), \quad S_1 = \int_v^{v_o} m_e v' dv' \frac{3v^2}{Ma^2}, \quad S_2 = \int_{v_o}^{v} \frac{dv}{a}\sqrt{3\nu_{ex}\nu_o},$$

and the energy derivations of these exponents are equal

$$S_1' = \frac{3v^2}{Ma^2}, \quad S_2' = \frac{4\nu_{ef}}{m_e v_o a}\left(\frac{\varepsilon - \Delta\varepsilon}{\varepsilon}\right)^{1/4}, \quad \nu_{ef} = \sqrt{3\frac{g_*}{g_o}\nu_q\nu_o} \tag{6.5.14}$$

A schematic character of the dependence (6.5.14) of the electron distribution function on the electron energy is given in Fig. 6.14. In this consideration we require that the characteristic electron energy ε_b defined as the derivations are equal $S_1'(\varepsilon_b) = S_2'(\varepsilon_b)$ to be close to the excitation energy $\Delta\varepsilon$. This corresponds to a small parameter

$$\delta = \left(\frac{\varepsilon_b - \Delta\varepsilon}{\Delta\varepsilon}\right)^{1/4} \ll 1 \tag{6.5.15}$$

If this criterion holds true, the distortion of the distribution function above the excitation threshold due to atom excitation becomes important. According to formula (6.5.14) this parameter is equal

$$\delta = \frac{3m_e}{4M}\frac{v_o v_o^2}{a v_{tf}} \sim \frac{1}{\kappa}\frac{m_e}{M}\frac{v_o^2}{v_{ex}^2} \tag{6.5.16}$$

In the case under consideration where the electron distribution function above the excitation threshold decreases sharply at removal from the threshold, the point of change of the distribution function derivation in Fig. 6.15 is located close to the excitation threshold.

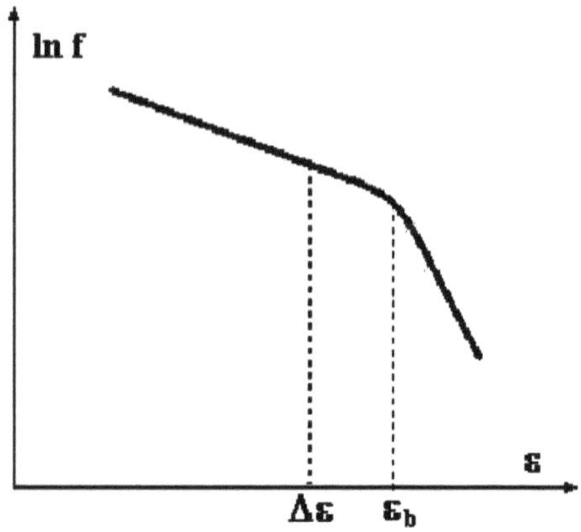

Fig. 6.15 The energy distribution function of electrons near the threshold of atom excitation

We now sum up the results of the above analysis and find the rate of atom excitation in a gas discharge plasma. In the case of small electric field strengths E the atom excitation threshold corresponds to a tail of the energy distribution function of electrons, and then the distribution function varies sharply at the atom excitation energy with energy variation. In addition, slow electrons resulted from atom excitation does not change the distribution function that allows us to ignore this process in determination of the distribution function of slow electrons. In this case the excitation rate is relatively small and it can be determined on the basis of the above model that is given by formula (6.5.14). We must modify formula (6.5.14) in this case. Indeed, represent the electron distribution function in the form

$$f_0(\varepsilon) = f_0(\Delta\varepsilon)\exp(-S_1),\ \ \varepsilon < \varepsilon_b;\ f_0(\varepsilon) = f_0(\Delta\varepsilon)\exp(-S_2),\ \ \varepsilon > \varepsilon_b$$

In the regime of a high electron number density, where the energy distribution function of electrons is the Maxwell one (2.1.2), it may be represented in the form [288]

$$f_0(v) \sim \left[\exp\left(-\frac{\varepsilon}{T_e}\right) - \exp\left(-\frac{\Delta\varepsilon}{T_e}\right)\right], \tag{6.5.17}$$

if we identify the boundary energy ε_b with the atom excitation energy $\Delta\varepsilon$.

The above form of the distribution function given in Fig. 6.14 divides the electron energy range in two ranges with the boundary $\varepsilon = \varepsilon_b$. At lower energies the distribution function is determined by elastic electron-atom collisions, while all electrons of the second group excite atoms and become slow electrons. Then the rate of atom excitation in this gas discharge plasma is determined by formula (6.5.13) and consists of two parts. The first part includes the electron flux $k_<$ in the electron energy space towards the boundary energy ε_b, and this flux is created by elastic electron-atom collisions. The second part is excitation of atoms that takes energy of fast electrons, i.e. fast electrons disappears as a result of this process. The total excitation rate constant is given by formula (6.5.13).

6.6 Electron Kinetics in Gas in Strong Field

The above expressions for the energy distribution function of electrons hold true if a typical electron energy is small compared to the excitation energy of atoms, and then inelastic electron-atom collisions do not influence on the electron distribution function. Let us consider the case of high electric field strengths at low electron number densities where electron-electron interactions are neglected. In this case each electron is accelerated under the action of the electric field, and when it reaches the atom excitation energy, the electron transfers energy on atom excitation and stops, starting to accelerate again (see Fig. 6.16).

We below determine the electron distribution function accounting for the recurring character of the energy of a test electron, as it shown in Fig. 6.16. Then the energy distribution function of electrons $f_0(\varepsilon)d\varepsilon$ is proportional to a time interval dt at which the electron energy ranges from ε to $\varepsilon + d\varepsilon$. Note that in determination the electron distribution function on the basis of dynamics of a test electron, we use the ergodic theorem [289–291] according to which after an average over a large time the probability that some parameter is located in a certain interval is proportional to a time during which this parameter is found in an indicated interval. Let us use the energy equation for a test electron

$$m_e \frac{d\varepsilon}{dt} = eEw_e, \tag{6.6.1}$$

where ε is a current electron energy, $w_e = w_{el}E/N_a$ is a current electron drift velocity when it has an energy ε. We assume that due to a small parameter m_e/M (the ratio of

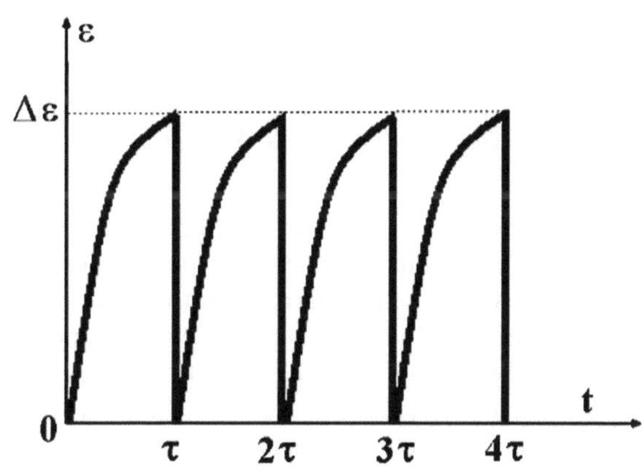

Fig. 6.16 Character of energy variation for a test electron in time. A test electron acquires an energy from an external electric field and an increase of the electron energy is restricted by excitation of gas atoms, so that the electron stops after each atom excitation. τ is a time of one period during which the electron energy varies from zero up to the excitation threshold

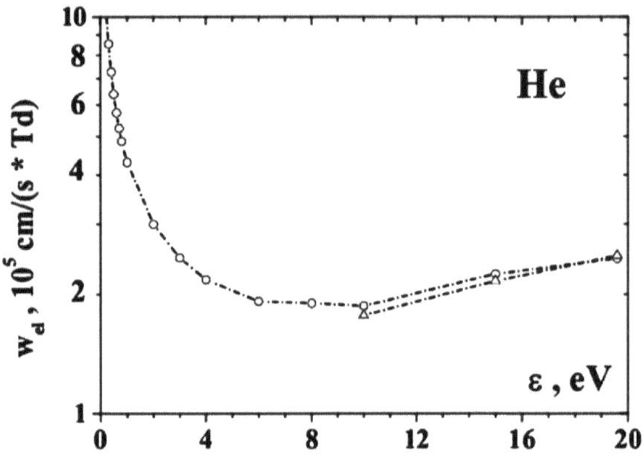

Fig. 6.17 Current reduced drift velocity of an electron in helium at a given electron energy in accordance with formula (6.6.2). The *curve* with *open circles* is constructed on the basis of cross sections of Table 3.1 [130], *open triangles* correspond to the approximation dependence $w_{el} = 5.6 \times 10^4 \varepsilon^{1/2}$, where the reduced drift velocity is expressed in cm/(s · Td), and the electron energy is given in eV

electron and atom masses) the electron distribution function is almost isotropic with a small width $\Delta\varepsilon \sim (m_e/M)\varepsilon$. Hence a current electron drift velocity is

$$w_e = \frac{eE}{m_e N_a v \sigma_{ea}(\varepsilon)} \tag{6.6.2}$$

Figure 6.17 contains the reduced current drift velocity of an electron $w_{el} = w_e N_a / E$ in helium as a function of its energy ε, and Fig. 6.18 contains this dependence in the argon case. Correspondingly, the energy distribution function of electrons in this case is [292]

$$f_0(\varepsilon)d\varepsilon = Cv\sigma_{ea}(\varepsilon)d\varepsilon, \ C = N_e \left(\int v\sigma_{ea}(\varepsilon)d\varepsilon \right)^{-1} \tag{6.6.3}$$

Note that this distribution function differs from the Maxwell one (6.2.4) if the energy change of a test electron results from collision with other electrons of a gas discharge plasma. In the case under variation an external electric field dominates in kinetics of the spherically symmetric distribution function of electrons rather than collisions with other electrons. In other words, in this case electron-atom collisions determine variation of the momentum of a test electron in electron kinetics in a space of electron velocities, whereas variation of a current energy of a test electron results from the action of an external electric field.

We assume in this formula that an electron partakes in many collisions with atom until it attains the energy of atom excitation. Using formula (6.1.7) for the current electron drift velocity and as early the dependence $\nu \sim v^k$ ($\sigma^*_{ea}(v) \sim v^{k-1}$) for the rate of electron-atom collisions, we obtain for the electron distribution function

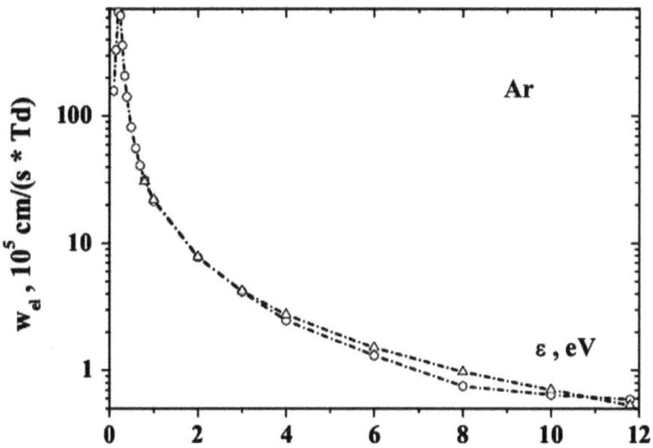

Fig. 6.18 Current reduced drift velocity of an electron in argon at a given electron energy in accordance with formula (6.6.2). The *curve* with *open circles* is constructed on the basis of cross sections of Table 3.1, *open triangles* correspond to the approximation dependence for a current electron drift velocity $w_{el} = 2.2 \times 10^6/\varepsilon^{3/2}$, where the current reduced drift velocity is expressed in cm/(s · Td), and the electron energy is given in eV: 1 optically thin plasma, 2 optically thick plasma

$$f_0(\varepsilon)d\varepsilon = C\varepsilon^{k/2}d\varepsilon, \quad \varepsilon \leq \Delta\varepsilon \tag{6.6.4}$$

Here C is the normalization coefficient that is equal to

$$C = N_e \frac{1 + k/2}{(\Delta\varepsilon)^{1+k/2}} \tag{6.6.5}$$

We have for a current electron drift velocity at the threshold electron energy $\Delta\varepsilon$

$$w_e = \frac{eE}{m_e \nu(\Delta\varepsilon)} \tag{6.6.6}$$

The energy distribution function of electrons (6.6.4) allows one to determine the average electron energy $\overline{\varepsilon}$ is an electron moves in a gas in an external electric field of a strength E that is equal to

$$\overline{\varepsilon} = \int_0^{\Delta\varepsilon} \varepsilon \frac{f_0(\varepsilon)d\varepsilon}{N_e} = \frac{1 + k/2}{2 + k/2}\Delta\varepsilon \tag{6.6.7}$$

Let us apply these results to helium and argon. Approximating the diffusion cross section of electron-atom scattering below the excitation energy in accordance with the data of Table 3.1, we obtain approximately $\sigma^*(\varepsilon) \sim 1/\varepsilon$ for helium and $\sigma^*(\varepsilon) \sim \varepsilon$ for argon. This gives $k \approx -1$ for helium and $k \approx 3$ for argon, and the energy distribution function has the following form in these cases

$$f_{He}(\varepsilon)d\varepsilon = N_e \frac{2d\varepsilon}{\sqrt{\varepsilon\Delta\varepsilon}}, \; f_{Ar}(\varepsilon)d\varepsilon = N_e \frac{2\varepsilon^{3/2}d\varepsilon}{5(\Delta\varepsilon)^{5/2}}, \; \varepsilon \leq \Delta\varepsilon, \tag{6.6.8}$$

where an index at the distribution function indicates a gas, and the excitation energy is $\Delta\varepsilon = 19.8\,\text{eV}$ for helium and $\Delta\varepsilon = 11.55\,\text{eV}$ for argon. The average electron energy according to formula (6.6.7) is equal in the helium and argon cases

$$\overline{\varepsilon}(He) = \frac{\Delta\varepsilon}{3} \approx 7\,\text{eV}, \quad \overline{\varepsilon}(Ar) = \frac{5}{7}\Delta\varepsilon \approx 8\,\text{eV} \tag{6.6.9}$$

The rate constant of ionization of excited atoms by electron impact k^*_{ion} is given by

$$k^*_{ion} = \int_{J_*}^{\Delta\varepsilon} f(\varepsilon)k_{ion}(\varepsilon)d\varepsilon, \tag{6.6.10}$$

where the rate of atom ionization $k_{ion}(\varepsilon)$ is determined by formula (3.6.17). Since the energy distribution function of electrons does not depend on the electric field strength in a range of strong fields under consideration, this rate constant is independent of the electric filed strength and $k^*_{ion}(He) = 2.5 \times 10^{-7}\,\text{cm}^3/\text{s}$ for ionization metastable helium atoms $He(2^3S)$, whereas $k^*_{ion}(Ar) = 3.6 \times 10^{-7}\,\text{cm}^3/\text{s}$ for ionization metastable argon atoms $Ar(^3P_2)$.

We also have from (6.6.1) for the distance Δx that is passed by electron during one period and the period time τ in accordance with Fig. 6.16 is given by

$$\Delta x = \frac{\Delta\varepsilon}{eE}, \quad \tau = \frac{1}{eE}\int_0^{\Delta\varepsilon} \frac{d\varepsilon}{w_e(\varepsilon)} \tag{6.6.11}$$

Formula (6.6.11) gives for the reduced way of an electron per one cycle

$$\Delta x(He)E \approx 20\,\text{V}, \quad \Delta x(Ar)E \approx 12\,\text{V} \tag{6.6.12}$$

Using the dependence of a current reduced drift velocity of electrons on its energy that is $w_{el}(He) \sim \varepsilon^{1/2}$ and $w_{el}(Ar) \sim \varepsilon^{-3/2}$, we find the average reduced drift velocity of electrons as

$$\overline{w_{el}} = \frac{1}{N_e}\int_0^{\Delta\varepsilon} w_{el}(\varepsilon)f_0(\varepsilon)d\varepsilon$$

This gives for the helium and argon cases

$$\overline{w_{el}}(He) = \frac{1}{2} w_{el}(He)|_{\varepsilon=\Delta\varepsilon} = 1.2 \times 10^5 \frac{\text{cm}}{\text{s} \cdot \text{Td}},$$
$$\overline{w_{el}}(Ar) = \frac{5}{2} w_{el}(Ar)|_{\varepsilon=\Delta\varepsilon} = 1.5 \times 10^5 \frac{\text{cm}}{s \cdot \text{Td}} \qquad (6.6.13)$$

We assume above that the main contribution to this value give electron energies where a power approximation holds true. As it follows from Figs. 6.16 and 6.17, this is fulfilled better for argon, whereas for helium it leads to an error. Next, the electron drift velocity w_e that is averaged over electron energies is expressed through the reduced one as

$$w_e = \overline{w_{el}} \frac{E}{N_a} \qquad (6.6.14)$$

Next, on the basis of formula (6.6.11) we have

$$\tau(He) = \frac{2\Delta x}{w_e(\Delta\varepsilon)}, \quad \tau(Ar) = \frac{2\Delta x}{5w_e(\Delta\varepsilon)}$$

Introducing the reduced time τ_o of one cycle as $\tau_o = \tau(E^2/N_a^2)$, we find for this parameter

$$\tau_o(He) = 1.6 \times 10^{-4}\text{s} \cdot \frac{\text{V}}{\text{cm}} \cdot \text{Td}, \quad \tau_o(Ar) = 7.9 \times 10^{-5}\text{s} \cdot \frac{\text{V}}{\text{cm}} \cdot \text{Td} \qquad (6.6.15)$$

The peculiarity of the electron drift process in strong electric fields is such, that an electron is accelerated and acquires an energy from the electric field, so that it is able to excite an atom. Hence, the parameters of electron drift do not depend on the mass ratio m_e/M because inelastic collisions determine the electron energy and this scheme works until $w_e \ll \sqrt{\Delta\varepsilon/m_e}$. This character of the electron process allows us to determine the rate constant k_{ex} of atom excitation in this regime. Since the power obtained from an external field by an electron is eEw_e and the atom excitation energy is $\Delta\varepsilon$, the excitation rate constant k_{ex} is given by

$$k_{ex}N_a = \frac{eEw_e}{\Delta\varepsilon} \qquad (6.6.16)$$

In derivation this formula we assume that the main resistance an electron consumes to attain the atom excitation energy, and after this the excitation process proceeds effectively. Since the electron drift velocity in this regime is proportional to E/N_a, one can represent the excitation rate constant as

$$k_{ex} = k_o \left(\frac{E}{N_a}\right)^2 \qquad (6.6.17)$$

where the reduced excitation rate constant k_o is independent of the electric field strength E and the atom number density N_a. One can express the excitation rate constant through a time τ that is acquired by an electron per one cycle and is spent on atom excitation. This gives

$$k_{ex} = \frac{1}{\tau N_a},$$

and $k_o \sim 1/\tau_o$. From this we have for the reduced excitation rate constants

$$k_o(He) = 6.2\times 10^{-14}\,\mathrm{cm}^3/(\mathrm{s}\cdot\mathrm{Td}^2),\ k_o(Ar) = 1.3\times 10^{-13}\mathrm{cm}^3/(\mathrm{s}\cdot\mathrm{Td}^2) \quad (6.6.18)$$

Let us determine the electron diffusion coefficient if an ionized gas is located in a strong electric field. As above we assume the electron momentum variation to be resulted in electron-atom collisions that gives an isotropic velocity distribution. We also consider the above regime where a tail of the distribution function at electron energies above the atom excitation energy $\varepsilon > \Delta\varepsilon$ gives a small contribution the the distribution function normalization. Then the electron diffusion coefficient is

$$D_e = \int_0^{\Delta\varepsilon} \frac{v^2}{3\nu} f_0(\varepsilon)d\varepsilon, \quad (6.6.19)$$

where v is a current electron velocity, $\nu = N_a v\sigma_{ea}$ is the rate of elastic electron-atom collisions, so that $v^2/(3\nu)$ is a current diffusion coefficient of a test electron. Using the energy distribution function (6.6.4), we obtain for the electron diffusion coefficient D_e averaged over electron energies

$$D_e = D_o \frac{1 + k/2}{2}, \quad (6.6.20)$$

where D_o is the diffusion coefficient of electrons at the energy $\varepsilon = \Delta\varepsilon$. These formulas give in the helium case at the excitation threshold $v\sigma_{ea} = 7.1\times 10^{-8}\,\mathrm{cm}^3/\mathrm{s}$ that gives $D_o N_a = 3.3\times 10^{23}\,\mathrm{cm}^{-1}\mathrm{s}^{-1}$ and $D_e N_a = 8.1\times 10^{22}\,\mathrm{cm}^{-1}\mathrm{s}^{-1}$ $(k = -1)$. In the same manner we have in the argon case $v\sigma_{ea} = 3.0\times 10^{-7}\,\mathrm{cm}^3/\mathrm{s}$ that gives $D_o N_a = 4.6\times 10^{22}\,\mathrm{cm}^{-1}\mathrm{s}^{-1}$ and $D_e N_a = 5.8\times 10^{22}\,\mathrm{cm}^{-1}\mathrm{s}^{-1}$ $(k = 3)$.

The above case holds true under the condition if

$$eEw_e \ll \nu_{ea}\delta\varepsilon, \quad (6.6.21)$$

where w_e is the current electron number density, $\nu_{ea} = N_a k_{ea}$ is the rate of elastic electron collision with atoms, k_{ea} is the rate constant of elastic electron-atom collisions, $\delta\varepsilon$ is a typical energy which characterizes a remarkable variation of this rate constant. Under these conditions an electron changes the motion direction many times until its energy varies remarkably, i.e. a current distribution function of a test electron at each time has the spherical form as it is given in Fig. 6.19. Hence criterion (6.6.21) takes the form

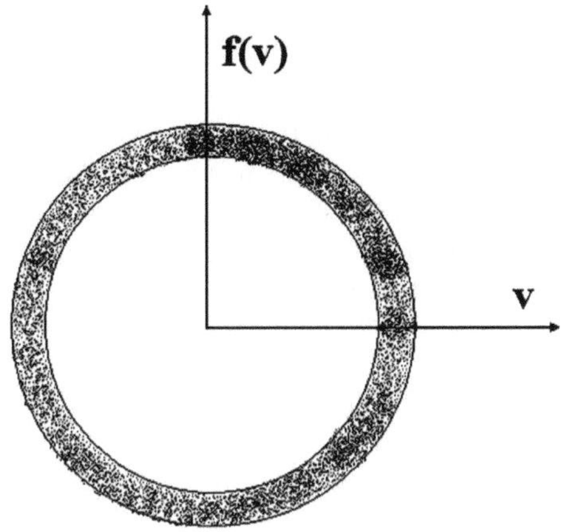

Fig. 6.19 Current energy distribution function of electrons in the course of electron motion in a gas in an electric field. Because of a large elastic cross section of electron-atom collisions compared with inelastic ones, the energy distribution function of electrons is close to an isotropic one: 1 optically thin plasma, 2 optically thick plasma

$$\left(\frac{eE}{N_a}\right)^2 \ll k_{ea}^2 m_e \delta\varepsilon \tag{6.6.22}$$

In particular, taking in the helium case at the atom excitation threshold $k_{ea} = 7.1 \times 10^{-8}\text{cm}^3/\text{s}$ and $\delta\varepsilon = 10\,\text{eV}$, we obtain the criterion (6.6.22) as $E/N_a \ll 500\,\text{Td}$. In the argon case at the atom excitation threshold we have $k_{ea} = 3.0 \times 10^{-7}\text{cm}^3/\text{s}$ and $\delta\varepsilon = 6\text{eV}$, that gives $E/N_a \ll 2{,}000\,\text{Td}$. In the opposite limiting case the energy distribution function is $f(\mathbf{v}) = C\delta(\mathbf{v} - \mathbf{i}w_x)$, where $\mathbf{i}$ is the unit vector along the electric field.

The expression (6.6.16) for the excitation rate constant k_{ex} holds true also if $k_< \ll k_>$, i.e. the atom excitation is determined by the electron flux in the electron energy space where the electron energy is small compared to the atom excitation energy. Let us analyze the atom excitation process in a gas discharge plasma located in a strong field from another standpoint considering this process when the electron energy exceeds the atom excitation energy $\varepsilon > \Delta\varepsilon$. Because the criterion holds true that is opposite with respect to criterion (6.5.4), the equation of variation of the electron energy ε if it moves in a gas in an external electric field of a strength E has the form

$$\frac{d\varepsilon}{dt} = eEw_e - N_a k_{ex} \Delta\varepsilon \tag{6.6.23}$$

One can use the principle of detailed balance (3.3.7) and express the excitation rate through the quenching rate constant k_q of the lowest atom excited state by electron impact which is independent on the electron energy at small energies. We have

$$\frac{d\varepsilon}{dt} = eEw_e - N_a k_q \frac{g_* \Delta\varepsilon(\varepsilon - \Delta\varepsilon)}{g_o \varepsilon} \tag{6.6.24}$$

Taking $t = 0$ when $\varepsilon = \Delta\varepsilon$, we solve (6.6.24) in the range $\varepsilon - \Delta\varepsilon \ll \varepsilon$, assuming that the atom excitation process proceeds mostly near the excitation threshold. In consideration this equation under the above conditions, we assume the electron drift velocity to be independent on time and take it at the excitation threshold according to formula (6.6.6). The solution of (6.6.24) has the form

$$\varepsilon - \Delta\varepsilon = \varepsilon_o \left[1 - \exp(-t/\tau_o)\right], \quad \varepsilon_o = \frac{3x^2}{m_e k_{ef}^2}, \quad \tau_o = \frac{g_o}{g_* N_a k_q}, \tag{6.6.25}$$

where the reduced electric field strength $x = eE/N_a$, and the rate constant k_{ef} is given by formula (6.5.3).

Note that (6.6.24) describes evolution of the average electron energy, and hence (6.6.24) may be used as estimations. In particular, the parameter ε_o is a typical excess of the threshold energy for the electron energy, and τ_o is a typical time

$$\varepsilon_o \ll \Delta\varepsilon \tag{6.6.26}$$

Let us find the reduced electric field strength at which $\varepsilon_o = \Delta\varepsilon/2$ in the helium and argon cases. In the helium case according to the data of Table 6.1 we obtain $k_q = 3 \times 10^{-9}\text{cm}^3/\text{s}$ and $k_{ef} = 1.4 \times 10^{-7}\text{cm}^3/\text{s}$, and this gives $x = 600$ Td. In the argon case we join 4 atomic excited levels of Table 6.1 with the electron shell $3p^5 4s$. The rate constant of the joined level with the statistical weight $g_* = 12$ is $1.4 \times 10^{-9}\,\text{cm}^3/\text{s}$, and this leads to the effective rate constant $k_{ef} = 1.2 \times 10^{-7}\text{cm}^3/\text{s}$. From this it follows $x = 400$ Td, if $\varepsilon_o = \Delta\varepsilon/2$.

Let us find the rate constant of ionization of excited atoms in a gas discharge plasma of helium and argon by electrons moved in a strong electric field. This determines stepwise ionization in this plasma. Taking formulas (3.6.15), (3.6.16) for the cross section of an atom with one valence s-electron by electron impact, we obtain for the ionization rate constant in a strong electric field

$$k_{ion}^* = \frac{k_o}{N_e} \int_J^{\Delta\varepsilon} \frac{(x-1) f(\varepsilon) d\varepsilon}{x^{1/2}(x+8)}, \quad k_o = \frac{10e^4}{J^2}\sqrt{\frac{2J}{m_e}}, \quad x = \frac{\varepsilon}{J}, \tag{6.6.27}$$

and the atom ionization potential for the lowest excited state is $J = 4.77$ eV in the helium case, and $J = 4.21$ eV in the argon case. We have $k_o(He) = 1.2 \times 10^{-6}\text{cm}^3/\text{s}$ and $k_o(Ar) = 1.4 \times 10^{-6}\,\text{cm}^3/\text{s}$. This leads to the following expressions for the ionization rate constants in the helium and argon cases

$$k_{ion}^*(He) = k_*(He) \int_1^{4.15} \frac{(x-1)dx}{x(x+8)}, \quad k_{ion}^*(Ar) = k_*(Ar) \int_1^{2.74} \frac{x(x-1)dx}{(x+8)} \tag{6.6.28}$$

with $k_*(He) = 1.2 \times 10^{-6}\mathrm{cm}^3/\mathrm{s}$ and $k_*(Ar) = 4.6 \times 10^{-8}\mathrm{cm}^3/\mathrm{s}$. From this we find the ionization rate constants of excited helium and argon atoms by electron impact for a gas discharge plasma located in a strong electric field as

$$k_{ion}^*(He) = 1.8 \times 10^{-7}\mathrm{cm}^3/\mathrm{s}, \quad k_{ion}^*(Ar) = 1.5 \times 10^{-8}\mathrm{cm}^3/\mathrm{s} \qquad (6.6.29)$$

Note that formula (6.6.3) for the energy distribution function of electrons corresponds to conditions of Fig. 6.16, where a test electron excites an atom of a gas discharge plasma near the threshold, and hence the tail of the energy distribution function of electrons at energies above the atom excitation energy does not give a contribution to the normalization of the distribution function. In deriving formula (6.6.3), we take into account that the inelastic cross section of electron-atom collisions is small compared to that for elastic collisions, and hence the velocity distribution function of electrons is close to an isotropic one. In addition, one can neglect by variation of an electron energy in elastic electron-atom collisions. We assume also

$$eE\lambda \ll \Delta\varepsilon, \qquad (6.6.30)$$

where $\lambda = 1/(N_a\sigma_{ea}^*)$ is the mean free path of electrons in a gas with respect to elastic collisions with electrons, $\Delta\varepsilon$ is the atom excitation energy. In the helium and argon case criterion (6.6.30) has the form $E/N_a \ll 10^3$ Td.

Let us determine the energy distribution function at energies above $\Delta\varepsilon$. For simplicity we assume that the distortion of the tail of the energy distribution function of electrons $f_o(\varepsilon)$ results for excitation of one level (may be, a joined level). From the statistical consideration, this distribution function follows from the relation

$$f_o(\varepsilon)d\varepsilon = W(t)dt,$$

where $W(t)$ is the probability of surviving of a test electron with respect to atom excitation, if at $t = 0$ the electron energy coincides with the atom excitation energy $\varepsilon = \Delta\varepsilon$. This probability satisfies to equation

$$\frac{dW}{dt} = -N_a k_{ex} W,$$

and its solution is

$$W = \exp(-\xi), \quad \xi(\varepsilon) = \int\limits_0^t N_a k_{ex} dt = N_a k_q \frac{g_*}{g_o} \int\limits_{\Delta\varepsilon}^{\varepsilon} \frac{\varepsilon - \Delta\varepsilon}{\varepsilon} \frac{d\varepsilon}{eEw_e(\varepsilon)}, \qquad (6.6.31)$$

where we use the principle of detailed balance (3.3.7), k_q is the rate constant of quenching of an excited atom. Using expression (6.6.2) for a current electron drift velocity $w_e(\varepsilon)$, we have

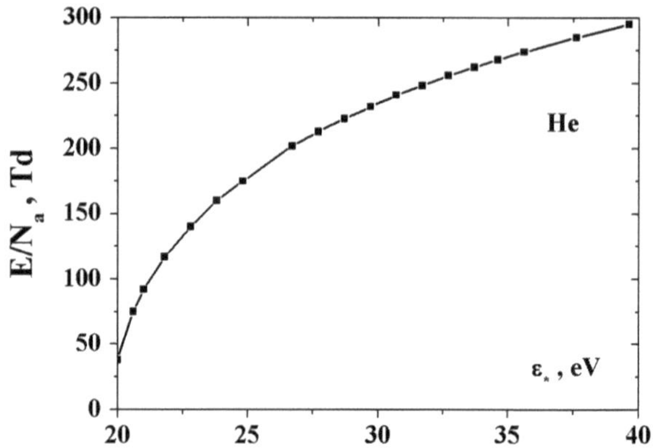

Fig. 6.20 Dependence of the reduced electric field strength E/N_a on the electron energy ε_* in a helium gas discharge plasma in accordance with equation $\xi(\varepsilon_*) = 1$

$$\xi(\varepsilon) = \left(\frac{N_a}{eE}\right)^2 k_q m_e \frac{g_*}{g_o} \int\limits_{\Delta\varepsilon}^{\varepsilon} \frac{d\varepsilon(\varepsilon - \Delta\varepsilon)}{\varepsilon} v\sigma_{ea}(\varepsilon) \qquad (6.6.32)$$

We give in Fig. 6.20 the dependence of the reduced electric field strength E/N_a on the electron energy ε_* which satisfies to equation $\xi(\varepsilon_*) = 1$.

We now formulate criteria of electron drift in a gas in an external electric field. Until the electron energy ε is small compared to the atom excitation energy $\Delta\varepsilon$, the rate of elastic electron-atom collisions dominates compared with other scattering mechanisms. This leads to the spherically symmetric velocity distribution function of electrons like to that of Fig. 6.19. In addition, the diffusion character of evolution of the electron distribution function over energies takes place at energies below the atoms excitation energy $\Delta\varepsilon$ that is determined by electron-electron collisions, electron-atom collisions and under the action of an external electric field. The corresponding diffusion coefficients B_{ee}, B_{ea} and B_E in an energy space are given by formulas (5.5.4), (5.3.8) and (6.4.10). In particular, the distribution function (6.6.3) satisfies to the criterion

$$B_E \gg B_{ea} \gg B_{ee} \qquad (6.6.33)$$

Chapter 7
Transport Processes in Gas Discharge Plasma

Abstract Transport processes involving atoms, electrons and ions are considered for various regimes of evolution of a gas discharge plasma with using appropriate models for collision processes. Ambipolar diffusion characterizes transport of charged particles in a plasma as a whole if the plasma expands in a surrounding space or attaches to walls. Electrons as more mobile particles remove outside a plasma region faster and create a field which acts on ions, and external fields influences on the ambipolar diffusion coefficient in a gas. Character of plasma properties as a result of its heating or under the action of a magnetic field is analyzed.

7.1 Transport Phenomena in Gases

Because atoms or molecules are usually the main component of ionized gases, their transport in gases may be of importance. We below formulate the definitions of basic kinetic (transport) coefficients which characterize transport phenomena in a weakly nonuniform gas. Then small gradients of certain quantities in a gas cause corresponding fluxes, and kinetic coefficients connects these fluxes with gradients of gas parameters. We first consider a gas consisting of several components, and the concentration c of a given component varies weakly in a space. Then the flux $\mathbf{j}$ of this component occurs that trends to eliminate the concentration gradient and is proportional to this gradient

$$\mathbf{j} = -DN_a\nabla c, \tag{7.1.1}$$

where N_a is the total number density of gas atoms or molecules, and D is the diffusion coefficient. If the concentration of a given component is small ($c \ll 1$), as it takes place when electrons, ions or excited atoms are an admixture to a neutral gas, one can rewrite this formula

$$\mathbf{j} = -D\nabla N, \tag{7.1.2}$$

B.M. Smirnov, *Theory of Gas Discharge Plasma*, Springer Series on Atomic, Optical, and Plasma Physics 84, DOI 10.1007/978-3-319-11065-3_7

where $N = N_a c$ is the number density for particle of a given component. One can see the analogy of this definition with (6.1.7). We consider a weakly nonuniform gas where the mean free path λ of particles is small compared to a typical dimension L of this plasma where the quantity under consideration varies remarkable, i.e.

$$L \gg \lambda \tag{7.1.3}$$

Along with diffusion of particles of another components in a gas, formulas (7.6.7) and (7.1.2) describe the self-diffusion process where diffusion of some probe gas atoms is studied.

Let us give the definitions for other transport phenomena which will be studied below. The thermal conductivity coefficient κ characterizes the connection between the heat flux $\mathbf{q}$ and temperature gradient ∇T, i.e.

$$\mathbf{q} = -\kappa \nabla T \tag{7.1.4}$$

The viscosity coefficient η is the proportionality coefficient between the friction force F per unit gas area and the gradient of the drift gas velocity. If the gas drift velocity $\mathbf{w}$ directs along the axis x, and its gradient directs along the axis z, the friction force acts on the plane xy and is proportional to the gradient of the drift gas velocity $\partial w_x/\partial z$. Then the gas viscosity coefficient is defined as

$$F = -\eta \frac{\partial w_x}{\partial z}, \tag{7.1.5}$$

In addition, formula (2.3.2) gives the definition of the mobility K of charged particles in a weak electric field. This definition requires a weak electric field strength, so that the energy transmitted to a probe particle from the field in a time between neighboring collisions must be relatively small. In particular, for ions whose mass is comparable with the mass of gas atoms this criterion has the form

$$eE\lambda \ll T, \tag{7.1.6}$$

where T is a typical thermal energy of ions or their temperature expressed in energetic units. This criterion means that the energy obtained from field between neighboring ion-atom collisions is small compared to a thermal ion energy.

Let us introduced the conductivity Σ of a weakly ionized quasineutral gas on the basis of the Ohm law that connects the electric current density $\mathbf{i}$ and the electric field strength $\mathbf{E}$ as

$$\mathbf{i} = \Sigma \mathbf{E} \tag{7.1.7}$$

Because the electric current in a weakly ionized gas is determined by electrons and ions, it is equal

$$\mathbf{i} = eN_e\mathbf{w}_e + eN_i\mathbf{w}_i,$$

where N_e, N_i are the number densities of electrons and ions, and $\mathbf{w}_e$, $\mathbf{w}_i$ are their drift velocities. From this on the basis of formula (2.3.2) we have for the conductivity of a weakly ionized gas located in a weak electric field

$$\Sigma = K_eN_e + K_iN_i \tag{7.1.8}$$

Transport coefficients of gases are determined by collisions between their atoms or molecules, namely, by elastic collisions between them because the cross sections of elastic collisions of atoms exceeds the cross sections of inelastic processes at thermal conditions. Hence, transport coefficients of gases are expressed through the cross sections of elastic collisions of gas atoms or molecules. The simple connection between these quantities is given by the practical Chapman-Enskog method [293–295] that is in reality the expansion of transport coefficients over a small numerical parameter, and we use below the first Chapman-Enskog approximation for transport coefficients. In this approximation the diffusion coefficient of a test atomic particle in a gas is given by [296–298]

$$D = \frac{3\sqrt{\pi T}}{8N_a\sqrt{2\mu\overline{\sigma}}}, \quad \overline{\sigma} \equiv \Omega^{(1,1)}(T) = \frac{1}{2}\int_0^\infty e^{-t}t^2\sigma^*(t)dt, \quad t = \frac{\mu g^2}{2T}. \tag{7.1.9}$$

Here T is the gas temperature expressed in energetic units, μ is the reduced mass of a test atomic particle and gas atom, $\sigma^*(g)$ is the diffusion cross section for collision of these particles with the relative velocity g, and the averaging $\langle\rangle$ is given with the Maxwell distribution function of gas atoms.

The thermal conductivity coefficient κ the first Chapman-Enskog approximation is determined by formula [296–298]

$$\kappa = \frac{25\sqrt{\pi T}}{32\overline{\sigma_2}\sqrt{m}}, \tag{7.1.10}$$

where m is the mass of gas atoms or molecules, and the average cross section of two gas atoms is given by

$$\overline{\sigma_2} \equiv \Omega^{(2,2)}(T) = \int_0^\infty t^2\exp(-t)\sigma^{(2)}(t)dt, \quad t = \frac{\mu g^2}{2T}, \quad \sigma^{(2)}(t) = \int (1-\cos^2\vartheta)d\sigma, \tag{7.1.11}$$

The viscosity coefficient η in the first Chapman-Enskog approximation is determined by formula [296, 297]

$$\eta = \frac{5\sqrt{\pi T m}}{24\overline{\sigma_2}}, \qquad (7.1.12)$$

where the average cross section $\overline{\sigma_2}$ for collision of two atoms is given by formula (7.1.11). As is seen, in the first Chapman-Enskog approximation the thermal conductivity and viscosity coefficients are connected by the relation

$$\kappa = \frac{15}{4m}\eta \qquad (7.1.13)$$

We also give the expression for the mobility of a single charged atomic particle that follows from formula (7.1.9) for its diffusion coefficient in the first Chapman-Enskog approximation and the Einstein relation (2.3.4)

$$K = \frac{3e\sqrt{\pi}}{8N\sqrt{2\mu T}\overline{\sigma}}, \quad \overline{\sigma} \equiv \Omega^{(1,1)}(T) = \frac{1}{2}\int_0^\infty e^{-t}t^2\sigma^*(t)dt, \quad t = \frac{\mu g^2}{2T} \qquad (7.1.14)$$

In reality the interaction potential of colliding gas atoms or molecules depends sharply on a distance between them. This provides the validity of the hard sphere model according to which atoms collide like to billiard balls. Note that just this model allows to Boltzmann and Maxwell to create the kinetics theory of gases [84, 256, 299, 300]. According to the hard sphere model, the differential cross section $d\sigma$ for particle scattering in the center of mass frame of reference is connected with the scattering angle ϑ in this frame of reference as

$$d\sigma = \sigma_o d\cos\vartheta, \quad \sigma_o = \pi R_o^2,$$

where R_o is the hard sphere radius. This simplifies the expressions (7.1.9), (7.1.10), and (7.1.12) since the average cross sections has now the following form

$$\overline{\sigma} = \sigma_o; \quad \overline{\sigma_2} = \frac{2}{3}\sigma_o, \qquad (7.1.15)$$

From this we obtain the following expressions for the above transport coefficients of gases

$$D = \frac{3\sqrt{\pi T}}{8\sqrt{2\mu}N_a\sigma_o}; \quad \kappa = \frac{75\sqrt{\pi T}}{64\sigma_o\sqrt{m}}; \quad \eta = \frac{15\sqrt{\pi T m}}{32\sigma_o} \qquad (7.1.16)$$

The basic component of a gas discharge plasma under consideration of helium and argon is a neutral atomic gas of helium or argon. Therefore, in the analysis of transport phenomena in a gas discharge plasma, we start from such processes in helium and argon. The diffusion coefficient of atoms in a gas is under normal conditions at temperature $T = 300\,\mathrm{K}$ [301, 302] $D(He - He) = 1.6\,\mathrm{cm^2/s}$ for helium atoms in helium, $D(He - Ar) = 0.76\,\mathrm{cm}^2/\mathrm{s}$ for helium atoms in argon or argon atoms in helium, and $D(Ar - Ar) = 0.16\,\mathrm{cm}^2/\mathrm{s}$ for argon atoms in argon. The hard sphere model (for example, [284]) describes well collision processes of atoms in these gases, so that colliding atoms in these processes are modelled by rigid balls. In that time, this model, the model of billiard balls, allowed to Maxwell and Boltzmann [84, 256, 299, 300] to construct the kinetic theory of gases. Within the framework of our consideration, the hard sphere model means that the cross section which determines a given transport coefficient is independent of the collision velocity of atoms.

Using the hard sphere model, we introduce the gas kinetic cross section σ_{gas}, defining it as the cross section that leads to a measured value of the diffusion coefficient under the condition that the diffusion cross section of atom collision is independent of the collision velocity. Then in the first Chapman-Enskog approximation [296, 297] for the diffusion coefficient, we obtain the following connection between the gas kinetic cross section σ_{gas} and the diffusion coefficient D

$$\sigma_{gas} = \frac{0.47}{N_a D}\sqrt{\frac{T}{\mu}}, \qquad (7.1.17)$$

where N_a is the number density of atoms, T is the gas temperature expressed in energetic units, μ is the reduced mass of colliding atoms. On the basis of experimental values of the diffusion coefficients [301, 302] we obtain from formula (7.1.17) for the gas kinetic cross sections: $\sigma_{gas} = 13\,\text{Å}^2$ for collision of two helium atoms, $\sigma_{gas} = 21\,\text{Å}^2$ for collision of helium and argon atoms, and $\sigma_{gas} = 37\,\text{Å}^2$ for collision of two argon atoms.

Table 7.1 contains the coefficients of thermal conductivity and viscosity of helium and argon [301, 303]. The thermal conductivity coefficient of argon in the temperature range $T = 300 - 1{,}000\,\mathrm{K}$ with the accuracy 2 % is approximated by the dependence

$$\kappa(T) = \kappa_o \left(\frac{T}{T_o}\right)^k \qquad (7.1.18)$$

Table 7.1 The thermal conductivity coefficients (κ) in $10^{-4}\mathrm{W}/(\mathrm{cm}\,K)$ and viscosity coefficients (η) in $10^{-5}\mathrm{g}/(\mathrm{cm\,s})$ for helium and argon [301, 303]

T	$\kappa(He)$	$\eta(He)$	$\kappa(Ar)$	$\eta(Ar)$
100	7.2	9.77	0.66	8.3
200	11.5	15.4	1.26	16.0
300	15.1	19.6	1.77	22.7
400	18.4	23.8	2.22	28.9
600	25.0	31.4	3.07	38.9
800	30.4	38.2	3.74	47.4
1,000	35.4	44.5	4.36	55.1

An argument in these coefficients indicates a gas, the gas temperature T is expressed in K

with the parameters $T_o = 300\,\mathrm{K}$, $k = 0.75$ $\kappa_o = 1.8 \times 10^{-4}\mathrm{W/(cm\,K)}$. In the temperature range $T = 1{,}000 - 8{,}000\,\mathrm{K}$ with the accuracy 5 % this formula is described by the parameters $T_o = 1{,}000\,\mathrm{K}$, $k = 0.68$, $\kappa_o = 4.2 \times 10^{-4}\mathrm{W/(cm\,K)}$.

Processes in a gas discharge plasma involving excited atoms can influence on ionization equilibrium in the plasma. On the other hand, important channel of destruction of excited atoms results from diffusion of atoms to the walls. The diffusion coefficient D of excited atoms in a gas is inversely proportional to the number density N_a of gas atoms and usually is reduced to the normal number density of atoms $N_a = 2.69\times 10^{19}\,\mathrm{cm}^{-3}$. The diffusion coefficients of excited helium atoms and molecules in helium averaged over measurements [167, 305–319] is [194] $0.59\,\mathrm{cm}^2/\mathrm{s}$ for a metastable atom $He(2^3S)$ in helium, $0.52\,\mathrm{cm}^2/\mathrm{s}$ for a metastable atom $He(2^1S)$ in helium, and $0.45\,\mathrm{cm}^2/\mathrm{s}$ for a metastable molecule $He(2^3\Sigma_u)$ in helium. The diffusion coefficient of metastable argon atoms $Ar(^3P_2)$, $Ar(^3P_0)$ in argon at room temperature averaged over measurements [317–325] is [194] $0.071\,\mathrm{cm}^2/\mathrm{s}$. Though the diffusion coefficients for atoms in metastable states $Ar(^3P_2)$ and $Ar(^3P_0)$ are different, a difference between their values is less than that from statistics due to a sum of measurements. Note that the diffusion coefficients of excited atoms in a gas are lower than those for atoms in the ground state.

7.2 Electron Drift in Gas in External Electric Field

The electron distribution function contains various information about the electron behavior in a gas discharge plasma including transport coefficients for electrons moved in a gas in an external field. In particular, the electron drift velocity according to its definition is

$$w_e = \int v_x f d\mathbf{v} = \int v_x^2 f_1 d\mathbf{v} = \frac{1}{3}\int v^2 f_1 d\mathbf{v},$$

where we use the expansion (6.1.3) for the distribution function and the spherical symmetry of functions f_o and f_1. From this on the basis of the first equation of the set (6.1.20) for the kinetic electron one can transform the expression for the electron drift velocity w_e to the following form [275] in accordance with formula (6.1.7)

$$w_e = \int v_x^2 f_1 d\mathbf{v} = \frac{eE}{3m_e}\left\langle \frac{1}{v^2}\frac{d}{dv}\left(\frac{v^3}{\nu}\right)\right\rangle,$$

where an average is made over the spherical distribution function f_0 and only elastic electron-atom collisions are taken into account. Formula (6.1.7) exhibits the dependence of the electron drift velocity on the reduced electric field strength E/N_a. This dependence in the case of electron drift in helium and argon is represented in Figs. 5.1 and 5.2. These Figures are based on experimental results [257–262] for helium and [6, 7, 258, 260, 261, 263–267] for argon. Though this information is obtained long ago, it determines contemporary knowing of this problem. Rich information for transport coefficients of electrons in gases is collected in review [43].

In the simple case where the rate of elastic electron-atom collisions $\nu(v)$ is independent of the electron velocity, we have from formula (6.1.7a) for the electron drift velocity and from formula (5.2.5) for the average electron energy

$$w_e = \frac{eE}{m_e \nu}, \quad \overline{\varepsilon} = \frac{3}{2}T + \frac{M}{2}w_e^2 \tag{7.2.1}$$

From this it follows that an external electric field influences significantly on an electron energy at large electric field strengths. An average in formula for the electron drift velocity is made over the spherical part of the electron distribution function. In the case $\nu(v) = const$ the drift velocity is proportional to the electric field strength. In other cases for a monotonic dependence of the rate of elastic electron-atom scattering on the electron velocity in the form $\nu(v) = bv^k$ dependence the electron drift velocity is given by formula (6.1.13).

The diffusion coefficient in the transverse direction with respect to the electric field according to its definition is equal to

$$D_\perp = \langle \frac{v^2}{3\nu} \rangle, \tag{7.2.2}$$

According to this formula, the diffusion coefficient is inversely proportional to the number density of gas atoms. Figures 7.1 and 7.2 contain experimental data for the reduced transverse diffusion coefficient $D_\perp N_a$ of electrons as a function of the reduced electric field strength E/N_a in helium and argon correspondingly. At a monotonic dependence of the rate of elastic electron-atom scattering on the electron

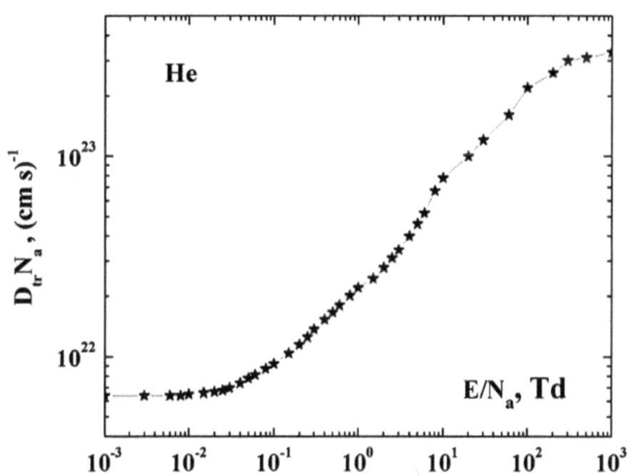

Fig. 7.1 Reduced diffusion coefficient for electrons in helium in an external electric field constructed on the basis of experimental data [43]. *Dotted curve* is calculated on the basis of formula (7.2.3)

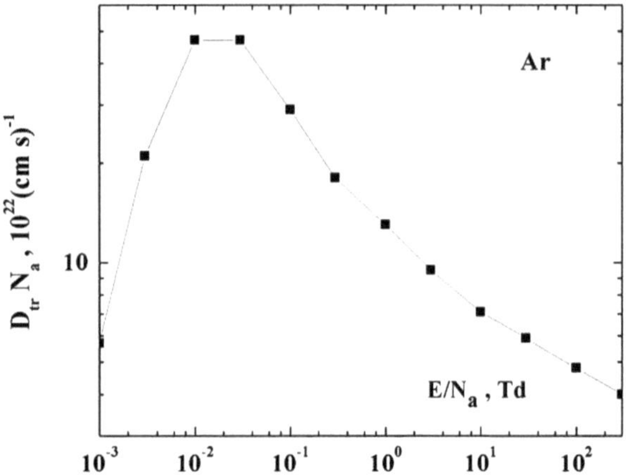

Fig. 7.2 Reduced diffusion coefficient for electrons in argon in an external electric field constructed on the basis of experimental data [43]

velocity if it has the form $\nu(v) = bv^k$ and large electron energy compared to a thermal energy of atoms the energy distribution function of electrons is given by (6.6.5). Then the diffusion coefficient of electrons in a gas discharge plasma in the transversal direction with respect to the electric field is given by

$$D_\perp = \left\langle \frac{v^2}{3\nu} \right\rangle = \frac{1}{3b} \cdot \left[\frac{(2k+2k)Ma^2}{3m_e b^2} \right]^{(2-k)/(2k+2)} \cdot \left[\frac{\Gamma(\frac{5-k}{2k+2})}{\Gamma(\frac{3}{2k+2})} \right] \tag{7.2.3}$$

A convenient parameter of electron drift in a gas in an external electric field at low electron number densities is the characteristic temperature T_{ef} [326] that is defined as

$$T_{ef} = \frac{eD_\perp}{\mathrm{K}} = \frac{eED_\perp}{w} \tag{7.2.4}$$

At small electric field strengths the characteristic temperature coincides with the gas temperature. If the rate of electron-atom collisions is independent of the electron velocity ($\nu(v) = const$), the electron distribution function conserves the form of the Maxwell distribution function, as it follows from (6.1.4), at any electric field strengths, i.e. $f_o(v) \sim \exp(-\varepsilon/T_{ef})$.

In the case where one can neglect inelastic processes and the dependence of the rate of elastic electron-atom scattering on the electron velocity has the form $\nu(v) = bv^k$, the characteristic electron temperature according to formula (7.2.4) is given by

$$T_{ef} = \frac{eED_\perp}{w_e} = \frac{m_e}{3-k} \left[\frac{(2k+2k)Ma^2}{3m_e b^2} \right]^{1/(k+1)} \cdot \left[\frac{\Gamma(\frac{5-k}{2k+2})}{\Gamma(\frac{3-k}{2k+2})} \right] \tag{7.2.5}$$

The diffusion coefficient of electrons in a gas discharge plasma is isotropic $D_\perp = D_\parallel$ at small electric field strengths when the average electron energy corresponds to their thermal energy. In strong fields the equality between the electron diffusion coefficients in transverse and longitudinal directions with respect to the electric field direction is violated. The ratio of these diffusion coefficients is determined by the dependence $\sigma_{ea}(v)$. In particular, in the case $\sigma_{ea}(v) = const$ in strong electric fields we have $D_\parallel/D_\perp = 0.49$. As is seen, transport coefficients of electrons in a gas discharge plasma in the case of elastic electron-atom collisions are expressed through the diffusion cross section of electron-atom collision. Note that in these collisions an electron velocity exceeds significantly an atom one, and hence an atom may be assumed to be motionless in such collisions.

Let us determine the characteristic temperature T_{ef} for a helium and argon gas discharge plasma if it is located in a strong electric field. In this case according to formula (6.6.20) we have $D_\perp N_a = 8.1 \times 10^{22}\,\mathrm{cm^{-1}s^{-1}}$ in the helium case and $D_\perp N_a = 5.8 \times 10^{22}\,\mathrm{cm^{-1}s^{-1}}$ in the argon case. The reduced electron mobility follows from formula (6.6.13) and is equal $K_e N_a/e = 1.2 \times 10^5\,\mathrm{cm/(s\,Td)}$ and $K_e N_a/e = 1.5 \times 10^5\,\mathrm{cm/(s\,Td)}$ in the argon case. As a result, we obtain for the characteristic energy in strong electric fields according to formula (7.2.4) $T_{ef} = 6.6\,\mathrm{eV}$ in the helium case and $T_{ef} = 3.9\,\mathrm{eV}$ in the argon case. Since energies of electrons $\varepsilon < \Delta\varepsilon$ and $\varepsilon \sim \Delta\varepsilon$, one can expect that $T_{ef} \sim \Delta\varepsilon$. In particular, in the helium and argon cases the average electron energy $3T_{ef}/2 \approx \Delta\varepsilon/2$.

The method of measurement of the characteristic temperature of electrons was suggested by Townsend [327, 328] 100 years ago and consists in determination of a space distribution of electrons in a plane that is perpendicular to the electron flux. This electron flux is generated by a point electron source, and a distance from this source exceeds significantly the mean free path of electrons in a gas, so that propagation of electrons in the transverse direction results from electron diffusion in this gas. Figures 7.3, 7.4 and 7.5 contain the characteristic electron energies in helium and argon correspondingly resulted from such measurements [7, 261, 262, 266].

The electron mobilities and the electron diffusion coefficients for helium and argon are given in Table 7.2 at room gas temperature and for small electric field strengths when the electron drift velocity is proportional to the electric field strength [329, 330]. The reduced mobilities are obtained on the basis of the Einstein relation (2.3.3).

Being guided by a gas discharge plasma of helium and argon, we give in Tables 7.3 and 7.4 the electron transport coefficients in helium and argon under the action of an electric field. These data are based on the measurements given in Figs. 5.1, 5.2, 7.1, 7.2, 7.3 and 7.4 and relate to the regime of low electron number densities (6.1.2). Old data are collected in review [43]. The error in these data is approximately 30 %. In the limit of low electric field strengths the reduced diffusion coefficient is equal to that given in Table 7.1. Note that this limit takes place at $E/N_a \ll 0.003\,\mathrm{Td}$ for heliumand at $E/N_a \ll 3 \times 10^{-4}\,\mathrm{Td}$ for argon. The limiting value of the electron drift

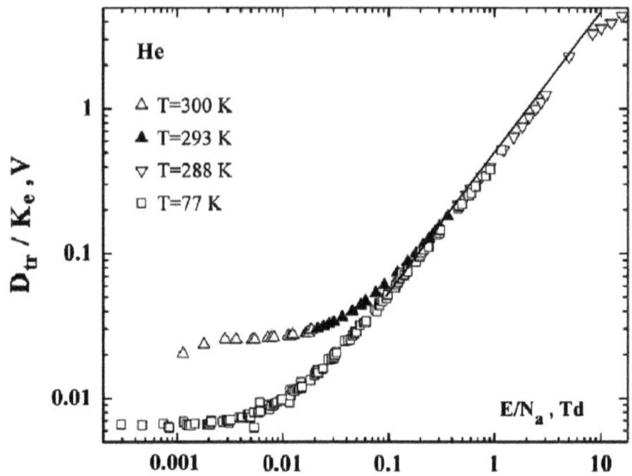

Fig. 7.3 Townsend characteristic energy of electrons in helium constructed on the basis of experimental data [43]. *Dotted curve* is calculated on the basis of formula (7.2.5)

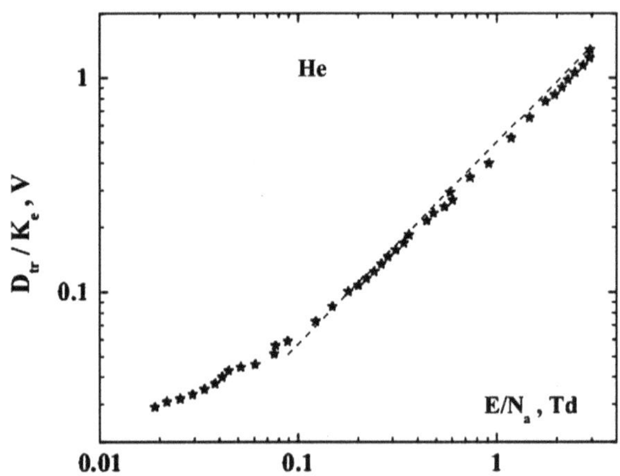

Fig. 7.4 Townsend characteristic energy of electrons in helium at room temperature and low electric field strengths constructed on the basis of experimental data [43]. *Dotted curve* is calculated on the basis of formula (7.2.5)

velocity for large electric field strengths is given by formula (6.6.8). The diffusion coefficient of electrons at large electric field strengths, as it follows from formula (7.2.2) with using the energy distribution function of electrons (6.6.8), is equal to

$$D_{\perp} = \langle \frac{v^2}{3\nu} \rangle = \frac{\Delta\varepsilon}{3m_e k_{el}(\Delta\varepsilon)} \tag{7.2.6}$$

This formula gives $D_{tr}N_a = 1.6 \times 10^{23}(\mathrm{cm\,s})^{-1}$ in the helium case and $D_{tr}N_a = 2.2 \times 10^{22}(\mathrm{cm\,s})^{-1}$ in the argon case.

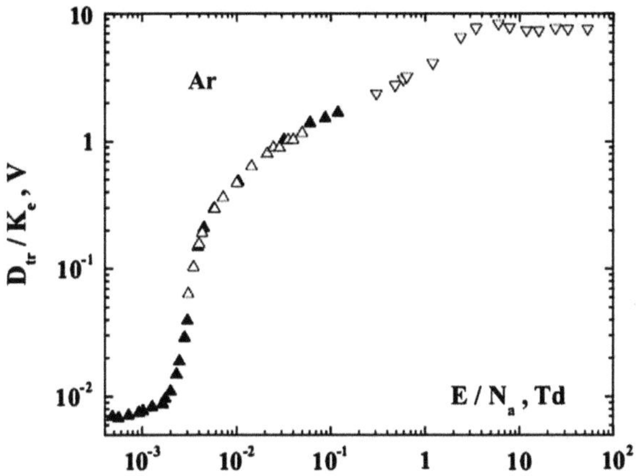

Fig. 7.5 Townsend characteristic temperature of electrons in argon constructed on the basis of experimental data [43]

Table 7.2 Reduced transport coefficients of electrons in helium and argon at room temperature and in the limit of small electric field strengths [329, 330]

Gas	He	Ar
D_eN_a, $10^{21}\,\mathrm{cm^{-1}s^{-1}}$	6.6	21
K_eN_a, 10^{23} $(\mathrm{cm\,s\,V})^{-1}$	2.6	8.1

Table 7.3 Transport coefficients for electrons in helium in the regime of a low electron number density

E/N_a(Td)	w_e(cm/s)	$D_{tr}N_a(\mathrm{cm\,s^{-1}})$	eD_{tr}/K(eV)
0.001	2.5×10^3	6.2×10^{21}	0.025
0.003	7.4×10^3	6.2×10^{21}	0.025
0.01	2.5×10^4	7.2×10^{21}	0.029
0.03	7.4×10^4	8.9×10^{21}	0.036
0.1	1.6×10^5 (1.7×10^5)	1.0×10^{22} (7.8×10^{21})	0.064 (0.046)
0.3	2.9×10^5 (2.9×10^5)	1.4×10^{22} (1.3×10^{22})	0.15 (0.14)
1	4.9×10^5 (5.3×10^5)	2.2×10^{22} (2.4×10^{22})	0.44 (0.46)
3	9.0×10^5 (9.1×10^5)	3.6×10^{22} (4.2×10^{22})	1.2 (1.4)
10	2.1×10^6 (1.7×10^6)	7.6×10^{22} (7.8×10^{22})	3.6 (4.6)
30	7.0×10^6 $[6.5 \times 10^6]$	1.2×10^{23}	5.3
100	2.50×10^7 $[2.2 \times 10^7]$	2.2×10^{23}	8.2
300	7.0×10^7 $[6.5 \times 10^7]$	3.0×10^{23}	11

Here E/N_a is the reduced electric field strength given in Td ($1\mathrm{Td} = 10^{-17}\,\mathrm{V\,cm^2}$), w_e is the electron drift velocity, $D_\perp$ is the coefficient of transverse diffusion, $K = w_e/E$ is the electron mobility [27, 43, 191]. The gas temperature is 300 K. Values in parentheses are calculated on the basis of formulas (7.3.3), and the values in square parentheses correspond to high electric field strengths in accordance with formula (6.6.8)

Table 7.4 Transport coefficients for electrons in argon in the regime of a low electron number density

E/N_a(Td)	$w_e(10^5$ cm/s)	$D_{tr}N_a(10^{22}$cm s$)^{-1}$	eD_{tr}/K(eV)
0.001	0.15	5.7	0.038
0.003	0.48	21	0.13
0.01	1.0	47	0.47
0.03	1.4	47	1.0
0.1	1.8	29	1.6
0.3	2.3	19	2.5
1	2.7	12	4.4
3	4.3	10	6.9
10	10	7.1	7.1
30	24 [18]	5.9	7.4
100	60 [62]	5.0	8.3
300	140 [180]	4.0	8.5

Here E/N_a is the reduced electric field strength given in Td (1Td $= 10^{-17}$V cm^2), w_e is the electron drift velocity, $D_\perp$ is the coefficient of transverse diffusion, $K = w_e/E$ is the electron mobility [43, 331–333]. The gas temperature is 300 K, and the values in square parentheses correspond to high electric field strengths in accordance with formula (6.6.8)

We also represent in Figs. 7.6 and 7.7 the rate constants of relaxation of the electron momentum k_P and energy k_ε which are defined by formulas (5.5.8). These values are obtained by substitution in formulas (5.5.8) experimental values of the dependencies on the reduced electric field strength E/N_a for the electron drift velocity w_e, the reduced diffusion coefficient $D_{tr}N_a$ and the characteristic energy T_{ef} which are given in Figs. 5.1, 5.2, 7.1, 7.2, 7.3, 7.4 and 7.5, and Tables 7.3 and 7.4. In the limit of low electron number densities one can express an average electron energy in balance

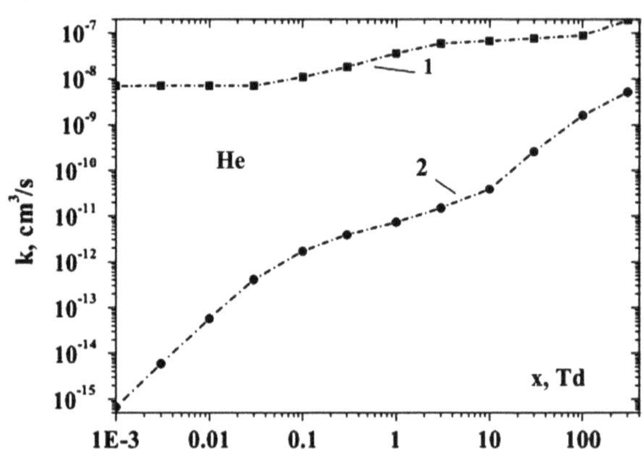

Fig. 7.6 Rate constants of momentum k_P *(1)* and energy k_ε *(2)* for relaxation of electrons located in helium in a constant electric field

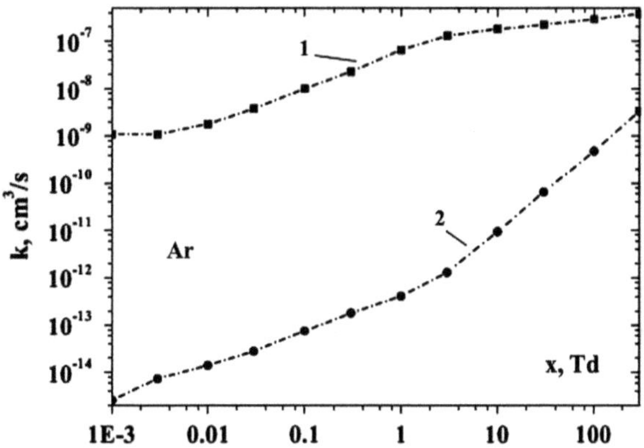

Fig. 7.7 Rate constants of momentum k_P *(1)* and energy k_ε *(2)* for relaxation of electrons located in argon in a constant electric field

(5.5.9) as $\overline{\varepsilon} = 3T_{ef}/2$. Then formula (5.5.8) for relaxation of the electron average energy takes the form

$$k_\varepsilon = \frac{2w_e^2}{3D_{tr}N_a} \tag{7.2.7}$$

Table 7.5 contains numerical values of the relaxation rate constants for the electron momentum and energy in helium and argon.

Table 7.5 Defined by formulas (5.5.9) and expressed in cm^3/s the relaxation rates for the average momentum k_P and average energy k_ε of electrons in helium and argon

Value	$k_P(He)$	$k_\varepsilon(He)$	$k_P(Ar)$	$k_\varepsilon(Ar)$
0.001	7.0^{-9}	6.7^{-16}	1.1^{-9}	2.6^{-15}
0.003	7.1^{-9}	5.9^{-15}	1.1^{-9}	7.4^{-15}
0.01	7.1^{-9}	5.8^{-14}	1.8^{-9}	1.4^{-14}
0.03	7.1^{-9}	4.1^{-13}	3.8^{-9}	2.8^{-14}
0.1	1.1^{-8}	1.7^{-12}	1.0^{-8}	7.5^{-14}
0.3	1.8^{-8}	3.9^{-12}	2.3^{-8}	1.8^{-13}
1	3.6^{-8}	7.4^{-12}	6.5^{-8}	4.1^{-13}
3	5.9^{-8}	1.5^{-11}	1.3^{-7}	1.3^{-12}
10	6.7^{-8}	3.9^{-11}	1.8^{-7}	9.4^{-12}
30	7.5^{-8}	2.6^{-10}	2.2^{-7}	6.5^{-11}
100	8.8^{-8}	1.6^{-9}	2.9^{-7}	4.8^{-10}
300	1.9^{-7}	5.1^{-9}	3.8^{-7}	3.3^{-9}

The value 7.0^{-9} means 7.0×10^{-9}

7.3 Electrons in Gas of Hard Spheres

Using the model of hard spheres for electron-atom collisions in a gas, we assume the diffusion cross section of such collisions to be independent on the electron velocity. Because this assumption holds true for collision of an electron with helium atoms (see Fig. 3.7), where the diffusion cross section is $\sigma^*_{ea} = 6\,\text{Å}^2$ with the accuracy of 20 % at collision energies below 10 eV, we consider this model in greater detail. In the limit of low electric field strengths the electron mobility K_e and the diffusion coefficient D_e are determined in this case by formulas (7.1.14) and (7.1.16) correspondingly, where $\overline{\sigma} = \sigma_o = \sigma^*_{ea}$. Taking the diffusion cross section of electron-atom scattering to be $\sigma^*_{ea} = 6\,\text{Å}^2$, we obtain for the reduced diffusion coefficient of electrons in helium at temperature $T = 300K$ $D_e N_a = 5.5 \times 10^{21}\,\text{cm}^{-1}\text{s}^{-1}$ instead of $(6.4 \pm 0.3) \times 10^{21}\,\text{cm}^{-1}\text{s}^{-1}$ according to experiments [329, 330].

We also consider large electron energies

$$T \ll eE\lambda \ll 10\,\text{eV}, \tag{7.3.1}$$

where T is the gas temperature expressed in energetic units, and $\lambda = 1/(N_a \sigma^*_{ea})$ is the mean free path of electrons, E is the electric field strength. In this limiting case the electron drift velocity w_e given by formula (6.1.10), the transverse diffusion coefficient D_{tr} according to formula (7.2.3) with $k = 1$, the characteristic temperature according to formulas (7.2.5), and the average electron energy given by formula (6.1.11) have the form

$$w_e = 0.90 \left(\frac{m_e}{M}\right)^{1/4} \left(\frac{eE\lambda}{m_e}\right)^{1/2}, \quad D_{tr} = 0.29 \left(\frac{M}{m_e}\right)^{1/4} \lambda \left(\frac{eE\lambda}{m_e}\right)^{1/2},$$

$$T_{ef} = 0.325\sqrt{\frac{M}{m_e}} eE\lambda, \quad \overline{\varepsilon} = 0.43\sqrt{\frac{M}{m_e}} eE\lambda \tag{7.3.2}$$

In the helium case these formulas take the form

$$w_e = w_o\sqrt{x}, \quad D_{tr}N_a = a\sqrt{x}, \quad T_{ef} = 0.76\overline{\varepsilon}, \quad \overline{\varepsilon} = bx \tag{7.3.3}$$

Here $x = E/N_a$ that is expressed in Td, the coefficients of this formulas are $w_o = 5.3 \times 10^5\,\text{cm/s}$, $a = 2.4 \times 10^{22}\,\text{cm}^{-1}\text{s}^{-1}$, $b = 0.6\,\text{eV}$. Applying these formulas to the helium case and taking the electron-atom cross section of elastic scattering $\sigma^*_{ea} = 6\,\text{Å}^2$, we obtain that the criterion (7.3.1) holds true more or less in the range electric field strengths $0.1\,\text{Td} < E/N_a < 10\,\text{Td}$. The values of the electron drift velocity w_e, the reduced transverse diffusion coefficient $D_{tr}N_a$ and the characteristic energy T_{ef} are given in Table 7.3 in parentheses.

One can join the cases of low and high electric field strengths in determination the electron drift velocity w_e and the electron temperature T_e in the regime of a large electron number density. In the limit of low electric field strengths one can ignore the second term in the denominator of the distribution function (6.1.15) compared with the first term. Correspondingly, the criterion of low fields is

$$x^2 F(T) \ll T$$

In the helium case where the mean free path λ of electrons in a gas is independent of the electron velocity, this criterion has the form

$$eE\lambda \ll \sqrt{\frac{m}{M}}\, T \tag{7.3.4}$$

Taking $\sigma^*_{ea} = 6\,\text{Å}^2$ for collision of an electron and helium atom and $T = 300\,\text{K}$, we have in the helium case

$$\frac{E}{N_a} \ll 0.02\text{Td} \tag{7.3.5}$$

The electron distribution function in this limiting case coincides with the Maxwell one

$$f_0 = \frac{2}{\pi^{1/2} T^{3/2}} \varepsilon^{1/2} \exp\left(-\frac{\varepsilon}{T}\right),$$

where this function is normalized to one. In the limit under consideration formulas (6.1.7) and (2.1.2) give for the drift velocity w_e and electron diffusion coefficient $D_\perp$

$$w_e = \frac{2eE\sqrt{2}}{3\sqrt{\pi m_e T^5}} \int_0^\infty \varepsilon\lambda(\varepsilon) d\varepsilon \exp\left(-\frac{\varepsilon}{T}\right), \quad D_\perp = \frac{1}{3}\sqrt{\frac{8}{\pi m_e T^3}} \int_0^\infty \varepsilon\lambda(\varepsilon) d\varepsilon \exp\left(-\frac{\varepsilon}{T}\right) \tag{7.3.6}$$

In the helium case, where the cross section of electron-atom collision σ^* and correspondingly the mean free path of electrons in a gas λ are independently of the electron energy, these expressions take the form

$$w_e = \frac{eE\lambda}{3}\sqrt{\frac{8}{\pi m_e T}}, \quad D_\perp = \frac{\lambda}{3}\sqrt{\frac{8T}{\pi m_e}} \tag{7.3.7}$$

As it follows from the above formulas, the drift velocity and the transverse diffusion coefficient of electrons are connected by the Einstein relation (2.3.3)

$$\frac{w_e}{D_\perp} = \frac{eE}{T}$$

Using the value $\sigma^* = 6\,\text{Å}^2$ for the diffusion cross sections for elastic scattering of an electron on the helium atom, we obtain in the zero field limit for the reduced electron coefficient diffusion coefficient $D_\perp N_a = 6.0 \times 10^{21}\,\text{cm}^{-1}\text{s}^{-1}$ in the helium case at $T = 300\,\text{K}$ and the reduced electron mobility equals to $K = 2.3 \times 10^{23}(\text{V cm s})^{-1}$.

The limit of high electric field strengths corresponds to the following criterion that is opposite to the criterion (7.3.4)

$$eE\lambda \gg \sqrt{\frac{m_e}{M}}T, \tag{7.3.8}$$

Taking the average diffusion cross section of electron-atom scattering to be $6\,\text{Å}^2$ in the helium case and the ratio of the nuclear mass to the electron one to be $M/m_e = 7{,}297$, we obtain on the basis of formula (6.1.10) for the electron drift velocity w_e

$$w_e = b\sqrt{x}, \tag{7.3.9}$$

where $b = 7.4 \times 10^5\,\text{cm/s}$.

The electron distribution function according to formula (6.1.5) allows us to evaluate the average electron energy that is equal in this limiting case

$$\overline{\varepsilon} = \frac{\Gamma(5/4)}{\Gamma(3/4)}\varepsilon_o = 0.427\sqrt{\frac{M}{m_e}}eE\lambda = 0.530Mw_e^2 \tag{7.3.10}$$

In the helium case this formula gives

$$\overline{\varepsilon} = 0.427\sqrt{\frac{M}{m_e}}eE\lambda = ax,$$

where $a = 0.6\,\text{eV}$, and x is taken here and below in Td.

The transverse electron diffusion coefficient $D_\perp$ according to formula (7.2.2) in the case $\lambda(v) = const$ is equal to

$$D_\perp = \frac{\lambda}{3}\cdot\overline{v} = \left(\frac{M}{m_e}\right)^{1/4}\lambda\sqrt{\frac{eE\lambda}{m_e}}\frac{\sqrt{2}}{3^{5/4}\Gamma\left(\frac{3}{4}\right)} = 0.292\left(\frac{M}{m_e}\right)^{1/4}\lambda\sqrt{\frac{eE\lambda}{m_e}} \tag{7.3.11}$$

In the helium case this formula for the transverse electron diffusion coefficient has the form

$$D_\perp = c\sqrt{x},$$

where $c = 1{,}500\,\text{cm/s}$ and the transverse electron diffusion coefficient relates to the normal number density of atoms $N_a = 2.69 \times 10^{19}\,\text{cm}^{-3}$.

The characteristic electron temperature according to formula (7.2.4) is equal to

$$T_{ef} = 0.325\sqrt{\frac{M}{m_e}}eE\lambda$$

The ratio of the effective temperature to the average kinetic energy $\overline{\varepsilon}$ of electrons equals to

$$\frac{3}{2}\frac{T_{ef}}{\overline{\varepsilon}} = 1.14$$

Using formula (6.2.1) for the electron drift velocity in this limiting case, we have

$$w_e = \frac{\sqrt{8}}{3\sqrt{\pi}}\left(\frac{12m_e}{M}\right)^{1/4}\sqrt{\frac{eE\lambda}{m_e}} = 0.99\left(\frac{m_e}{M}\right)^{1/4}\sqrt{\frac{eE\lambda}{m_e}} \qquad (7.3.12)$$

This formula leads to the same dependence on the problem parameters, as formula (6.1.10) that relates to another character of equilibrium for the electron component in the gas. The ratio of the electron drift velocities according to formulas (7.3.12) and (6.1.10) is 1.1.

Substituting the electric field strength from formula (7.3.20) in formula (7.3.12), we express the electron drift velocity through the electron temperature

$$w_e = \sqrt{\frac{32T_e}{3\pi M}} = w_o\sqrt{T_e} \qquad (7.3.13)$$

From this it follows, that the electron drift velocity is small compared to their typical velocity. In the helium case $w_o = 9.0 \times 10^5\,\text{cm/s}$ if the electron temperature is expressed in eV.

The transverse electron diffusion coefficient $D_\perp$ is expressed through the electron drift velocity (7.3.12) on the basis of the Einstein relation. Then, using the connection (7.3.20) between the electric field strength and the electron temperature, we obtain

$$D_\perp = \frac{T_e w_e}{eE} = \sqrt{\frac{M}{12m_e}}w_e\lambda = \sqrt{\frac{8T_e}{9\pi m_e}}\lambda \qquad (7.3.14)$$

Note that at the normal number density of atoms the mean free path of electrons is $\lambda = 0.62\,\mu\text{m}$ that gives $D_\perp = 24\,\text{cm}^2/\text{s}$ at the normal number density of helium atoms or $D_\perp N_a = 6.2 \times 10^{20}(\text{cm s})^{-1}$.

One can combine the expressions (7.3.7) and (6.1.10) for the electron drift velocity in the limit of low and high electric field strengths, that gives for the electron drift velocity in a wide range of electric field strengths in the regime of a low electron number density

$$w_e = \frac{0.53eE\lambda}{\sqrt{m_e T}\left[1 + 0.59\left(\frac{M}{m_e}\right)^{1/4}\sqrt{\frac{eE\lambda}{T}}\right]} \qquad (7.3.15)$$

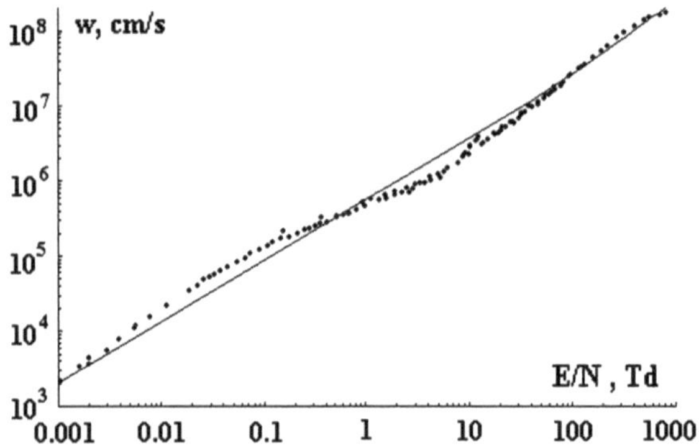

Fig. 7.8 The electron drift velocity in helium. *Solid curve* is given by formula (7.3.16), signs are experimental points according to [43]

In the helium case ($\sigma_{ea}^* = 6\,\text{Å}^2$) this formula gives

$$w_e = \frac{v_o}{1 + C\sqrt{x}}, \tag{7.3.16}$$

where $v_o = 2.8 \times 10^6\,\text{cm/s}$, $C = 3.7$, and the reduced electric field strength x is expressed in Td. This dependence in Fig.7.8 is compared with this experimental dependence [43].

We now consider the regime of a high electron number density where the energy distribution function of electrons is the Maxwell one (2.1.2). Then formula (6.2.5) gives for the drift electron velocity in the case where the mean free path of electrons in a gas is independent of the electron velocity

$$w_e = \frac{\sqrt{8}}{3\sqrt{\pi}} \frac{eE\lambda}{\sqrt{m_e T_e}} = 0.532 \frac{eE\lambda}{\sqrt{m_e T_e}}, \tag{7.3.17}$$

and formula the difference of electron and gaseous temperatures is equal on the basis of formula (6.2.7)

$$T_e - T = \frac{M}{m_e} \frac{(eE\lambda)^2}{12 T_e}$$

This leads to the following connection between the electron temperature and the reduced electric field strength

$$T_e = \frac{T}{2} + \sqrt{\frac{T^2}{4} + \frac{M}{12 m_e}(eE\lambda)^2} \tag{7.3.18}$$

Taking the average diffusion cross section of elastic electron collision with a helium atom to be $\sigma_{ea}^* = 6\,\text{Å}^2$, we get

$$T_e = 0.0129 + \sqrt{1.67 \times 10^{-4} + 0.169x^2}, \tag{7.3.19}$$

where the electron temperature T_e is expressed in eV and the electric field strength $x = E/N_a$ is given in Td. In the limit $T_e \gg T$ the electron temperature is equal to

$$T_e = \sqrt{\frac{M}{12m_e}}eE\lambda \tag{7.3.20}$$

We also determine within the framework of the hard sphere model the relaxation rate constants for electrons in a gas which are given by formulas (5.5.8). Using expressions for w_e and $\overline{\varepsilon}$ according to formulas (7.3.2), we have for these rate constants

$$k_P = \frac{eE}{N_a w_e m_e} = 1.1\left(\frac{M}{m_e}\right)^{1/4}\sqrt{\frac{eE\sigma_{ea}^*}{N_a m_e}}, \quad k_\varepsilon = \frac{eEw_e}{N_a \overline{\varepsilon}} = 2.1\left(\frac{m_e}{M}\right)^{3/4}\sqrt{\frac{eE\sigma_{ea}^*}{N_a m_e}} \tag{7.3.21}$$

As it follows from this formula, the ratio of the relaxation rates in the field range under consideration is given by

$$\frac{k_P}{k_\varepsilon} = 0.5\frac{M}{m_e} \tag{7.3.22}$$

Thus, in the range (7.3.1) of electric field strengths, where a typical electron energy is large compared to a thermal atom energy and is less than the atom excitation energy, i.e. electron relaxation is determined by elastic electron-atom collisions, a typical relaxation rate of the electron momentum exceeds that for a typical electron energy in M/m_e times. Note that the relaxation rate constants k_P and k_ε evaluated on the basis of formula (7.3.21) for helium with $\sigma_{ea}^* = 6\,\text{Å}^2$ in the range of the reduced electric field strengths $E/N_a = 0.3 - 3\,\text{Td}$ coincide with those given in Fig.7.6 within 10 %. Since at large electric fields the average electron energy cannot exceed the excitation energy, the difference between the rate constants k_P and k_ε decreases with an increasing electric field strength.

Let us join now the results for the electron drift velocity in helium obtained for different ranges of the electric field strengths in which this system is located. At low electric field strengths according to the definition the electron drift velocity is $w_e = EK$, where the reduced electron mobility is given in Table 7.2. At higher electric field strengths, where the electron drift velocity is determined by elastic electron-atom collisions and the cross section of this collision is independent of a collision energy (we approximate it as $\sigma_{ea} = 6\,\text{Å}^2$), the electron drift velocity is given by formula (6.1.10) [or by formula (7.3.2)]. At more higher electric field strengths electron braking is determined by the excitation process and the electron drift velocity is given by formula (6.6.8). The results of this joining is given in Fig.7.9, where theoretical values with using of indicated formulas are given by straightforward lines, and experimental data are taken from Fig. 5.1.

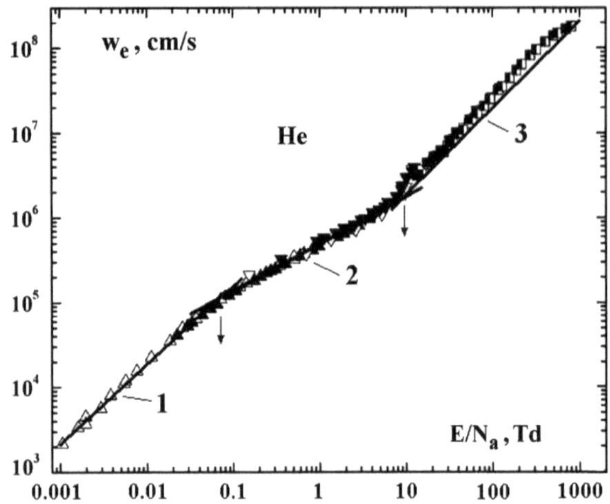

Fig. 7.9 Electron drift velocity in helium depending on the reduced electric field strength in neglecting electron-electron collisions. Range 1 [formula (2.3.2)] corresponds to low electric field strengths where the action of the electric field leads to a relatively small change of the electron energy; in the range 2 [formula(6.1.10)] this action is strong, but the energy distribution function of electrons is determined by elastic electron-atom scattering, excitation of atoms dominates in formation of the energy distribution function of electrons in range 3 [formula (6.6.6)], but the electron drift velocity is small compared to its typical isotropic velocity. Signs include experimental data [43] taken from Fig. 5.1

7.4 Conductivity of Weakly Ionized Gas

The electron drift velocity organizes the conductivity Σ of a weakly ionized gas including a gas discharge plasma. The plasma conductivity is defined as the ratio of the current density $i = eN_e w_e$ to the electric field strength $\Sigma = eN_e w_e/E$ that is limit of a low electric field strengths is given by

$$\Sigma = \frac{N_e e^2}{3m_e}\left\langle \frac{1}{v^2}\frac{d}{dv}\left(\frac{v^3}{\nu}\right)\right\rangle \qquad (7.4.1)$$

In the case if the rate $\nu = 1/\tau$ of electron-atom collisions is independent of the electron velocity v, this formula is reduced to a widely used form

$$\Sigma = \frac{N_e e^2 \tau}{m_e} \qquad (7.4.2)$$

Let us consider now the case if the mean free path of an electron in a gas is independent of the electron velocity ($\lambda(v) = 1/(N_a \sigma^*_{ea}) = const$) in the regime of a low electron number density where the electron drift velocity is given by formula (6.1.10). This formula relates to the limit of high electric field strengths when a typical electron energy ε_o in formula (6.1.10) exceeds significantly a typical atom thermal energy T ($eE\lambda \gg T$). Then the conductivity of an ionized gas is given by

$$\Sigma = 0.90 \frac{N_e e^{3/2} \lambda^{1/2}}{(m_e M)^{1/2} E^{1/2}} \quad (7.4.3)$$

As is seen, in this case the conductivity of an ionized gas depends on the electric field strength E.

Let us consider the regime of a high electron number density where we have the Maxwell distribution function of electrons over electron velocities, and this distribution function is characterized by the electron temperature T_e that may differ from the gas temperature T. Then the drift velocity of electrons is given by formula (6.2.5), and formula (6.2.7) determines the electron temperature T_e. In particular, in the case where the mean free path of electrons is independent of the electron velocity these formulas give for the conductivity Σ of an ionized gas

$$\Sigma_{ea} = \frac{N_e}{N_a} \frac{2\sqrt{2}}{3\sqrt{\pi}} \frac{e^2}{\sigma_{ea}\sqrt{m_e T_e}} \frac{N_e}{N_a} \Sigma_o; \quad \Sigma_o = \frac{2\sqrt{2}}{3\sqrt{\pi}} \frac{e^2}{\sigma_{ea}\sqrt{m_e T_e}}, \quad (7.4.4)$$

where $\sigma_{ea} = 6\,\text{Å}^2$ in the helium case. One can represent formula (7.4.4) in the helium case in the form

$$\Sigma_{ea} = \frac{N_e}{N_a} \frac{C_o}{\sqrt{T_e}},$$

where the electron temperature T_e is expressed in eV, and $C_o = 1.1 \times 10^8$ $\text{eV}^{1/2}/(\Omega\,\text{cm})$.

Above we consider the limiting case where the conductivity of an ionized gas is determined by elastic electron-atom scattering. But electron-ion scattering becomes significant at not large electron number densities, and this takes place at low degrees of gas ionization because of a long-range character for the Coulomb electron-ion interaction. Let us determine the conductivity of an ionized gas in the limiting case where it is determined by electron-ion scattering. Then on the basis of formula (3.1.7) for the electron-ion diffusion cross section we obtain for the electron-ion scattering rate

$$\nu_{ei} = N_i \frac{4\pi e^4 \ln \Lambda}{m_e^2 v^3},$$

where N_i is the number density of ions. Substituting this in formula (6.2.5) for the drift velocity with the Maxwell distribution function (6.2.4) of electrons, we obtain for the electron drift velocity

$$w_e = \frac{2^{5/2}}{\pi^{3/2}} \frac{E T_e^{3/2}}{m_e^{1/2} N_i e^3 \ln \Lambda} \quad (7.4.5)$$

From this we find the plasma conductivity $\Sigma = eN_e w_e$ for a quasineutral plasma $N_e = N_i$ that is given by Spitzer formula [334] for the plasma conductivity

$$\Sigma_{ei} = \frac{2^{5/2} T_e^{3/2}}{\pi^{3/2} m_e^{1/2} e^2 \ln \Lambda} \tag{7.4.6}$$

This formula is valid at low degree of ionization $N_e \ll N_a$. Indeed, because the Coulomb electron-ion interaction potential exceeds that for electron-atom interaction, electron-ion scattering becomes dominate at low ionization degree, and then the Spitzer formula (7.4.6) holds true.

The Spitzer formula holds true in the limit of a high electron concentration of electrons, where the plasma conductivity is given by

$$\Sigma_{ei} = C_{sp} T_e^{3/2}$$

Expressing the electron temperature T_e in eV and taking a typical value of the Coulomb logarithm $\ln \Lambda = 7$, we get the proportionality coefficient in the Spitzer formula in the helium case $C_{sp} = 8.8 \times 10^6\,\text{eV}^{-3/2}\Omega^{-1}\text{cm}^{-1}$. As it follows from this, at the electron temperature the contribution to the plasma conductivity is identical for atoms and ions at the electron temperature $T_e = 1\,\text{eV}$, if the ionization degree is several percent.

In a general case we take into account both electron-atom and electron-ion scattering, so that the scattering rate is given by

$$\nu = N_a \sigma_{ea}^* v + N_i \frac{4\pi e^4 \ln \Lambda}{m_e^2 v^3}$$

Here N_a, N_i are the number densities of atoms and ions correspondingly, and formula (3.1.7) is used for the diffusion cross section of electron-ion collision. Let us assume for definiteness that the mean free path of an electron in a gas is independent of the electron velocity ($\lambda(v) = 1/(N_a \sigma_{ea}^*) = const$). We then obtain for the gas conductivity

This gives for the electron drift velocity

$$\Sigma = \Sigma_{ea} \Phi(\xi), \quad \xi = \frac{N_i}{N_a} \frac{\pi e^4 \ln \Lambda}{T_e^2 \sigma_{ea}^*}, \quad \Phi(\xi) = \int_0^\infty \frac{e^{-x} x dx}{1 + \xi/x^2}, \tag{7.4.7}$$

where the conductivity of an ionized gas Σ_{ea} due to electron-atom scattering is given by formula (7.4.4). As is seen, the competition of contributions for electron-atom and electron-ion scattering in an ionized gas is determined by the parameter ξ. The function $\Phi(\xi)$ has the following limiting expressions

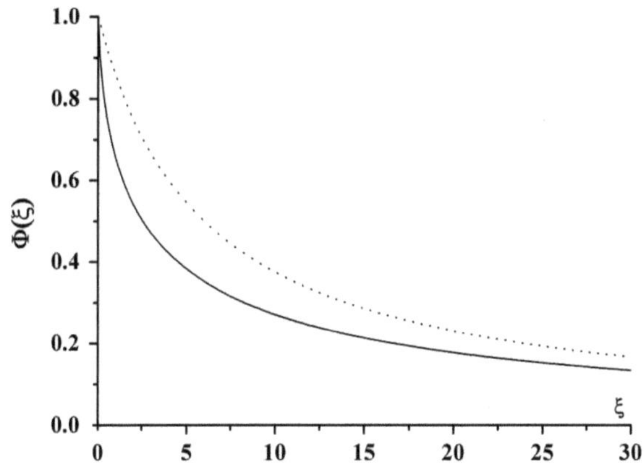

Fig. 7.10 Function $\Phi(\xi)$ that characterizes the contribution of the electron-ion scattering to the conductivity of an ionized gas. The *Dotted curve* corresponds to the approximation $\Phi(\xi) = (1 + \xi/6)^{-1}$

$$\Phi(\xi) = 1 - \frac{\xi}{2} \ln \frac{1}{e^{2C}\xi}, \quad \xi \ll 1; \quad \Phi(\xi) = \frac{6}{\xi}\left(1 - \frac{20}{\xi}\right), \quad \xi \gg 1 \qquad (7.4.8)$$

Figure 7.10 represents the dependence $\Phi(\xi)$, and also its simple approximation $(1 + \xi/6)^{-1}$ which gives the same limiting results.

The parameter ξ in formula (7.4.7) for the helium case, where we take $\sigma_{ea} = 6\,\text{Å}^2$, is given by

$$\xi = \frac{N_i}{N_a} \frac{\pi e^4 \ln \Lambda}{T_e^2 \sigma_{ea}^*} = \frac{N_i}{N_a} \frac{1{,}600}{T_e^2}$$

Here the electron temperature T_e is taken in eV, and $\ln \Lambda = 7$. In particular, if $N_i/N_a = 10^{-4}$ and the electron temperature $T_e = 1\,\text{eV}$, the electron drift velocity decreases due to electron-ion scattering by 6 %, at $N_i/N_a = 10^{-5}$ a decrease of the electron drift velocity is about 2 % due to electron-ion scattering, and in the case $N_i/N_a = 10^{-6}$ this drift velocity decrease is approximately 0.3 %.

7.5 Electron Thermal Conductivity of Helium Arc Plasma

Transport of electrons gives a certain contribution to heat transport in a gas discharge plasma, and we analyze this below. For definiteness, we consider the helium case of electron transport in the regime of a high number density of electrons, when this gas is located in an external electric field with a weak gradient ∇T_e of the electron temperature that causes a heat flux. We are guided by typical conditions in a gas discharge plasma when the electron concentration is low, but due to a high mobility electrons creates a heat flux $\mathbf{q}_e$ that is independent of the gas heat flux and can be included in equation (8.6.1) of heat transport. Correspondingly the total heat flux is a sum of these fluxes.

Let us introduce the coefficient of electron thermal conductivity κ_e according to the heat transport equation

$$\mathbf{q}_e = -\kappa_e \nabla T_e, \tag{7.5.1}$$

where the electron temperature T_e characterizes the Maxwell distribution function of electrons $\varphi(v)$. The gradient of the electron temperature induces the additive non-symmetric distribution function, and the total velocity distribution function of electrons has the form

$$f(\mathbf{v}) = \varphi(v) + (\mathbf{v}\nabla \ln T_e) f_1(v)$$

Using the stationary kinetic equation for electrons $\mathbf{v}\nabla f = I_{ea}(f)$ that takes into account a change of the electron momentum as a result of their collisions with atoms, together with the gas state equation for the electron pressure $p_e = N_e T_e$, we have from the kinetic equation for the electron distribution function

$$\varphi(v)\left(\frac{m_e v^2}{2T_e} - \frac{5}{2}\right)\mathbf{v}\nabla T_e = I_{ea}(f)$$

From this we have on the basis of the expression (5.3.4) for the collision integral for the antisymmetric part of the electron distribution function

$$f_1(v) = -\frac{\varphi(v)}{\nu}\left(\frac{m_e v^2}{2T_e} - \frac{5}{2}\right)$$

From this we find the heat flux due to electron transport

$$(\mathbf{q}_e)_x = \int \frac{m_e v^2}{2} v_x f(\mathbf{v}) d\mathbf{v} = \int \frac{m_e v^2}{2} v_x^2 \nabla \ln T_e f_1(v) d\mathbf{v}$$

From this and the definition (7.8.2) for the electron thermal conductivity coefficient, we have

$$\kappa_e = N_e \left\langle \frac{v^2}{3\nu_{ea}} \frac{m_e v^2}{2T_e} \left(\frac{m_e v^2}{2T_e} - \frac{5}{2}\right)\right\rangle, \tag{7.5.2}$$

where an average is made over the Maxwell distribution function of electrons, and the coefficient of the electron thermal conductivity is proportional to the electron concentration in a gas.

We first consider a general case where the dependence of the rate of electron-atom collisions $\nu_{ea}(v)$ on the electron velocity v is approximated by the dependence

$\nu_{ea} \sim v^k$. At this approximation including a variable $z = m_e v^2/(2T_e)$, we obtain for the electron thermal conductivity

$$\kappa_e = \frac{4}{3\sqrt{\pi}} \cdot \frac{T_e N_e}{\nu_o m_e} \left(1 - \frac{k}{2}\right) \Gamma\left(\frac{7-k}{2}\right)$$

In the case $\nu_{ea} = const$ this gives

$$\kappa_e = \frac{5T_e N_e}{2\nu_o m_e},$$

and in the case $k = 1$ it follows from the above formula

$$\kappa_e = \frac{2}{3\sqrt{\pi}} N_e \lambda \sqrt{\frac{2T_e}{m_e}}, \tag{7.5.3}$$

where the relation $\nu_{ea} = v/\lambda$ is used with the expression $\lambda = 1/(N_a \sigma_{ea})$ for the mean free path of electrons in a gas.

In the helium case, taking the diffusion cross section of electron-atom collisions to be independent of the electron velocity, we transform formula (7.5.3) to the form

$$\kappa_e = \frac{\sqrt{8}}{3\sqrt{\pi}} N_e \lambda \sqrt{\frac{T_e}{m_e}} = 0.532 c_e \frac{1}{\sigma_{ea}^*} \sqrt{\frac{T_e}{m_e}},$$

where $c_e = N_e/N_a$ is the electron concentration. Taking the average cross section of electron-atom collisions as $\sigma_{ea} = 6\,\text{Å}^2$, we obtain for the electron thermal conductivity coefficient in helium

$$\kappa_e = 3.08 c_e \sqrt{T_e}, \tag{7.5.4}$$

where the electron temperature T_e is expressed in eV, and the thermal conductivity coefficient κ_e is given in W/(cm K).

One can compare the electron thermal conductivity coefficient κ_e in helium with the gas thermal conductivity coefficient κ that in helium at room temperature is equal to $\kappa_a = 1.51 \times 10^{-3}$W/(cm K). Taking the diffusion cross section of electron-atom scattering in helium to be $\sigma_{ea} = 6\,\text{Å}^2$, we obtain for the electron thermal conductivity coefficient $\kappa_e/c_e = 0.616\sqrt{T_e/T_o}$W/(cm K), where $T_o = 1$ eV, and $c_e = N_e/N_a$ is the electron concentration. From this we obtain for the ratio of the electron and gaseous thermal conductivity coefficients

$$\frac{\kappa_e}{\kappa_a} \approx 410 c_e, \quad T_e = 1\,\text{eV}, \quad \frac{\kappa_e}{\kappa_a} \approx 580 c_e, \quad T_e = 2\,\text{eV}, \tag{7.5.5}$$

One can see that the contribution of electron thermal conductivity to the heat transport is significant at a high electron concentration $c_e \sim 0.001 - 0.01$.

7.6 Ion Drift and Diffusion in Gas in External Electric Field

Because of a small electron mass, electrons are more mobile than ions. But transport of electrons outside a gas discharge plasma creates electric fields that keep electrons and cause transport of ions together with electrons conserving the plasma to be quasineutral. In other words, transport of a plasma as a whole proceeds in this case, and this transport is determined by ion rates. Therefore drift of ions in an external field determines transport of a plasma as a whole.

In considering the ion drift in an external field, we use the macroscopic equation (5.2.2) for the mean electron momentum **P** when charged particles are located in a gas in an external electric field [190]

$$\frac{d\mathbf{P}}{dt} = e\mathbf{E} - \frac{\mu}{N_i N_a}\int \mathbf{g} g d\sigma^*(g) f\varphi d\mathbf{v} d\mathbf{v}_a,$$

In the simplest case when the rate of ion-atom collisions $k_{ia} = \int g d\sigma^*(g)$ is independent of the collision velocity, this equation takes the form

$$m\frac{d\mathbf{w}}{dt} = e\mathbf{E} - \mu\mathbf{w}k_{ia}N_a, \quad (7.6.1)$$

where **w** is the ion drift velocity, and the mean ion momentum is $\mathbf{P} = m\mathbf{w}$. In the stationary case this equation gives

$$\mathbf{w} = \frac{e\mathbf{E}}{\mu k_{ia} N_a} \quad (7.6.2)$$

This case relates to motion of ions in a foreign gas with the polarization ion-atom interaction potential $U(R)$ that corresponds to ion-atom interaction at large distances R and has the form

$$U(R) = -\frac{\alpha e^2}{2R^4}$$

where α is the atom polarizability, e -is the ion charge. Then the ion mobility K is given by the Dalgarno formula [113]

$$K = \frac{36}{\sqrt{\alpha\mu}}, \quad (7.6.3)$$

Here the polarizability is given in $\mathrm{cm}^2/(\mathrm{V\,s})$, the atom polarizability α is expressed in a_o^3, the reduced mass μ of colliding particles is given in a.u.m. It is of importance that the ion drift velocity w according to formula (7.6.2) is proportional to the electric field strength E at any strengths, including large E where the ion energy exceeds a thermal ion energy.

If an atomic ion is located in a parent gas, ion drift in a gas in an external electric field results from the resonant charge exchange process (3.5.4). In the absence of elastic scattering charge transfer has a relay character [226], i.e. a forming ion obtains the velocity of an atom that partakes in the charge exchange event. This character of the charge exchange process is called the Sena effect. Note that at room temperature the contribution of elastic ion-atom scattering in the mobility of helium atomic ions in helium is 6 %, and in the case of argon atomic ions in argon this contribution is 11 % [190].

The ion mobility in a small electric field may be evaluated on the basis of the Chapman-Enskog method [294–297], that corresponds to expansion of transport coefficients over a small numerical parameter. The ion mobility in the first Chapman-Enskog approximation is given by

$$K = \frac{eD}{T} = \frac{3\sqrt{\pi}}{16\sqrt{MT}N_a\overline{\sigma^*}} \tag{7.6.4}$$

where M is the ion mass, $\overline{\sigma^*}$ is the diffusion cross section of ion-atom scattering, and the connection between the mobility K and the diffusion ion coefficient D follows from the Einstein relation (2.3.3). In the absence of elastic scattering the diffusion cross section of ion-atom scattering is expressed through the cross section of resonant charge exchange σ_{res} as $\sigma^* = 2\sigma_{res}$ [335]. In addition, the contribution of the second approximation to the first one is about 2 % [336]. Accounting for a weak dependence on the collision velocity for the cross section $\sigma_{res}(v)$ of resonant change exchange, we obtain for the mobility of atomic ions in a parent gas [190]

$$K_o = \frac{1342}{\sqrt{MT}\sigma_{res}(\sqrt{9T/M})}, \tag{7.6.5}$$

where the ion mobility K is expressed in $\mathrm{cm^2/(V\,s)}$ and relates to the normal number density of atoms ($2.69 \times 10^{19}\,\mathrm{cm^{-3}}$), the ion and atom mass M is given in a.u.m., the gas temperature T is given in K, the cross section of resonant charge exchange σ_{res} is expressed in $10^{-15}\,\mathrm{cm^2}$, and an argument in the cross section indicates a velocity at which it is taken with accounting for a weak velocity dependence of the cross section of resonant charge exchange. Elastic ion-atom scattering is ignored in this formula. Accounting for a small contribution of elastic scattering into the ion mobility due to ion-atom polarization interaction, we have for the atomic ion mobility in a parent gas at small electric field strength [190]

$$K = \frac{K_o}{1 + x + x^2}, \quad x = \frac{\sqrt{\alpha e^2/T}}{\sigma_{res}(\sqrt{7.5T/M})} \tag{7.6.6}$$

where K_o is the ion mobility in absence of elastic scattering according to formula (7.6.3), α is the atom polarizability. Note that accounting for elastic scattering leads to a mobility decrease.

The diffusion coefficient of atomic ions in a parent atomic gas in weak electric fields is given by formula (7.1.16) with the diffusion cross section σ_o of ion-atom collisions $\sigma_o = 2\sigma_{res}$ [335]. In the first approximation the mean free path of ions in the expression for the ion diffusion coefficient is independent of the ion velocity. In the second approximation this dependence is such as in formula (7.6.5)

$$D_i = \frac{3\sqrt{\pi}}{8}\lambda\sqrt{\frac{T}{M}} = \frac{3\sqrt{\pi}}{16}\sqrt{\frac{T}{M}}\frac{1}{N_a\sigma_{res}(\sqrt{9T/M})} \tag{7.6.7}$$

Note that all the above expressions relate to low electric field strengths, i.e. between neighboring collisions an ion obtains from the electric field a small energy compared to a thermal one that corresponds to criterion

$$eE\lambda \ll T, \tag{7.6.8}$$

where E is the electric field strength, $\lambda = 1/(N_a\sigma_{res})$ is the ion mean free path in a gas, and N is the number density of atoms in a gas. In this limiting case the ion drift velocity w in a parent gas is proportional to the electric field strength

$$w = EK \tag{7.6.9}$$

In the limit of high electric field strengths, when the criterion

$$eE\lambda \gg T, \tag{7.6.10}$$

holds true, the velocity ion distribution function in the field direction (the axis x) has the form

$$f(v_x) = C\exp\left(-\frac{Mv_x^2}{2eE\lambda}\right), \quad v_x > 0, \tag{7.6.11}$$

where C is the normalization constant, and the cross section of resonant charge exchange assumes to be independent of the collision velocity. In this case the ion distribution function is the Maxwell one with the gas temperature in directions perpendicular to the field. We have for the ion drift velocity and ion energy

$$w_i = \overline{v_x} = \sqrt{\frac{2eE\lambda}{\pi M}}, \quad \overline{\varepsilon} = \frac{M\overline{v_x^2}}{2} = \frac{eE\lambda}{\pi} \tag{7.6.12}$$

In the next approximation one can include a weak velocity dependence for the cross section σ_{res} of resonant charge exchange, and then the cross section σ_{res} of resonant charge exchange in formula (7.6.12) is taken at the collision velocity $w_i = \sqrt{2eE\lambda/(\pi M)}$.

In an intermediate range of electric field strength, when the ion drift velocity is comparable with a thermal ion velocity, it can be found from solution of the kinetic equation (5.1.1) for the ion distribution function [216, 337, 338]. A simple approximation for the ion drift velocity for an intermediate range of the electric field strength takes the form [190, 339], if we account for a weak velocity dependence of the cross section of resonant charge exchange

$$w = \sqrt{\frac{2T}{M}} \cdot \frac{0.48\beta}{(1 + 0.22\beta^{3/2})^{1/3}}, \quad \beta = \frac{eE}{2TN_a\sigma_{res}\left[\sqrt{\frac{2T}{M}}\,(4.5 + 1.6\beta)\right]}, \quad (7.6.13)$$

where an argument indicates a velocity at which the charge exchange cross section is taken. The dependence of the reduced drift velocity $u = w_i/\sqrt{2T/M}$ on the parameter β that characterizes the electric field strength is given in Fig. 7.11.

Being guided by a gas discharge plasma of helium and argon, we give in Table 7.5 the values of the mobilities K and diffusion coefficients D of helium and argon ions in parent gases at temperatures $T = 300$ K and $T = 800$ K and in zero electric field. At temperature $T = 300$ K are used experimental values of mobilities averaged over various measurements [190]. The dependence of the cross section of resonant charge exchange σ_{res} on the collision energy ε is taken in the form $\sigma_{res} \sim \varepsilon^k$, where $k = 0.084$ for helium and $k = 0.082$ for argon [339, 229]. Transport coefficients of molecular ions in parent gases are taken at temperature $T = 300$ K by averaging on various measurements, and the rate constant of ion elastic scattering on atoms assumes to be independent of the collision energy, as it takes place for the polarization ion-atom interaction.

The contribution of elastic scattering to the mobility of atom ions in a parent gas that is determined by formula (7.6.6) decreases with an increase of an average ion

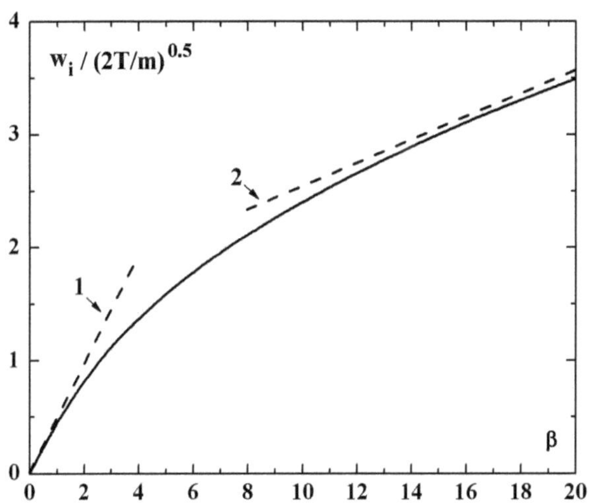

Fig. 7.11 The ion drift velocity in a gas in an electric field in accordance with formula (7.6.13). Lines 1, 2 correspond to the limits of low and high electric field strengths

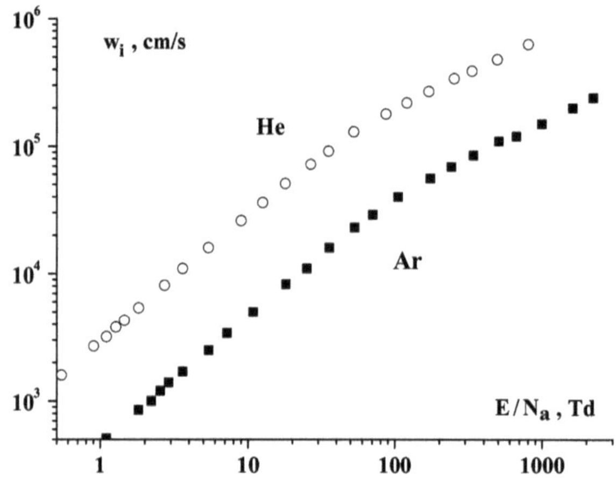

Fig. 7.12 The drift velocity of the atomic helium and argon ions in the parent gas in a constant electric field

energy, and this contribution at room temperature and in low fields is 6 % in the helium case and 11 % in the argon case. The dependence on the reduced electric field strength E/N_a for the drift velocity of atomic ions of helium and argon in parent gases is given in Fig. 7.12.

One can separate the average ion energy $\overline{\varepsilon}$ in a thermal and field parts basing on formula (5.2.6). Requiring the accurate asymptotic behavior of the average energy in the limit small and large electric field strengths, we represent formula (5.2.6) for the average energy of the average ion energy in the form

$$\overline{\varepsilon} = \sqrt{\left(\frac{3T}{2}\right)^2 + \left(\frac{eE\lambda}{2}\right)^2} \tag{7.6.14}$$

where the mean free path of ions in a parent gases at large electric field strengths is given by

$$\lambda = \frac{1}{N_a\sigma_{res}\left(\sqrt{\frac{4\overline{\varepsilon}}{M}}\right)} \tag{7.6.15}$$

Figure 7.13 gives the mean energy of helium and argon atomic ions in parent gases depending on the reduced electric field strength E/N_a on the basis of formulas (7.6.14) and (7.6.15).

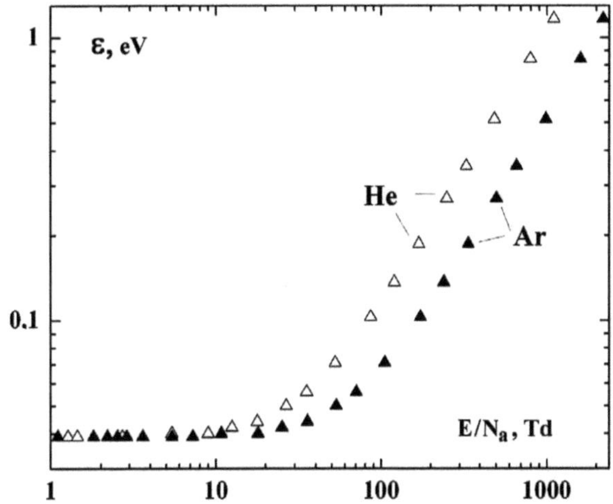

Fig. 7.13 The average energy of helium and argon ions in parent gases in accordance with formulas (7.6.14) and (7.6.15)

The electric current in a gas discharge plasma is determined by electrons and ions give a small contribution to this. Indeed, the electric current density in an ionized gas under the action of the electric field is given by

$$\mathbf{i} = -eN_e\mathbf{w}_e + eN_i\mathbf{w}_i,$$

where N_e, N_i are the number densities of electrons and ions, $\mathbf{w}_e$, $\mathbf{w}_i$ are their drift velocities, and the electron drift velocity exceeds significantly that for ions because of a less mass. Since a plasma is quasineutral $N_e = N_i$, the contribution of ions into the plasma conductivity is w_i/w_e. We give this ratio in Fig. 7.14 for the case

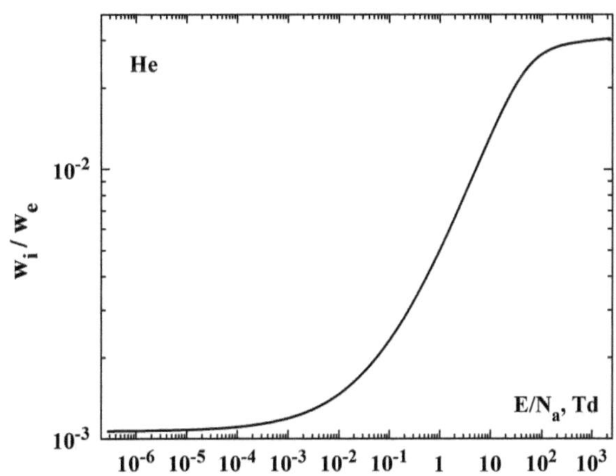

Fig. 7.14 The ratio of the drift velocities of atomic helium ions and electrons in helium

where atomic ions are located in a gas discharge plasma. Note that this contribution is higher for helium compared to other inert gases because of a less atom mass.

We have two limiting cases of ion drift with respect to the electric field strengths, and the ion drift velocity is determined by formula (7.6.5) at low electric field strengths and by formula (7.6.12) at high field strengths. The parameter β defined by formula (7.6.13) determines the transition between these limiting cases. In particular, in the case of helium atomic ions transition between these limiting cases takes place at $\beta = 2.8$ where the above formulas give identical values for the ion drift velocity. According to formula (7.6.13) at this β the ion drift velocity exceeds a typical thermal ion velocity $\sqrt{2T/M}$ by 6 %. Next, $\beta = 2.8$ corresponds to $E/N_a = 50\,\mathrm{Td}$ for helium and $E/N_a = 100\,\mathrm{Td}$ for argon. These reduced electric field strengths are intermediate for transition from low to high electric field strengths.

Let us make a similar analysis in the case where molecular ions are realized in a gas discharge plasma. One can use formula (7.6.2) for the drift velocity of ions in a foreign gas, and then the ion mobility is determined by the polarization interaction (7.6.3) of ions and gas atoms. But, it is convenient to use the experimental values of the ion drift velocity (in particular, [8, 190, 340]) and determine the ion drift velocity in the limit of low electric field strengths. At an intermediate value of the electric field strength the ion drift velocity is equal to a typical thermal velocity. The mobility of helium molecular ions He_2^+ in helium at the normal number density of atoms is $18\,\mathrm{cm^2/(V\,s)}$ (with the accuracy of 10 %) and for molecular argon ions Ar_2^+ in argon this value is $1.9\,\mathrm{cm^2/(V\,s)}$. This gives the intermediate electric field strength approximately 20 Td for helium and 50 Td for argon. These values are lower than those in the case of atomic ions.

One can check the accuracy of the polarization theory [113] for the mobility of molecular ions basing on experimental data which according to the data of Table 7.6 are equal $K = (18 \pm 2)\,\mathrm{cm^2/(V\,s)}$ for helium molecular ions He_2^+ in helium and $K = (1.86 \pm 0.05)\,\mathrm{cm^2/(V\,s)}$ for the mobility of molecular ions in helium

Table 7.6 The zero-field mobilities K and diffusion coefficients D for helium and argon ions in parent gases

	$T = 300\,\mathrm{K}$	$T = 800\,\mathrm{K}$
$K(He^+ - He)$	10.4 ± 0.3	6.9
$K(He_2^+ - He)$	18 ± 2	18
$D(He^+ - He)$	0.27	0.40
$D(He_2^+ - He)$	0.46	1.2
$K(Ar^+ - Ar)$	1.55 ± 0.1	1.0
$K(Ar_2^+ - Ar)$	1.86 ± 0.05	1.9
$D(Ar^+ - Ar)$	0.040	0.060
$D(Ar_2^+ - Ar)$	0.048	0.13

An argument indicates the ion and gas sort; the mobilities are given in $\mathrm{cm^2/(V\,s)}$, the diffusion coefficients are expressed in $\mathrm{cm^2/s}$ and reduce to the normal number density of atoms $2.69 \times 10^{19}\,\mathrm{cm^{-3}}$

and molecular ions Ar_2^+ in argon in zero field at the normal number density of atoms. In the case of the ion-atom polarization interaction the ion mobility in an atomic gas in zero field is given by formula (7.6.3). The polarizability of a helium atom is 1.383 a_o^3, and the polarizability of an argon atom is 11.1 a_o^3. On the basis of formula (7.6.3) we obtain the zero field mobility of molecular ions He_2^+ in helium $K = 19\,\mathrm{cm}^2/(\mathrm{V\,s})$, and for molecular argon ion Ar_2^+ in argon $K = 2.1\,\mathrm{cm}^2/(\mathrm{V\,s})$. As is seen, these values are in accordance with experimental data with the accuracy 10–20 %.

One can compare the drift velocity of electrons and ions. In particular, at the pressure of 10 Torr and the electric field strength of 50 V/cm we have the electron drift velocity in argon is 1.5×10^6 cm/s, the drift velocity of atomic ions Ar^+ in argon is 6×10^3 cm/s (their mobility equals to $120\,\mathrm{cm}^2/(\mathrm{V\,s})$) and the drift velocity of molecular ions Ar_2^+ is 8×10^3 cm/s (their mobility in argon equals to $160\,\mathrm{cm}^2/(\mathrm{V\,s})$). As is seen, the drift velocity of electrons exceeds that of ions by more than two orders of magnitude. Therefore the contribution of ions to the conductivity of a quasineutral argon ionized gas under indicated conditions is below 1 %.

The ion diffusion coefficient of atomic ions in a parent gas in the longitudinal direction at high electric field strengths $D_\parallel$ is given by [254]

$$D_\parallel = 0.137\lambda w \tag{7.6.16}$$

In derivation the expression for the transverse diffusion coefficient $D_\perp$ of atomic ions in a strong field $eE\lambda \gg T$ we start from a general expression for the diffusion coefficient

$$D_\perp = \left\langle \frac{v_z^2}{\nu} \right\rangle,$$

where v_z is an ion velocity in the transverse direction, ν is the rate of variation of the transverse ion velocity in collisions with atoms $\nu = N_a w_i \sigma_{res}$, and an average is made over ion-atom collision velocities. Since $\langle v_z^2 \rangle = T/M$, where M is the ion mass, we obtain for the ion diffusion coefficient in a transverse direction to the field [254]

$$D_\perp = \frac{T}{M}\frac{\lambda}{w_i} = \frac{T\sqrt{\pi\lambda}}{\sqrt{2eEM}}, \tag{7.6.17}$$

One can join the expressions for the ion transverse diffusion coefficient in low (7.6.7) and strong (7.6.17) electric fields. Let us construct the expression for the transverse diffusion coefficient of atomic ions in a parent gas assuming that the cross section of resonant charge exchange in collisions of an atomic ion with an own atom is independent of the collision velocity. We require this expression to be transferred into formulas (7.6.7) and (7.6.17) in the limits of low and strong electric field strengths. One can combine these formulas as

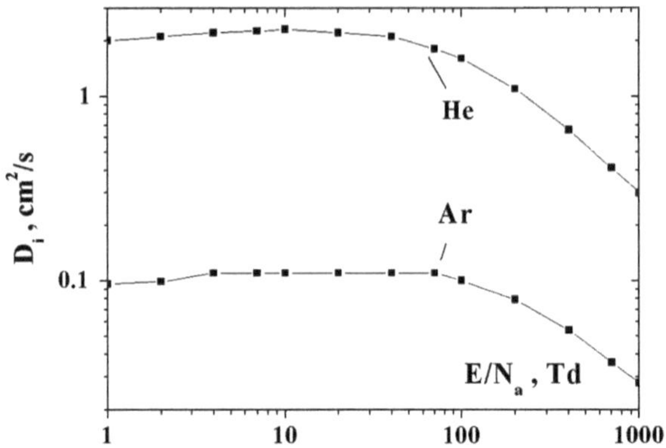

Fig. 7.15 Reduced to the normal density of atoms the diffusion coefficient of helium and argon ions in a parent gas as a function of the reduced electric field strength according to formula (7.6.18) at the gas temperature $T = 400\,\mathrm{K}$

$$D_\perp = \frac{3\sqrt{\pi}}{16}\lambda\sqrt{\frac{T}{M}}\left(1 + \frac{3}{8\sqrt{2}}\sqrt{\frac{eE\lambda}{T}}\right)^{-1} \qquad (7.6.18)$$

Figure 7.15 gives the dependence of the diffusion coefficient of atomic helium and argon atoms in a parent gases at the gas temperature $T = 400\,\mathrm{K}$ in accordance with formula (7.6.18). A weak increase of the ion diffusion coefficient with an increasing electric field strength proceeds due to a decrease of the cross section of resonant charge exchange with an increased energy of ion-atom collision.

7.7 Ambipolar Diffusion of Plasma in Gas in Electric Field

If a plasma is propagated in an empty space, its quasi neutrality is violated due to different velocities of electrons and ions. This leads to creation of an electric field that brakes electrons and accelerated ions, and this plasma moves as a whole. If this self-consistent regime of plasma motion proceeds in a dense gas, where electron and ion displacement is accompanied by their diffusion in a gas, this regime of plasma motion is called ambipolar diffusion and proceeds near the walls of gas discharge of a not low pressure [18, 29].

We now consider such a self-consistent regime of transport of a gas discharge plasma near its boundary, in particular, plasma departure to electrodes or walls of a chamber where this plasma is located. Because separation of a plasma charge creates high electric fields, the plasma tries to move as a whole, conserving the quasineutrality in the course of propagation. Nevertheless, because of different electron and ion masses and hence due to different mobilities of electrons and ions, electrons turn away from plasma ions and create in this way an electric field that pulls and accelerates the ions. Assuming the number density of electrons and ions in a gas discharge plasma with a high density of atoms to be small compared the atom number density, we have for the flux of electrons $\mathbf{j_e}$ and ions $\mathbf{j_i}$

$$\mathbf{j_e} = -K_e N_e \mathbf{E_a} - D_e \nabla N_e, \quad \mathbf{j_i} = K_i N_i \mathbf{E_a} - D_i \nabla N_i \tag{7.7.1}$$

Here N_e, N_i are the number densities of electrons and ions, K_e, D_e are the mobility and diffusion coefficient of electrons, K_i, D_i are the mobility and diffusion coefficient of ions, $\mathbf{E}_a$ is the electric field strength due to separation of electron and ion charges, and we take into account that the force from this electric field acts on electrons and ions in different directions. Since the quasineutrality of this plasma conserves in the course of its evolution, the electron and ion fluxes are equal $\mathbf{j_e} = \mathbf{j_i}$. From this it follows that because $K_e \gg K_i$ and $D_e \gg D_i$, in scales of electron values we have $\mathbf{j_e} = 0$. From this we find the electric field strength created by a charge separation

$$\mathbf{E_a} = \frac{D_e}{K_e} \cdot \frac{\nabla N_e}{N_e} \tag{7.7.2}$$

Taking the plasma flux as

$$\mathbf{j} = \mathbf{j_i} = -D_a \nabla N \tag{7.7.3}$$

and defining in this manner the coefficient D_a of ambipolar diffusion, one can find for the coefficient of ambipolar diffusion with accounting for the plasma quasineutrality $N_e = N_i = N$

$$D_a = D_i + \frac{D_e}{K_e} K_i \tag{7.7.4}$$

Thus, propagation of this plasma reduces to the diffusion regime of ion motion in a dense gas. This regime, ambipolar diffusion, is typical for a gas discharge plasma [18, 29].

In the case of the Maxwell distribution function of electrons and ions this formula have the form

$$D_a = D_i \left(1 + \frac{T_{ef}}{T_i}\right), \tag{7.7.5}$$

where T_{ef} is the effective temperature of electrons, and T_i ia the temperatures of ions. In a general case of a strong electric field the electron temperature must be replaced by the effective temperature which coincides with the characteristic temperature (7.2.4)

$$T_{ef} = \frac{eD}{K}$$

for electrons and ions separately.

The effective temperature of electrons coincides with the electron temperature if the electron distribution function has the Maxwell form (6.2.4). In other cases it requires a more precise. Let us consider a spread case of a gas discharge plasma when it is located in a cylinder tube along which a voltage is applied and a plasma goes to walls in the transverse direction due to ambipolar diffusion. Because of high electric field strengths E the dependence $w_e(E)$ for the electron drift velocity is nonlinear. We find in this case the effective electron temperature T_e that is included in formula (7.7.4) and according to its definition it is equal to

$$T_{ef} = eD_{\perp}^{e}/\frac{w_a}{E_a},$$

where $D_{\perp}^{e}$ is the electron diffusion coefficient in the transverse direction with respect to a discharge electric field $\mathbf{E}_o$, E_a is the electric field strength in the transverse direction resulted from ambipolar diffusion, i.e. separation of electrons and ions, and w_a is the electron drift velocity under the action of this field. Denoting by w_o the electron drift velocity due to the discharge electric field strength E_o, we find a change of the electron drift velocity δw resulted from inclusion of ambipolar diffusion that corresponds to addition of the electric field strength δE in the transverse direction

$$\delta w = \frac{\partial w}{\partial E}\delta E = \sqrt{w_o^2 + w_a^2} - w_o = \frac{w_a^2}{2w_o}, \quad \delta E = \frac{E_a^2}{2E_o}$$

From this it follows

$$\frac{w_a}{E_a} = \sqrt{\frac{w}{E}\frac{\partial w}{\partial E}}$$

This gives for the effective electron temperature in the transverse direction

$$T_{ef} = eD_{\perp}^{e} \cdot \left(\frac{w}{E}\frac{\partial w}{\partial E}\right)^{-1/2} \tag{7.7.6}$$

Evidently, this expression gives a correct transition to the limit of low electric field strengths.

Let us consider one more peculiarity of ambipolar diffusion at high electric field strengths where the average electron energy exceeds its significantly a thermal energy of heavy atomic particles. A weak change of an energy in elastic electron-atom collisions leads to a dominate role of the second term of formula (7.7.4) in a wide range of electric field strengths. Let us demonstrate this for the case when the rate of electron-atom collisions ν is independent of the collision velocity. Then according to formulas (2.3.2) and (7.2.2) we have

$$\frac{T_{ef}}{T} = 1 + \frac{Mw_e^2}{3T}$$

Taking the same formula for ions, we obtain that since the electron drift velocity exceeds remarkably the ion drift velocity, we obtain a wide range of electric field strengths, where the ion and gas temperatures coincide, while a typical electron energy exceeds significantly a thermal atom energy.

Let us formulate the criterion of the ambipolar diffusion regime for plasma propagation. Denote by L a typical distance on which plasma parameters vary remarkably, and this distance exceeds significantly he mean free path of charged particles in the gas. From the Poison equation we have for a typical difference between the electron and ion number densities

$$\Delta N \equiv\mid N_i - N_e \mid\sim \frac{E}{4\pi eL}$$

An estimate for the electric field strength resulted from charge separation has the form

$$\frac{\Delta N}{N} \sim \left(\frac{r_D}{L}\right)^2,$$

where r_D is the Debye-Hückel radius. From this it follows that a quasineutrality of this plasma $\Delta N \ll N$ corresponds to the criterion $r_D \ll L$, that is the condition of plasma existence.

Note that a typical electron temperature exceeds significantly a thermal ion energy because a small part of an electron energy transfers to atoms in single collisions with atoms. Therefore in a wide range of electric field strengths a typical electron energy exceeds a thermal one, whereas the energy distribution function of ions coincides with that at zero field. As it follows from the data of Table 7.3 and Fig. 7.3, a typical electron energy in helium exceeds a thermal one at the electric field strengths $E/N_a > 0.1$ Td. In the argon case it takes place at $E/N_a > 0.001$ Td due to the data of Table 7.4 and Fig. 7.4. At low electric field strengths at which the electron drift velocity is small compared to a thermal electron velocity, electrons are found in equilibrium with a gas and the effective electron temperature T_{ef} in formula (7.7.5) is equal to the gas temperature $T_{ef} = T$. In this limit $D_a = 2D_i$, i.e. the contribution to the ambipolar diffusion coefficient is equal due to ion diffusion and their drift under the action of an electric field that is created by electrons. But in the course of an increasing electric field strength the contribution to the ambipolar diffusion coefficient from ion drift owing to an electron electric field increases. There is a wide range of electric field strengths where a typical electron energy is large compared with a thermal energy of atoms, whereas the electric field acts weakly on the energy distribution of ions. In this range the ion diffusion coefficient D_i in formula (7.7.5) for the ambipolar diffusion coefficient is independent of the electric field strength, whereas as the second term in the parentheses of this formula varies with an increasing electric field strength.

Figure 7.16 contains the coefficients of ambipolar diffusion in a helium gas discharge plasma as a function of the reduced electric field strengths for two regimes where atomic or molecular ions dominate. In this case the diffusion coefficient of helium atomic ions is $0.27\,\mathrm{cm}^2/\mathrm{s}$ and the diffusion coefficient of helium molecular

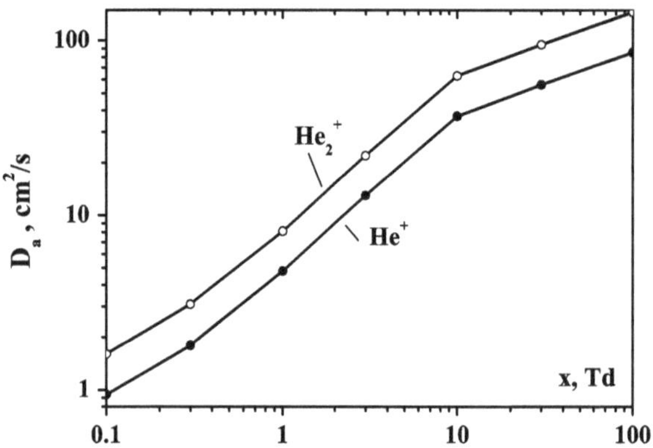

Fig. 7.16 The ambipolar diffusion coefficient of a helium plasma in accordance with formula (7.7.5) if the basic ion sort in the plasma is He^{+} or He_2^{+}

ions is $0.46\,\mathrm{cm}^2/\mathrm{s}$ in accordance with the data of Table 7.6. The effective electron temperature according to formula (7.2.4) is $T_{ef} = eD_e/K_e$ in helium in an external electric field is taken from Table 7.3, so that D_e is the transverse diffusion coefficient of electrons and $K_e = W_e/E$ is the electron mobility at this electric field strength (w_e is the electron drift velocity, E is the electric field strength). Note that these values of the ambipolar diffusion coefficient relate to the normal number density of gas atoms $N_n = 2.87 \times 10^{19}\,\mathrm{cm}^{-3}$. If the number density of atoms is N_a, these diffusion coefficients must be multiplied by the factor N_n/N_a.

The coefficients of ambipolar diffusion given in Fig. 7.17 for an argon gas discharge plasma are obtained in the same method. The diffusion coefficient of argon atomic ions Ar^{+} in argon is $0.040\,\mathrm{cm}^2/\mathrm{s}$ and the diffusion coefficient of argon

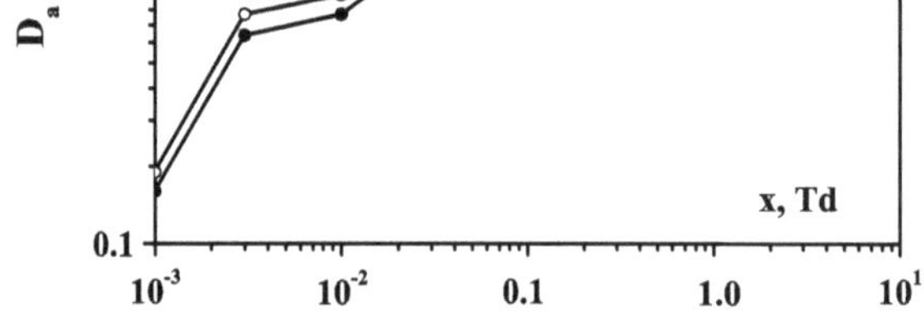

Fig. 7.17 The ambipolar diffusion coefficient of an argon plasma in accordance with formula (7.7.5) if the basic ion sort in the plasma is Ar^{+} or Ar_2^{+}

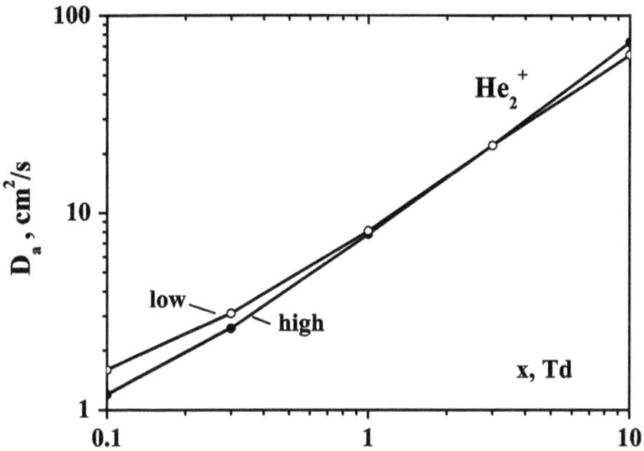

Fig. 7.18 The ambipolar diffusion coefficient in a helium gas discharge plasma for regimes of low and high electron number densities if the basic ion sort is He_2^+

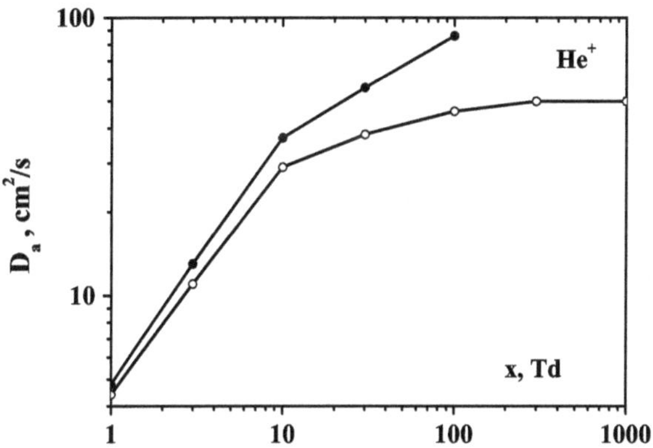

Fig. 7.19 The ambipolar diffusion coefficients in accordance with formula (7.7.5) for the regime of a low electron number density in a helium plasma where the basic ion sort is He^+. *Open circles* correspond to low electric field strengths where the ion diffusion coefficient in formula (7.7.5) does not depend on the electric field strength and these values are represented also in Fig. 7.16. *Closed circles* take into account the field dependence of the ion diffusion coefficient in accordance with formula (7.7.1)

molecular ions Ar_2^+ in argon is 0.048 cm^2/s according to the data of Table 7.6. The effective electron temperatures $T_{ef} = eD_e/K_e$ in argon in an electric field are taken from Table 7.4. One can see that the data of Figs. 7.16 and 7.17 relate to the regime of a low electron number density where one can ignore electron-electron collisions (Fig. 7.18).

In the regime of a high number density of electrons the coefficient of ambipolar diffusion in a plasma is given by formula (7.7.5) and the electron temperature in the helium case is determined according to formula (6.2.7) that is valid for $E/N_a <$ 50 Td. Figure 7.19 contains the coefficients of ambipolar diffusion in a helium plasma

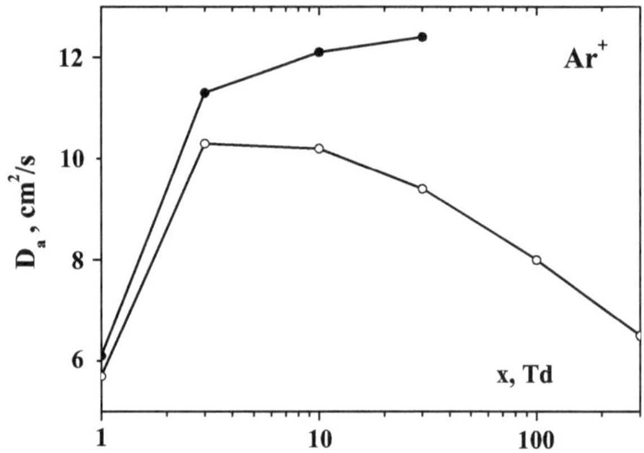

Fig. 7.20 The ambipolar diffusion coefficients in accordance with formula (7.7.5) for the regime of a low electron number density in an argon plasma where the basic ion sort is Ar^+. *Open circles* correspond to low electric field strengths where the ion diffusion coefficient in formula (7.7.5) does not depend on the electric field strength and these values are represented also in Fig. 7.17. *Closed circles* take into account the field dependence of the ion diffusion coefficient in accordance with formula (7.7.7)

if the basic ion sort is He^+, and the same results are given in Fig. 7.20 for the argon case with Ar^+ as the basic ion sort in a gas discharge plasma. In these figures the values of ambipolar diffusion are compared for the regimes of low and high electron number density. In reality the difference between these values is comparable with the accuracy of measurements for the effective temperature of electrons (Tables 7.3 and 7.4).

We now consider the case of large electric field strengths where an average ion energy differs from a thermal atom energy. For definiteness we consider the situation in a cylinder discharge tube where an electric field directs along the tube axis, whereas plasma drift proceeds to walls, i.e. in the perpendicular direction. Therefore it is necessary to use the transverse diffusion coefficient of ions in formula (7.7.5) for the diffusion coefficient of ions. At large electric field strengths we use formula (7.6.17) for the transverse diffusion coefficient of ions. Being guided by a typical ion energy of 0.1 eV, we take the cross section of resonant charge exchange to be $3.5 \times 10^{-15}\,\mathrm{cm}^2$ for the helium case and $7.0 \times 10^{-15}\,\mathrm{cm}^2$ for the argon case according to Table 3.4 data. At the gas temperature $T = 300\,\mathrm{K}$ formula (7.6.18) for the transverse diffusion coefficient of ions has the following form for atomic helium ions He^+ in helium and for atomic argon ions Ar^+ in argon correspondingly

$$D_i(He) = \frac{0.27}{1 + 0.088\sqrt{x}}, \quad D_i(Ar) = \frac{0.040}{1 + 0.060\sqrt{x}}, \qquad (7.7.7)$$

where the diffusion coefficient is given in cm^2/s and is reduced to the normal number density of atoms, $x = E/N_a$ is the reduced electric field strength expressed in Td.

The values of the ambipolar diffusion coefficient are evaluated according to formula (7.7.5) where the dependence of the diffusion coefficients of ions are connected with the resonant charge exchange process and their values are determined by formula (7.7.7). Values of the effective temperature of electrons T_{ef} are taken from Table 7.3 for a helium gas discharge plasma and from Table 7.4 for an argon gas discharge plasma. The latter means that these data relate to the regime of a low electron number density.

7.8 Heat Processes in Gas Discharge Plasma

Heat processes in a gas discharge plasma accompany the processes of energy transfer from an external electric field to a gas through electrons. As a result of heat processes, a gas is heated, and the gas temperature at the axis of a discharge tube is higher than that near its walls. Therefore the reduced electric field E/N_a is higher at the axis, and electric properties of a plasma inside a cylinder gas discharge tube at axis differs from those near the walls. This requires the analysis of heat processes in a gas discharge plasma. Below we analyze them for a gas discharge plasma in discharges of a restricted power where the temperature at the tube axis is comparable with that at its walls.

Let us consider first a gas discharge plasma in the regime of a high electron number density and determine the rate of relaxation of the electron temperature. For this goal we transform the balance equation for the average electron energy (6.2.9) to the equation for the electron temperature $T_e = 2\overline{\varepsilon}/3$

$$\frac{3dT_e}{2dt} = eEw_e - \frac{m_e^2}{M} \cdot \left(1 - \frac{T}{T_e}\right)\left\langle v^2 \nu_{ea}\right\rangle, \qquad (7.8.1)$$

where ν_{ea} is the rate of electron-atom collisions. In particular, in the case $\nu_{ea} = \nu = const$ the balance Equation (7.8.1) takes the form

$$\frac{dT_e}{dt} = -\nu_\varepsilon (T_e - T_o),$$

where T_o is the equilibrium temperature for a given electric field strength, T_e is a current electron temperature, ν_ε is the rate of relaxation of the electron temperature that is connected with the rate of electron-atom collisions by the relation

$$\nu_\varepsilon = 2\frac{m_e}{M}\nu,$$

From this it follows that relaxation of the electron temperature proceeds longer in $M/(2m_e)$ times than relaxation of the electron momentum. For helium the ratio of these relaxation times is 4×10^3, and in the argon case this ratio is $\sim 4 \times 10^4$.

The heat flux in a gas discharge plasma is a sum of fluxes due to transport of atoms and electrons because these fluxes are independently. We now determine the partial thermal conductivity of an ionized gas due to electron transport in the regime of a high number density of electrons. This flux occurs due to a weak gradient ∇T_e of the electron temperature. Let us introduce the coefficient of electron thermal conductivity κ_e according to the heat transport equation

$$\mathbf{q}_e = -\kappa_e \nabla T_e, \tag{7.8.2}$$

where the electron temperature T_e characterizes the Maxwell distribution function of electrons $\varphi(v)$. The gradient of the electron temperature induces the additive non-symmetric distribution function, and the total velocity distribution function of electrons has the form

$$f(\mathbf{v}) = \varphi(v) + (\mathbf{v}\nabla \ln T_e) f_1(v)$$

Using the stationary kinetic equation for electrons $\mathbf{v}\nabla f = I_{ea}(f)$ that takes into account a change of the electron momentum as a result of their collisions with atoms, together with the gas state equation for the electron pressure $p_e = N_e T_e$, we have from the kinetic equation for the electron distribution function

$$\varphi(v)\left(\frac{m_e v^2}{2T_e} - \frac{5}{2}\right)\mathbf{v}\nabla T_e = I_{ea}(f)$$

From this we have on the basis of the expression (5.3.4) for the collision integral for the antisymmetric part of the electron distribution function

$$f_1(v) = -\frac{\varphi(v)}{\nu}\left(\frac{m_e v^2}{2T_e} - \frac{5}{2}\right)$$

From this we find the heat flux due to electron transport

$$(\mathbf{q}_e)_x = \int \frac{m_e v^2}{2} v_x f(\mathbf{v}) d\mathbf{v} = \int \frac{m_e v^2}{2} v_x^2 \nabla \ln T_e f_1(v) d\mathbf{v}$$

From this and the definition (7.8.2) for the electron thermal conductivity coefficient, we have

$$\kappa_e = N_e \left\langle \frac{v^2}{3\nu_{ea}} \frac{m_e v^2}{2T_e} \left(\frac{m_e v^2}{2T_e} - \frac{5}{2}\right)\right\rangle, \tag{7.8.3}$$

where an average is made over the Maxwell distribution function of electrons, and the coefficient of the electron thermal conductivity is proportional to the electron concentration in a gas.

We note that because the electron thermal conductivity κ_e in a gas discharge plasma results from electron-atom collisions, its value is connected with the dependence of the rate of electron-atom collisions $\nu_{ea}(v)$ depends on the electron velocity v that we take below in the form $\nu_{ea} \sim v^k$. Using this dependence, we determine the coefficient of the electron thermal conductivity in κ_e a gas discharge plasma. Formula (7.8.3) with accounting for an indicated approximation for the electron-atom collisions in the form $\nu_{ea}(v) = \nu_o z^{k/2}$, where $z = m_e v^2/(2T_e)$, gives

$$\kappa_e = \frac{4}{3\sqrt{\pi}} \cdot \frac{T_e N_e}{\nu_o m_e} \left(1 - \frac{k}{2}\right) \Gamma\left(\frac{7-k}{2}\right)$$

In the case $\nu_{ea} = const$ this gives

$$\kappa_e = \frac{5 T_e N_e}{2 \nu_o m_e},$$

and in the case $k = 1$ it follows from formula (7.8.3)

$$\kappa_e = \frac{2}{3\sqrt{\pi}} N_e \lambda \sqrt{\frac{2T_e}{m_e}}, \tag{7.8.4}$$

where the relation $\nu_{ea} = v/\lambda$ is used with the mean free path $\lambda = 1/(N_a \sigma_{ea})$ of electrons in a gas.

7.9 Plasma Transport in Magnetic Field

Magnetic fields influence on the behavior of charged particles of a gas discharge plasma, and we below consider properties of this plasma located in a magnetic field. A magnetic field compels a free charged particle to move by circle trajectories with the Larmor frequency ω_H that is given by

$$\omega_H = \frac{eH}{mc} \tag{7.9.1}$$

Here H is the magnetic field strength, m is the particle mass, c is the light speed, e is the electron charge that assumes to be the particle charge. Electrons of a gas discharge plasma are magnetized, if the Larmor frequency for electrons

$$\omega_H = \frac{eH}{m_e c}, \tag{7.9.2}$$

is large compared to the rate of collision involving electrons. We assume this criterion to be not valid for atomic ions, i.e. a magnetic field acts only the electron component of a gas discharge plasma.

Let us consider the behavior of electrons in crossed electric and magnetic fields. Then in the regime of low density of electrons the kinetic (5.1.1) takes the form

$$\left(e\mathbf{E} + \frac{e}{c}[\mathbf{vH}]\right)\frac{\partial f}{\partial \mathbf{v}} = I_{ea}(f)$$

Take the direction of the electric field $\mathbf{E}$ along the axis x, and take the direction of the magnetic field $\mathbf{H}$ along the axis z. Then by analogy with formula (6.1.3) the distribution function of electrons has the form

$$f(\mathbf{v}) = f_0(v) + v_x f_1(v) + v_y f_2(v),$$

and the collision integrals from the symmetric and asymmetric distribution functions are given by expressions (5.3.4) and (5.3.9). Substituting this in the kinetic equation, we find by analogy with (6.1.4) the following connection between partial distribution functions

$$vf_1 = \frac{a\nu}{(\nu^2 + \omega_H^2)}\frac{df_0}{dv}, \quad vf_2 = \frac{a\omega_H}{(\nu^2 + \omega_H^2)}\frac{df_0}{dv},$$

where $a = eE/m_e$, ω_H is the Larmor frequency (7.9.2) for electrons, and $\nu = N_a v\sigma^*_{ea}$ is the rate of electron-atom elastic collisions. This gives for the electron drift velocity

$$w_x = \frac{eE}{3m_e}\left\langle\frac{1}{v^2}\frac{d}{dv}\left(\frac{\nu v^3}{\nu^2 + \omega_H^2}\right)\right\rangle, \quad w_y = \frac{eE}{3m_e}\left\langle\frac{1}{v^2}\frac{d}{dv}\left(\frac{\omega_H v^3}{\nu^2 + \omega_H^2}\right)\right\rangle, \tag{7.9.3}$$

where the average is made on the basis of the spherical distribution function f_0 of electrons. In the limit of a small magnetic field $\omega_H \ll \nu$ the first formula is converted into (6.1.7), and in the limit of a rareness gas

$$\omega_H \gg \nu \tag{7.9.4}$$

this plasma is magnetized. In this limiting case the electrons move in the direction that is perpendicular both to the electric and to the magnetic field, and the electron drift velocity is

$$w_y = \frac{eE}{m_e\omega_H} = c\frac{E}{H} \tag{7.9.5}$$

The drift velocity in the direction of the electric field is equal in the limit (7.9.4) of magnetized electrons

$$w_x = \frac{eE}{3m_e\omega_H^2}\left\langle\frac{d(v^3\nu)}{v^2 dv}\right\rangle \tag{7.9.6}$$

In particular, if the cross section of electron-atom collision σ_{ea}^* is independent of the collision velocity v, formula (7.9.6) gives

$$w_x = \frac{4eE\overline{v}}{3m_e\lambda\omega_H^2}, \tag{7.9.7}$$

where $\overline{v}$ is the average electron velocity, $\lambda = 1/(N_a\sigma_{ea}^*)$ is the mean free path of electrons in a gas. From this it follows for the plasma conductivity (7.4.2) if it is located in a strong magnetic field

$$\Sigma = \frac{eN_e w_e}{E} = \frac{4N_e e^2\overline{v}}{3m_e\lambda\omega_H^2}, \tag{7.9.8}$$

In the regime of a high electron number density with the Maxwell distribution function of electrons, we determine the electron temperature T_e from the energy balance equation

$$eEw_x = \int \frac{m_e v^2}{2} I_{ea}(f_0)d\mathbf{v}$$

This equation takes into account that electrons obtain energy from the electric field and transfer it to atoms in electron-atom elastic collisions. This equation with using formula (7.9.3) for the electron drift velocity and formula (5.3.9) for the integral of electron-atom collisions gives for the difference of electron T_e and gaseous T temperatures by analogy with expression (6.2.7)

$$T_e - T = \frac{Ma^2}{3}\frac{\langle\frac{v^2\nu}{\nu^2+\omega_H^2}\rangle}{\langle v^2\nu\rangle} \tag{7.9.9}$$

In the limiting case of a strong magnetic field $\omega_H \gg \nu$ this formula takes a simple form

$$T_e - T = \frac{Ma^2}{3\omega_H^2} = \frac{Mc^2E^2}{3H^2} = \frac{M}{3}w_y^2, \quad \omega_H \gg \nu \tag{7.9.10}$$

We consider the coefficient of the electron transverse diffusion in a strong magnetic field $\omega_H \gg \nu$ in a gas discharge plasma in the absence of an electric field (or an electric field is directed along the magnetic field). Then the trajectory of a test electron consists of many circles whose centers are displaced after each electron-atom collision. Then according to definition of the diffusion coefficient we have

$$D_\perp = \frac{\langle x^2\rangle}{t},$$

where $\langle x^2 \rangle$ is the square of the electron displacement for time t in the direction x that is perpendicular to the magnetic field. Because

$$x - x_o = r_H \cos \omega_H t$$

where x_o is the projection on the axis x for a current circle of electron rotation, $r_H = v_\rho / \omega_H$ is the Larmor radius, and v_ρ is the electron velocity component in the direction perpendicular to the magnetic field. We have for the square of the electron trajectory

$$\langle x^2 \rangle = n \langle (x - x_o)^2 \rangle = \frac{n v_\rho^2}{2\omega_H^2},$$

where n is a number of collision during a time $t = n/\nu$. This gives

$$D_\perp = \left\langle \frac{v_\rho^2 \nu}{2\omega_H^2} \right\rangle = \left\langle \frac{v^2 \nu}{3\omega_H^2} \right\rangle, \quad \omega_H \gg \nu, \tag{7.9.11}$$

where an average is made over electron velocities. This formula for the electron diffusion coefficient may be combined with that (7.2.2) in the absence of the magnetic field. This gives

$$D_\perp = \frac{1}{3} \left\langle \frac{v^2 \nu}{\omega_H^2 + \nu^2} \right\rangle \tag{7.9.12}$$

Part III
Processes in Gas Discharge

Chapter 8
Ionization Equilibrium in Gas Discharge Plasma

Abstract Various forms of ionization equilibrium in a plasma are considered for atom ionization in single collisions with electrons, stepwise atom ionization and various regimes of loss of electrons and ions.

8.1 Townsend Scheme for Self-maintaining of Gas Discharge

A gas discharge plasma is a weakly ionized gas in an external field that occupies a volume restricted by a discharge construction, and this plasma is supported by processes in it. As a result of leaving of charged particles for electrodes and walls, this plasma perishes, and within the framework of the Townsend scheme plasma reproducing proceeds due to processes of atom ionization by electron impact. Self-maintaining of gas discharge is its principal property, and now the rate of this process is characterized by the first Townsend coefficient α that is expressed through the ionization cross section σ_{ion} of an atom by electron impact with the formula

$$\alpha(E) = N_a \frac{\langle v\sigma_{ion}\rangle}{w_e}, \tag{8.1.1}$$

where N_a is the number density of atoms, v is the electron velocity, w_e is the electron drift velocity, and an average is made over the energy distribution of electrons.

In considering a gas discharge plasma, we will guided mostly by glow discharge where gas heating under the action of the electric discharge current does not act significantly on the gas properties. Figure 2.6 represents basic regions of gas discharge that is supported in a cylinder tube. In addition to this, Fig. 8.1 gives the space distribution for the electric field strength and voltage along the discharge tube. Note that our consideration relates to gas discharge of a not low pressure, so that the mean free path of atomic particles in the gas is relatively small. The cathode region where a gas discharge plasma is created and the positive column where the ionization equilibrium is supported by ionization of atoms by electron impact and plasma attachment to the tube walls, are of principle for this plasma. The dark region is an intermediate region between the cathode region and positive column, and fast electrons penetrate there from the cathode region and these electrons are responsible for ionization in this

B.M. Smirnov, *Theory of Gas Discharge Plasma*, Springer Series on Atomic, Optical, and Plasma Physics 84, DOI 10.1007/978-3-319-11065-3_8

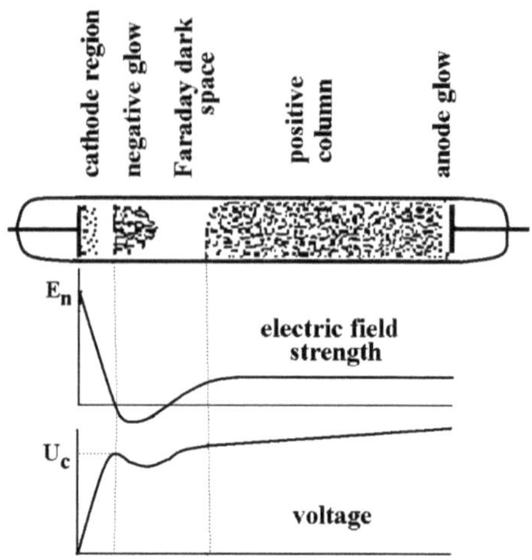

Fig. 8.1 The character of plasma parameters in glow gas discharge

region. The attachment of ions to the anode increases slightly the electric field near the anode, but its contribution to the discharge voltage is not of principle. Therefore we focus below on two principal regions of gas discharge, the cathode region and the positive column. Note that if we increase a length of the discharge tube, an additional tube length will be occupied by the positive column.

Let us derive the condition of self-maintaining of discharge within the framework of the Townsend scheme [2, 3, 5]. Along with ionization in a volume in electron-atom collisions, this scheme uses the process of formation of secondary electrons on the cathode as a result of cathode bombardment by ions accelerated in an electric field. This process is characterized by the second Townsend coefficient γ that is the probability of formation of a secondary electron per ion contacted with the cathode. The balance equation for the electron number density N_e near the cathode and its solution has the form

$$\frac{dN_e}{dz} = \alpha N_e; \ N_e = N_o \exp\left(\int \alpha dz\right), \tag{8.1.2}$$

where z is the distance from the cathode, and N_o is the electron number density at the cathode. Let us consider a gas discharge plasma between two plane electrodes with a distance L between them. Then the condition of self-maintaining of this plasma has the form

$$\int_0^L \alpha dz = \ln(1 + 1/\gamma) \tag{8.1.3}$$

This Townsend scheme of ionization balance was introduced a century ago and conserves its actuality now.

In the Townsend scheme of ionization balance we use the first Townsend coefficient as the single parameter to describe the atom ionization by electron impact. This is simplification of a real character for ionization of gas atoms in an electric field because it requires direct atom ionization in pair collision with an electron. In reality other ionization channels exist, in particular, ionization processes involving excited atoms. Location of excited atoms in a gas discharge plasma acts on the ionization balance twofold. First, excitation of atoms leads to a loss of fast electrons and causes a decrease of the ionization rate in a gas discharge plasma. Second, the ionization channel involving excited atoms may give a contribution to the total ionization rate under certain conditions. According to this example, the Townsend scheme requires to use caution. Therefore, based on this scheme we will use only experimental values of the first Townsend coefficient assuming the conditions analyzed to be identical to the experimental conditions.

If we assume a pair character of atom ionization and assume that the rate of the ionization process (and α) do not depend on the electron number density, a general form of the first Townsend coefficient as a function of the atom number density is

$$\frac{\alpha}{N_a} = F\left(\frac{E}{N_a}\right), \tag{8.1.4}$$

Figure 8.2 gives experimental values of the reduced first Townsend coefficient α/N_a [43, 341–343] as a function of the reduced electric field strength eE/N_a, and Fig. 8.3 represents these values for argon [43, 137, 331, 344–346]. Though the used measurements relate to a half-century period, the data of different time are in accordance. It is convenient to use a simple approximation [4, 12] for the function $F(x)$ of formula (8.1.4) that is

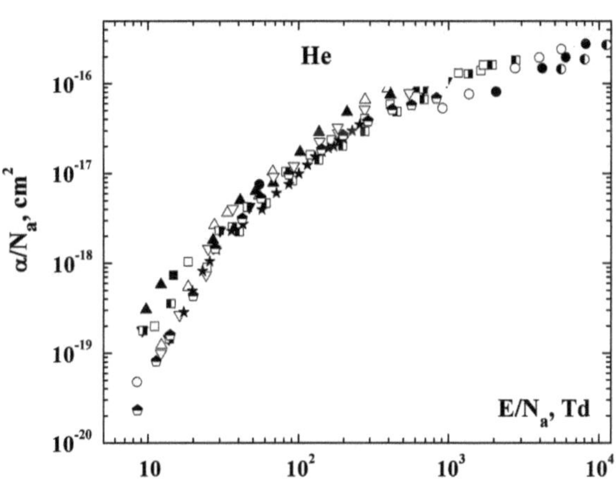

Fig. 8.2 The reduced first Townsend coefficient for helium as a function of the reduced electric field strength [43, 341–343]

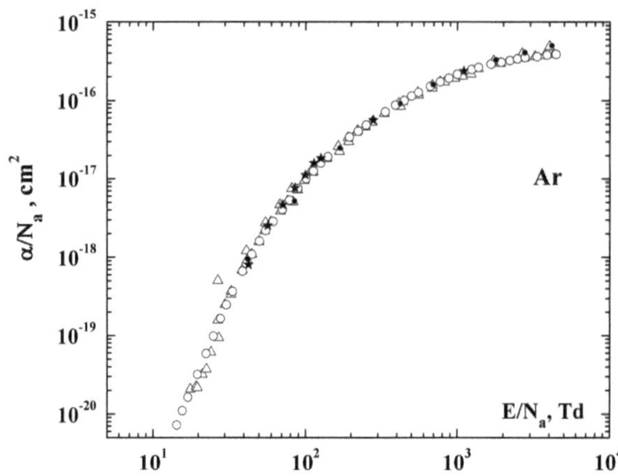

Fig. 8.3 The reduced first Townsend coefficient for argon as a function of the reduced electric field strength [43, 137, 331, 344–346]

$$F(x) = A \exp\left(-\frac{B}{x}\right), \qquad (8.1.5)$$

It is suggested [4, 12] to use the parameters of this formula for helium $A = 0.85 \times 10^{-16}\,\text{cm}^2$, $B = 96\,\text{Td}$ in a range of reduced electric field strengths $E/N_a = 60 - 400\,\text{Td}$, and for argon these parameters are $A = 4.0 \times 10^{-16}\,\text{cm}^2$, $B = 500\,\text{Td}$ in a range $E/N_a = 300 - 1{,}700\,\text{Td}$. But this approximation is not well in the total indicated range. In order to convince in the validity of the approximation (8.1.5), it is convenient to give these experimental dependencies of the reduced first Townsend coefficient α/N_a on the reciprocal electric field strength N_a/E as it is represented in Figs. 8.4 and 8.5 in the helium and argon cases. Then in logarithmic axes these dependence are expressed by straightforward lines. From Figs. 8.4 and 8.5 it follows that it is convenient to divide the range of E/N_a into some intervals where these dependencies may be approximated by straightforward lines in Figs. 8.4 and Fig. 8.5, and then the parameters of formula (8.1.5) relate to a corresponding range of electric field strengths. In particular, in a range of high fields $10^4 > E/N_a > 10^3\,\text{Td}$ we have for the parameters of formula (8.1.5) $A = 2.5 \times 10^{-16}\,\text{cm}^2$, $B = 1{,}100\,\text{Td}$ for helium and $A = 9.0 \times 10^{-16}\,\text{cm}^2$, $B = 1{,}540\,\text{Td}$ for argon. This approximation is represented in Figs. 8.4 and Fig. 8.5 that shows a restricted accuracy of this approximation.

At lower electric field strengths it is convenient to approximate $F(x)$ by a power dependence on the reduced electric field strength taking it in the form

$$\frac{\alpha}{N_a} = a \cdot (E/N_a)^k \qquad (8.1.6)$$

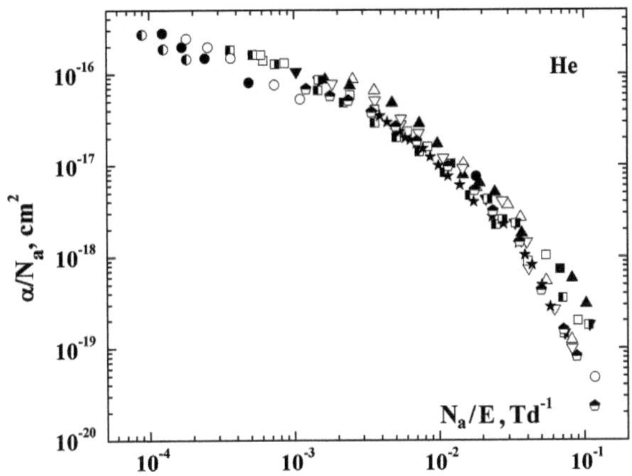

Fig. 8.4 Constructed on the basis of Fig. 8.2, the reduced first Townsend coefficient for helium as a function of a reciprocal reduced electric field strength

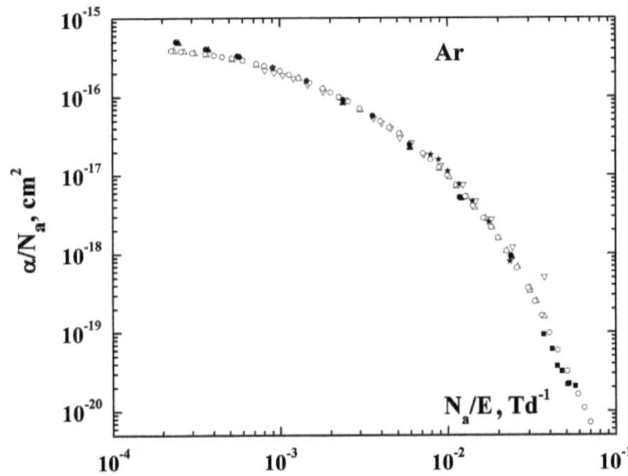

Fig. 8.5 Constructed on the basis of Fig. 8.3 data, the reduced first Townsend coefficient for argon as a function of a reciprocal reduced electric field strength

The parameters of this formula in the argon case are $a = 9.2\times 10^{-25}\,\text{cm}^2, k = 3.67$ in the range $E/N_a \leq 50\,\text{Td}, a = 8.6\times 10^{-23}\,\text{cm}^2, k = 2.54$ in the range $50 < E/N_a \leq 100\,\text{Td}$, and $a = 1.2\times 10^{-19}\,\text{cm}^2, k = 1.09$ in the range $100 < E/N_a \leq 1{,}000\,\text{Td}$.

We also give in Table 8.1 the values of the first Townsend coefficient α and the rate constant of atom ionization k_{ion} that according to the relation (8.1.1) is determined by the formula

$$k_{ion} = \frac{\alpha w_e}{N_a} \qquad (8.1.7)$$

Table 8.1 Ionization parameters of helium and argon atoms by an electron moved in the gas in an electric field

E/N_a (Td)	$\alpha(He)/N_a$ (cm^2)	$k_{ion}(He)$ (cm^3/s)	$\alpha(Ar)/N_a$ (cm^2)	$k_{ion}(Ar)$ (cm^3/s)
10	4×10^{-20}	1×10^{-13}	1×10^{-21}	1×10^{-15}
30	2×10^{-18}	4×10^{-11}	2×10^{-19}	5×10^{-13}
100	1×10^{-17}	2×10^{-10}	1×10^{-17}	8×10^{-11}
300	5×10^{-17}	4×10^{-9}	6×10^{-17}	1×10^{-9}
1,000	9×10^{-17}	3×10^{-8}	2×10^{-16}	1×10^{-8}
3,000	4×10^{-16}	2×10^{-7}	2×10^{-16}	3×10^{-8}

E/N_a is the reduced electric field strength in Td (1Td= 10^{-17} V · cm^2), α is the first Townsend coefficient, k_{ion} is the atom ionization rate constant by electron impact according to formula (8.1.7)

Table 8.2 Parameters of the approximation (8.1.5) of the first Townsend coefficient

Gas	Range E/N_a (Td)	A 10^{-17} (cm^2)	B (Td)
He	100–300	8.7	200
He	300–1,000	14	340
He	1,000–3,000	22	850
Ar	100–300	1.3	230
Ar	300–1,000	3.4	520
Ar	1,000–3,000	9.1	1,600

The data of Table 8.1 use a sum of measurements according to review [43] and are based on average values of the first Townsend coefficient and electron drift velocities in helium and argon given in Figs. 5.1 and 5.2. In addition, the approximation (8.1.5) of experimental data for the first Townsend coefficient is valid in a not wide range of electric field strengths. Table 8.2 contains parameters of this approximation in an indicated range of the reduced electric field strengths.

We note one more peculiarity of the first Townsend coefficient α. According to its definition (8.1.2) of the first Townsend coefficient, an electron is moving along the electric field, so that a random diffusion motion of electrons is ignored. Because due to drift electron motion its removal from an initial electron position through time t is $w_e t$ and due to diffusion electron motion this removal is $\sim\sqrt{D_e t}$, the drift electron motion dominates through times

$$t \gg \frac{D_e}{w_e^2}$$

This criterion must be valid for a time of electron doubling $t \sim (w_e\alpha)^{-1}$, so that the criterion of the Townsend mechanism of ionization has the form

$$\alpha \ll \frac{w_e}{D_e}$$

Introducing the electron mobility $K_e = w_e/E$ and replacing above the longitudinal diffusion coefficient D_e for electrons in a gas by the transverse diffusion coefficient, we use the characteristic or the Townsend temperature $T_{ef} = eD_e/K_e$ in accordance with formula (7.2.4). Then the criterion of the drift character of electron propagation in a gas in an external electric field is

$$\xi = \frac{\alpha T_{ef}}{eE} \ll 1 \tag{8.1.8}$$

This criterion holds true usually.

Based on the above experimental data for $\alpha(E/N_a)/N_a$, we assume experimental conditions to be analogous to those under consideration. But at low reduced electric field strengths this quantity is sensitive to other ionization channels. In particular, a low concentration of excited atoms in a gas discharge plasma may change remarkable the value of α. Therefore experimental values of the first Townsend coefficient can be used at not very small values of α.

It should be noted that the Townsend concept [42] with using the first Townsend coefficient α as the characteristic of gas ionization is analogous to ionization of atoms or molecules in collisions with electrons in gases. Along with the first Townsend coefficient α, one can introduced the effective cross section of atom excitation σ_{ex}, so that the quantity $1/\beta$ is the electron mean free path in a gas with respect to excitation of gas atoms. One can determine this value at high electric field strengths where the electron distribution function is given by formula (6.6.4). In this case the energy acquired by an electron from an external electric field is consumed on excitation of gas atoms. The power P transmitted by one electron is eEw_e, where E is the electric field strength, w_e is the electron drift velocity. This gives for the excitation rate constant by electron impact in this regime according to formula (6.6.17)

$$k_{ex} = \frac{eEw_e}{\Delta\varepsilon N_a} = k_o \left(\frac{E}{N_a}\right)^2,$$

and the values of k_o are given by formula (6.6.18) and in Table 8.2. In this limit the effective excitation cross section σ_{ex} is equal

$$\sigma_{ex} = \frac{k_{ex}}{w_e} = \frac{eE}{\Delta\varepsilon N_a} = \sigma_o \cdot \frac{E}{N_a} \tag{8.1.9}$$

The effective cross section of atom excitation σ_{ex} in the lowest excited state is compared with the reduced first Townsend coefficient α/N_a in Fig. 8.6 for helium and in Fig. 8.7 for argon in the range of high electric field strengths. In addition, Table 8.2 contains values of the parameters k_o and σ_o for helium and argon.

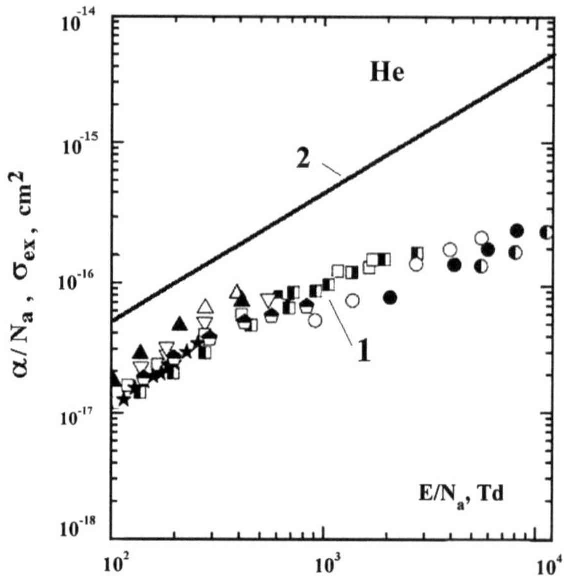

Fig. 8.6 Reduced first Townsend coefficient α/N_a *(1)* in helium constructed on the basis of experimental data of Fig. 8.2 and the effective cross section of atom excitation σ_{ex} *(2)* in the limit of high electric field strengths

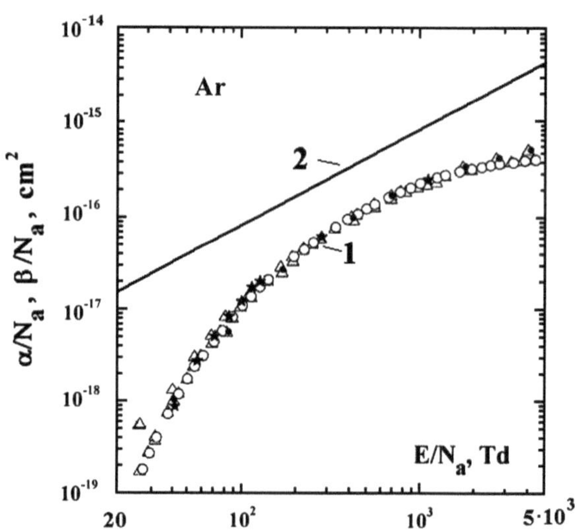

Fig. 8.7 Reduced first Townsend coefficient α/N_a *(1)* in argon constructed on the basis of experimental data of Fig. 8.3 and the effective cross section of atom excitation σ_{ex} *(2)* in the limit of high electric field strengths

8.2 Ionization Equilibrium in Positive Column of Gas Discharge

In addition to the regime of self-maintaining ionization along a tube with a gas where electric current of gas discharge propagates, a self-maintaining regime is realized in a transverse direction of a long gas discharge tube. We below consider such a self-consistent regime in a long cylinder tube where an ionization equilibrium is supported inside it due to ionization processes in a gas under the action of an electric field of gas discharge (Table 8.3).

Thus, we analyze the ionization equilibrium in a long tube with a gas and a longitudinal electric field, so that electrons and ions are formed in pair electron-atom collisions and perish as a result of attachment to the tube walls. The gas distribution over the tube cross section assumes to be uniform, and the mean free path of electrons is small compared to a tube radius. Such conditions are typical for glow discharge, and this regime of ionization equilibrium is called the Schottky regime [347, 348]. This regime relate to a small electron number density, where electrons do not influence on the distribution of atomic particles in the gas discharge tube.

The equation of ionization balance under the above conditions has the following form for the electron number density N_e

$$D_a \Delta N_e + k_{ion} N_e N_a = 0, \tag{8.2.1}$$

where D_a is the coefficient of ambipolar diffusion, N_a is the number density of gas atoms, k_{ion} is the rate constant of atom ionization in collisions with electrons. We use the cylinder symmetry of the electron distribution if a gas discharge plasma is located in a cylinder tube of a radius ρ_o. Then (8.2.1) has the form

$$\frac{D_a}{\rho}\frac{d}{d\rho}\left(\rho\frac{dN_e}{d\rho}\right) + k_{ion} N_e N_a = 0, \tag{8.2.2}$$

Solution of this equation with the boundary condition $N_e(\rho_o) = 0$ is given by

$$N_e(\rho) = N_o J_0\left(\frac{2.4\rho}{\rho_o}\right), \tag{8.2.3}$$

where J_0 is the Bessel function, and N_o is the electron number density at the tube axis. This dependence is called the Schottky distribution. The boundary condition $N_e(\rho_o) = 0$ leads to the following relation between parameters of processes

Table 8.3 Parameters of excitation of helium and argon atoms by electrons in the limit of low electron number densities and high electric field strengths

Parameter	He	Ar
k_o, $10^{-14}\,\mathrm{cm^3/(Td^2 s)}$	6.2	13
σ_o, $10^{-19}\,\mathrm{cm^2/Td}$	5.1	8.5

$$N_a \frac{k_{ion}\rho_o^2}{D_a} = 5.78 \qquad (8.2.4)$$

As a matter, this relation accounts for the balance between the process of destruction of charged particles as a result of ambipolar diffusion and the process of their formation in electron-atom ionization collisions. In particular, this balance equation gives for the flux of electrons and ions to walls

$$j_e = -D_a \frac{dN_e}{d\rho}|_{\rho=\rho_o} = \frac{1.25 D_a N_o}{\rho_o}, \qquad (8.2.5)$$

where N_o is the electron number density at the axis.

Let us consider the limit of low electron number densities for the Schottky regime of ionization equilibrium, where the ionization balance results from formation of electrons and ions in collisions with electrons and loss of electrons and ions owing to attachment to walls of a gas discharge tube. In this case one can neglect by collisions between electrons, and the rate of atom ionization is expressed through the first Townsend coefficient α in accordance with formula (8.1.1). As a result, equation of ionization equilibrium (8.2.4) takes the form

$$\alpha w_e = \frac{5.78 D_a}{\rho_o^2} \qquad (8.2.6)$$

We now estimate the parameter $\alpha\rho_o$ that is the ratio of a tube radius to the mean free path of electrons with respect to ionization. From (8.2.6) of ionization equilibrium it follows

$$(\alpha\rho_o)^2 \sim \alpha \frac{D_a}{w_e}$$

According to its structure, the right hand side of this relation is the ratio of the mean free path of elastic scattering of electrons in a gas to that due to atom ionization. Evidently, this ratio is small, i.e.

$$\alpha\rho_o \ll 1 \qquad (8.2.7)$$

This means that existence of the positive column requires that the length of a discharge tube exceeds significantly its radius.

The distribution of electrons (8.2.3) over the cross section of the discharge tube allows one to determine various parameters of the positive column for the Schottky regime of gas discharge. In particular, we below find the electric potential between the axis and walls of a cylinder discharge tube for this regime. Indeed, as a result of ambipolar diffusion, an electric field occurs that compels ions to move towards the walls. According to formula (7.7.2) this electric field strength E is equal for a dense plasma which is characterized by the electron temperature T_e

$$E = \frac{T_e}{e} \frac{\partial N_e}{N_e \partial\rho},$$

and correspondingly the electric potential ΔU between the axis and walls is

$$\Delta U = \int_0^{\rho_*} \frac{T_e}{e} \frac{\partial N_e}{N_e \partial \rho} d\rho,$$

where $\rho_* \approx \rho_o$, i.e. this point is located near the walls and the diffusion character of plasma motion takes place in this point. Hence $\rho_o - \rho_* \sim \lambda$, where λ is the mean free path of electrons or ions in a gas. This gives

$$\Delta U = \frac{T_e}{e} \ln \frac{N_o}{N_e(\rho_*)},$$

where N_o is the electron number density at the tube axis.In considering the Schottky regime, we have on the basis of formula (8.2.4)

$$\Delta U \approx -\frac{T_e}{e} \ln J_0 \left[2.405(1 - \lambda/\rho_o)\right] \approx \frac{T_e}{e} \ln \frac{8\rho_o}{\lambda} \qquad (8.2.8)$$

As is seen, the energy that is transferred to an ion during its motion from the axis to walls exceeds significantly a typical thermal electron energy.

8.3 Stepwise Ionization of Atoms in Positive Column

We consider above the Schottky regime of ionization equilibrium in the positive column of gas discharge. This case corresponds to low electron number densities at which excited atoms do not partake in the ionization balance. At higher number densities of electrons ionization of metastable or excited atoms may give contribution to ionization processes, and we below consider the ionization regime where the ionization process proceeds through formation of a metastable state, and the ionization process corresponds to the following scheme

$$e + A \to e + A^m, \quad e + A^m \to 2e + A^+ \qquad (8.3.1)$$

Here free electrons result from ionization of metastable atoms by electron impact, and destruction of metastable atoms in this regime proceeds due to travelling of electrons in a space and due to attachment of metastable atoms to the walls. According the scheme (8.3.1), instead of direct ionization of atoms by electron impact that corresponds to the Schottky regime we are based on stepwise ionization through a metastable state, and decay of ions A^+ and excited atoms A^m in this case results from diffusion to walls. The set of balance equations which describe the scheme the ionization scheme (8.3.1) has the following form for the number densities of electrons N_e, metastable atoms N_m and atoms N_a in the ground state

$$4D_a\frac{d}{dx}\left(x\frac{dN_e}{dx}\right)+k_{ion}^m N_e N_m \rho_o^2=0,\ 4D_m\frac{d}{dx}\left(x\frac{dN_m}{dx}\right)+k_{ex}N_e N_a \rho_o^2=0 \tag{8.3.2}$$

Here we use the reduced distance from the axis $x = \rho^2/\rho_o^2$, where ρ is a distance from the axis, and ρ_o is a radius of the discharge tube, and the reduced variable x varies from 0 to 1. In addition, we denote above the rate constant for atom excitation in the metastable state by electron impact as k_{ex}, the rate constant for ionization of metastable atoms as k_{ion}^m, the diffusion coefficient for metastable atoms as D_m, and the ambipolar diffusion coefficient for a gas discharge plasma as D_a. The boundary conditions $N_e(\rho_o) = N_m(\rho_o) = 0$ must be added to these equations.

By analogy with the Schottky regime, where we find the parameters of the ionization balance as the eigen values of (8.2.2), we now make the same operation for the set of (8.3.2). In order to simplify this operation, we use the following space distributions for the number densities of electrons and metastable atoms

$$N_e = N_e(0)(e^{-ax} - e^{-a}),\quad N_m = N_m(0)(e^{-bx} - e^{-b}) \tag{8.3.3}$$

Determination of the parameters a and b for these expressions will allow us to find the eigen values of the equation set (8.3.2). In order to determine the accuracy of this operation, we apply it first to the balance (8.2.2) of the Schottky regime. Let us set the above expression for the electron number density N_e into (8.2.2) and take it at $x = 0$. In addition, integrate this equation over the tube cross section. On the basis of these operations we obtain $a = 0.842$ and

$$\frac{N_a k_{ion}\rho_o^2}{D_a} = \frac{4a}{1-e^{-a}} = 5.9$$

instead of the numerical parameter 5.78 for the precise solution of (8.2.2) according to (8.2.4). The difference of these values characterizes the accuracy of the used operation.

We now return to the set (8.3.2) of equations apply to that the above operations. As a result, we obtain

$$a = 1.435,\ b = 0.988,\ k_{ion}^m N_m(0) = \frac{7.55D_a}{\rho_o^2},\ \ N_a N_e(0)k_{ex} = \frac{6.28D_m N_m(0)}{\rho_o^2} \tag{8.3.4}$$

These parameters characterize the distribution of electrons and ions, as well as metastable atoms over the cross section of a discharge tube for this ionization mechanism through metastable atoms. One can exclude the number density of metastable atoms from two last relations, and this gives

$$k_{ex}k_{ion}^m c_e(0) = \frac{7.55D_a N_a}{\rho_o^2}\cdot\frac{6.28D_m N_a}{\rho_o^2}, \tag{8.3.5}$$

where $c_e = N_e/N_a$ is the electron concentration. This equation of ionization balance describes the voltage-current characteristic of the positive column of gas discharge for the mechanism (8.3.1) for ionization equilibrium where destruction of electrons and metastable atoms is determined by diffusion of these particles toward walls and subsequent their attachment to the walls.

We now consider another regime of ionization equilibrium where destruction of metastable atoms within the framework of the scheme (8.3.1) results in collisions with electrons rather than as a result of excitation transfer to the walls. We than obtain the following set of balance equations instead of the set (8.3.2)

$$4D_a\frac{d}{dx}\left(x\frac{dN_e}{dx}\right) + k^m_{ion}N_eN_m\rho_o^2 = 0\,, \quad k_QN_mN_e = k_{ex}N_eN_a, \qquad (8.3.6)$$

where k_Q is the rate constant of metastable atom destruction by electron impact that includes quenching of the metastable atom with its transition in the ground and other excited states and also its ionization. As early, we ignore gas heating due to discharge currents, so that $N_a = const$ over the tube cross section. Then the second equation of the set (8.3.3) gives $N_m = const$ over the tube cross section except a wall boundary layer, and the first equation of the set (8.3.6) is analogous to the balance (8.2.2). Therefore using formula (8.2.4), we obtain now

$$N_a\frac{k^m_{ion}k_{ex}}{k_Q} = 5.78\frac{D_a}{\rho_o^2} \qquad (8.3.7)$$

One can see that this regime of ionization equilibrium dominates if $N_aN_ek_Q \gg 6D_mN_m/\rho_o^2$ (we replace the factor 5.78 by 6), or

$$(N_a\rho_o)^2\frac{c_e}{c_m} \gg \frac{6D_mN_a}{k_Q} \qquad (8.3.8)$$

From this it follows that the diffusion mechanism of destruction of metastable atoms takes place at a low reduced number density of atoms or radius of a discharge tube.

8.4 Plasma of Positive Column of Low Pressure Gas Discharge

In the positive column of gas discharge of low pressure the mean free path of charged particles exceeds a size of a gas discharge chamber. Because plasma quasineutrality is conserved except a narrow region near walls, a specific distribution of the plasma potential is established that equalizes fluxes of electrons and ions on walls. This equilibrium determines the regime of ionization equilibrium in this plasma [89, 349] that is considered below on the basis of the Firsov method [350]. This method allows one to obtain the expression for the ion current density to walls for the Boltzmann space distribution of electrons and ions in a plasma irrespectively to the rate constant of atom ionization, though the ionization character is used there.

We first consider the geometry of this positive column when a gas discharge plasma is located between two parallel plane electrodes. Taking the direction x perpendicular to electrodes and the origin in the middle between electrodes, we have for the plasma electric potential $\varphi(x)$ because of the symmetry $\varphi(x) = \varphi(-x)$. This potential drops monotonically from the center to walls, and we take for definiteness $\varphi(0) = 0$. Assuming a thermodynamic equilibrium of locked in a gap electrons with the electron temperature T_e, we have for the number density of electrons $N_e(x)$ in a space

$$N_e(x) = N_o \exp\left[\frac{e\varphi(x)}{T_e}\right],$$

where N_o is the number density of electrons in the middle between electrodes. Ignoring collisions of ions with gas atoms, we have that an ion formed in a point ξ has a velocity $v_x = \sqrt{2e[\varphi(\xi) - \varphi(x)]/M}$ in point x. Denoting by $g(\xi)$ a number of ions formed in point ξ per unit time and unit volume, we have for the number density of ions $N_i(x)$ in point x

$$N_i(x) = N_e(x) = N_o \exp\left[\frac{e\varphi(x)}{T_e}\right] = \int_0^x \frac{g(\xi)d\xi}{\sqrt{2e[\varphi(\xi) - \varphi(x)]/M}}, \qquad (8.4.1)$$

Let us introduce the dimensionless potential $\eta(x) = -e\varphi(x)/T_e$ and a typical electron flux $j_o = N_o\sqrt{2T_e/M}$. In these variables (8.4.1) takes the form

$$j_0 e^{-\eta} = \int_0^x \frac{g(\xi)d\xi}{\sqrt{\eta(x) - \eta(\xi)}}, \qquad (8.4.2)$$

where $\eta > 0$. In solving (8.4.2), we use the relation

$$\int_0^x \frac{d\eta(x)}{dx} dx \frac{1}{\sqrt{[\eta(y) - \eta(x)][\eta(x) - \eta(\xi)]}} = \int_{\eta(\xi)}^{\eta(y)} \frac{d\eta}{\sqrt{[\eta(y) - \eta][\eta - \eta(\xi)]}} = \pi$$

Multiplying (8.4.2) by $(d\eta/dx)\,[\eta(y) - \eta(x)]^{-1/2}$ and integrating over dx in the limits between ξ and y, we obtain the ion (electron) flux $j(y)$ at a distance y from the origin with using the above relation

$$j(y) = \int_0^y g(\xi)d\xi = \frac{j_0}{\pi}\int_0^{\eta(y)} \exp(-\eta)\frac{d\eta}{\sqrt{\eta(y) - \eta}} \qquad (8.4.3)$$

Let us represent the ion (electron) flux to the electrode in the form

$$j(\eta) = \frac{j_o}{\pi}\int_0^{\eta} \exp(-\eta')\frac{d\eta'}{\sqrt{\eta-\eta'}} = \frac{2j_o}{\pi}\sqrt{\eta} + \frac{2j_o}{\pi}\int_o^{\eta} \exp(-\eta')\sqrt{\eta-\eta'}d\eta',$$

and this flux increases at moving toward the electrode. The boundary condition at the electrode is $dj/d\eta = 0$ and gives the equation for the reduced electric potential $\eta_o = \eta(0)$ at the electrode

$$\sqrt{\eta_0}\int_0^{\eta_o} \exp(-\eta)\frac{d\eta}{\sqrt{\eta_o-\eta}} = 1 \tag{8.4.4}$$

Solution of this equation gives $\eta_o = 0.855$, that gives for the flux of charged particles on the electrode

$$j = \frac{j_o}{\pi\sqrt{\eta_o}} = 0.344 j_o = 0.344 N_o\sqrt{\frac{2T_e}{M}} \tag{8.4.5}$$

From this it follows the ionization balance where charged particles are formed by atom ionization in collisions with electrons. This balance equation has the form [38, 170]

$$N_o k_{ion}(T_e) N_a \int_0^{x_0} e^{-\eta}dx = 0.344 N_0\sqrt{\frac{2T_e}{M}}, \tag{8.4.6}$$

where N_a is the number density of atoms, k_{ion} is the rate constant of atom ionization by electron impact. This balance equation corresponds to the following character of processes. Ions move to electrodes freely, while the electric plasma potential hampers pass of electrons to electrodes and equalizes the fluxes of electrons and ions. Reproduction of charged particles results from atom ionization by electron impact.

Evidently, the same situation takes place in a cylinder discharge tube, where the electric potential of walls with respect to the tube center is equal to [38]

$$eU(\rho_o) = 1.145 T_e, \tag{8.4.7}$$

where ρ_o is a tube radius. Figure 8.8 gives the distribution of the transverse potential over the tube radius.

Correspondingly, the flux of ions j_i or electrons on walls is equal [38, 350]

$$j_i = 0.272 N_o\sqrt{\frac{2T_e}{M}} \tag{8.4.8}$$

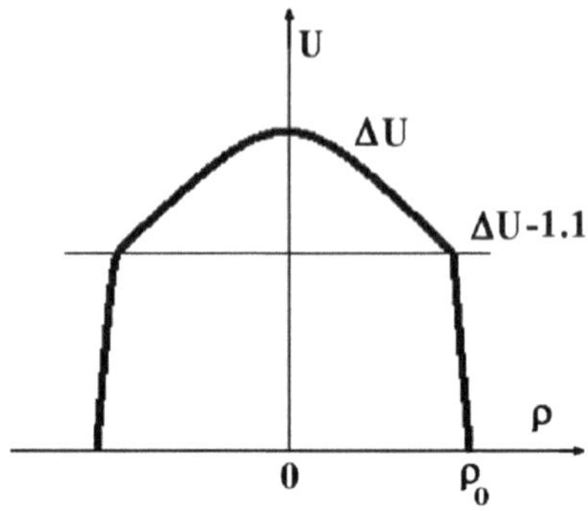

Fig. 8.8 The electric potential distribution over a radius of a *cylinder* discharge tube with a gas discharge plasma of a low pressure

By analogy with (8.4.6), the ionization balance equation for an individual electron at this geometry has the form

$$N_a k_{ion}(T_e) = \frac{0.76}{\rho_o}\sqrt{\frac{2T_e}{M}} \tag{8.4.9}$$

Thus, the nature of this regime is that electrons and ions of a gas discharge plasma are formed due to atom ionization by electron impact and are lost as a result of electron and ion attachment to walls where they recombine. A charge on walls causes repulsion of electrons during their motion towards the walls and accelerate ions that equalizes currents of electrons and ions attached to walls. This regime is realized if the following criterion holds true

$$\lambda_i \gg \rho_o \gg r_D, \tag{8.4.10}$$

so that the regime under consideration is determined by plasma separation over all the tube cross section rather than in a near-wall layer where the wall field acts.

Note that formula (8.4.8) for the ion flux to walls coincides almost with the ion flux to the wall surface if these ions are located in a space which is free from fields. Indeed, if the ion number density in this space is N_o and the ion temperature is T_e, their flux to a surface is

$$j_i = N_o \cdot \frac{1}{4}\sqrt{\frac{8T_e}{\pi M}} = 0.282 N_o \sqrt{\frac{2T_e}{M}}, \tag{8.4.11}$$

that coincides with (8.4.8) within the accuracy of several percent.

8.5 Heat Processes in Positive Column of Gas Discharge

In considering the Schottky regime of the positive column for a low-current discharge (8.2.4), we assume that electrons do not change the atom distribution on the cross section of a discharge tube, and hence the number density of atoms is constant over the cross section. But this influence starts at relatively small currents. Indeed, propagation of a discharge current causes a heat release, that in turn leads to an increase of the gas temperature at the tube axis. Because the gas temperature in the tube center is stronger than that at walls, and the gas pressure is constant over the tube cross section, the number density of atoms N_a at the tube center is lower than in the tube center. The electric field strength E is constant over the cross section, and therefore the reduced electric field strength E/N_a is higher at the center. Because of a strong dependence of the ionization rate on E/N_a, a region of ionization tightens to the axis. In the end, another ionization regime of gas discharge occurs that differs from the Schottky one. As it follows from this, the heat regime of the positive column requires a special analysis.

Let us consider the heat regime of the positive column of gas discharge in a cylinder tube if heat release results from a gas thermal conductivity. The heat balance equation has the form

$$\frac{1}{\rho}\frac{d}{d\rho}\left[\rho\kappa(T)\frac{dT}{d\rho}\right]+p(\rho)=0, \tag{8.5.1}$$

where ρ is a distance from the tube axis, κ is the thermal conductivity coefficient of a gas, $p(\rho)=iE$ is the specific power of heat release, where $i=eN_e w_e$ is the electric current density, E is the electric field strength, w_e is the electron drift velocity, N_e is the electron number density.

In determination the temperature profile $T(\rho)$ over the cross section, it is convenient to introduce new variables $x=\rho^2/\rho_o^2$ and $Z=\int_T^{T_o}\kappa(T)dT$, where T is a current temperature, T_o is the temperature at the center. In these variables (8.5.1) takes the form

$$\frac{d}{dx}\left(x\frac{dZ}{dx}\right)+\frac{p\rho_o^2}{4}=0 \tag{8.5.2}$$

Approximating the specific power as $p(\rho)=p_o(1-x^n)$, we have for solution of this equation

$$Z(x)=\frac{P}{4\pi}\cdot\frac{n+1}{n}\left[x-\frac{x^{n+1}}{(n+1)^2}\right],$$

where P is the heat release power per unit tube length that is given by

$$P=2\pi\rho_o\left[-\kappa\frac{dT(r_o)}{d\rho}\right]=4\pi\frac{dZ(x=1)}{dx}=\pi p_o\rho_o^2\cdot\frac{n}{n+2}$$

Thus, we have

$$Z(1)\equiv\int_{T_w}^{T_o}\kappa(T)dT=\frac{P}{4\pi}\frac{n+2}{n+1}, \tag{8.5.3}$$

where T_w is the wall temperature. In particular, the Schottky regime (8.2.3) corresponds to $n = 0.67$, and then (8.5.3) gives

$$\int_{T_w}^{T_o} \kappa(T)dT = 0.13P$$

Thus, introducing the function

$$g(n) = \frac{Z(1)}{P} = \frac{n+2}{4\pi(n+1)}, \tag{8.5.4}$$

we obtain that this function varies weakly in the range of the parameter n that corresponds to real space distributions of heat release ($n < 0.67$). The dependence of this function on n is given in Fig. 8.9.

One can expect that the space distribution of the electron number density in reality may be sharper only than that in the Schottky limit, i.e. $n < 0.67$. Hence, the ratio $g(n)$ of the thermal transport function $Z(1)$ and the released power P per unit tube length lies in the limits from $g = 0.159$ for the constriction form of the electron distribution and $g = 0.127$ in the Schottky limit. One can average formally this quantity over possible distributions

$$g = 0.143 \pm 0.016 \tag{8.5.5}$$

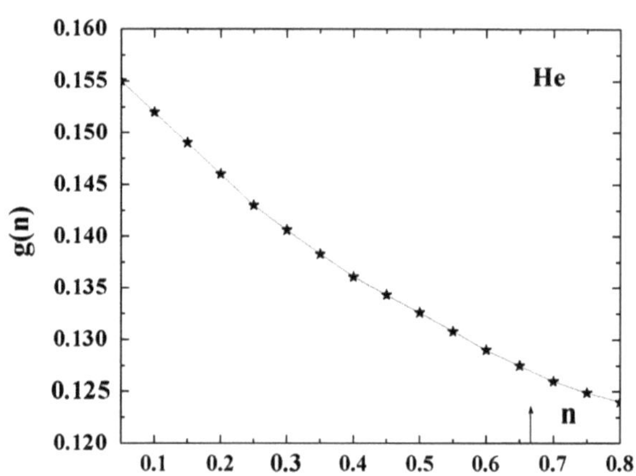

Fig. 8.9 The ratio g according to formula (8.5.4). An *arrow* corresponds to the case where the heat release is proportional to the electron number density in the Schottky case according to formula (8.2.3)

From this we obtain on average

$$\int_{T_w}^{T_o} \kappa(T)dT = 0.14P \tag{8.5.6}$$

with the accuracy of 10 %. In particular, in the limit of low discharge power ($\Delta T = T_o - T_w \ll T_o$) we have for the temperature difference

$$\Delta T = 0.14\frac{P}{\kappa} \tag{8.5.7}$$

If we take the temperature dependence for the coefficient of gas thermal conductivity in the form of a power dependence $\kappa(T) \sim T^{\alpha}$, (8.5.6) takes the form

$$T_o\kappa(T_o) - T_w\kappa(T_w) = 0.13(1+\alpha)P$$

In particular, in the helium and argon cases $\alpha \approx 0.71$ at strong gas heating $T_o \gg T_w$ we have

$$T_o\kappa(T_o) = 0.22EI, \tag{8.5.8}$$

where E is the electric field strength, I is a current strength.

Let us analyze the constructed form of the distribution of the electron number density in a cylinder tube. We introduce a typical radius a occupied by an electric current inside a cylinder discharge tube of a radius ρ_o such that the released specific power P is connected with the released power p_o per unit volume at the tube axis as

$$P = \pi a^2 p_o \tag{8.5.9}$$

Hence

$$\frac{a^2}{\rho_o^2} = \frac{n}{n+2} \tag{8.5.10}$$

and for a constricted discharge we have

$$a = \rho_o\sqrt{\frac{n}{2}}$$

In this limiting case instead of formula (8.5.6) we have

$$\int_{T_w}^{T_o} \kappa(T)dT = 0.16P \tag{8.5.11}$$

In addition, we consider heat balance for a plasma of the positive column of gas discharge located in a cylinder tube if the Schottky regime is realized, so that the electron number density is distributed over the tube cross section according to formula (8.2.3). Then (8.5.2) takes the form

$$\frac{d}{\rho d\rho}\left[\kappa(T)\rho\frac{dT}{d\rho}\right]+p_oJ_0\left(\frac{2.4\rho}{\rho_o}\right)=0,$$

where κ is the gas thermal conductivity coefficient, $p(\rho)=i(\rho)E$ is the power of heat release per unit volume, p_o is the specific discharge power at the tube axis. Using the variables $x=\rho^2/\rho_o^2$ and $Z=\int_T^{T_o}\kappa(T)dT$, where T is a current temperature, T_o is the temperature at the axis, and approximating the Bessel function as $J_0(2.405\sqrt{x})=1-1.38x+0.38x^2$ that is valid within the accuracy of 1 %, we obtain after the first integration of equation of heat transport over the tube cross section

$$\frac{d}{dx}\left(x\frac{dZ}{dx}\right)=\frac{p_o\rho_o^2}{2}(x-0.69x^2+0.13x^3)$$

We here use that the heat flux is zero at the axis. For the power P of heat release per unit tube length that is expressed through the heat flux to the walls

$$P=-2\pi\rho_o\cdot\frac{dT}{d\rho}|_{\rho=\rho_o}=4\pi\frac{dZ}{dx}|_{x=1}$$

This reduces equation of heat transport to the form

$$\frac{dZ}{dx}=0.182P(1-0.69x+0.13x^2)$$

Solution of this equation by analogy with formula (8.5.6) gives

$$\int\limits_{T_w}^{T_o}\kappa(T)dT=0.13P \tag{8.5.12}$$

where T_w, T_o is the gas temperature at walls and axis respectively. This coincides practically with the relation (8.5.7) that corresponds to an average over various distributions of the electron number density over the tube cross section.

Note that the gas temperature $T(\rho)$ as a function from a distance from the tube axis has the form

$$T(\rho)=T_o-\beta\rho^2, \tag{8.5.13}$$

as it follows from the symmetry of a cylinder tube. One can find the parameter β from two considerations. In the first case we solve (8.5.1) near the axis that gives

$$T(\rho) = T_o - \beta_1\rho^2 = T_o - \frac{p_o\rho^2}{4\kappa(T_o)} \tag{8.5.14}$$

In the second case we require relation (8.5.13) to be valid at any distances from the axis that gives

$$T(\rho) = T_o - \beta_2\rho^2 = T_o - \frac{(T_o - T_w)\rho^2}{\rho_o^2}, \tag{8.5.15}$$

where T_w is the wall temperature. We compare these distributions assuming the specific power is proportional to the number density of electrons $N_e(\rho)$ that is given by formula (8.2.3). From this it follows for the total power per unit length

$$P = \int_0^{\rho_o} p(\rho) \cdot 2\pi\rho d\rho = 1.36 p_o\rho_o^2$$

Being guided by helium and argon, we have the temperature dependence for the thermal conductivity coefficient

$$\kappa(T) \sim T^{0.7}$$

From this it follows

$$Z = \int_{T_w}^{T_o} \kappa(T)dT = \frac{[T_o\kappa(T_o) - T_w\kappa(T_w)]}{1.7} = \frac{T_o\kappa(T_o)}{1.7}\left[1 - \left(\frac{T_w}{T_o}\right)^{1.7}\right]$$

Hence on the basis of relation (8.5.12) we obtain

$$\frac{Z}{P} = 0.13 = \int_{T_w}^{T_o} \kappa(T)dT = \frac{[T_o\kappa(T_o) - T_w\kappa(T_w)]}{1.7} = \frac{T_o\kappa(T_o)}{1.7}\left[1 - \left(\frac{T_w}{T_o}\right)^{1.7}\right]$$

This gives

$$\frac{\beta_1}{\beta_2} = \frac{4\kappa(T_o)(T_o - T_w)}{p_o\rho_o^2} = 1.2\frac{1 - T_w/T_o}{1 - (T_w/T_o)^{1.7}} \tag{8.5.16}$$

Figure 8.10 represents this ratio at $T_w = 300K$.

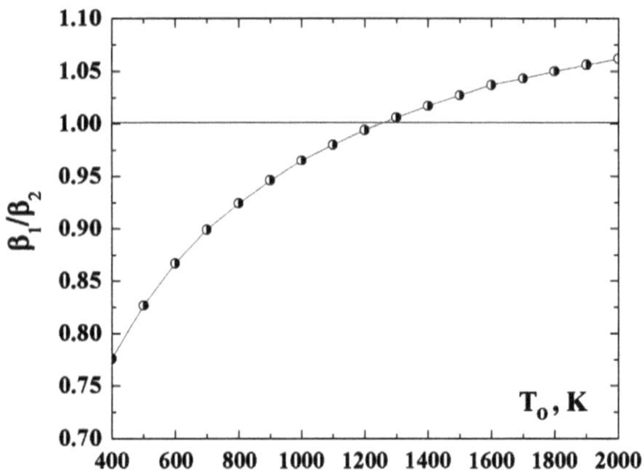

Fig. 8.10 The ratio of the parameters for the temperature space distribution in accordance with formula (8.5.16)

8.6 Local Thermodynamic Equilibrium in Arc Plasma of High Pressure

We now consider the heat regime of the positive column of high pressure arc discharge when heat release result form gaseous and electron thermal conductivity. the heat balance equation has the following form instead of (8.5.1)

$$\frac{1}{\rho}\frac{d}{d\rho}\left[\rho\kappa(T)\frac{dT}{d\rho}\right]+\frac{1}{\rho}\frac{d}{d\rho}\left[\rho\kappa_e(T_e)\frac{dT_e}{d\rho}\right]+p(\rho)=0, \qquad (8.6.1)$$

where $\kappa_e(T_e)$ is the coefficient of electron thermal conductivity that is proportional to the electron number density N_e, $p(\rho)$ the specific heat release. This equation is called the Elenbaas-Heller equation, if $p(\rho)$ is represented in the form $p(\rho) = \Sigma E^2$ (Σ is the plasma conductivity, E is the electric field strength). Though the number density of electrons is small compared to the atom number density, heat transport due to electrons may be significant because of a high electron mobility.

In considering the heat balance in the arc positive column, we will be guided the case of a dense plasma with thermodynamic equilibrium for electrons, so that the number density of electrons N_e is connected with the electron temperature T_e by the Saha formula (2.1.8)

$$N_e \sim \exp\left(-\frac{J}{T_e}\right),$$

where J is the atom ionization potential, and the criterion $J \gg T_e$ provides a sharp dependence on the electron temperature both the electron number density and the parameters $\kappa_e(T_e)$ and $p(\rho)$. Because of a sharp dependence of the electron number density on the electron temperature, electrons may be concentrated mostly in the region near the tube axis. For the analysis of the electron distribution near the axis, it is convenient to use the variable

$$y = \frac{[T_e(0) - T_e(\rho)]\,J}{2T_e^2(0)}, \tag{8.6.2}$$

Assuming thermodynamic equilibrium for electrons, we have on the basis of the Saha formula (2.1.8)

$$N_e(\rho) = N_e(0)e^{-y}$$

Because the specific power of heat release and the coefficient of electron thermal conductivity are proportional to the electron number density, we have for these quantities near the axis

$$p(\rho) = p(0)e^{-y}, \quad \kappa_e \sim N_e \sim e^{-y}$$

Correspondingly, the heat balance (8.6.1)with using the variable $x = \rho^2/\rho_o^2$ takes the form

$$\frac{d}{dx}\left[x\left(e^{-y} + \zeta\right)\frac{dy}{dx}\right] - Ae^{-y} = 0, \tag{8.6.3}$$

and the parameters of this equations are given by

$$\zeta = \frac{T\kappa(T)}{\kappa_e(T_e)}\alpha, \quad \alpha = \frac{dT(\rho)}{dT_e(\rho)}, \quad A = \frac{p_o\rho_o^2 J}{8T_e^2\kappa_e(T_e)} \tag{8.6.4}$$

Thus, a sharp dependence of the electron number density on the electron temperature allows us to obtain the balance equation with plasma parameters on the axis of a discharge tube.

Let us consider the limiting case where heat transport is determined by the gas thermal conductivity $\zeta \gg 1$. Then equation (8.6.3) has the Fock solution

$$y = 2\ln\left(1 + \frac{Ax}{2\zeta}\right) \tag{8.6.5}$$

This solution leads to the following distribution of the electron number density over the cross section of a discharge tube

$$N_e(\rho) = N_e(0)e^{-y} = N_e(0)\left(1 + \frac{\rho^2}{r_o^2}\right)^{-2} \tag{8.6.6}$$

If a typical size r_o of a region occupied by a plasma is small compared to the radius of a discharge tube ρ_o, the discharge contraction takes place that is determined by heat processes in a plasma.

The heat release power per unit length of a discharge tube is connected with the rate of heat release and is given by

$$P = IE = \int p_o e^{-y} \cdot 2\pi\rho d\rho = \frac{16T_e^2 T\kappa(T)\alpha}{J} \tag{8.6.7}$$

Expressing this quantity through the electron number density with the distribution (8.6.6) over the tube cross section, one can determine a size of a region that is occupied by the current.

In other limiting case $\zeta \ll 1$ heat transport is determined by electron thermal conductivity, and equation (8.6.3) in a region $y < \ln(1/\zeta)$, where the plasma is concentrated, is transformed to the form

$$\frac{d}{dx}\left(x\frac{dY}{dx}\right) + AY = 0$$

where the variable $Y = N_e(\rho)/N_e(0) = e^{-y}$. The solution of this equation

$$Y = J_0\left(2\sqrt{Ax}\right) \tag{8.6.8}$$

leads to the following expression for the electric power per unit tube length

$$P = IE = \int p_o Y \cdot 2\pi\rho d\rho = 1.36 p_o r_o^2,$$

where

$$r_o^2 = \frac{5.78\rho_o^2}{A} = \frac{12T_e^2\kappa_e(T_e)}{p_o J},$$

and the specific power in this regime is

$$P = IE = \frac{16T_e^2\kappa_e(T_e)}{J} \tag{8.6.9}$$

Joining formulas (8.6.7) and (8.6.9), we have for the specific power per unit length of the positive column of arc discharge

$$P = IE = \frac{16T_e^2\kappa_e(T_e)(1+3.2\zeta)}{J} \tag{8.6.10}$$

In the same manner we obtain for the radius of a region occupied by a current in the case of the Schottky regime

$$\int N_e \cdot 2\pi\rho d\rho = 1.36 N_o r_o^2$$

This gives in a general case for the radius of a region occupied by a plasma

$$r_o^2 = \frac{12T_e^2\kappa_e(T_e)(1+3.2\zeta)}{p_o J} \qquad (8.6.11)$$

We now analyze equation (8.6.1) in the form (8.6.3) from the standpoint of current contraction. We have that the heat release power depends sharply on the electron temperature. Heat transport due to electron thermal conductivity also depends strongly on the electron temperature, while this dependence due to the atom thermal conductivity is weak. Therefore if heat transport is determined by the electron thermal conductivity, i.e. $\zeta \ll 1$, according to solution (8.6.8) an arc plasma is spread over all the cross section, and contraction of the positive column is absent. In the case $\zeta \gg 1$ with the plasma distribution (8.6.6) contraction of the electric current in the arc positive column is possible in the case $r_o \ll \rho_o$.

In order to find the criterion when contraction starts, we analyze the heat balance equation (8.6.1) for the total tube cross section with ignoring the electron thermal conductivity that can eliminate the arc contraction. Counting the gas temperature T out the wall temperature T_w, we introduce the reduced temperature as

$$\Theta = \frac{T - T_w}{T_w},$$

and (8.6.1) in this case takes the form

$$\frac{1}{x}\frac{d}{dx}\left(x\frac{d\Theta}{dx}\right) + B\exp(b\Theta) = 0$$

where we ignore the temperature dependence $\kappa(T)$ for the thermal conductivity coefficient, and the parameters of this equation are

$$B = \frac{r_o^2 p(T_w)}{4T_w\kappa(T_w)}; \quad b = T_w \frac{d\ln p(T)}{dT}_{|T=T_w}$$

This equation is added by the boundary conditions $\Theta(1) = 0$, $d\Theta(0)/dx = 0$ according to its physical nature. In contrast to (8.6.3) that describes a region near the axis $x \ll 1$, this equation relates to all the distances from the axis $0 < x < 1$. Next, the variable of this equation is connected with the gas temperature.

One can represent formally the Fock solution of this equation

$$\Theta = \frac{1}{b}\ln\left[\frac{2\gamma}{Bb(1+\gamma x)^2}\right], \tag{8.6.12}$$

where γ is a variable parameter, and the validity of this solution may be checked by the substitution of this solution into equation. This solution takes into account the boundary condition $d\Theta(0)/dx = 0$ at the axis, and on the basis of the other boundary condition $\Theta(1) = 0$ one can find the equation for the variable parameter γ

$$2\gamma = Bb(1+\gamma)^2 \tag{8.6.13}$$

From this it follows that the real solution of the heat balance equation takes place for the following relation between parameters

$$Bb \leq \frac{1}{2} \tag{8.6.14}$$

Violation of this condition leads to a thermal instability. Its physical sense is such that an exponential growth of the rate of heat release is not compensated by heat transport which rate depends weakly on the temperature. This instability leads to contraction that increases the rate of heat transport.

The threshold of this instability is given by $Bb = 1/2$, $\gamma = 1$, that gives $b\Theta = \ln[2/(1+\kappa)]$. This leads to the following relation between the specific power of heat release at the axis and on the walls for the threshold of thermal instability

$$p(T_o) = 4p(T_w), \tag{8.6.15}$$

where T_o is the temperature at the axis.

On the basis of the above analysis one can estimate a size r_o of the region that is occupied by the discharge electric current which follows from the balance of heat release and heat transport. This gives

$$r_o^2 \sim \frac{\kappa(T_o)}{dp/dT} \tag{8.6.16}$$

Thus, heat transport in arc of high pressure is of importance and determines the temperature distribution in the positive column. In addition, due to the character of ionization equilibrium in this gas discharge plasma, the ionization processes proceed mostly in a region near the axis of the gas discharge tube. This leads to two regimes of ionization equilibrium for the plasma over all the tube. In the regime of local thermodynamic equilibrium, the number density of electrons and ions at a given distance from the tube axis is determined by the temperature of this point. On contrary, in the Schottky regime formed electrons and ions propagate over all the tube cross section, i.e. in contrast to the case of local thermodynamic equilibrium, the number density of electrons and ions has the same order of magnitude for any tube cross section, and its distribution over the cross section is determined by the boundary condition.

Let us give the criterion of local thermodynamic equilibrium in a discharge tube with an arc plasma of high pressure. Taking ionization equilibrium according to the scheme (3.8.1)

$$2e + A^+ \leftrightarrow e + A, \qquad (8.6.17)$$

where A and A^+ are an atom and atomic ions respectively, we have the rate of electron decay in this scheme $1/\tau = KN_a$. On the other hand, according to formula (8.4.1) a typical rate for electron travel to walls is approximately $6D_a/\rho_o^2$, where D_a is the ambipolar diffusion coefficient, ρ_o is the tube radius. From this it follows that local thermodynamic equilibrium is not realized inside the discharge tube if the following criterion holds true

$$KN_aN_e \ll \frac{6D_a}{\rho_o^2} \qquad (8.6.18)$$

If this criterion is violated, plasma transport takes place on a distance $\Delta\rho$ from the point of its formation in accordance with

$$\Delta\rho \sim \sqrt{\frac{6D_a}{KN_a}} \ll \rho_o \qquad (8.6.19)$$

Because the ionization rate $\sim \exp(-J/T_e)$, where J is the atom ionization potential, it is necessary for local thermodynamic equilibrium that the electron temperature T_e varies on a distance $\Delta\rho$ on a value that is small compared with T_e^2/J.

8.7 Ionization Equilibrium in Arc Plasma of High Pressure

In considering a dense arc plasma, one can introduce separately the gaseous T and electron T_e temperatures because of a high electron number density. On the other hand, because of high electric currents, heat processes are of importance for parameters of this plasma. We below analyze properties of a dense arc plasma located in a cylinder tube where the criterion (8.6.18) holds true, i.e. this plasma is characterized by the gaseous $T(\rho)$ and electron $T_e(\rho)$ temperatures and is supported by an external electric field of a strength E, but the local thermodynamic equilibrium is violated in this regime. Evidently, both temperatures depend on the reduced electric field strength $x = eE/N_a$. We first find the connection between the gaseous $T(\rho)$ and electron $T_e(\rho)$ temperatures in this regime in the case if the cross section of elastic electron-atom collision is independent on the electron velocity. Then according to formula (6.2.7) we have

$$(T_e - T)T_e = const \cdot \left(\frac{eE}{N_a}\right)^2$$

This balance equation takes into account the energy transfer from an electric field to gas atoms through collisions with electrons and uses the assumption that the diffusion cross section of elastic electron-atom collisions is independent of the electron energy. Since the electric field strength is constant over the tube cross section, we have

$$\frac{T_e}{T}\left(\frac{T_e}{T} - 1\right) = const, \qquad (8.7.1)$$

that is the electron temperature varies over the tube cross section proportional to its gaseous temperature.

Since the ionization rate depends on the electron temperature as $\sim \exp(-J/T_e)$, where J is the atom ionization potential, one can find from this that in a region which is responsible for the ionization balance, variations of the electron ΔT_e and gaseous ΔT are given by relations

$$\Delta T_e \sim \frac{T_e^2}{J}, \ \frac{\Delta T}{T} \sim \frac{T_e}{J}, \qquad (8.7.2)$$

There are two regimes of the positive column of gas discharge, diffusive and contracted ones, and the Schottky case relates to the diffusive regime, where an electron current and discharge glowing occupy all the tube cross section. On contrary, in the contracted regime an electric current possesses a small part of the tube cross section. Transition between these forms of gas discharge takes place if the difference of helium gaseous temperatures ΔT between center and walls of the discharge tube satisfy to formula (8.7.2). Taking a typical electron temperature for a helium plasma of the positive column to be $T_e \sim 3eV$, obtain that this transition proceeds at $\Delta T \sim 40K$.

We now analyze ionization equilibrium for the positive column of a dense arc plasma in this ionization regime where formation of electrons and ions proceeds from atom ionization by electron impact, and plasma decay results from ambipolar diffusion of this plasma to tube walls. Under given conditions, electrons and ions are formed in a narrow region near the tube axis and then are spread over the total tube cross section by ambipolar diffusion with recombination on tube walls. Let us assume for simplicity that ionization proceeds in a region of a radius a that is small compared with the tube radius ρ_o, i.e. $a \ll \rho_o$. Because the total electron flux toward the walls $j = -D_a \cdot 2\pi\rho dN_e/d\rho$ is independent of a distance ρ from the axis aside the ionization boundary $\rho \geq a$, we have for the electron number density in this region

$$N_e = N_o \frac{\ln(\rho_o/\rho)}{\ln(\rho_o/a)}, \ \rho > a \qquad (8.7.3)$$

where we account for the boundary condition $N_e(\rho_o) = 0$.

Under this condition equation of ionization equilibrium takes the form

$$\int k_{ion}(\rho)2\pi\rho d\rho = 2\pi\rho D_a \frac{dN_e}{d\rho}_{|\rho=a} = \frac{2\pi D_a(\rho_o)N_o}{\ln(\rho_o/a)}, \quad (8.7.4)$$

where we use the above expression for the electron number density $N_e(\rho)$. Assuming the ionization rate constant k_{ion} and the ambipolar diffusion coefficient to be independent of a distance from the axis ρ, we assume the ionization balance equation in the form

$$N_a k_{ion} = \frac{2D_a}{a^2 \ln(\rho_o/a)} \quad (8.7.5)$$

This equation corresponds to the criterion $\rho_o \gg a$ and is valid within the accuracy of a constant factor under the logarithm. Take this factor such that the balance (8.7.5) is transformed into (8.2.4) in the limit $\rho_o = a$. Then we obtain the equation of ionization equilibrium in the form

$$N_a k_{ion} = \frac{2D_a}{a^2 \left[0.35 + \ln(\rho_o/a)\right]} \quad (8.7.6)$$

Chapter 9
Cathode and Wall Processes

Abstract Properties of the cathode region of glow discharge are analyzed depending on the character of cathode processes. Principles of magnetron discharged are described.

9.1 Electric Breakdown of Gases

Let us define the breakdown electric potential in gases as the voltage between two electrodes which provides a stationary electric current in a gas located in a gap between two electrodes in accordance with the relation (8.1.3). Basing on the condition of existence of an electric current in a gas between two electrodes separated a distance L between them, we leave aside dynamics of origin of an electric current that may have various mechanisms [12, 17–19, 21, 22]. So, the Townsend electric breakdown under consideration is multiplication of electron avalanches where an individual avalanche is a circle of processes of electron and ion production starting from formation of a secondary electron at the cathode and finishing by attachment of resulted electrons to the anode. In this consideration we suppose each avalanche to be developed independently on other ones and are based on the relation (8.1.3) as the condition of a self-maintaining process. Along with this mechanism of electric breakdown, the streamer mechanism [21, 351–357] is possible and dominates at large distances between electrodes.

Within the framework of the Townsend mechanism of electric breakdown for a gas located in a gap between two parallel electrodes, it is convenient to use the approximation (8.1.5) for the reduced first Townsend coefficient (8.1.4) that leads to the following expression for the breakdown electric potential $U_b = EL$ of a gaseous gap [4]

$$U_b = \frac{B(N_a L)}{\ln\left[AN_a L/\ln(1 + 1/\gamma)\right]} \qquad (9.1.1)$$

The dependence of the breakdown electric potential U_b on the reduced gas pressure or on the reduced number density of atoms $N_a L$, the Pashen curve [358] is given in Fig. 9.1 for helium and argon.

B.M. Smirnov, *Theory of Gas Discharge Plasma*, Springer Series on Atomic, Optical, and Plasma Physics 84, DOI 10.1007/978-3-319-11065-3_9

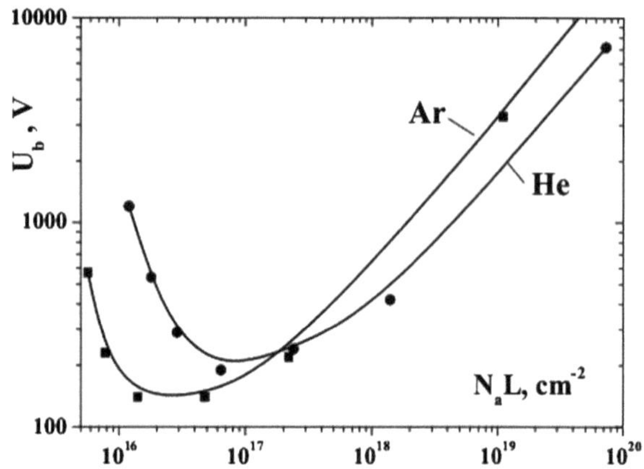

Fig. 9.1 Pashen curves for helium and argon calculated on the basis of formula (9.1.1) and data of Table 8.1

Within the framework of the approximation (8.1.5) for the reduced first Townsend coefficient (8.1.4), one can find that the minimum value of the breakdown electric potential corresponds to the following value of the reduced number density of atoms (or the gap size) [4]

$$(N_a L)_{\min} = \frac{e}{A} \ln \left(1 + \frac{1}{\gamma}\right) \qquad (9.1.2)$$

This leads to the following expression for the minimum breakdown electric potential [4]

$$U_{\min} = B(N_a L)_{\min} = \frac{B}{A} \ln \left(1 + \frac{1}{\gamma}\right) \qquad (9.1.3)$$

Let us use formulas (9.1.2) and (9.1.3) for electric breakdown of helium and argon if this gas is located between plane copper electrodes. The results are given in Table 8.2. Note that the accuracy of the approximation (8.1.5) influences on the values of parameters $(N_a L)_{\min}$ and $U_{\min}$, as it exhibits in Table 8.2. Note that the experimental values of the cathode voltage are 177 V for helium and 130 V for argon [12], and they conform to the minimum voltage of the Townsend discharge form in accordance with Table 8.2 data. One more notation relates to the second Townsend coefficient that accounts for generation of secondary electrons from the cathode which may be resulted both from ion bombardment the cathode and also from collisions excited atoms and photons with the cathode surface. These excited atoms and photons are formed in electron collisions with atoms. In the geometry of measurement of the Pashen curves a radius of the electric current is small compared with a size of a region occupied by a gas, and hence a small part of formed excited atoms or photons loses on the cathode. Hence the second Townsend coefficient for the Pashen curve is determined by ion collisions with the cathode, and under conditions of Table 9.3 for a copper cathode it is equal $\gamma = 0.19$ in helium and $\gamma = 0.08$ in argon.

Parameters of the minimum of the Pashen curves for a gap with helium and argon between copper electrodes evaluated on the basis of formulas (9.1.2) and (9.1.3).

These parameters are given if the approximation (8.1.5) is obtained in different ranges of the reduced electric field strength, that leads to different parameters A and B of formula (8.1.5) and respectively to different values of yield parameters.

Introducing a new variable $\xi = (N_a L)/(N_a L)_{\min}$, one can represent formula (9.1.1) in the form

$$U = U_{\min} \frac{\xi}{\ln \xi + 1}$$

From this it follows that large breakdown potential corresponds to a large distance between electrodes. Therefore another mechanism of breakdown corresponds to large values of the parameter $N_a L$, when the plasma uniformity is violated, and breakdown is realized in the streamer form [21, 351–357] that is a non-uniform plasma propagated between electrodes.

Note that we use above the condition (8.1.3) as the condition of electric breakdown of a gas. In reality it is the condition of self-maintaining of an electric current in a gas, and within the framework of electric breakdown means that an increase of the electric field strength compared with that followed to the condition (8.1.3) leads to an exponential current growth in time in some range of currents.

9.2 Electron Emission from Cathode in Ion Collisions

The Townsend scheme of ionization (1.01) relates to some conditions of gas discharge, in particular, to glow gas discharge. In this scheme, reproduction of electrons which attach to the anode and walls of a discharge tube, takes place in ion collisions with the cathode surface (the second (1.0.1)). The ion velocity in this process is small compared to a typical electron velocity, and hence the process of formation of secondary electrons at the cathode surface has the potential character that is represented in Fig. 9.2 [359–361]. As is seen, two channels may be responsible for formation of secondary electrons. The first channel (Fig. 9.2a) is similar to the Auger process [362, 363] in an atom (or ion) with a removed electron from an internal electron shell. This hole is filled in electron transition from a higher electron shell, and an electron energy difference is transferred to another electron of higher electron shell that leads to its release. The Auger effect is used for diagnostics of some materials in the form of Auger spectroscopy [364, 365].

Another channel of generation of free electrons at the cathode corresponds to collisions of excited atoms with the cathode surface (Fig. 9.2b). As a result of this process, an atom transits in the ground state and a released energy is consumed on ionization of a metal electron. An analog of this process is the Penning process (3.7.3) [192, 193] in collisions involving an excited atom. It is possible that ion collision with the cathode surface has a two-stage character and corresponds to the scheme [366, 367]:

$$A^+ + M \to A^* + M^+, \;\; A^* + M^+ \to A + M^{++} \tag{9.2.1}$$

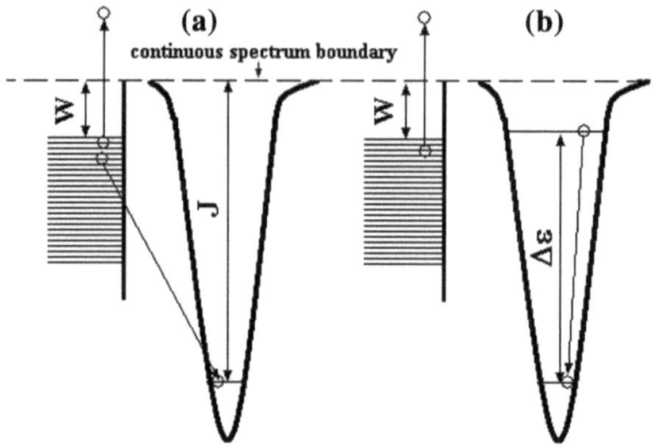

Fig. 9.2 Two mechanisms of electron liberation from a cold cathode. **a** Auger process as a result of ion interaction with metal electrons; **b** Penning process in interaction of an excited atom with metal electrons

This mechanism of electron emission is confirmed experimentally [368]. Note that this process is effective if the cathode surface absorbs atoms of a surrounding gas, so that the resonant charge exchange process includes incident ions and bound atoms.

In consideration of these processes a simple model is used for metal electrons (Fig. 9.2) where electrons occupy a band in the energetic space, and the minimum electron binding energy is the metal work function W. Hence the process (Fig. 9.2a) is possible if the criterion $J > 2W$, and the process (Fig. 9.2b) requires the criterion $\Delta\varepsilon > W$, and the ionization potentials J for helium and argon atoms are equal to 24.59 and 15.76 eV correspondingly, and the excitation energies $\Delta\varepsilon$ for metastable helium and argon atoms are equal to 19.82 and 11.55 eV. In addition, Table 2.1 contains the average work functions of metals which are used as the cathode. As is seen, the criteria of processes of Fig. 9.2 are fulfilled in the helium and argon cases.

The second Townsend coefficient which is responsible for reproduction of electrons at the cathode, is different for breakdown in a gas gap located in a space between two electrodes and that in the cathode layer. Indeed, in the case of the cathode layer because of its small depth compared to a current radius the second Townsend coefficient γ may be represented in the form [24, 369]

$$\gamma = \gamma_i + \gamma_m + \gamma_p, \tag{9.2.2}$$

where γ_i, γ_m, and γ_p correspond to emission of secondary electrons due to ions, excited atoms and photons. Excited atoms may be also an intermediate channel of formation of secondary electrons, and the photoionization process is shown to be important for emission of secondary electrons from the cathode [370]. One can separate these processes in a special experimental setup which concept is given in Fig. 9.3. In this case a beam of ions of a certain type and energy is directed to the cathode and formed secondary electrons are collected. The second Townsend coefficient γ_i due

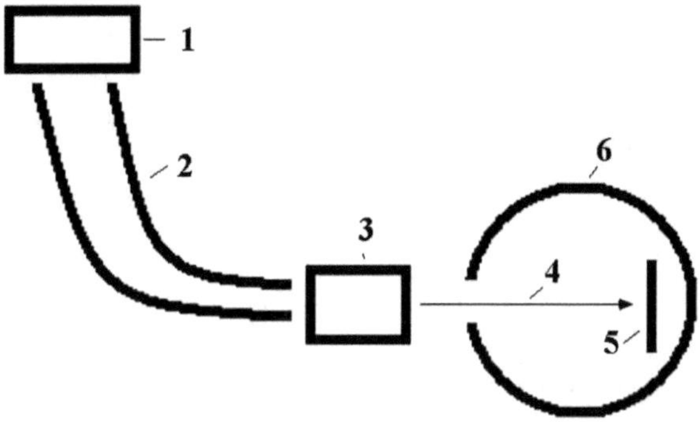

Fig. 9.3 Character of measurement of the second Townsend coefficient γ_i [371]. *1* source of ions, *2* mass-spectrometer, *3* accelerator of ions, *4* beam of ions, *5* cathode, *6* collector of electrons

Table 9.1 The partial second Townsend coefficient γ_i due to ions [373–378]

Ion/Cathode	Mo	Ta	W
He^+	0.25	0.14	0.295
He_2^+	-	0.10	-
He^{++}	0.72	0.6	0.74
Ar^+	0.083	-	0.095
Ar^{++}	-	-	0.42

to ions is determined as the ratio of the current of collected electrons to the current of incident ions. Measurements in the range of ion energies 10–1000 eV exhibits a weak dependence of γ_i on the ion energy that confirms the potential character of formation of secondary electrons in accordance with the scheme of Fig. 9.3. Some results of these measurements [371–375] are given in Table 9.1 together with [376]. One can see that if ions in the experimental scheme of Fig. 9.3 can reflect from the surface, the measured second Townsend coefficient may result from many ion collisions with the surface until the ion attaches to this surface.

The second Townsend coefficient depends on the cathode surface and is changed if the cathode surface absorbs gaseous atoms or molecules. This is confirmed by data given in Fig. 9.4 where the results are given for second Townsend coefficients for the Mo-cathode if its surface is pure or is covered by a monolayer of absorbed atoms. An additional information gives the energy distribution of electrons resulted from collisions of helium ions with the cathode surface measured in [374, 375, 377]. According to the Auger mechanism of electron emission, the energy of released electrons ε exceeds $J - 2W$ (J is the ionization potential of gaseous atoms, W is the working function of a cathode material). Figures 9.5 and 9.6 give the measured distribution functions of electrons if a cathode material is a semiconductor [377]. The boundary electron energy in these distributions is close to $J - 2W$.

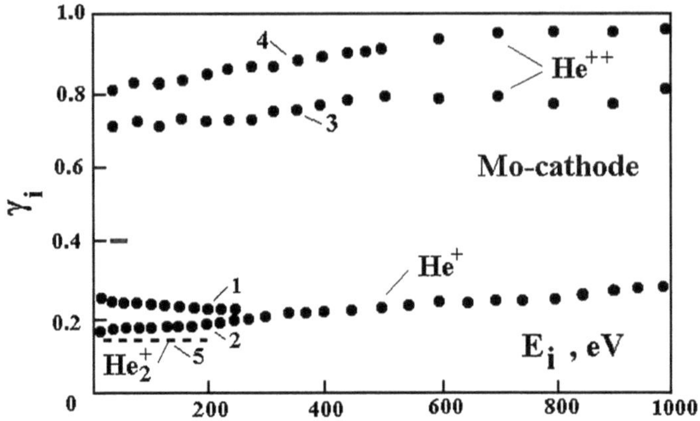

Fig. 9.4 The partial second Townsend coefficient γ_i for a Mo-cathode due to helium ions in helium. The cases *1* and *3* correspond to atomically clean molybdenum, while in the cases *2*, *4*, *5* the molybdenum cathode is covered by a monolayer of gas atoms. Ion sorts are indicated [371]

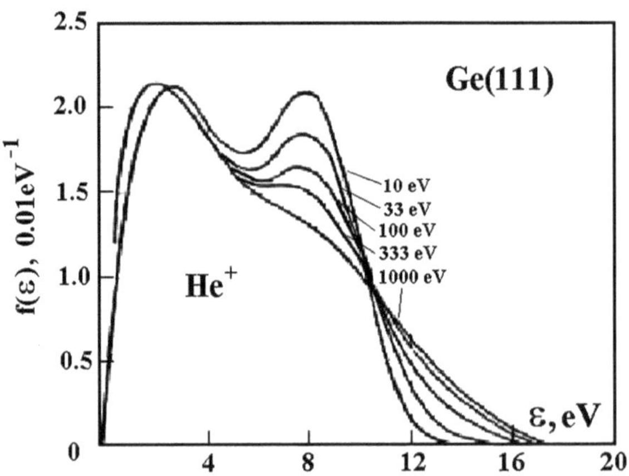

Fig. 9.5 The energy distribution function of released electrons in collisions of He^+-ions with the germanium cathode surface of the direction 111 some energies of incident ions [377]

9.3 Properties of Cathode Region of Glow Discharge

Let us consider glow discharge of high pressure where the mean free path of electrons and ions in a gas is small compared to typical sizes of their space distribution. Two elements of gas discharge are of principle from the standpoint of discharge self-maintaining, that are the cathode region and the positive column. The cathode region is responsible for reproduction of charged particles that are destructed on electrodes, i.e. for self-maintaining of the gas discharge current. The positive column contains an uniform plasma in the longitudinal direction, and reproduction of charged particles in the positive column compensates their destruction on walls of the discharge tube.

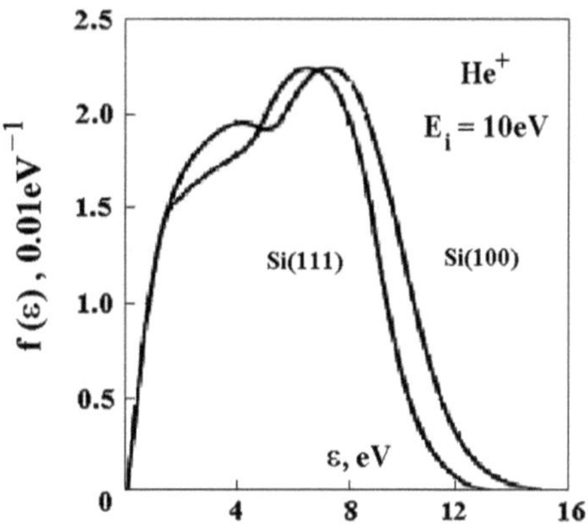

Fig. 9.6 The energy distribution function of released electrons in collisions of the 10 eV He^+-ions with the silicon surface of different directions [377]

The condition of self-maintaining of gas discharge in the cathode region is given by formula (9.1.1) that is generalized to a non-uniform electric field

$$\int \alpha dz = \ln(1 + 1/\gamma) \tag{9.3.1}$$

Here z is the distance from the cathode, and we assume the cathode region depth to be small compared to the cathode radius, that allows us to reduce the problem to a one-dimensional case.

We first consider in accordance with [4] the cathode layer of glow discharge where the ionization balance in the cathode region is given by (9.3.1) and the parameters of these processes are the first and second Townsend coefficients. In addition to (9.3.1) that describes reproduction of charged particles at the cathode, we have the Poisson equation

$$\frac{dE}{dz} = 4\pi e(N_i - N_e), \tag{9.3.2}$$

that describes a non-uniformity of the electric field strength E near the cathode and where N_i, N_e are the number densities of ions and electrons. Because the mean free path of ions and electrons is large compared to a size of the cathode region, the ion current densities for electrons i_e and ions i_i according to definition are given by

$$i_i = eK_iN_iE, \; i_e = -eK_eN_eE$$

where K_e, K_i are the mobilities of electrons and ions. Since $K_e \gg K_i$, the number density of electrons in the cathode region is respectively small $N_e \ll N_i$, and the Poisson equation (9.3.2) is reduced to the form

$$\frac{dE}{dz} = 4\pi eN_i = -4\pi \frac{i_i}{K_iE}$$

A sign of the last equation accounts for a drop of the electric field strength with removal from the cathode.

Note that the total current density $i = i_i + i_e$, is conserved in the cathode region, and accounting for generation of secondary electrons by ion impact, we have the boundary condition at the cathode $i_e(0) = \gamma i_i(0)$, or $i_i(0) = i/(1+\gamma)$ as the relation between the electron i_e and ion i_i current densities. Because ions are not lost in the cathode region, their current is constant along x, and the last relation is valid for the total cathode region

$$i_i = \frac{i}{(1+\gamma)}$$

Accounting for this in solution of the Poisson equation, we have [4]

$$E^2 = E_c^2 - \frac{8\pi i}{K_i(1+\gamma)} z, \tag{9.3.3}$$

Here z is a distance from the cathode, and $E_c = E(0)$ is the electric field strength at the cathode. Introducing a length l of the cathode region such that the electric field strength on the boundary is zero $E(l) = 0$ and assuming the ion mobility to be independent of the electric field strength, we find from the solution of this equation [4]

$$i = \frac{E_c^2 K_i(1+\gamma)}{8\pi l} \tag{9.3.4}$$

Note that the electric field strength is nonzero at any discharge point, and the boundary condition means only that outside the cathode region the electric field strength is small compared with that at the cathode E_c. The voltage drop U_c in the cathode region is expressed through the electric field strength at the cathode E_c, in turn according to formula (9.3.4) E_c depends on a size of the cathode region. Evidently, optimal parameters of the cathode region corresponds to the minimum of the voltage drop. Hence, by variation of the current density i and the length l of the cathode layer we choose the optimal conditions which lead to the minimum of the cathode voltage drop U_c. This corresponds to the normal regime of glow discharge. The values of the cathode voltage drop U_c depend on the cathode material and gas sort. Table 9.2 contains experimental values of U_c for gas discharge in helium and argon and some cathode materials.

The values of the cathode voltage drop U_c may be found from the minimum condition for the voltage drop. But as it follows from the relations obtained, the electric field strength E_c as well as the cathode voltage drop U_c are expressed through a current density. Therefore the condition of the minimum of the cathode voltage drop chooses an optimal value of the current density.

In the normal regime of gas discharge the electric current density i and the cathode region length l under optimal conditions do not vary at change of the discharge electric current. Hence the regime under consideration corresponds to low discharge currents, when a discharge current occupies a restricted part of the cathode. This gives the following criterion of the normal regime of glow gas discharge

Table 9.2 Values of the cathode voltage drop U_c, V [12]

Cathode	He	Ar	Cathode	Hc	Ar
$Mg(3.7)$	125	119	$Sr(2.6)$	86	93
$Al(4.2)$	140	100	$Ag(4.6)$	162	130
$K(2.3)$	59	64	$Cd(4.1)$	167	119
$Ca(2.9)$	86	93	$Ba(2.5)$	86	93
$Fe(4.7)$	150	165	$Pt(5.6)$	165	131
$Ni(5.2)$	158	131	$Au(5.4)$	165	131
$Cu(4.8)$	177	130	$Pb(4.2)$	177	124
$Zn(3.6)$	143	119	$Bi(4.3)$	137	136

The work functions of cathode materials [452] are given in parentheses

$$I < i \cdot \pi r_o^2 \qquad (9.3.5)$$

Here I is the total discharge current, i is the current density, r_o is a cathode radius that assumes to be equal to the tube radius. This is the normal discharge regime. At large currents when the criterion (9.3.5) is violated (abnormal discharge regime) the current number density exceeds the optimal value. As a discharge current increases, the cathode is heated, and in the end this regime is violated. In particular, we have the arc discharge regime when the cathode temperature becomes high, and electrons are formed at the cathode as a result of thermoemission.

The peculiarity of the cathode region is violation of plasma quasineutrality in it, and an external voltage is required to support this unipolar plasma. The properties of this plasma are determined by the boundary conditions at the cathode, that in turn are connected with processes of formation of charged particles at the cathode. In the case of glow discharge secondary electrons are formed as result of cathode bombardment by ions and then electrons are multiplied in the cathode region. We now consider one more type of the cathode region as it takes place in diodes. Then electrons are formed at the cathode and reach another electrode without collisions. Our goal is to connect the voltage U_o between electrodes, the electric current density i and a distance L between electrodes. For definiteness, we assume that electrons are located in a space between electrodes as it takes place in an electron lamp. If ions determine a charge in a space near the cathode, as it takes place in the cathode layer of glow discharge and magnetron discharge, it is necessary to substitute the electron mass in subsequent formulas by the ion mass.

In a collisionless plasma the electron velocity at a distance x from the electrode equals to $v_e = \sqrt{2eU(x)/m_e}$, where m_e is the electron mass, $U(x)$ is the plasma potential that is zero at a corresponding electrode. The Poisson equation in the gap is

$$\frac{dE}{dz} = -4\pi e N_e(z) = -4\pi i \sqrt{\frac{m_e}{2eU}},$$

where N_e is the electron number density. Because the electrons are not formed and perish in the gap, the electron current density is conserved $i = eN_e(x)v(x) = const$ in the gap or the cathode region. Let us multiply the Poisson equation by $E = -dU/dx$, and using the boundary condition $E(0) = 0$, $U(0) = 0$, we obtain

$$E^2(z) = 16\pi i\sqrt{\frac{m_e U(z)}{2e}}$$

Counting off the plasma electric potential U from the cathode, we find from solution of this equation ($E = -dU/dz$)

$$U(z) = \left(9\pi i\sqrt{\frac{m_e}{2e}}\right)^{2/3} z^{4/3}, \tag{9.3.6}$$

This gives for the discharge current density under the action of the voltage U_o between electrodes

$$i = \frac{2}{9\pi}\sqrt{\frac{e}{2m_e}}\frac{U_o^{3/2}}{l^2}, \tag{9.3.7}$$

where l is a distance between electrodes. This relation is known as the three halves law [378–381] or the Child law. If the voltage $U(l)$ is given together with the current density i, the above formulas give the size l of the gap through which the electric current passes in accordance with the expression

$$l = \left(\frac{2}{9\pi}\right)^{1/2}\left(\frac{e}{2m_e}\right)^{1/4}\frac{U_o^{3/4}}{i^{1/2}} \tag{9.3.8}$$

We above ignore electron-atom collisions using the expression $v_e = \sqrt{2eU(x)/m_e}$ for the electron velocity. Hence these formulas require the validity of the following criterion

$$l \ll \lambda, \tag{9.3.9}$$

where λ is the mean free path of electrons in a gas. Next, ions are absent in a gap in the above consideration. In the same way one can apply the above formulas to the case where only ions are located inside the gap. Then the electron mass m_e in these formulas must be changed by the ion mass M and λ in criterion (9.3.9) is the mean free path of ions. Such a situation takes place in magnetron discharge where electrons do not penetrate to the cathode due to an external magnetic field, and ions are accelerated in the gap, so that their bombardment of the cathode causes formation of secondary electrons that provides self-maintaining of this discharge.

The above formulas describe the electron behavior in the electron lamp where electrons are formed at a hot cathode as a result of thermoemission and are accelerated in a vacuum under the action of the net electric potential. Another example of this relates to gas discharge with a diode configuration (for example [382]), where ions are formed at the anode and accelerating under the action of an electric field of gas discharge. These ions bombard the cathode causing its sputtering. This phenomenon is known from 19 century [383–386]. This allows one to form a flux of metal atoms which are deposited onto a surface and form there films. But magnetron discharge [387–389] becomes more effective for this goal because identical sputtering parameters are achieved there at voltages in several times lower.

Let us apply the relation (9.3.6) to the cathode region of glow discharge assuming the criterion is fulfilled that is opposite with respect to criterion (9.3.9). Taking into account that the space charge in the cathode region is created by ions as a more slow component,we assume atomic ions to be located in the cathode region. Then the drift ion velocity w_i is determined by the resonant charge process and is given by formula (7.6.12) that has the form

$$w_i = \sqrt{\frac{2eE\lambda}{\pi M}}, \quad eE\lambda \gg T,$$

where λ is the mean free path of ions in the cathode region, and M is the ion or atom mass. In determination the parameters of the cathode layer, we repeat operation used for deduction of formula (9.3.6). Indeed, substituting the expression (7.6.12) for the ion drift velocity into the Poisson equation, we reduce the latter to the form

$$\frac{dE}{dz} = 4\pi e N_i = -\frac{4\pi i}{w_i} = -\frac{4\pi i\sqrt{\pi M}}{\sqrt{2eE\lambda}},$$

where x is a distance from the cathode, N_i is the ion number density, i is the ion current density to the cathode, and a sign "minus" account for drift of ions in a negative direction. Solving this equation, obtain instead of (9.3.3) the following expression for the electric field strength in the cathode region

$$E^{3/2} = E_n^{3/2} - \frac{6\pi i\sqrt{\pi M}}{\sqrt{2e\lambda}} z, \quad eE_n\lambda \gg T \tag{9.3.10}$$

where E_c is the electric field strength at the cathode which we denote by E_n for the normal regime of the cathode region. From this we obtain instead of formula (9.3.4) the following formula for the current density in the normal regime of glow discharge ($\gamma \ll 1$)

$$i_n = \frac{E_n w_n}{6\pi l} = \frac{E_n^{3/2}(e\lambda)^{1/2}}{3\pi l(2\pi M)^{1/2}}, \tag{9.3.11}$$

where w_n is the ion drift velocity near the cathode, the electric fields strength is E_c, and l is the depth of the cathode layer. Correspondingly, the electric field strength distribution over the cathode layer has the form

$$E = E_n(1 - z/l)^{2/3}, \tag{9.3.12}$$

that gives for the cathode voltage U_c in this case

$$U_c = \frac{3E_n l}{5} \tag{9.3.13}$$

The electric field strength near the cathode E_c follows from the condition of discharge self-maintenance (9.3.1). Introduction the reduced distance from cathode $x = z/l$ leads to the relation

$$J(b) = \int_0^1 \exp\left(-\frac{b}{z^{2/3}}\right) dz = \frac{\ln(1+1/\gamma)}{AN_a l}, \tag{9.3.14}$$

where $b = BN_a/E_c$ and the first Townsend coefficient α is given by formula (8.1.4). Taking the cathode voltage (9.3.13) as a function of the parameter b, we require this voltage to be minimal with respect to this parameter that is equivalent to the maximum of the function $bJ(b)$ and leads to the following value of this parameter

$$b = \frac{BN_a}{E_n} = 0.665 \tag{9.3.15}$$

This gives $E_n/N_a = 1650\,\mathrm{Td}$ for helium and $E_n/N_a = 2300\,\mathrm{Td}$ for argon. From this we have the following relations for other quantities

$$y = AN_a l = 3.24\ln(1+1/\gamma), \ U_c = 2.92\frac{B}{A}\ln(1+1/\gamma), \tag{9.3.16}$$

and these relations allow one to determine the parameters of the cathode region on the basis of parameters A and B of the first Townsend coefficient and the value γ of the second Townsend coefficient. In using of these relations and experimental values of parameters of the cathode layer, one can determine the second Townsend coefficient. In particular, the second relation (9.3.16) gives

$$\gamma = \frac{1}{\exp\left(\frac{AU_c}{B}\right) - 1} \tag{9.3.17}$$

Table 9.3 contains values of the second Townsend coefficient in helium and argon obtained on the basis of formula (9.3.17). Of course, the accuracy of these data is less, the less value γ is. Assuming the reduced electric field strength at the cathode to be in the range $10^3\,\mathrm{Td} < E_n/N_a < 3\cdot 10^3\,\mathrm{Td}$, we obtain on the basis of Table 9.2 data and formulas (9.3.16) for the cathode voltage U_c in helium and argon

$$U_c(He) = 97\ln(1+1/\gamma), \ U_c(Ar) = 48\ln(1+1/\gamma) \tag{9.3.18}$$

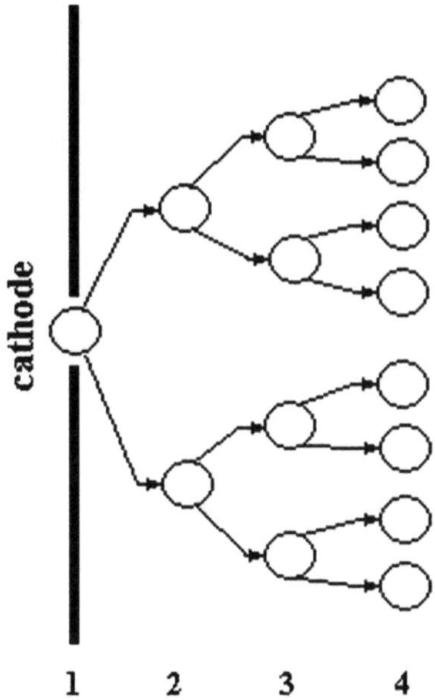

Fig. 9.7 Local character of atom ionization by electron impact in the cathode layer. In this case neglecting by energy losses as a result of electric electron-atom collisions, one can assume that each electron produces one ionization act if it acquires from the electric field on average an energy ε_o, the energetic cost of an electron-ion pair. Values mean a number of electron generation

where the cathode voltage U_c is expressed in V. Correspondingly, on the basis of formulas (9.3.16) and (9.3.18) with using $A = 2.2\,\text{Å}$ for helium and $A = 0.91\,\text{Å}$ according to formula (8.1.5) we obtain

$$N_a l(He) = 1.5 U_c(He), \ N_a l(Ar) = 7.4 U_c(Ar), \qquad (9.3.19)$$

where U_c is the cathode voltage given in Table 9.2 and expressed in this formula in V, and $N_a l$ is expressed in $10^{14}\,\text{cm}^{-2}$.

In the above consideration we are based on the local character of ionization in the cathode layer, as it is shown in Fig. 9.7. Introducing an energy ε_o that is consumed on one ionization act on average, we obtain a number of ionization acts $k = U_c/\varepsilon_o$ for a secondary electron which is formed at the cathode surface. As a result, we have for the total number of electrons formed from one secondary electron which is emitted from the cathode surface by ion bombardment

$$n = 2^k,$$

that is valid in the limit $k \to \infty$. Since the condition of electron reproduction has the form $n = 1/\gamma$ (for simplicity we assume $\gamma \ll 1$), we obtain on the basis of formulas (9.3.18)

$$k = \frac{\ln(1/\gamma)}{\ln 2} = \frac{U_c}{U_o \ln 2} = \frac{U_c}{\varepsilon_o}$$

Table 9.3 Values of the second Townsend coefficient obtained on the basis of formula (9.3.17)

Cathode/Gas	He	Ar	Cathode/Gas	He	Ar
Mg	0.38	0.10	Sr	0.70	0.18
Al	0.31	0.16	Ag	0.23	0.08
K	1.2	0.38	Cd	0.22	0.10
Ca	0.70	0.18	Ba	0.70	0.18
Fe	0.27	0.038	Pt	0.22	0.079
Ni	0.24	0.079	Au	0.22	0.079
Cu	0.19	0.08	Pb	0.19	0.091
Zn	0.30	0.10	Bi	0.32	0.071

From this we obtain for the energetic cost ε_o of one electron-ion pair

$$\varepsilon_o = U_o \ln 2 \tag{9.3.20}$$

This gives $\varepsilon_o = 67\,\text{eV}$ in the helium case and $\varepsilon_o = 33\,\text{eV}$ in the argon case. Correspondingly, the efficiency of ionization J/ε_o (J is the atom ionization potential) is equal to 0.36 in the helium case and 0.47 in the argon case. We thus conclude within the framework of the local model that the energetic cost ε_o of formation of one electron-ion pair in the cathode layer, as well as the efficiency J/ε_o of this act, is independent of the cathode material. One can explain this by facts that the electric field strength E_c near the cathode and ionization processes in the cathode layer are independent of the cathode material.

By analogy with formulas (9.3.18) and (9.3.19) we obtain for the reduced normal current density i_n at the cathode on the basis of formulas (9.3.11) and (9.3.15)

$$\frac{i_n}{N_a^2} = \frac{E_n^{3/2}(e\lambda)^{1/2}}{3N_a^2 \pi l (2\pi M)^{1/2}} = \frac{0.078 B^{3/2}}{N_a l \sqrt{\sigma_{res} M}}, \tag{9.3.21}$$

where σ_{res} is the cross section of resonant charge exchange, M is the ion mass. Tables 9.4 and 9.5 contain the values of $N_a l$ according to formula (9.3.19) and the reduced normal current density i_n/N_a^2 according to formula (9.3.21) for helium and argon, where the cross sections of the resonant charge exchange process are taken from [228, 229] at the collision energy 10 eV.

It should be noted a restricted accuracy of data of Tables 9.3, 9.4, and 9.5. The basis of determination of these data is the used scheme of processes in the cathode layer. We assume the number density of atoms to be large, and the ion mean free path as a result of the resonant charge exchange process is small compared to the depth of the cathode layer. In addition, we assume the Townsend mechanism of electron reproduction at the cathode as a result of ion bombardment and also the number density of electrons to be low that leads to a direct character of atom ionization in

Table 9.4 The reduced cathode layer depth $N_a l$ calculated on the basis of formula (9.3.19), and the reduced electric current density i_n/N_a^2 in accordance with formula (9.3.21) for helium

Cathode	γ	$N_a l,\ 10^{16}\,\text{cm}^{-2}$	$i_n/N_a^2,\ 10^{-38} A\cdot\text{cm}^4$
Mg	0.38	1.9	4.5
Al	0.31	2.1	4.0
K	1.2	0.9	9.5
Ca	0.70	1.3	6.5
Fe	0.27	2.3	3.5
Ni	0.24	2.4	3.5
Cu	0.19	2.7	3.2
Zn	0.30	2.2	3.9
Sr	0.70	1.3	6.5
Ag	0.23	2.5	3.4
Cd	0.22	2.5	3.4
Ba	0.70	1.3	6.5
Pt	0.22	2.5	3.4
Au	0.22	2.5	3.4
Pb	0.19	2.7	3.2
Bi	0.32	2.1	4.1

Table 9.5 The reduced cathode layer depth $N_a l$ calculated on the basis of formula (9.3.19), and the reduced electric current density i_n/N_a^2 in accordance with formula (9.3.21) for argon

Cathode	γ	$N_a l,\ 10^{16}\,\text{cm}^{-2}$	$i_n/N_a^2,\ 10^{-39} A\cdot\text{cm}^4$
Mg	0.10	8.6	2.5
Al	0.16	7.4	2.9
K	0.38	4.8	4.5
Ca	0.18	6.9	3.1
Fe	0.038	12	1.7
Ni	0.079	12	2.2
Cu	0.08	9.6	2.2
Zn	0.10	8.8	2.4
Sr	0.18	6.9	3.1
Ag	0.08	9.6	2.2
Cd	0.10	8.8	2.4
Ba	0.18	6.9	3.1
Pt	0.079	9.7	2.2
Au	0.079	9.7	2.2
Pb	0.091	9.2	2.3
Bi	0.071	10	2.1

pair collisions of electrons with atoms. If any of these assumptions violates, the data of Tables 9.3, 9.4, and 9.5 becomes wrong.

The above regime of the cathode region is typical for atomic gases or vapors where ion transport is determined by the resonant charge exchange process. The electric current in this regime, the normal regime of glow discharge, occupies only a part of the cathode, and the depth l of the cathode region is less than a distance L between electrodes. Under these conditions, this form of the cathode region is independent of the geometry of a discharge chamber. In particular, it is suitable for gas discharge between two parallel electrodes with a small distance L between them compared to their radius R or if gas discharge is burnt in a cylinder tube.

Let us consider also another case where the ion drift velocity w_i is proportional to the electric field strength E. This holds true at low electric field strength ($eE\lambda \ll T$) or at any electric field strengths for the polarization ion-atom interaction potential that is suitable more or less for molecular ions located in inert gases. In this case the Poisson equation gives for the electric field strength in the cathode region

$$E = E_n\sqrt{1 - z/l}, \tag{9.3.22}$$

where l is the cathode layer length, E_c is the electric field strength at the cathode. According to the definition, the electric field strength is zero at the cathode layer boundary, that means in reality that the electric field strength at the cathode layer boundary is lower than that near the cathode, because a plasma becomes quasineutral there. Formula (9.3.22) gives for the cathode drop

$$U_c = \frac{2}{3}E_n l \tag{9.3.23}$$

Next, equation (9.3.1) of ionization balance takes the form now

$$y\int_0^1 \exp\left(-\frac{b}{x^{1/2}}\right)dx = \ln(1 + 1/\gamma),$$

where $y = AN_a l,\ x = \sqrt{1 - x/l}$. From this it follows

$$b = \frac{BN_a}{E_n} = \frac{2BN_a l}{3U_c} = \frac{2By}{3AU_c},$$

and this equation connects parameters y and b. One more equation of the minimum of the cathode voltage $dU_c/dy = 0$ has the form

$$\frac{db}{dy} = \frac{b}{y}$$

These two equations with the dependence (8.1.4) for the first Townsend coefficient on the electric field strength give

$$b = B\frac{N_a}{E_c} = 0.71 \tag{9.3.24}$$

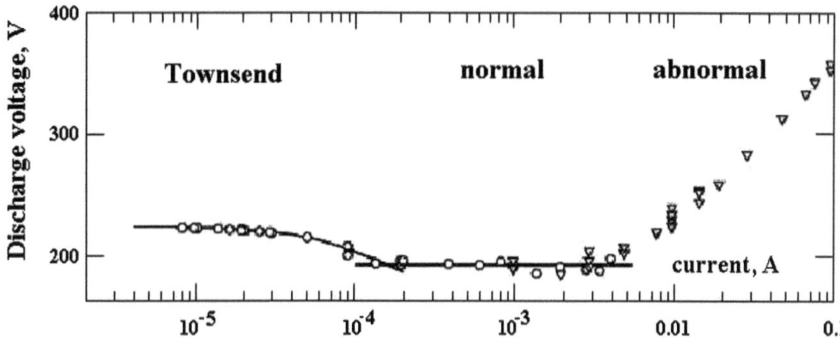

Fig. 9.8 Dependence of the voltage on the discharge current for a system with plane copper electrodes of a diameter 78 mm and the product $pl = 2\,\mathrm{cm}\cdot\mathrm{Torr}$ ($N_a l = 7\cdot 10^{15}\,\mathrm{cm}^{-2}$), where l is a distance between plates and p is the the pressure of argon that fills a space between electrodes [390]

From this we have the following relations for parameters of the cathode layer

$$y = AN_a l = 3.06\ln(1+1/\gamma),\ \ U_c = 2.87\frac{B}{A}\ln(1+1/\gamma) \qquad (9.3.25)$$

These relations are analogous to (9.3.16) for the regime of the cathode region with atomic ions. They allow us to connect the electric parameters of the cathode region with the reduced first and second Townsend coefficients α/N_a, γ and the ion drift velocity w_i.

9.4 Transition from Glow Discharge to Arc and Townsend Discharges

As a result of this analysis, one can understand the role of the cathode region of gas discharge consists in self-maintaining of this gas discharge. This is based on the ionization balance in the cathode region, i.e. the rate of electrons attached to the cathode is equal to the rate of electron formation as a result of ionization processes in a gas. According to the character of ionization processes in the cathode plasma one can separate three range of cathode currents which are represented in Fig. 9.8 for certain parameters of discharge in argon between two parallel electrodes in the form of disks. At low currents which we call the Townsend discharge form, the ionization equilibrium is determined by equation (8.1.3). In this limit the electron concentration is low and electrons of this discharge do not interact with each other. the electric field strength in a space between electrodes is uniform in this case.

As the discharge current increases, i.e. the electron number density in a space between electrodes increases, the space distribution of the electric field becomes nonuniform owing to noncompensated space charge between electrodes. Moreover,

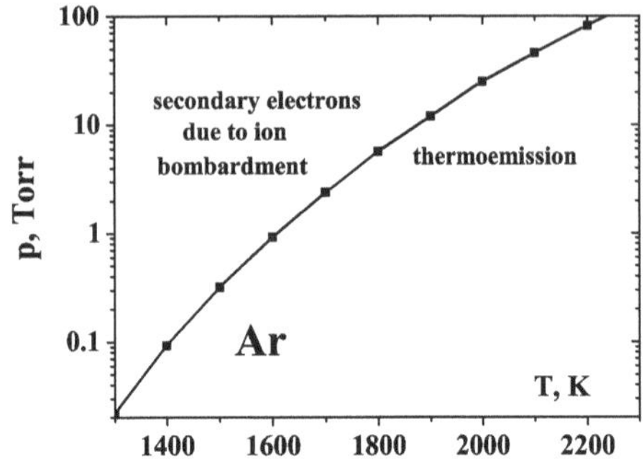

Fig. 9.9 The boundary between normal glow discharge and arc in argon and for an aluminum cathode

the optimal current density is realized in this regime that allows one to chose the minimum cathode drop which is independent of the electric current. In this normal regime of the cathode region electrons and ions occupy only a part of the cathode surface, and this part increases with an increasing current. Therefore the voltage-current characteristic in this current range is given by a horizontal line, as it takes place in Fig. 9.8. At large currents the abnormal regime of the cathode region occurs. In this regime the current covers all the cathode region that does not allow one to reach the optimal voltage, and an additional space charge is compensated by an increasing voltage of the cathode region in accordance with Fig. 9.8.

A subsequent increase of the discharge current leads to transition from glow discharge to arc. It should be noted that this transition results from change of electron emission from the cathode, while properties of the positive column does not vary at this transition. Indeed, emission of electrons as a result of cathode bombardment by ions for glow discharge is changed by thermoemission of electrons for arc. In this case heat release from the cathode is restricted and the cathode temperature becomes enough large. In arc the thermoemission current exceeds remarkable that due to emission of second electrons as a result of cathode bombardment of ions. As example, in the case of the Al cathode and argon as a gas the normal current density is $i_n = 0.22\,\mathrm{mA/(cm^2Torr^2)}$ [393] for glow discharge in argon. Figure 9.9 shows the dependence of the argon pressure p on the argon temperature T if the normal current density i_n in glow discharge is equal to the current density of thermoemission emission of electrons i_{th} given by formula (2.5.3). Transition from glow discharge to arc takes place at cathode temperatures of such of an order of magnitude.

The above analysis of the cathode region of normal glow discharge allows us also to estimate at which parameters the transition to Townsend or dark gas discharge occurs. In analyzing the transition between the Townsend and glow regimes, we use that reproduction of electrons at the cathode is the same in both cases, so that the character of this transition is determined by the second Townsend coefficient γ and the number density N_a of gas atoms in the standard consideration that was analyzed both experimentally [392, 393] and theoretically (for example [394–397]. Now we only compared parameters of Townsend and glow discharges and a boundary between these discharge regimes.

Let us consider for simplicity glow gas discharge between two parallel electrodes which are disks, and the current occupies a small part of the electrode surface in the normal regime. The current may travel along the electrode surface, and in order to exclude this effect, let us place in the electrode center small discs that decreases a distance between electrodes in this region and compels the current to pass through this electrode region. The parameters of the cathode region for glow discharge are determined by formulas (9.3.13) and (9.3.16). Since the current occupies a small part of the electrode surface, parameters of a gas discharge plasma in the cathode region are independent of the discharge geometry.

Based on the data for helium and argon, we obtain two regimes of the electric current between electrodes - disks, so that for the Townsend regime the current is distributed more or less uniformly over the electrode surface, in the glow regime the current occupies a small part of the electrode area. One can compare these regimes if $L > l$, i.e. a distance between electrodes L exceeds the cathode region depth l of glow discharge. Practically this corresponds to the right part of the Pashen curves given in Fig. 9.1. On the basis of data of Tables 9.2 and 9.4 we have $U_c > U_{min}$ so that near the minimum of the Pashen curve the Townsend regime of gas discharge is favorable. At larger number densities of atoms or a distance between electrodes glow discharge form becomes preferable. It should be noted that the accuracy of data of Table 9.2 and 9.4 is restricted because of errors in experimental data and in approximation of the first Townsend coefficient. But the general conclusion of existence of two regimes of gas discharge holds true. At given parameters one of these regimes is stable, and the other one is metastable.

We note that the parameters $(N_aL)_{min}$ and U_{min} of the Townsend discharge given by formulas (9.1.2) and (9.1.3) are in conformity with the parameters N_al and U_c according to formula (9.3.16) for glow discharge. Indeed, if the first Townsend coefficient is approximated by formulas (8.1.4) and (8.1.5) and the parameters A and B of this approximation are the same in all the range of electric field strengths, the ratios of these parameter are

$$\frac{(N_aL)_{min}}{N_al} = \frac{e}{3.24} = 0.84, \quad \frac{U_b}{U_c} = \frac{e}{2.92} = 0.93 \tag{9.4.1}$$

Nevertheless, these two forms of gas discharge differ significantly and the character of electric currents in them is not the same. Transition to glow discharge corresponds to plasma shrinking and charge separation in it, so that the electric field strength near the cathode increases by an order of magnitude under optimal conditions for Townsend discharge.

Thus, we have two forms of gas discharge in a gap between two parallel electrodes filled by a gas, if reproduction of electrons near the cathode results from ion bombardment of the cathode. A gas discharge plasma occupies uniformly all the space between electrodes for Townsend discharge and the number density of electrons and ions is small, i.e. an electric field created by electrons and ions is weak. In glow discharge the electric current occupies a small part of the electrode area and a gas discharge plasma is concentrated in a small part of a space between electrodes.

At low currents the latter becomes unstable because transverse diffusion of a gas discharge plasma in a surrounding space. But the transition between Townsend and glow discharge requires redistribution of a gas discharge plasma in a space and may be dependent of certain discharge conditions.

9.5 Plasma Sheath at Walls

If a plasma is a restricted by boundaries, one can extract two regions [46, 47, 90]. The main part of this is uniform, and just this part obtained the name "plasma" [46]. This plasma is almost neutral and drift of charged particles, electrons and positive ions, to the plasma boundary results in this region from ambipolar diffusion [18, 29]. The plasma near boundaries is nonuniform and according to the Langmuir analysis of the positive column of mercury arc in 1923 [48], this plasma region forms "a sheath of definite thickness containing only positive ions and neutral atoms". This structure of the boundary region equalizes the electron and ion currents to the plasma boundaries.

We first consider the case where the thickness Δ of the plasma sheath region is small compared to the mean free path of ions λ_i [398–400], i.e. a plasma of the double layer is collisionless. We take the electron temperature T_e to be large compared to the ion temperature that is equal to the gaseous temperature T, i.e. $T_e \gg T$. This criterion leads to a large drift velocity of ions compared to that in the presheath region. If ions penetrate in the sheath region with the velocity u_o, the conservation of the ion flux gives

$$N_i(z) = \frac{N_o}{\sqrt{1 - -2eU(z)/(Mu_o^2)}}, \tag{9.5.1}$$

where $U(z)$ is the electric potential of the sheath plasma on a distance z from walls. The Poisson equation has the following form in this case

$$\frac{d^2U}{dz^2} = 4\pi e(N_i - N_i) = 4\pi e N_o \left[\frac{1}{\sqrt{1 - 2eU(z)/(Mu_o^2)}} - \exp\left(-\frac{eU}{T_e}\right)\right], \tag{9.5.2}$$

where we use the Boltzmann formula for the electron number density

$$N_e(z) = N_o \exp\left[-\frac{eU(z)}{T_e}\right]$$

One can see that in the region $eU \gg T_e$ this problem may be reduced to the Child-Langmuir problem [378–381] if atomic charged particles are moving through a plane gap, and the charge of these particles created the electric field which acts on motion of these charged particles. In particular, we encounter with this problem in the analysis of the cathode layer, and then the electric voltage $U(x)$ for the plane layer, the current density i of ions in this layer and the layer thickness L are connected by formulas (9.3.6), (9.3.7), and (9.3.8). These relations corresponds to the criterion (9.3.9). More precise solution of (9.5.2) is given in [401].

Let us consider one more peculiarity of interaction of the ion flux with a plasma through which it propagates [402–404]. Oscillations may be developed in a uniform plasma in the form of the ion sound with the phase velocity $\sqrt{T_e/M}$, if the electron temperature is large compared to the ion temperature $T_e \gg T_i$. Therefore strong interaction takes place of an ion flux with the ion sound if the flux velocity exceeds weakly the phase velocity $\sqrt{T_e/M}$ of the ion sound. But though the velocity of an ion flux in the double layer may exceed the phase velocity of the ion sound, interaction of this ion flux with ion sound is weak because of a low number density of electrons.

We now consider another limiting case if the double layer thickness Δ is large compared to the mean free path λ of ions. In this case motion of electrons and ions in a gas is determined by their kinetic coefficients in gas, and assuming the double layer to be narrow, one can describe the distribution of the number densities of electrons $N_e(x)$ and ions $N_i(x)$ within the framework of a one-dimensional problem by equations of the electron and ion balance and Poisson equation for the electric field strength $E(x)$ as

$$j = -D_e \frac{dN_e}{dz} - EK_eN_e, \; j = -D_i \frac{dN_i}{dz} + EK_iN_i, \; \frac{dE}{dz} = 4\pi e(N_i - N_e), \quad (9.5.3)$$

Here z is a distance from walls, j is the flux of electrons and ions toward walls, D_e, D_i are the diffusion coefficients of electrons and ions in a gas, K_e, K_i are their mobilities, and we take the drift velocities of electrons w_e and ions w_i as $w_e = EK_e$, $w_i = EK_i$. The boundary conditions for (9.5.3) are

$$N_e(0) = 0, \; N_i(0) = 0, \; E(0) = E_o, \quad (9.5.4)$$

and E_o is the electric field strength at walls. Equations (9.5.3) with boundary conditions (9.5.4) describes a general concept of the double layer and its transition to a uniform plasma with the ambipolar mechanism of plasma drift to walls. Take for definiteness that a gas discharge plasma is restricted by dielectric walls. Then a negative charge is accumulated on the walls which repel electrons and equalizes currents of electrons and ions toward the walls. Attaching to walls, electrons and ions recombine of this surface subsequently.

In considering the double layer as a transient layer between the positive column and walls, let us describe the positive column of gas discharge within the framework of the Schottky model [347, 348], if the number density of electrons is relatively small and the electric current does not influence on its temperature distribution along the cross section. For a gas discharge plasma inside a cylinder tube of a radius ρ_o the number density of electrons and ions along the tube cross section is equal [347, 348]

$$N_e(\rho) = N_o J_0 \left(2.405 \frac{\rho}{\rho_o}\right),$$

where N_o is the number density of electrons at the tube axis, ρ is a distance of a given point from the axis. This formula gives near walls $\rho \approx \rho_o$

$$N_e(\rho) = 1.25N_o\left(1 - \frac{\rho}{\rho_o}\right) \tag{9.5.5}$$

Let us use an approximated solution of the set (9.5.3) of equations assuming inside the double layer $N_e \ll N_i$. Solving the Poisson equation (9.5.3) for the electric field strength with the boundary conditions (9.5.4) $E(0) = E_o$ and $E(\Delta) = 0$ we have

$$E(x) = E_o\left(1 - \frac{x^2}{\Delta^2}\right), \tag{9.5.6}$$

where $x = \rho_o - \rho$ is a distance from walls. Taking the distribution of ions (9.5.5) near walls, we have the relation between the double layer thickness Δ and the electric field strength E_o at walls

$$\Delta = \sqrt{\frac{E_o\rho_o}{2.5\pi N_o e}} \tag{9.5.7}$$

It should be noted that a general form of the electric field strength E_o has the form [221]

$$E_o \sim \frac{T_e}{r_D}$$

because this field is screened at a distance that is of the order of the Debye-Hückel radius r_D that increases with a decreasing number density of electrons and ions. Since we take the Debye-Hückel radius r_D that corresponds to the number density of electrons and ions N_o at the axis, the expression for the double layer thickness contains the tube radius ρ_o.

We note that the electron temperature inside the double layer coincides with that in the positive column because the mean free path of electrons in inert gases with respect to change their energy exceeds that for change of the direction of electron motion roughly by two orders of magnitude [76]. Hence the energy distribution function of electrons located inside the double layer coincides with that in the positive column. Introducing the effective electron temperature T_e according to the Einstein relation [92, 93] as well as the ion temperature T_i

$$T_e = \frac{eD_e}{K_e}, \quad T_i = \frac{eD_i}{K_i},$$

we have for the number densities of electrons $N_e(x)$ and ions $N_i(x)$ inside the double layer with accounting for the boundary conditions (9.5.4)

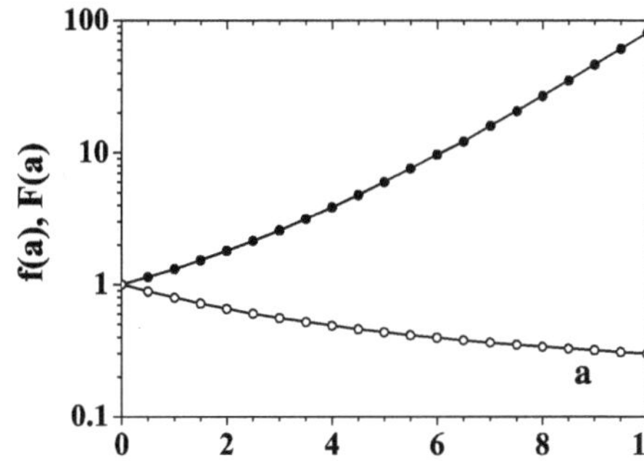

Fig. 9.10 Dependencies $f(a)$ (*the upper curve*) and $F(a)$ (*the lower curve*) according to formulas (9.5.10)

$$N_e(x) = \frac{j}{D_e}\exp\left[\frac{eU(x)}{T_e}\right]\int_0^x \exp\left[-\frac{eU(x')}{T_e}\right]dx',$$

$$N_i(x) = \frac{j}{D_i}\exp\left[\frac{eU(x)}{T_i}\right]\int_0^x \exp\left[-\frac{eU(x')}{T_i}\right]dx', \qquad (9.5.8)$$

and the electric voltage $U(x)$ is

$$U(x) = E_o x\left(1 - \frac{x^2}{3\Delta^2}\right), \quad U_o = U(\Delta) = \frac{2eE_o\Delta}{3} \qquad (9.5.9)$$

We define the electric voltage $U(0) = 0$ and take at the double layer boundary $N_e(\Delta) = N_i(\Delta)$. From this relationship we have on the basis of formulas (9.5.8) for the number densities of electrons and ions

$$\frac{D_e(T_e)}{D_i(T_i)} = \frac{f(a_i)}{F(a_e)}, \quad a_i = \frac{eE_o\Delta}{T_i}, \quad a_e = \frac{eE_o\Delta}{T_e},$$

$$f(a) = \int_0^1 dt\exp\left[\tfrac{a}{3}(1-t)(2-t-t^2)\right],$$

$$F(a) = \int_0^1 dt\exp\left[-\tfrac{a}{3}(1-t)(2-t-t^2)\right] \qquad (9.5.10)$$

The dependencies $f(a)$ and $F(a)$ are given in Fig. 9.10. In the limit of high a these functions are

$$f(a) = \frac{\exp(2a/3)}{a}, \quad F(a) = \sqrt{\frac{\pi}{4a}} \qquad (9.5.11)$$

We note that for typical electric field strengths in the positive column of glow gas discharge the ion temperature coincides with the gas temperature T, i.e. the velocity distribution function of electrons is close to the Maxwell one, and an electric field of glow gas discharge acts weakly on ions. On contrary, action of electric field on electrons is strong, and the effective electron temperature exceeds strongly the ion and atom temperature.

Let us consider the Townsend regime of the positive column where the number density of electrons and ions is small and does not influence on plasma properties. Evidently, this regime correspond to criteria

$$r_D \gg \rho_o, \ \lambda \gg \Delta \tag{9.5.12}$$

Under these criteria we have for the fluxes of charged particles to the walls for a two-temperature gas discharge plasma

$$j_e = N_o \sqrt{\frac{T_e}{2\pi m_e}} \exp\left(-\frac{eU}{T_e}\right), \ j_i = N_o \sqrt{\frac{T_i}{2\pi m_i}}, \tag{9.5.13}$$

where N_o is the number densities of electrons and ions in the quasineutral plasma region, m_e and m_i are the electron and ion masses, and U is the wall electric potential with respect to the plasma. These formulas are based on the Maxwell distribution function of electrons and ions far from the walls. From equality of the electron and ion fluxes to walls we obtain for the wall potential

$$eU_o = \frac{T_e}{2} \ln\left(\frac{T_e m_i}{T_i m_e}\right) \tag{9.5.14}$$

9.6 Principles of Magnetron Discharge

The magnetic field inserted in a plasma may create magnetic traps for electrons and in this manner to influence the parameters of gas discharge. In magnetron discharge that we consider below a magnetic trap varies cathode processes because it does not allow for electrons to penetrate in the cathode region from a plasma. This leads to an increase the cathode voltage, and ions accelerate in the cathode region up to high energies in order to maintain the ionization equilibrium. As the secondary phenomenon, sputtering of the cathode proceeds due to a high ion energy, and such a generator of metal atoms is used widely in applications.

A concept of the magnetron discharge was suggested by Penning [387], and later this scheme was modified [388, 389, 405, 406] that provides using the magnetron discharge for applications. A typical scheme of magnetron discharge [407] is given in Fig. 9.11 where the magnetic field is created by two coaxial magnets with opposite positions of poles. At these positions of magnets the total magnetic field strength has a maximum near the cathode and is directed parallel to its radius, as it is shown in Fig. 9.12 [408]. Under the action of the electric and magnetic fields, the electron trajectory in absence of collisions takes place along the circle of the maximum mag-

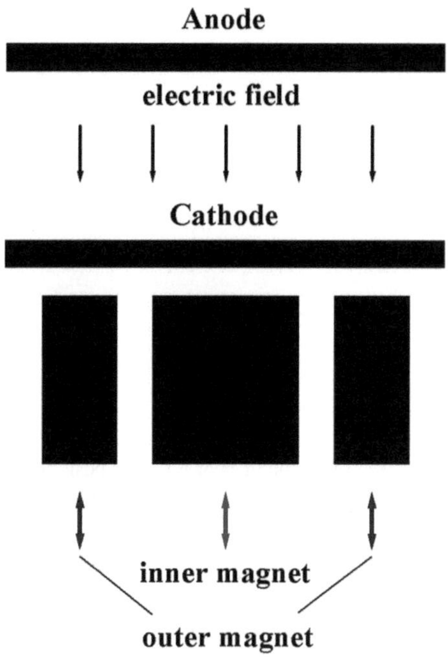

Fig. 9.11 Scheme of typical magnetron discharge [408]

netic field according to formula (7.9.5). As a result, we obtain a magnetic trap which captures slow electrons. Captured electrons move in the region of a high magnetic field and can leave this region in collisions with other electrons if an exchange by energy is comparable with the energy of a magnetic well that is of the order of μH, where H is the magnetic field strength, μ is the magnetic moment for a captured electron. In the scheme of (Fig. 9.11) the electric field directs along the axis and has the axial symmetry. Though in reality the axial symmetry of the electric field may be violated, we conserve it subsequently for simplification of the analysis.

Let us determine the depth of the magnetic trap well eU_{max} that is equal to $eU_{max} = \mu H$, and the magnetic moment μ for this trap is

$$\mu = \frac{JS}{c},$$

where $J = ev_e/2\pi r$ is the electric current for a test electron, v_e is its drift velocity according to formula (7.9.5), r is the radius of a circle - the electron trajectory that is determined by the geometry of the magnetic field, and $S = \pi r^2$ is the area restricted the electron trajectory. On the basis of these expressions we obtain

$$eU_{max} = \mu H = \frac{eEr}{2} \qquad (9.6.1)$$

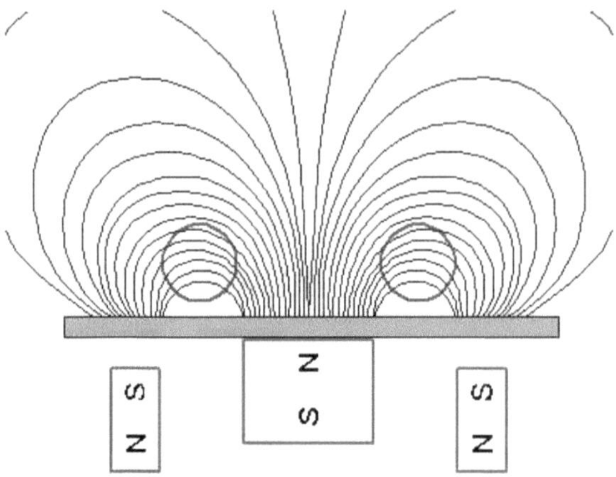

Fig. 9.12 Magnetic lines of force for positions of magnets given in Fig. 9.11. The regions of the maximum magnetic fields is marked [408]

From this it follows that the depth of the potential well for a captured electron is independent of the magnetic field strength and is determined by the electric field strength. The criterion of the magnetic trap

$$\mu H > T_e, \tag{9.6.2}$$

where T_e is the electron temperature, requires that the number density of electrons increases in the region of the magnetic trap and is connected with a large electric field strength. Note that we assume a uniform electric field in the scheme of Fig. 9.11, while in reality it is essential the electric field strength in the region of the magnetic trap. Hence, a nonuniform electric field distribution in a space is used usually [408–411]. Nevertheless, we will use a uniform electric field, as it is given in Fig. 9.11 since this simplifies the analysis.

Assuming the magnetic trap to be located near the cathode, we have that electrons are captured mostly by the magnetic trap and cannot not to reach the cathode. This means that ionization in the cathode region proceeds due to ions only which bombard the cathode that leads to formation secondary electrons. These secondary electrons compensate losses due to departure of ions to the cathode and in this manner maintain the magnetron discharge. Hence, the condition of maintaining of magnetron discharge has the form

$$\gamma(n+1) = 1, \tag{9.6.3}$$

where γ is the probability of formation of a secondary electron as a result of cathode bombardment by an ion, and n is a number of electrons formed in the cathode region due to one secondary electron. This criterion holds true if the cathode layer dimension L is small compared with the mean free path λ of ions

$$\lambda \gg L, \tag{9.6.4}$$

Table 9.6 Energetics of typical processes including cathode processes for argon sputtering in magnetron discharge [414, 415]

Cathode	ε	E_{kin}	E_{rad}	E_{heat}
Al	3.33	6	4	13
Ti	4.86	8	9	21
V	5.29	7	8	20
Cr	4.11	8	4	18
Fe	2.26	9	4	18
Ni	4.45	11	4	17
Cu	3.50	6	2	14
Zr	6.34	13	7	34
Nb	6.50	13	8	28
Mo	6.88	13	6	36
Rh	5.60	13	4	33
Cd	1.16	4	1	7
In	2.52	4	2	15
Hf	6.33	20	7	48
Ta	8.10	21	9	43
W	8.80	22	9	56
Au	3.92	13	2	21

The specific energies of corresponding processes are expressed in $eV/atom$ and relate to the binding energy in solid metals ε, the average kinetic energy of sputtered atoms E_{kin}, the radiation cathode energy E_{rad}, and a typical specific heat energy transferred to argon and to the cathode E_{heat}

and ions pass the cathode region without collisions. In addition, the current density of ions is restricted, and the voltage drop (9.3.6) due to the ion charge is small compared to the cathode potential U_c

$$U_c \gg U \tag{9.6.5}$$

Because the cathode voltage drop for magnetron discharge is several hundreds of eV, ion bombardment of the cathode along with formation of secondary electrons leads to other phenomena, in first turn, to cathode erosion and formation of free metal atoms that is the reason of applications of magnetron discharge. Typical values for energetic processes on the cathode of magnetron discharge are given in Table 9.6 [412, 413]. These values are guided by axial magnetron discharge in argon of a pressure 0.13-1.3 Pa, the magnetic fields strength of 100–200 Gs near the cathode, the discharge voltage of 500–1000 V and electric current densities of 30–120 A/m^2.

In the end of this paragraph we compare the cathode layers of glow gas discharge and magnetron discharge as regions where electrons are reproduced. In glow discharge the reproduction of electrons results from ionization electron-atom collisions in the cathode layer, and this region length exceeds the mean free path of electrons by some orders of magnitude. In addition, reproduction of electrons results from plasma electrons located in the cathode layer, while in magnetron discharge these electrons are locked in the magnetic trap and cannot partake in the reproduction process. Hence in magnetron discharge the process of discharge self-maintenance is owing to secondary electrons, and the ionization process proceeds outside the cathode layer where secondary electrons lose their energy. Thus, introduction of a magnetic field change the mechanism of electron reproduction, and hence parameters of magnetron discharge depend on the position of the magnetic trap.

In this consideration, an additional element, the magnetic trap, is introduced in magnetron gas discharge compared with glow discharge, and hence the position of the magnetic trap with respect to position of the cathode acts on the parameters of magnetron discharge. Indeed, the electric potential that act on electrons is a sum of the cathode and magnetic trap [414], and the cathode voltage becomes large, if the magnetic trap is located far from the cathode (until magnetron discharge transfers into glow discharge), and action of the magnetic filed decreases if the magnetic trap approaches to the cathode. Evidently, the optimal distance of the magnetic trap from the cathode is of the order of the mean free path of ions, and detailed analysis of the cathode region (for example, [411, 414–417]) testifies the complexity of processes which provide self-maintenance of magnetron discharge.

Part IV
Helium and Argon Gas Discharge Plasmas

Chapter 10
Atom Excitation in Helium and Argon Uniform Plasma

Abstract Atom excitation in a uniform gas discharge plasma of helium and argon is analyzed under various conditions. Fast electrons under consideration determine the rates of atom excitation and ionization in a gas discharge plasma. The specific power of plasma radiation in a continuous spectrum range resulted from the photorecombination process is given.

10.1 Excitation of Metastable State in Helium Plasma at Low Electron Concentrations

Taking the diffusion cross section for electron scattering on the helium atom according to formula (3.2.8), we obtain for the parameter χ defined by formula (6.5.3)

$$\chi = \frac{350}{x^2}, \tag{10.1.1}$$

where the reduced electric field strength $x = E/N_a$ is given in Td. We assume the average electron energy to be small compared to $\varepsilon_c = 10\,\mathrm{eV}$, and the normalization of the distribution function coincides with that for the Druyvesteyn distribution function (6.1.9). Note that in the Druyvesteyn case with $\sigma_* = 6\,\text{Å}^2$ we have $\chi = 580/x^2$, i.e. the transition from the constant cross section to the cross section (3.2.8) leads to a change of the distribution function at the excitation threshold for any electric field strength. This shows that the behavior of the electron-atom elastic cross section is important for atom excitation by electron impact in a gas discharge plasma.

We now analyze the validity of the criterion (6.5.4) that corresponds to a relative smallness of the parameter $1/\kappa$, where κ is given by formula (6.5.3). Substituting in formula (6.5.3) parameters of excitation of the metastable helium atom $He(2^3S)$, namely, $v_o = 2.64 \times 10^8\,\mathrm{cm/s}$, $k_q = 3.1 \times 10^{-9}\,\mathrm{cm^3/s}$, $k(v_o) = 7.1 \times 10^{-8}\,\mathrm{cm^3/s}$, $k_{ef} = 4.5 \times 10^{-8}\,\mathrm{cm^3/s}$, we obtain for excitation of the 2^3S state

$$\kappa = \frac{268}{x} \tag{10.1.2}$$

B.M. Smirnov, *Theory of Gas Discharge Plasma*, Springer Series on Atomic, Optical, and Plasma Physics 84, DOI 10.1007/978-3-319-11065-3_10

As is seen, in the range of the electric field strengths $2Td < x < 10Td$ this parameter is large. This means a strong distortion of the tail of the electron distribution function due to atom excitation.

Let us use the parameter δ in accordance with formulas (6.5.15) and (6.5.16) which characterizes the closeness of the effective energy of atom excitation and the atom excitation energy $\Delta\varepsilon$. For excitation of the state $He(2^3S)$ the parameter (6.5.16) is equal

$$\delta = \left(\frac{\varepsilon_b - \Delta\varepsilon}{\Delta\varepsilon}\right)^{1/4} = \frac{3m_e}{4M}\frac{v_o \nu_o^2}{a\nu_{ef}} = \frac{0.23}{x}, \qquad (10.1.3)$$

and in the range of the electric field strengths $2Td < x$ under consideration this parameter is small.

Thus, there are two limits for atom excitation by electron impact in a gas discharge plasma. In the first case the rate of excitation is relatively small, so that the excitation process does not influence on the energy electron distribution function at energies above the atom excitation threshold. Then in the regime of a high electron number densities we have the Boltzmann distribution function of electrons above the excitation threshold, and the rate constant of atom excitation is given by formula (3.3.8). In the other limit, where the criterion (6.5.4) holds true, atom excitation by electron impact is a self-consistent process, so that this process leads to a decrease of the electron distribution function and this, in turn, influences on the excitation rate. Hence the rate of atom excitation by electron impact is given by formula (6.5.13). The latter takes place in the case of excitation of the helium atom into the metastable state.

We first consider the regime of a low number density of electrons and determine the rate constants of excitation of the metastable $He(2^3S)$ state $k_<$ and $k_>$ according to formulas (6.3.8) and (6.5.7). Using the energy distribution function (6.1.9), we obtain the rate constant of formation of fast electrons in an ionized helium under the action of the electric field

$$k_< = \frac{3.8 \cdot 10^{-9}}{x^{3/2}} \exp\left(-\frac{350}{x^2}\right), \qquad (10.1.4)$$

where the rate constant is given in cm^3/s and the reduced electric field strength $x = E/N_a$ is expressed in Td.

In the same manner on the basis of formulas (6.1.9) and (6.5.9) we find the rate constant of excitation of helium atoms $k_>$ which corresponds to assumption that a weak process is atom excitation above the excitation threshold, but the excitation process disturbs significantly the electron distribution function above the excitation threshold ($\kappa \gg 1$). We have

$$k_> = 1.6 \cdot 10^{-9} x^{-0.3} \exp\left(-\frac{350}{x^2}\right) \qquad (10.1.5)$$

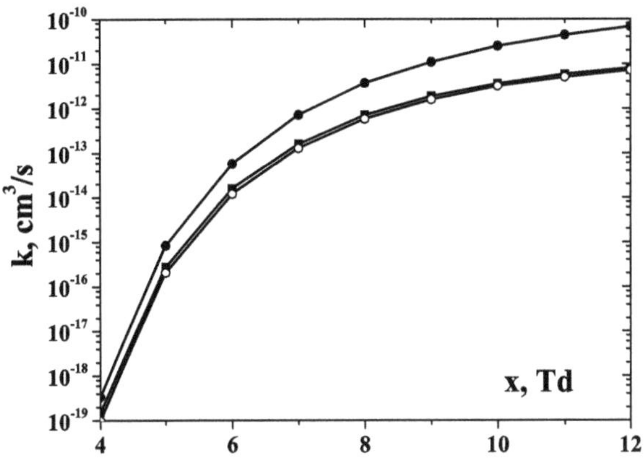

Fig. 10.1 Rate constants of helium atom excitation to the metastable state $He(2^3S)$ by electron impact for weakly ionized helium located in a constant electric field in the regime of low electron number densities. The rate constants $k_>$ calculated on the basis of formula (10.1.6) are given by *closed circles*, the rate constants $k_<$ determined by to formula (10.1.5) are represented by *closed squares*, and *open circles* correspond to the total rate constants k_{ex} of atom excitation according to formula (10.1.7)

From this on the basis of formula (6.5.13) we obtain for the rate constant of excitation of the metastable state $He(2^3S)$ by electron impact without assumption about the ratio between $k_<$ and $k_>$

$$k_{ex} = \frac{1.6 \cdot 10^{-9} \exp(-350/x^2)}{x^{0.3}(1 + 0.43x^{1.2})} \qquad (10.1.6)$$

Figure 10.1 contains dependencies (10.1.6), (10.1.6), and (10.1.6) in the appropriate range of electric field strengths.

Using formulas (10.1.4) and (10.1.5) for the rate constants $k_<$ and $k_>$ of atom excitation in the state $He(2^3S)$ by electron impact in an external electric field in the regime of low electron number densities, we obtain for the ratio of these rate constants

$$\zeta = \frac{k_>}{k_<} = \frac{x^{1.2}}{2.3} \qquad (10.1.7)$$

Figure 10.2 contains this ratio as a function of the reduced electric field strength $x = E/N_a$ in the appropriate range of the electric field strengths. These rate constants coincide at the reduced electric field strength $x = 2Td$, and in the range under consideration atom excitation above the excitation threshold dominates.

From this one can determine the efficiency of atom excitation in a gas discharge plasma, i.e. the part of energy transferred from an electric field to electrons which is consumed on atom excitation. We have for the parameter γ defined according to formula (6.4.3)

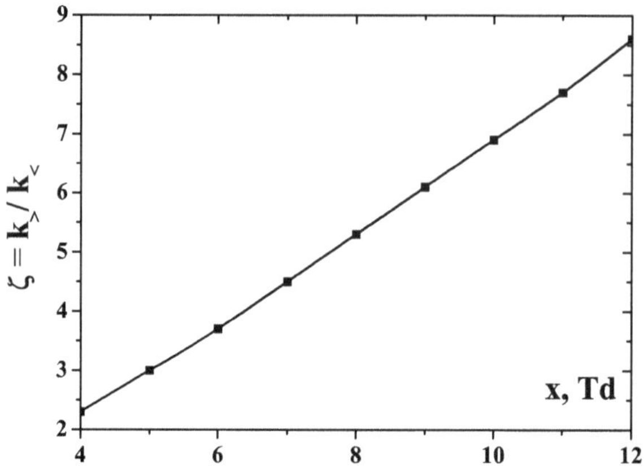

Fig. 10.2 The ratio for the atom excitation rates by electron impact between pair electron-atom collisions for the tail of the electron distribution function and that resulted from diffusion electron motion in an energy space for excitation of the helium atom in the metastable state $He(2^3S)$ by electron impact in weakly ionized helium located in a constant electric field for the Druyvesteyn case of electron-atom scattering in the regime of a low electron number density. The reduced electric field strength $x = E/N_a$ is expressed in Td

$$\gamma = \frac{p_{ex}}{p_{el}}, \tag{10.1.8}$$

where $p_{ex} = \Delta\varepsilon k_{ex}$ is the specific power that is consumed on atom excitation, and $p_{el} = eEw_e/N_a$ is the specific power that is consumed on elastic electron scattering on atoms. Using formula (6.1.10) for the electron drift velocity in the Druyvesteyn case, where the diffusion cross section of electron-atom scattering is independent of the collision velocity, we have for the specific power due to elastic electron-atom scattering in helium

$$p_{el} = 5.3 \cdot 10^{-12} x^{3/2}, \tag{10.1.9}$$

where the specific power is expressed in $\mathrm{eV} \cdot \mathrm{cm}^3/\mathrm{s}$, and the reduced electric field strength is given in Td.

The efficiency of atom excitation is given by formula (6.4.4)

$$\xi = \frac{\gamma}{1+\gamma}$$

Figure 10.3 contains the efficiency of excitation of the helium atom in the metastable state $He(2^3S)$ by electron impact in a helium gas discharge plasma that follows from formulas (10.1.7), (10.1.8) and (10.1.9).

Note that according to formulas (10.2.1) and (10.2.2), at large electric field strengths $k_> \gg k_<$, and $k_{ex} \approx k_<$ for specific powers of processes involving electron-atom collisions. This gives for the parameter γ

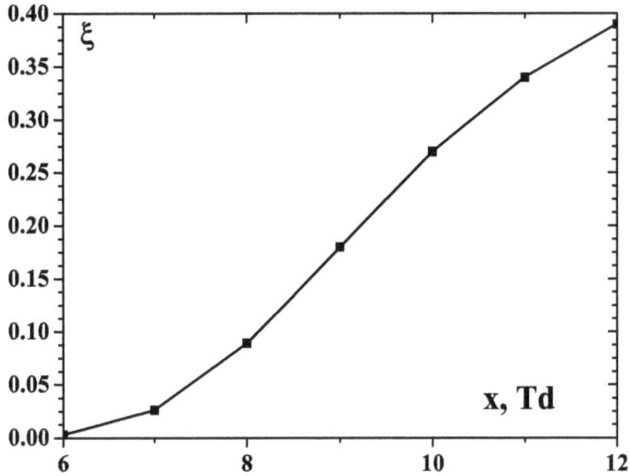

Fig. 10.3 Given by formulas (6.4.7) and (6.4.4) the efficiencies of excitation of the helium atom in the metastable state $He(2^3S)$ by electron impact in weakly ionized helium located in a constant electric field for the Druyvesteyn case of electron-atom scattering and in the regime of a low electron number density. The reduced electric field strength $x = E/N_a$ is expressed in Td

$$\gamma = \frac{1.4 \cdot 10^4}{x^3} \exp\left(-\frac{350}{x^2}\right)$$

This parameter has a maximum at $x = 15$, where $\gamma = 0.88$, and the maximum efficiency of atom excitation is $\xi = 0.47$. But this electric field strength corresponds to the boundary of stability of ionized helium.

10.2 Excitation of Metastable State in Helium Plasma at High Electron Concentrations

We now determine the rate constants of excitation of helium atoms in the metastable $He(2^3S)$ state by electron impact in the regime of a high electron number density. Using the Maxwell distribution function (2.1.3) for electrons and formula (6.3.17) for the rate constant of atom excitation by electron impact in an ionized gas in the regime of high electron number densities, we have for excitation of metastable helium atoms in the 2^3S state [287]

$$k_< = \frac{8.1 \cdot 10^{-4}}{T_e^{5/2}} \exp\left(-\frac{19.8}{T_e}\right)(c_e + 2.0 \cdot 10^{-7}x^2) \qquad (10.2.1)$$

Here $c_e = N_e/N_a$ is the electron concentration, we take a typical value $\ln\Lambda = 7$, the rate constant is expressed in cm^3/s, the reduced electric field strengths x are given in Td, the electron temperature is measured in eV. At low electric field strengths, i.e.

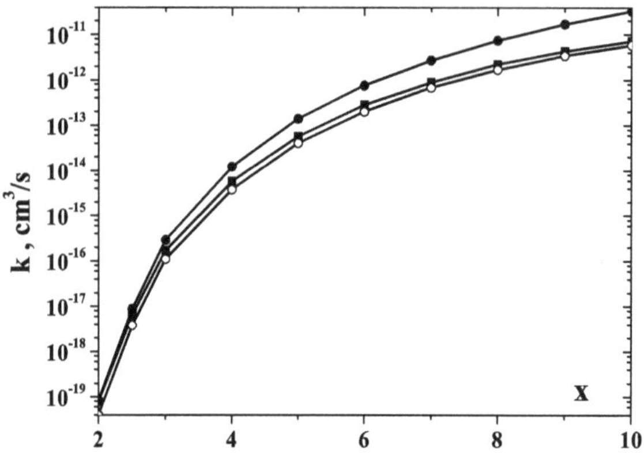

Fig. 10.4 The rate constants of excitation of the helium atom in the state 2^3S by electron impact if this process proceeds in weakly ionized helium located in a constant electric field in the regime of high electron number densities. The excitation rate constant $k_<$ is given by formula (10.2.1) and corresponds to the electron concentration $c_e = 10^{-6}$, and the excitation rate constant $k_>$ is determined by formula (10.2.2). The values $k_<$ are given by *open circles*, the values of $k_>$ are marked by *filled squares*, and the total excitation rate constants k_{ex} in accordance with formula (10.2.3) are labelled by *filled circles*

for the Druyvesteyn case, the connection between the electron temperature T_e and the reduced electric field strength $x = E/N_a$ is given by formula (6.2.12) ($T_e = 0.41x$) at low electric fields strengths, and at higher electric field strengths it is determined from equation (6.2.14).

In the case, where atom excitation is determined by electron acceleration in helium below the excitation threshold dominates, we have for the excitation rate constant on the basis of formula (6.3.8) and he Maxwell distribution function (6.2.4) for electrons [287]

$$k_> = 8.3 \cdot 10^{-10} \frac{x^{1.2}}{T_e^{1.5}} \exp\left(-\frac{19.8}{T_e}\right) \tag{10.2.2}$$

Figure 10.4 gives the rate constants $k_<$ and $k_>$ according to formulas (10.2.1) and (10.2.2) in the range $T_e \ll \Delta\varepsilon$, and also the excitation rate constant that is the combination of these quantities according to formula (6.5.13) [287]

$$k_{ex} = 8.3 \cdot 10^{-10} \frac{(c_e + 2.0 \cdot 10^{-7}x^2)}{(c_e + 2.0 \cdot 10^{-7}x^2 + 1 \cdot 10^{-6}x^{1.2}T_e)} \frac{x^{1.2}}{T_e^{1.5}} \exp\left(-\frac{19.8}{T_e}\right) \tag{10.2.3}$$

In addition, we compare the excitation rate constants for the metastable helium atom by electron impact in a gas discharge plasma in the regime of low and high electric field strengths. This comparison is made in Fig. 10.5.

One can evaluate the efficiency of excitation of the helium atom in the regime of a high electron number density in accordance with formula (6.4.4) by analogy with that given in Fig. 10.3. As above, the parameter γ characterized the efficiency

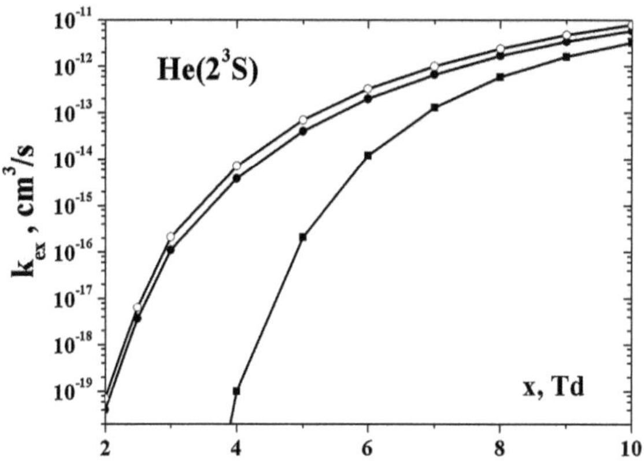

Fig. 10.5 The rate constants k_{ex} of atom excitation to the metastable state $He(2^3S)$ by electron impact in a constant electric field. *Closed squares* correspond to the regime of low electron number densities in accordance with formula (10.1.6), the rate constants for the regime of high electron number densities are determined by formula (10.2.3). *Closed circles* relate to the electron concentration $c_e = 10^{-6}$ and *open circles* correspond to the electron concentration $c_e = 10^{-5}$

of atom excitation is defined according to formula (10.1.8)

$$\gamma = \frac{p_{ex}}{p_{el}},$$

where $p_{ex} = \Delta\varepsilon k_{ex}$ is the specific power that is consumed on atom excitation, so that the rate constant of atom excitation in a helium gas discharge plasma is given by formula (10.2.3), and $p_{el} = P_{el}/N_a = eEw_e/N_a$, where p_{el} is the power per electron that is consumed on elastic electron scattering on atoms. Using formula (6.2.10) for the electron drift velocity in a gas, we obtain for the specific power p_{el} due to elastic electron scattering on helium atoms in the regime of high electron number densities

$$p_{el} = 3.7 \cdot 10^{-12} \frac{x^2}{\sqrt{T_e}} \qquad (10.2.4)$$

where p_{el} is measured in eV $\cdot$ cm^3/s, the reduced electric field strength x is given in Td, and the electron temperature T_e is expressed in eV. In the limit of low electric field strengths, where the Druyvesteyn case is valid and the electron temperature of electrons in helium is connected with the reduced electric field strength x through formula (6.2.12) ($T_e = 0.41x$), formula (10.2.4) takes the form

$$p_{el} = 5.8 \cdot 10^{-12} x^{3/2},$$

The efficiency of atom excitation by electron impact in an ionized gas is the part of energy that is consumed on atom excitation. Figure 10.6 contains the efficiencies

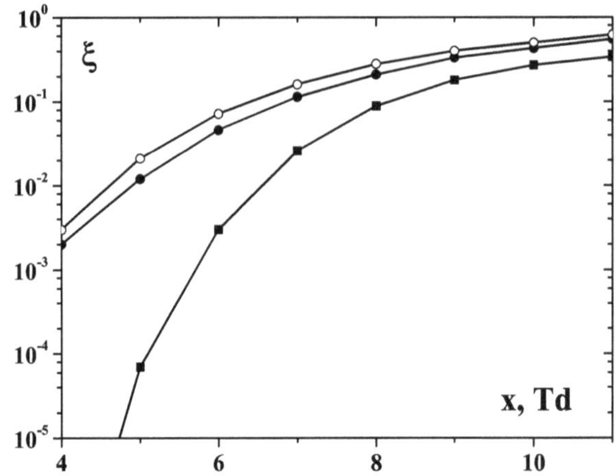

Fig. 10.6 The efficiency of excitation of the helium atom in the metastable state $He(2^3S)$ by electron impact in weakly ionized helium located in a constant electric field. *Filled squares* correspond to the regime of low densities of electrons, *filled circles* relate to the regime of high densities of electrons with the electron concentration $c_e = 10^{-6}$, and *open circles* correspond to the regime of high densities of electrons with the electron concentration $c_e = 10^{-5}$

of atom excitation in the regimes of low and high electron number densities which are obtained on the basis of the above formulas.

Let us compare the rate constant of atom excitation by electron impact in the regime of a high electron number density under conditions of thermodynamic equilibrium. This limit corresponds to the criterion which is opposite to the criterion (6.5.4)

$$\kappa \ll 1 \tag{10.2.5}$$

In addition, one more criterion must be fulfilled

$$k_< \gg k_> \tag{10.2.6}$$

These criteria do not hold true in the case of excitation of the metastable state 2^3S by electron impact. Nevertheless, under conditions of thermodynamic equilibrium the rate constant of atom excitation by electron impact in a plasma is given by formula (3.3.8), and in the case of excitation of the metastable state 2^3S by electron impact in a helium gas discharge plasma has the form

$$k_{ex}^B = 9.3 \cdot 10^{-9} \exp\left(-\frac{19.8}{T_e}\right), \tag{10.2.7}$$

where the rate constant is measured in cm^3/s, and the electron temperature is expressed in eV.

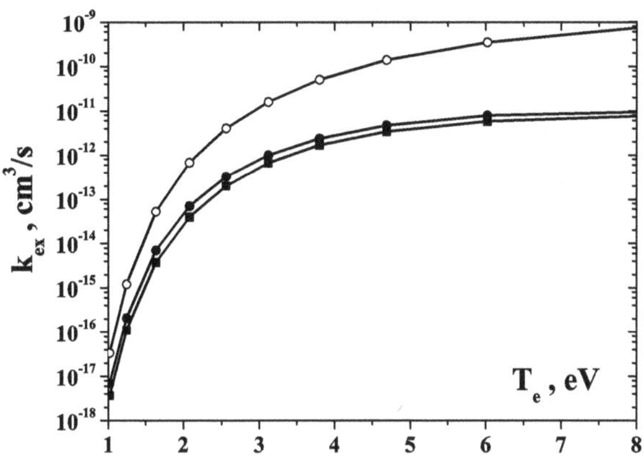

Fig. 10.7 The rate constants k_{ex} of atom excitation to the metastable state $He(2^3S)$ by electron impact in a constant electric field. *Open circles* correspond to thermodynamic equilibrium according to formula (10.2.7). The excitation rates for the regime of high electron number densities in accordance with formula (10.2.3) are represented by *closed squares* for the electron concentration $c_e = 10^{-6}$ and by *filled circles* for the electron concentration $c_e = 10^{-5}$

Note that the excitation rate constant (10.2.7) under conditions of thermodynamic equilibrium exceeds that (10.2.3) under the condition $\kappa \gg 1$ because as a result of violation of the criterion (10.2.6) becomes the electron energy distribution function is small compared to the Maxwell distribution function, and violation of the criterion (10.2.5) leads to a sharp decrease of the electron distribution function with an increasing electron energy above the excitation threshold. Figure 10.7 represents the comparison of the thermodynamic rate constant of helium atom excitation in the state 2^3S given by formula (10.2.7) with those for the case where an ionized helium is located in an external electric field for the regime of high electron number densities. This excitation rate constant in ionized helium is given by formula (10.2.3), and the connection between the electron temperature T_e and the reduced electric field strength x follows from (6.2.14).

10.3 Inelastic Electron Collisions with Excited Helium Atoms

We above consider excitation of helium atoms in a gas discharge plasma from the ground state that leads to formation the metastable atoms $He(2^3S)$ mostly. Excitation of higher excited states and formation of ions in a gas discharge plasma proceeds through metastable 2^3S atoms. Figure 10.8 gives the rates of transitions between the lowest excited helium states which are grouped around the metastable state $He(2^3S)$.

There are two channels of quenching of the metastable state 2^3S by electron impact. The first channel results from transition in the ground state and

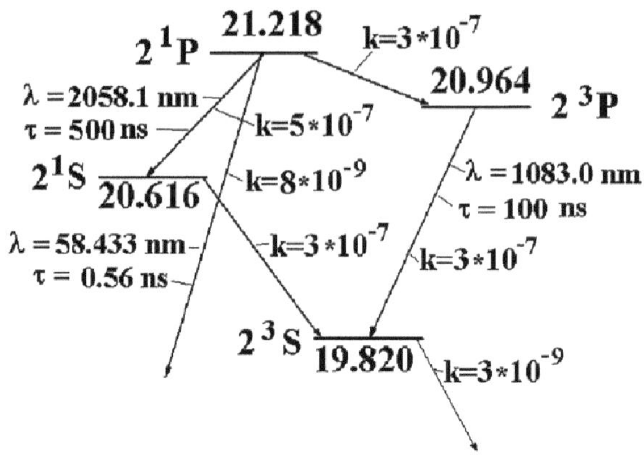

Fig. 10.8 The lowest excited states of a helium atom and the rates of transitions involving these states [40]

is characterized by the quenching rate constant $k_q = 3.1 \times 10^{-9}\,\mathrm{cm}^3/\mathrm{s}$. The second channel consists in excitation the metastable state 2^3S in the states 2^1S and 2^3P in collisions with electrons. The first channel dominates if the following criterion holds true

$$k_q \gg k_{ms} + k_{mp} \tag{10.3.1}$$

Here indices $m,\ s,\ P,\ p$ denote the states $2^3S,\ 2^1S,\ 2^1P$ and 2^3P correspondingly, k_q is the rate constant of quenching of the metastable state by electron impact with transition in the ground state, other rate constants correspond to transitions in accordance with indices. The criterion (10.3.1) means that decay of metastable atoms in collisions with electrons results in atom transition in the ground state rather than transition in excited states.

For determination the population of excited states we use the principle of detailed balance that connects the rates of excitation and quenching according to formula (3.3.7). Because the equilibrium between atoms in the metastable and connected with it states, the number density of atoms in an excited state N_{ex} which is in the equilibrium with the state 2^3S is given by

$$N_{ex} = N_m \frac{g_{ex}}{g_m} \left\langle \sqrt{\frac{\varepsilon - \Delta\varepsilon}{\varepsilon}} \right\rangle,$$

where $g_m = 3$ is the statistical weight of the state 2^3S, g_{ex} is the statistical weight for a given excited state (2^1S or 2^3P), $\Delta\varepsilon$ is the excitation energy of this state, and an average in the above formula is made with the electron distribution function. We use that in the helium case the electron-atom diffusion cross section of scattering σ^*_{ea} is independent of the electron velocity. Then the electron distribution is given by the Druyvesteyn distribution function [269, 285, 286] according to formula (6.1.9).

Therefore the above formula for the number density N_{ex} of atoms in a certain excited state that is connected with the metastable state by electron transitions, takes the form

$$N_{ex} = N_m \frac{g_{ex}}{g_m} J(\frac{\Delta\varepsilon}{\varepsilon_o}), \quad J(a) = \frac{2}{\Gamma(3/4)} \int_0^\infty \sqrt{z} \exp\left[-(z+a)^2\right] dz, \qquad (10.3.2)$$

For simplicity we construct the function $J(a)$ from its limiting values, so that $J(0) =$ 1 and

$$J(a \gg 1) = \frac{\sqrt{\pi} \exp(-a^2)}{2^{3/2}\Gamma(3/4)a^{3/2}} = 0.51a^{-3/2} \exp(-a^2)$$

Taking the function $J(a)$ such that it obtains the precise limiting values, we have

$$J(a) = \frac{1}{1 + 1.96a^{3/2} \exp(a^2)} \qquad (10.3.3)$$

We now compare the contribution of two channels for decay of metastable helium atoms, so the the first channel corresponds to quenching of this state with atom transition in the ground state, while the second channel of decay of the metastable $He(2^3S)$ state relates to excitation in nearby excited states. According to the criterion (10.3.1), the first channel dominates if

$$\frac{1}{3} J(a) + J(1.44a) \ll 0.01 \qquad (10.3.4)$$

The boundary of the transition between two channels corresponds to $a = 1.6$ or $\varepsilon_o = 0.5\,\text{eV}$. Taking the diffusion cross section of electron scattering on a helium atom to be $6\,\text{Å}^2$, we find according to formula (6.1.9) the boundary reduced electric field strength $E/N_a = 0.6Td$. We are guided by the range $2Td < x < 10Td$ for the specific electric field strength where quenching of metastable atoms in a helium gas discharge plasma results from transitions in higher excited states in collision with electrons.

Let us determine the number density of helium atoms in the metastable $He(2^3S)$ state for a gas discharge plasma in the regime of high electron number densities being guided by the electron number densities $N_e \sim 10^{12}\,\text{cm}^{-3}$, as it takes place in glow discharge. We use formula (10.2.3) for the rate constant of excitation of helium atoms in a gas discharge plasma by electron impact. Connecting the electron temperature T_e and the reduced electric field strength x by the relation $T_e = 0.41x$, that holds true at low electron temperatures, we obtain instead of formula (10.2.3) for the rate constant of atom excitation in the metastable state $He(2^3S)$ by electron impact

$$k_{ex} = \frac{3.1 \cdot 10^{-9}(c_e + 2.0 \cdot 10^{-7}x^2)}{x^{0.3}(c_e + 2.0 \cdot 10^{-7}x^2 + 4.2 \cdot 10^{-7}x^{2.2})} \cdot \exp\left(-\frac{48}{x}\right), \qquad (10.3.5)$$

where the rate constant k_{ex} is given in cm^3/s. The quenching rate for the metastable state $He(2^3S)$ in the ground electron state is equal to $k_q = 3.1 \times 10^{-9}\,\mathrm{cm}^3/\mathrm{s}$ and is independent on the electron energy near the excitation threshold. This rate constant characterizes quenching of the metastable state if it is in thermodynamic equilibrium with neighboring excited states. This equilibrium is supported by collisions of electrons with excited atoms, but along with collision transitions, radiative transitions proceed in a uniform ionized gases and they violate the thermodynamic equilibrium between excited atoms.

In this analysis we are guided by the scheme of processes involving the lowest excited states of the helium atom which are given in Fig. 10.8. The violation of the thermodynamic equilibrium proceeds due to radiation of atoms in the state 2^1P and is governed by the criterion $N_e k_q(2^1P)\tau_{ef} \ll 1$, where $k_q(2^1P) = 8 \times 10^{-7}\,\mathrm{cm}^3/\mathrm{s}$ is the total rate constant of transitions in lower states as a result of collisions with electrons τ_{ef} is the effective radiative time of the helium atom state 2^1P with accounting for reabsorption processes. We are guided by the case where a gas discharge plasma of helium is located in a cylinder tube of a radius 1 cm, and then $\tau_{ef} = 1.1\,\mu\mathrm{s}$ as it is obtained in Chap. 4. Then the criterion of violation of thermodynamic equilibrium for excited atoms $He(2^1P)$ has the form

$$N_e \ll 10^{12}\,\mathrm{cm}^{-3} \qquad (10.3.6)$$

We are guided by a helium plasma of glow gas discharge with

$$10^{10}\,\mathrm{cm}^{-3} \ll N_e \ll 10^{12}\,\mathrm{cm}^{-3}, \qquad (10.3.7)$$

where the criterion (10.3.6) holds true. Hence the population of the state 2^1P is small compared with thermodynamic one, and one can neglect the process of atom quenching with transitions between 2^1P and 2^1S states. In the analysis of these transitions, we restrict by the left branch of processes given in Fig. 10.8, and the notations of these processes are represented in Fig. 10.9.

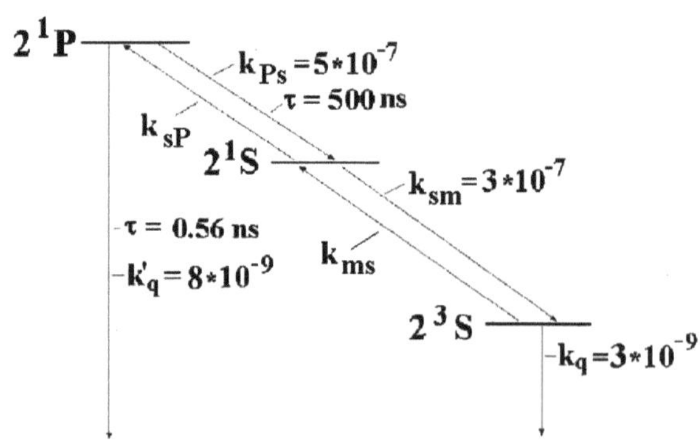

Fig. 10.9 The scheme of transitions between the lowest excited states of a helium atom which are taken into account in the balance of metastable helium atoms and notations of the rate constants of corresponding transitions. This scheme is a truncated scheme of Fig. 10.8

Thus, basing on the scheme of transitions between excited states of the helium atom according to Fig. 10.9, we have the following balance equations for the number density of atoms N_m in the metastable state 2^3S and the number density of atoms N_s in the metastable state 2^1S

$$\frac{1}{N_e}\frac{dN_m}{dt} = k_{ex}\,N_a - k_q\,N_m - k_{ms}\,N_m + k_{sm}\,N_s,\ \frac{1}{N_e}\frac{dN_s}{}dt = k_{ms}\,N_m - k_{sm}\,N_s - k_{sp}\,N_s$$

This allows us to introduce the effective rate constant k_{ef} of quenching of the metastable atom state 2^3S by electron impact in accordance with the balance equation for the number density of metastable atoms

$$\frac{1}{N_e}\frac{dN_m}{dt} = k_{ex}N_a - k_Q N_m,\ k_Q = k_q + \frac{k_{ms}k_{sp}}{k_{sm} + k_{sp}}, \tag{10.3.8}$$

where k_Q is the total rate constant of quenching of the metastable state 2^3S by electron impact. Solution of balance equations gives for the number densities of metastable atoms

$$N_m = \frac{k_{ex}}{k_Q}N_a,\ N_s = N_m\frac{k_{ms}}{k_{sm} + k_{sp}} \tag{10.3.9}$$

where N_a is the number density of atoms in the ground state. Note we use the stepwise character of excitation of the metastable state 2^1S and other states because the electron distribution function decreases sharply at energies above the threshold of excitation of the 2^3S state (see Fig. 6.14).

We use the scheme of processes with transitions between lower excited states of the helium atom given in Fig. 10.10 and represent in this figure notations of the rate constants for processes which partake in excitation of the resonantly excited state $He(2^1P)$ through the state $He(2^3P)$. From this scheme we have the following balance equation for the number density of atoms N_P located in the state $He(2^1P)$

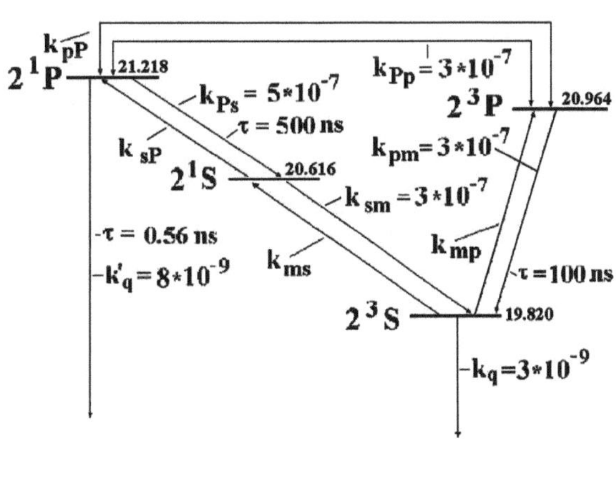

Fig. 10.10 Transitions between the lowest excited states of the helium atom and notations of the corresponding rate constants within the framework of the scheme of Fig. 10.8

$$\frac{dN_P}{dt} = N_e N_m k_{mP} - N_P\left(\frac{1}{\tau} + N_e k_{Pm} + N_e k_{Pp}\right)$$

The solution of this equation gives for the relative number density of atoms N_P in the state $He(2^1P)$

$$\frac{N_P}{N_m} = \frac{k_{mP}}{(\tau N_e)^{-1} + k_{Pm} + k_{Pp}} \approx N_e \tau k_{mP}, \qquad (10.3.10)$$

since decay due to radiation dominates in detachment of the state $He(2^3P)$ under given conditions.

From this we find an addition to the effective quenching rate constant due to excitation of atoms in the state $He(2^1P)$ through the state $He(2^3P)$

$$\Delta k = \frac{N_P}{N_m} k_{pP} \approx \tau N_e k_{Pm} k_{Pp} \qquad (10.3.11)$$

This formula is valid for $N_e \ll 3 \times 10^{13}\,\mathrm{cm}^{-3}$ and at $N_e = 10^{12}\,\mathrm{cm}^{-3}$ gives

$$\Delta k \approx 1 \cdot 10^{-8} \exp(-1.398/T_e),$$

where the electron temperature is expressed in eV, and the rate constant is given in cm^3/s. Comparing this with the effective quenching rate constant k_Q of the metastable atom $He(2^3S)$ according to formula (10.3.13), we find that under given conditions $\Delta k \ll k_Q$. In particular, at the electron temperature $T_e = 3\,\mathrm{eV}$ we have $\Delta k/k_Q = 0.14$.

We now determine the total rate constant for quenching of metastable atoms $He(2^3S)$ by electron impact in a gas discharge plasma for the regime of low electron number densities by analogy with that in the regime of a high number density of electrons. This rate constant k_Q for quenching of metastable atoms $He(2^3S)$ by electron impact is determined by formula (10.3.8)

$$k_Q = k_q + \frac{k_{ms} k_{sp}}{k_{sm} + k_{sp}},$$

where the rate constants of transition in lower states by energy are given in Fig. 10.8, and the rate constant for transitions in higher states are calculated on the basis of formula (6.3.2) with using the distribution function (6.1.9). Note that in contrast to excitation of atoms in the ground state in a plasma, now due to a not large excitation energy these rate constants are determined by electron energies where the Druyvesteyn distribution function (6.1.9) is valid. According to formula (6.3.2) we have

$$k_{ms} = \frac{\exp(-\frac{0.94}{x^2})}{1 + \frac{1.86}{x^{3/2}}}, \quad k_{sp} = 5\frac{\exp(-\frac{0.54}{x^2})}{1 + \frac{1.23}{x^{3/2}}} \qquad (10.3.12)$$

where the rate constants are expressed in 10^{-7} cm^3/s and the the specific electric field strength $x = E/N_a$ is given in Td. The results are given in Fig. 10.11.

Let us determine the concentrations of metastable atoms $He(2^3S)$ and $He(2^1S)$ located in the helium gas discharge plasma in the regime of high electron number densities as a function of the electron temperature T_e being guided by the electron number densities $N_e \sim 10^{12}\,\mathrm{cm}^{-3}$ that is typical for glow gas discharge. In this consideration we restrict ourselves by the scheme of processes of Fig. 10.9, so that the concentration $c_m = N_m/N_a$ of atoms in the metastable state (2^3S) is given by formula (10.3.9)

$$c_m = \frac{k_{ex}}{k_Q},$$

where the rate constant of atom excitation by electron impact in an external electric field is determined by formula (10.2.3). We use the connection (6.1.10) between the electron temperature T_e and the reduced electric field strength $x = E/N_a$ in the helium case where the diffusion cross section of electron-atom collisions is independent on the electron velocity, and this connection has the form $T_e = 0.41x$, if we express the electron temperature in eV and the reduced electric field strength in Td. Figure 10.12 contains the dependence of the concentration of the metastable atoms in the state 2^3S in a gas discharge plasma, where atom excitation proceeds through formation of fast electrons in the plasma due to electron kinetics in a space of electron energy and subsequent excitation of atoms by fast electrons. Data of Fig. 10.12 corresponds to the regime of low electron number densities where electron-electron collisions are not important for establishment the energy distribution function of

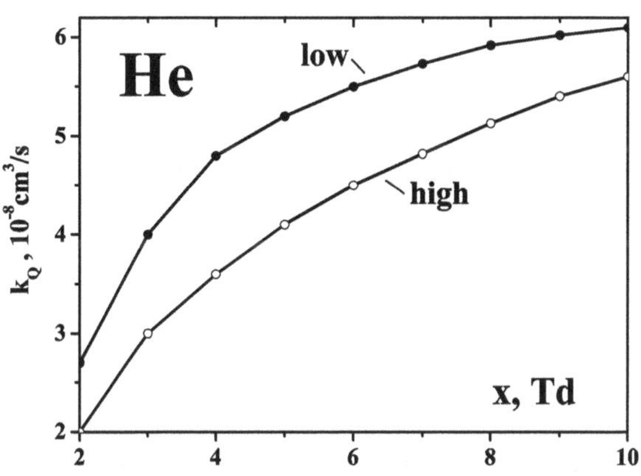

Fig. 10.11 The rate constant of quenching of metastable atoms $He(2^3S)$ by electron impact in a gas discharge plasma. *Filled circles* correspond to the regime of low electron number densities where the rate constant is calculated on the basis of formulas (10.3.8) and (10.3.12). The rate constants for the regime of high electron number densities are indicated by *open circles* and are evaluated according to formula (10.3.13)

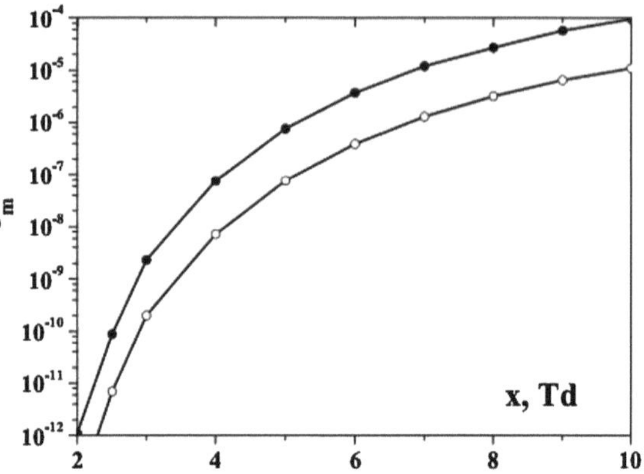

Fig. 10.12 Concentration of metastable helium atoms in a gas discharge plasma of helium in an external electric field as a function of the reduced electric field strength. Concentrations of $He(2^3S)$ are given by *filled circles*, and concentrations of $He(2^1S)$ are labelled by *open circles*. These data are based on formulas (10.3.9) and correspond to the regime of low electron number densities

electrons. These data are obtained on the basis of formulas (10.3.9), (10.3.5) and (10.3.12).

We now consider the regime of a high electron number density with the Maxwell distribution function of electrons by energy which results from electron-electron collisions. In considering as early the range of not high electron temperature, we take the rate constant of quenching of an excited state by electron impact to be independent of the electron energy for slow electrons that simplifies our analysis. Since this holds true for low electron energies, the results will be valid for low electric field strengths and will contain an error for middle electric field strengths. Using the principle of detailed balance, we then obtain for the excitation rate constants on the basis of the relation (3.3.7)

$$k_{ex} = k_q \frac{g_*}{g_o} \exp\left(-\frac{\Delta\varepsilon}{T_e}\right)$$

On the basis of parameters of Fig. 10.9, we obtain for the effective rate constant of quenching of the metastable state that is expressed in cm^3/s

$$k_Q = 3 \cdot 10^{-9} + \frac{5 \cdot 10^{-7} \exp(-1.398/T_e)}{1 + 5\exp(-0.602/T_e)}, \qquad (10.3.13)$$

where the electron temperature T_e is expressed in eV. Figure 10.13 gives the ratio of the effective quenching rate constant k_Q of the metastable atom $He(2^3S)$ by electron impact to the quenching rate constant. As is seen, excitation of the metastable atom $He(2^3S)$ by electron impact in higher excited states with subsequent radiation of the $He(2^1P)$ gives the main contribution to quenching of the $He(2^3S)$ atom at temperatures $T_e \sim 1\,\mathrm{eV}$ which are of interest for these transitions. Note the connection between the electron temperature T_e expressed in eV and the reduced electric field strength $x = E/N_a$ expressed in Td that according to formula (6.2.10) gives $T_e = 0.41x$.

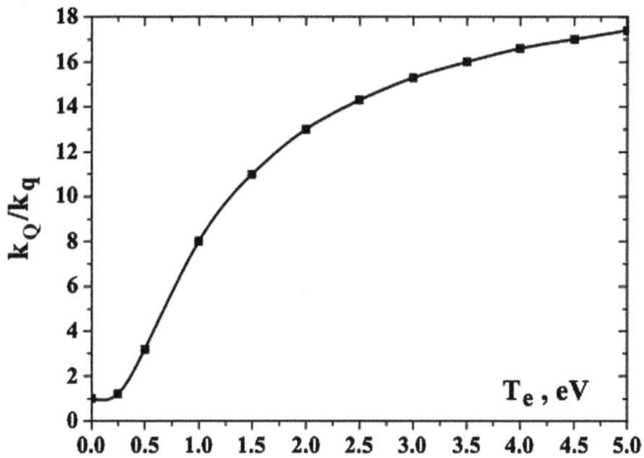

Fig. 10.13 The ratio of the total rate constant of quenching of the metastable state $He(2^3S)$ of the helium atom by electron impact in a helium gas discharge plasma to the rate constant of quenching with atom transition in the ground state as a function of the electron temperature T_e

Note that in the regime of a high electron number density of a gas discharge plasma with the Maxwell distribution function of electrons (2.1.2) an equilibrium between electrons means also electron equilibrium with excited atoms until other mechanisms of destruction of excited atoms are weaker than that as a result of collision with electrons. This leads to the thermodynamic equilibrium between nearby excited atom states in the case of a helium gas discharge plasma, and the number densities of excited helium states is connected with the number density of metastable helium atoms $N_m = N(2^3S)$ by relations

$$N_s \equiv N(2^1S) = \frac{N_m}{3}\exp\left(-\frac{0.794}{T_e}\right)/(1+N_e/N_1),\ N(2^3P) = N_m\exp\left(-\frac{1.144}{T_e}\right), \tag{10.3.14}$$

$$N_P \equiv N(2^1P) = \frac{N_m}{3(1+N_e/N_2)}\exp\left(-\frac{1.398}{T_e}\right) \tag{10.3.15}$$

Here N_e is the electron number density, $N_1 = 4\times 10^{12}\,\mathrm{cm}^{-3}$, $N_2 = 2\times 10^{15}\,\mathrm{cm}^{-3}$, the electron temperature T_e is expressed in eV, and for simplicity we assume the gas discharge plasma to be optically transparent for radiation with the wavelengths $\lambda = 2.058\,\mu\mathrm{m}$ and $\lambda = 1.083\,\mu\mathrm{m}$.

The criterion of the thermodynamic equilibrium between electrons and metastable helium atoms requires that quenching of metastable atoms by electron impact transfer these atoms in the ground state. This criterion has the form

$$k_q \gg k_o\left[\frac{1}{3}\exp\left(-\frac{0.794}{T_e}\right)+\exp\left(-\frac{1.144}{T_e}\right)\right], \tag{10.3.16}$$

where $k_o = 3 \times 10^{-7}\,\mathrm{cm}^3/\mathrm{s}$. One can see that the criterion (10.3.16) is fulfilled at $T_e < 0.2\,\mathrm{eV}$. But under conditions of a self-maintained gas discharge plasma the second channel of quenching of the metastable state $He(2^3S)$ by electron impact dominates, i.e. quenching of metastable atoms leads to their excitation.

Above in determination the number density of helium atoms in the metastable $He(2^3S)$ state for an ionized helium located in a constant electric field in the regime of high electron number densities we are restricted ourself by the left branch of quenching processes given in Fig. 10.8. We estimate the contribution of transitions in the 2^1P state through 2^3P state to the effective rate constant of quenching of the metastable state $He(2^3S)$ being guided by for the electron number density $N_e = 10^{12}\,\mathrm{cm}^{-3}$ which are typical for glow gas discharge.

We give in Fig. 10.14 the concentration of metastable atoms $He(2^3S)$ in a helium gas discharge plasma as a function of the electron temperature dependence for the regime of high electron number densities, and Fig. 10.15 represents the same data foe metastable atoms $He(2^1S)$. As is seen, at not small electric field strengths or electron temperatures the concentration of metastable atoms may exceed the electron concentration.

Summing up the analysis of kinetics of excited states in a helium gas discharge plasma, we conclude that the first stage of this process is excitation of metastable states $He(2^3S)$ by electron impact. The rate constant of this process is determined both by formation of fast electrons as a results of their diffusion in a space of electron energies and by excitation of helium atom in collisions with fast electrons. The latter is a self-consistent process that leads to a sharp decrease of the energy distribution function of electrons with an increasing electron energy above the excitation threshold. As a result of a sharp decrease of the energy distribution function of electrons

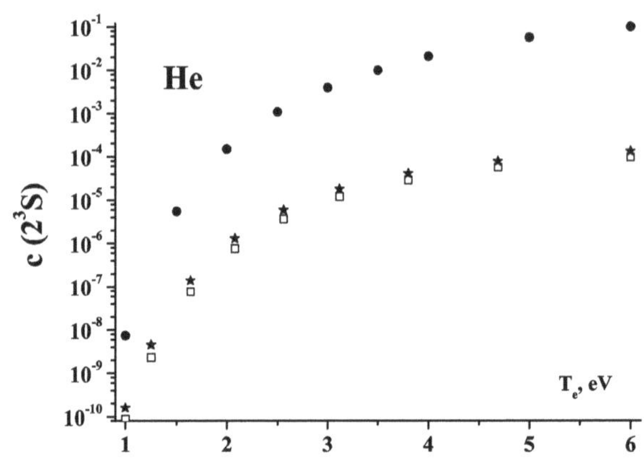

Fig. 10.14 Concentration of metastable helium atoms $He(2^3S)$ in a gas discharge plasma of helium as a function of the electron temperature which are based on formulas (10.3.9) and correspond to the regime of high electron number densities. *Open squares* and *stars* relate to the electron concentrations $c_e = 10^{-6}$ and $c_e = 10^{-5}$ respectively, *filled circles* correspond to thermodynamic equilibrium in accordance with formula (10.2.7)

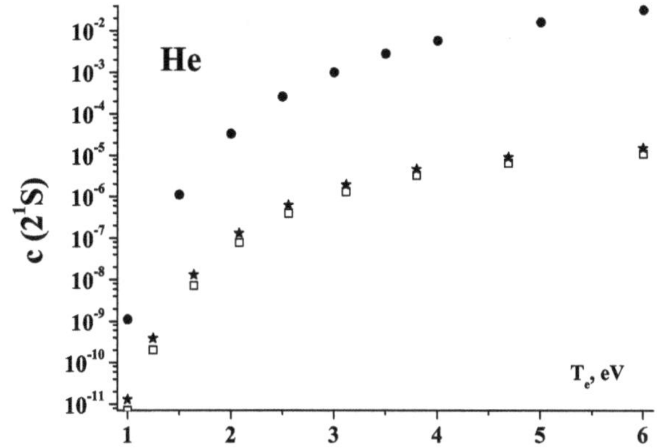

Fig. 10.15 Concentration of metastable helium atoms $He(2^1S)$ in a gas discharge plasma of helium as a function of the electron temperature which are based on formulas (10.3.9) and correspond to the regime of high electron number densities. *Open squares* and *stars* relate to the electron concentrations $c_e = 10^{-6}$ and $c_e = 10^{-5}$ respectively, *filled circles* correspond to thermodynamic equilibrium for excited atoms and electrons

in this energy range excitation of higher excited states of the helium atom proceeds through the lowest metastable state $He(2^3S)$, rather than by direct electron impact. One more peculiarity of kinetics of excited states in a helium gas discharge plasma is that quenching of the lowest metastable state $He(2^3S)$ in collisions with electrons proceeds through transitions in higher levels, rather than transition in the ground state of the helium atom. This character of processes improves the efficiency to use metastable atoms for plasma radiation and plasma ionization.

As it follows from the above analysis, quenching of metastable atoms proceeds through excitation and radiation of more excited atom states. Hence the basic part of the electric energy consumed on atom excitation will be transformed in the energy of emitted photons. Let us determine the rate of emission of resonant photons in a uniform helium gas discharge plasma as a result of the radiative transition

$$He(2^1P) \to He(1^1S) + \hbar\omega$$

We are guided by not high electron number densities $N_e \ll 10^{13}\,\text{cm}^{-3}$ at which decay of atoms $He(2^1P)$ results from emission of resonant photons which leave the plasma region.

Under given conditions formation of resonant photons proceeds through a chain of processes given in Fig. 10.10, and because decay of all the atoms in the state 2^1P results from their radiation, the rate of this process is determined by the following formula

$$\frac{N_p}{dt} = N_s N_e k_{sp}$$

Let us define the rate of formation of resonant photons k_{rad} in this case as

$$\frac{dN_p}{dt} = N_e N_a k_{rad},$$

and the rate constant of photon formation is

$$k_{rad} = k_{sp}\frac{N_s}{N_a} \tag{10.3.17}$$

On the basis of formula (10.3.9) we have

$$k_{rad} = \frac{k_{sp}k_{sm}}{k_{sm}+k_{sp}}\frac{N_m}{N_a},$$

and according to the expression (10.3.8) for k_Q with accounting for formula (10.3.9) for the concentration of atoms in the 2^1S state we obtain

$$k_{rad} = \frac{k_{ex}}{k_Q}\frac{k_{sp}k_{sm}}{k_{sm}+k_{sp}}k_{ex}\left(1-\frac{k_q}{k_Q}\right) \tag{10.3.18}$$

Since according to the data of Fig. 10.9 and formula (10.3.12) $k_Q \gg k_q$ in the range of the electron temperatures under consideration, the main channel of decay of metastable atoms is excitation in the 2^1P state with subsequent photon emissions, all the excitations of helium metastable states are finished by emission of resonant photons. We also give the criterion of this scenario where quenching of resonantly excited atoms results in emission of photons rather than quenching by electron impact. The corresponding criterion has the form

$$N_e k_Q \tau_{ef} \ll 1, \tag{10.3.19}$$

and the effective lifetime of resonant excitation with respect to its leaving the plasma region is given by formula (4.5.16) $\tau_{ef} = 2.6\tau_r\sqrt{k_o R}$. Taking for the radiative transition $He(2^1P) \to He(1^1S) + \hbar\omega$ the effective lifetime for a typical plasma size $R \sim 1\,\text{cm}$ to be $\tau_{ef} = 1.1\,\mu\text{s}$ and the rate constant of quenching of an excited state by electron impact $k_Q < 10^{-7}\,\text{cm}^3/\text{s}$, we have the criterion (10.3.19) in the form $N_e \ll 10^{13}\,\text{cm}^{-3}$.

10.4 Excitation of Atoms in Argon Gas Discharge Plasma

We above analyze the character of excitation of the lowest excited states of the helium atom in a gas discharge plasma by electron impact. This process results from the sum of the process of formation of fast electrons which are able to excite atoms and the process of atom excitation by fast electrons which leads to a sharp decrease of the electron distribution function of electrons with an increasing electron energy above the excitation threshold. The latter influences the resulted rate constant of excitation, i.e. the process of atom excitation in a gas discharge plasma is a self-consistent process. One can expect that the same situation takes place in excitation of argon atoms by electron impact in an argon gas discharge plasma. We below analyze the process

$$e + Ar(3p^6) \rightarrow e + Ar(3p^5 4s) \qquad (10.4.1)$$

in the limit of high electron concentrations (see Fig. 10.1) where the distribution function is the Maxwell one (2.1.3) for the most part of electrons. We represent in Fig. 10.16 positions of excited levels for processes (10.4.1).

In considering processes (10.4.1), we construct the parameter κ which according to formula (6.5.3) is equal

$$\kappa = \frac{2v_o \nu_{ef}}{5a} = \frac{\alpha}{x}, \quad \nu_{ef} = N_a \sqrt{3\frac{g_*}{g_o} k_o k_q} \qquad (10.4.2)$$

Here $k_o = v\sigma^*_{ea}(v_o)$ is the rate constant of elastic electron-atom scattering at the excitation threshold, k_q is the rate constant of quenching of an excited atom by electron impact, $a = eE/m_e$, g_o, g_* are the statistical weights for the ground and excited atom states, $x = E/N_a$ is the reduced electric field. strengths expressed in Td. Table 10.1 contains experimental rate constants [197, 199, 418] of quenching of metastable state $Ar(^3P_2)$ by electron impact. We assume an identical exchange interaction between an incident electron and metastable atom in $^3P_2$ and $^3P_0$ states that leads to identical rate constants of these states. Given in Table 10.1 the rate

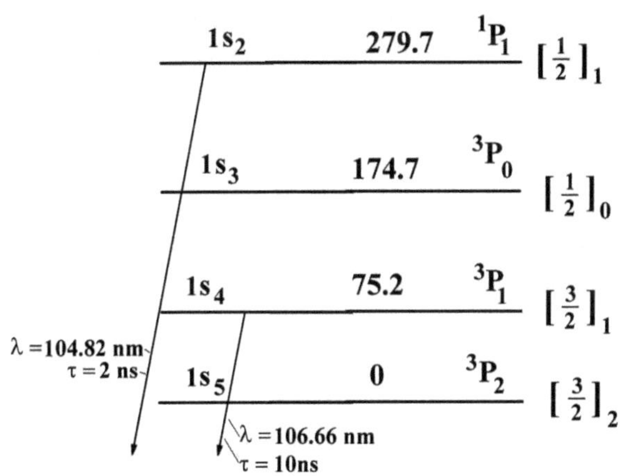

Fig. 10.16 Energies of the lowest levels (in cm^{-1}) for states involving in the process (6.5.15) and notations of these states

Table 10.1 Rate constants of quenching of the argon atom with the electron shell $3p^5 4s$ in collisions with a slow electron if the atom transfers in the ground state

State	$^3P_2$	$^3P_1$	$^3P_0$	$^1P_1$
k_q, $10^{-10}\,\mathrm{cm}^3/\mathrm{s}$	4	8.2	4	39

Table 10.2 Values of the parameter α in formula (6.5.16) for excitation of argon atom states with the electron shell $3p^5 4s$ if the reduced electric field strength is measured in Td

State	$^3P_2$	$^3P_1$	$^3P_0$	$^1P_1$
α	190	220	87	470

constant of quenching of resonantly excited $^3P_1$ and $^1P_1$ states we determine on the basis of formula (3.4.4). Note that the average rate constant of excited state quenching for an argon gas discharge plasma with the electron temperature 1-3eV [137] corresponds to these data averaged over the electron shell $3p^5 4s$.

Table 10.2 contains the values of the coefficient α defined by formula (10.4.2) which are obtained on the basis of data of Table 10.1. As it follows from data of Table 10.2, in the range of electric field strengths under consideration $2Td < x < 20Td$ the criterion (6.5.4) $\kappa \gg 1$ holds true.

The rate constant of atom excitation in a gas discharge plasma is formed from origin of fast electrons with the energy $\Delta\varepsilon$, the threshold energy of atom excitation, and its rate constant is according to formula (6.3.17)

$$k_< = \frac{8\sqrt{2\pi}}{3}\frac{e^4\varepsilon_o \ln\Lambda}{m_e^{1/2}T_e^{5/2}}\exp\left(-\frac{\varepsilon_o}{T_e}\right)\left(c_e + \frac{x^2}{4\pi e^2\sigma_{ea}(v_o)\ln\Lambda}\right) \tag{10.4.3}$$

Another part of the rate constant of atom excitation in a gas discharge plasma by electron impact as a result of atom excitation by electrons with the energy above the excitation threshold one $\Delta\varepsilon$ is given by formula (6.5.7) that now has the form

$$k_> = \frac{4.6\Delta\varepsilon^{3/2}}{(\pi T_e)^{3/2}}\frac{g_*}{g_o}\frac{k_q}{\kappa^{1.2}}\exp\left(-\frac{\varepsilon_o}{T_e}\right) \tag{10.4.4}$$

The total rate constant of atom excitation by electron impact is expressed through the above ones on the basis of formula (6.5.13)

$$k_{ex} = \frac{k_< k_>}{k_< + k_>} \tag{10.4.5}$$

Figure 10.17 represents the dependencies on the electron temperature T_e for the rate constants given by formulas (10.4.3), (10.4.4) and (10.4.5) in the case of excitation of the state $Ar(3p^5 4s^3P_2)$ by electron impact in an argon gas discharge plasma.

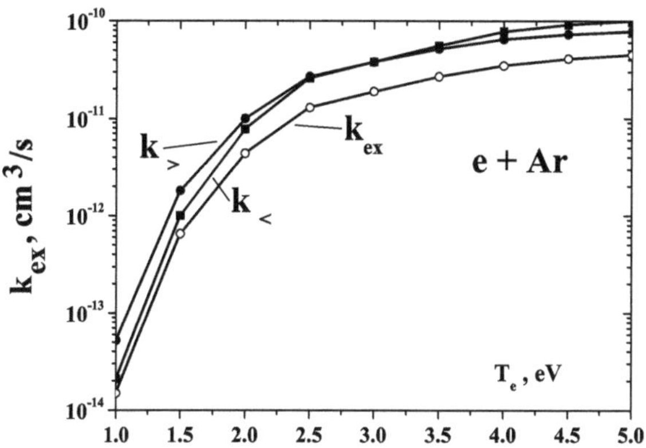

Fig. 10.17 Rate constants of excitation of an the argon atom by electron impact in the state $^3P_2$ as a function of the electron temperature in an argon gas discharge plasma in the regime of high electron number densities in accordance with formulas (10.4.3), (10.4.4), and (10.4.5) at the electron concentration $c_e = 10^{-6}$

Connection between the reduced electric field strength $x = E/N_a$ and the electron temperature T_e is given by the lower curve of Fig. 10.3.

It should be noted that the energy distribution function of electrons in an argon gas discharge plasma varies weakly in a range of excited levels with an electron shell $3p^54s$. Indeed, according to formulas (6.5.2) and (6.5.3) the electron distribution function above the excitation threshold of the state $Ar(3^3P_2)$ has the form

$$f_0(\varepsilon) = f(\Delta\varepsilon)\exp(-\varepsilon/T_e - S), \;\; S = \sum_i \kappa_i \left(\frac{\varepsilon - \Delta\varepsilon_i}{\Delta\varepsilon_i}\right)^{5/4} \eta(\varepsilon - \Delta\varepsilon_i), \quad (10.4.6)$$

where $f(\Delta\varepsilon)$ is the distribution function at the excitation threshold for the lowest excited level $Ar(3^3P_2)$, κ_i is given by formula (6.5.16), $\Delta\varepsilon_i$ is the excitation energy for i−th level, $\eta(z)$ is the unit function, so that it is equal to zero at $z < 0$ and one at $z > 0$. Using parameters of formula (6.5.16) given in Table 10.1, we represent in Fig. 10.18 the character of variation of the energy distribution function of electrons above the threshold of excitation of the lowest excited level. On the basis of this one can assume that the electron distribution function varies not strongly in the range of excitation of states with the electron shell $3p^54s$, whereas in the range of excitation of states with electron shell $3p^54p$ this variation is strong. From this one can conclude that excitation of state with other electron shells than $3p^54s$ can proceed only in a stepwise manner through the states with the electron shell $3p^54s$.

In determination the concentration of excited argon atoms in a gas discharge plasma where excitation and quenching processes in collisions with electrons are of importance, we ascertain first the mechanism of decay of excited atoms. In the case of metastable atoms it is necessary to compare the rate of mixing of states which is given by formula (3.5.5) and with the rate of excitation of metastable atoms

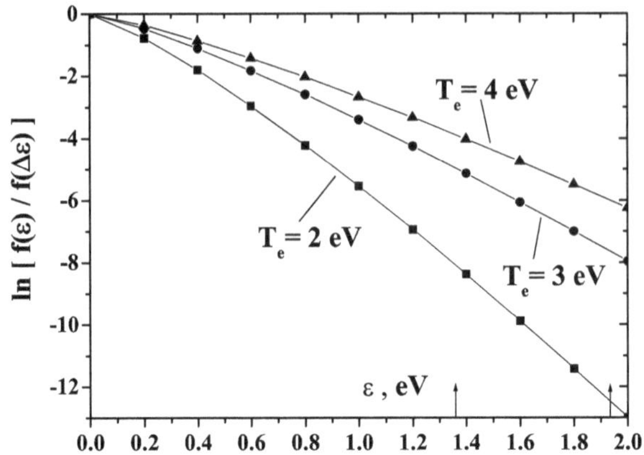

Fig. 10.18 Logarithm of the energy distribution function of electrons according to formula (10.4.6) above the threshold of excitation of the lowest excitation state which is taken as zero for the electron energy. *Arrows* indicate the boundaries of excitation of states with electron shell $3p^54p$

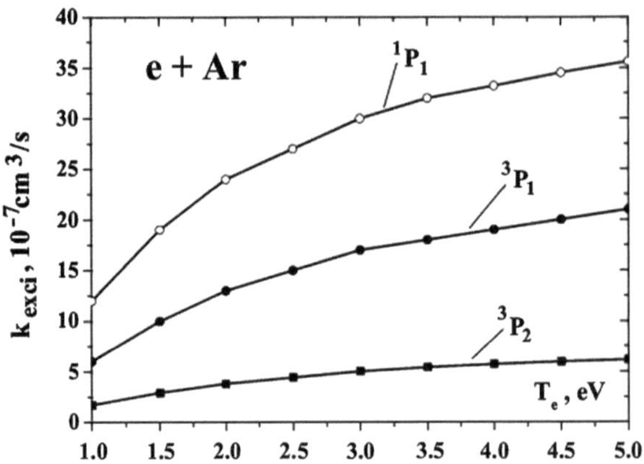

Fig. 10.19 The rate constants of transitions of argon atoms from the states $^3P_2$, $^3P_1$ and $^1P_1$ to the electron shell $Ar(3p^54p)$ as a result of collisions with electrons. These rate constants are evaluated on the basis of formula (3.4.4)

$k_{exci} = k(^3P_2 \to 3p^54p)$ in states with the electron shell $Ar(3p^54p)$. According to formula (3.4.4) which has the form

$$k_{exci} = \frac{k_o}{(\Delta\varepsilon)^{7/2}\tau_r},$$

the rate constants for excitation of $Ar(^3P_2)$ and $Ar(^3P_0)$ are nearby, and Fig. 10.17 gives the excitation rate for the state $^3P_2$. The rate constants of excitation of excited argon atoms with the electron shell $3p^54s$ in states with the the electron shell $3p^54p$ k_{exci} are represented in Fig. 10.19.

Table 10.3 The rate constants of electron-atom inelastic collisions involving the lowest resonantly excited states $Ar(3p^54s(^3P_1)$ and $Ar(3p^54s(^1P_1)$ calculated on the basis of formula (3.4.4)

Process	Rate constant, cm^3/s
$e + Ar(3p^54s\ ^3P_1) \to e + Ar(3p^6\ ^1S_0)$	8.2×10^{-10}
$e + Ar(3p^54s\ ^1P_1) \to e + Ar(3p^6\ ^1S_0)$	3.9×10^{-9}
$e + Ar(3p^54p) \to e + Ar(3p^54s\ ^3P_1)$	2.8×10^{-7}
$e + Ar(3p^54p) \to e + Ar(3p^54s\ ^1P_1)$	4.6×10^{-7}

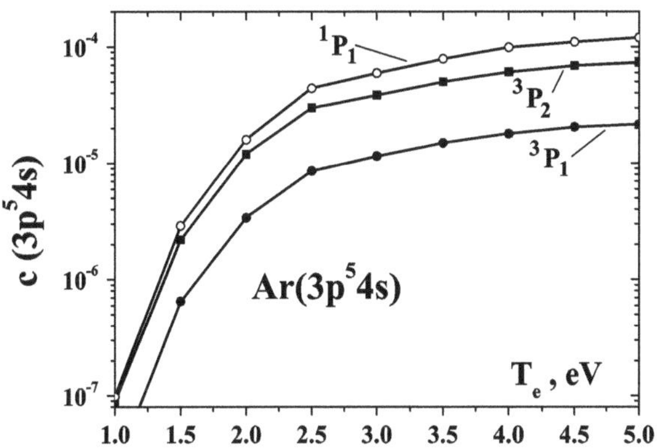

Fig. 10.20 Dependence on the electron temperature for the concentration of excited argon atoms $^3P_2, ^3P_1$ and $^1P_1$ of the electron shell $3p^54s$ in a gas discharge plasma in the regime of high electron number densities in accordance with formula (10.4.8)

In evaluation the concentrations for the lowest resonantly excited argon atoms $Ar(3p^54s)(^3P_1)$ and $Ar(3p^54s)(^1P_1)$, we first find the rate constants of quenching processes in collisions with electrons which involve these states. These rate constants are independent of the electron energy for a slow electron. Evaluated on the basis of formula (3.4.4) and are given in Table 10.3. The excitation rate constants from the states $Ar(3p^54s)(^3P_1)$ and $Ar(3p^54s)(^1P_1)$

Figure 10.20 contains the concentration of metastable atoms located in the lowest excited state $Ar(^3P_2)$ which is given by

$$c(^3P_2) = \frac{k_{ex}}{k_{exci}} \tag{10.4.7}$$

Correspondingly, in accordance with the statistical weights of metastable states, $c(^3P_0)$=$0.2c(^3P_2)$. Next, the excitation rate constants from the states $Ar(3p^54s(^3P_1)$ and $Ar(3p^54s(^1P_1)$ into states with the electron shell $3p^54p$ are determined by formulas (10.4.3), (10.4.4) and (10.4.5) where the quenching rate constant in the ground state of the argon atom is represented in Table 10.3. The rate constants of excitation of argon atoms by electron impact from the ground state $Ar(3p^6\ ^1S_0)$ to

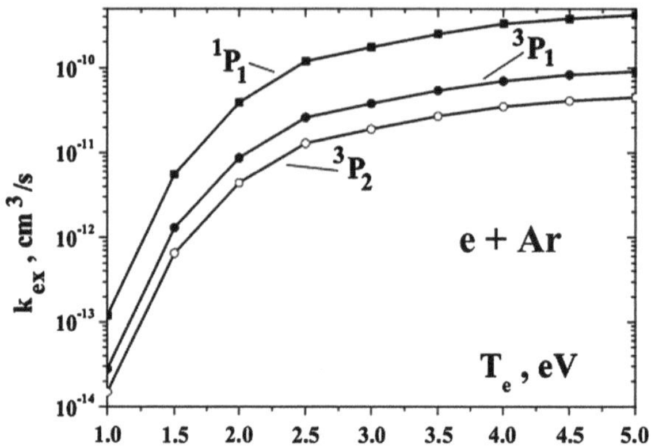

Fig. 10.21 The rate constants of excitation of argon atoms from the ground state $^1S_0$ to excited states of the electron shell $Ar(3p^54p)$ as a result of collision with electrons. These rate constants are evaluated on the basis of formulas (10.4.3), (10.4.4) and (10.4.5) (3.4.4) and Table 10.3 data

the lowest excited states with the electron shell $Ar(3p^54s)$ are given in Fig. 10.21. We consider first the limit of high electron number densities where the concentration of excited atoms is given by formula (10.4.6) like to the case of metastable atoms

$$c_o = \frac{k_{ex}}{k_{exci}}$$

Concentrations of resonantly excited states of argon atoms $Ar(^1P_1)$ and $Ar(^3P_1)$ are given in Fig. 10.20 in the limit of high electron densities if this concentration is established as a result of electron-atom collisions. As early, we assume that excited argon atoms $Ar(3p^54p)$ decay as a result of ionization and transitions in more excited states, whereas transitions to the electron shell $Ar(3p^54s)$ give a small contribution to quenching of these states. Note that all the results of Fig. 10.20 correspond to conditions where decay of states of the group $Ar(3p^54p)$ results from electron collisions with these excited atoms.

Including in the balance equation radiation of excited atoms, one can rewrite equation for the concentration c of atoms in states $^3P_1$ and $^1P_1$ in the form

$$c = \frac{k_{ex}N_o}{k_{exci}(N_e + N_o)}, \tag{10.4.8}$$

where

$$N_o = \frac{1}{\tau_{ef}k_q(4p)}$$

Here τ_{ef} is the lifetime of excited atom with respect to its location inside a gas discharge plasma. If a uniform plasma is located in a cylinder of a radius 1 cm, this effective lifetime is equal to 2.6 μs for the state $Ar(^3P_1)$ and to 3.0 μs for the state $Ar(^1P_1)$, as it is given in Chap. 4. Using data of Table 10.3, we obtain $N_o = 1.4 \times 10^{12}\,\text{cm}^{-3}$ for the state $Ar(^3P_1)$ and $N_o = 7.2 \times 10^{11}\,\text{cm}^{-3}$ for the state $Ar(^1P_1)$.

We now analyze the above results from the standpoint of general balance equations for the number density of excited argon atoms which have the following form

$$\frac{dN(4s)}{dt} = N_e k_{ex} N(3p) - N_e k_{exci} N(4s) - \frac{N(4s)}{\tau_r(4s \to 3p)} + N(4p)\left[\frac{1}{\tau_r(4p \to 4s)} + N_e k_q\right],$$
$$\frac{dN(4p)}{dt} = N_e k_{exci} N(4s) - N_e k_q N(4p) - \frac{N(4p)}{\tau_r(4p \to 4s)} - N_e k_{dec} N(4p) \tag{10.4.9}$$

Here $N(3p)$ is the number density of argon atoms in the ground state $Ar(3p^6\,{}^1S_0$, $N(4s)$ and $N(4p)$ are the number densities of argon atoms in one or group of states with the electron shell $3p^54s$ and $3p^54p$ respectively, τ_r is a time of radiative transitions between indicated states, k_{ex} is the excitation rate constant for transition $3p \to 4s$, k_{exci} is the rate constant of transition $4s \to 4p$, k_q is the rate constant of transition $4p \to 4s$ by electron impact, k_{dec} is the summary rate constant of transitions from the states with the electron shell $3p^54p$ in higher excited states and in ionized states as a result of collisions with electrons. Other notations are explained in previous formulas.

One can see that the concentrations of excited atoms in states $3p^54s$ given in Fig. 10.20 correspond to an equilibrium that takes into account two first terms of the first balance equation (10.4.9). The third term of the first balance equation (10.4.9) may be taken into account by using formula (10.4.8) for the concentration of resonantly excited argon atoms instead of formula (10.4.7). In addition, this consideration is based on fast destruction of excited $Ar(4p)$ states that holds true under the criterion

$$k_{dec} \gg \frac{1}{N_e\tau_r(4p \to 4s)} + k_q(4p \to 4s) \tag{10.4.10}$$

10.5 Continuous Spectrum of Radiation of Equilibrium Plasma

A dense gas discharge plasma emits both broaden resonant radiation that was considered above and radiation in continuous spectrum due to process (4.2.2). The cross section of this process with formation of an excited atom with n principal quantum number according to Kramers formula (4.2.11) is given by

$$\sigma_{rec} = \frac{16\pi}{3\sqrt{3}}\frac{e^{10}}{m_e c^3 v^2 n^3 \hbar^4 \omega}, \tag{10.5.1}$$

where v is the velocity of an incident electron, and the energy balance for this process has the form

$$\hbar\omega = \frac{m_e v^2}{2} + \frac{m_e e^4}{2\hbar^2 n^2},$$

if photorecombination leads to formation of a bound electron state with the principal quantum number n. One can rewrite this formula for the differential cross section of scattering $d\sigma$ of a classical electron on an ion with emitting of a photon with frequencies in a range from ω to $\omega + d\omega$ [149, 419]

$$d\sigma = \frac{16\pi}{3\sqrt{3}} \frac{e^6}{m_e^2 c^3 v^2} \frac{d\omega}{\hbar\omega} \tag{10.5.2}$$

Let us determine from this the power of radiation per unit volume P_{rad} of a uniform gas discharge plasma that is given by

$$P_{rad} = \langle N_e N_i \hbar\omega v \sigma_{rec} \rangle = \frac{16\pi}{3\sqrt{3}} \left(\frac{e^2}{\hbar c}\right)^3 \frac{e^4}{m_e n^3} \left\langle \frac{1}{v} \right\rangle \tag{10.5.3}$$

Here $N_e = N_i$ is the electron or ion number density, v is the electron velocity, and an average is made over the Maxwell distribution function (2.1.3) for electrons, where the electron energy is $\varepsilon = m_e v^2/2$. Since for the Maxwell distribution function we have

$$\left\langle \frac{1}{v} \right\rangle = \sqrt{\frac{2m_e}{\pi T_e}},$$

this leads to the following expression for the specific power of plasma radiation

$$\frac{P_{rad}}{N_e^2} = \frac{16\sqrt{2\pi}}{3\sqrt{3}} \left(\frac{e^2}{\hbar c}\right)^3 \frac{e^4}{(m_e T_e)^{1/2} n^3}, \tag{10.5.4}$$

where the final atom state is a bound electron state with the principal quantum number n. As is seen, the main contribution to the specific power of plasma radiation follows from low n.

Above we are based on the Kramers formula (4.2.11) that assumes the validity for a classical electron description for initial and final transition states. But the main contribution to the specific power of plasma radiation gives low n, where the classical approach is violated. Nevertheless, one can use the fact that Kramers formulas for the ground state of the hydrogen atom leads to an error of 25 % [149]. Hence we will use formula (10.5.4) for the ground atom state, and then the total specific power of plasma radiation is given by

$$\frac{P_{rad}}{N_e^2} = \frac{64\sqrt{\pi}}{3\sqrt{3}} \left(\frac{e^2}{\hbar c}\right)^3 \frac{\hbar^3 J^{3/2}}{m_e^2 e^2 T_e^{1/2}}, \tag{10.5.5}$$

where we use the following connection between the atom ionization potential J and the principal quantum number n in accordance with the energy balance equation for an electron and photon in this process

$$J = \frac{m_e e^4}{2\hbar^2 n^2}$$

Let us represent formula (10.5.4) in the form

$$\frac{P_{rad}}{N_e^2} = \frac{\kappa_o J^{3/2}}{T_e^{1/2}}, \qquad (10.5.6)$$

and if we express the atom ionization potential J and the electron temperature T_e in eV, the proportionality coefficient is $\kappa_o = 8.3 \times 10^{-33}\, W \cdot \mathrm{cm}^3$. Within the framework of the classical approach and based on the Kramers formulas, one can use formula (10.5.2) for both bremsstrahlung electron scattering on ions and photorecombination of electrons and ions with formation of highly excited atoms. On the basis of Kramers formula (10.5.2) one can find also spectrum of formed photons. Indeed, until the classical description holds true, we have on the basis of formula (10.5.2)

$$dP_{rad}(\omega) \sim d\omega, \qquad (10.5.7)$$

and large photon energies give the main contribution to the radiation power. Transferring to the ground atom state, we neglect the discrete structure of atom levels. Nevertheless, it is convenient to use the above formulas for estimations.

Let a uniform gas discharge plasma be located in a cylinder volume of a radius R that is small compared to a cylinder length. Its total radiation power is equal

$$P = \pi R^2 N_e^2 \frac{\kappa_o J^{3/2}}{T_e^{1/2}}$$

As is seen, the radiation power increases with an increasing number density of electrons and ions N_e and a cylinder radius. If this power becomes very large, plasma radiation takes place only from the plasma surface and its power per unit length of the cylinder is equal in this limit according to the Stephan-Boltzmann formula

$$P_S = 2\pi R \sigma T_e^4,$$

where $\sigma = 5.67 \times 10^{-12}\, \mathrm{W/(m^2K^4)} = 1.03 \times 10^5\, \mathrm{W/(cm^2eV^4)}$ is the Stephan-Boltzmann constant. Figure 10.22 contains the dependence of the reduced radius on its temperature T_e if the equality $P = P_S$ is fulfilled. As one can conclude from this figure, a laboratory gas discharge plasma is optically thin.

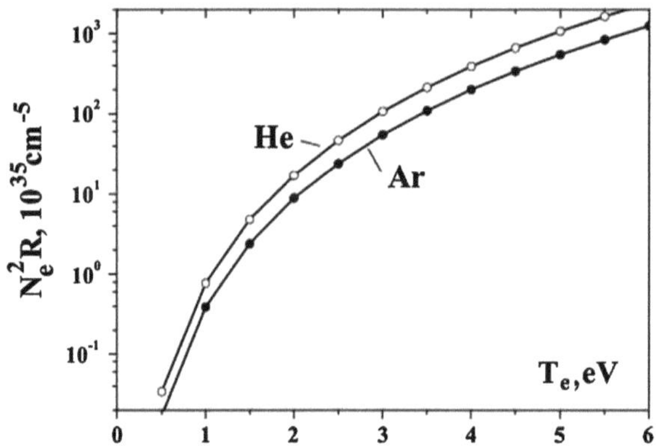

Fig. 10.22 Reduced number density of electrons in an uniform gas discharge plasma of helium and argon where the optical plasma thickness is compared to one

10.6 Tail of the Energy Distribution Function of Electrons

In consideration electron kinetics in a gas discharge plasma, we above neglect the influence of excited atoms on the energy distribution function of electrons, and the latter has the form as it is shown in Fig. 6.12. This corresponds to zero concentration of excited atoms. In reality, there is an additional contribution to the energy distribution function of fast electrons due to processes of quenching of excited atoms by electron impact. As a result of this process, a drop of the energy distribution function of electrons with an increasing electron energy, and Fig. 10.23 gives the electron distribution function in a gas discharge plasma with accounting for this process.

Let us find the tail part of the energy distribution function of electrons in the case of the Maxwell distribution function $\varphi(\varepsilon)$ according to formula (2.1.2) before the atom excitation energy ($\varepsilon < \Delta\varepsilon$), and if the drop of the distribution function in the range 2 of Fig. 10.23 is sharp, i.e. the electron energy ε_* is close to the atom excitation energy $\Delta\varepsilon$. In this case in accordance with the character of equilibrium as a result of electron-atom collisions the tail part of the electron distribution function is given by

$$f_0(\varepsilon) = \frac{k_q}{k_Q}\varphi(\varepsilon) \tag{10.6.1}$$

Here k_q is the rate constant of electron collision with an excited atom with atom transition in the ground state, $k_Q = k_{exci} + k_{ion}$ is the rate constant of quenching of an excited atom with its transition in other states, so that k_{exci} is the rate constant of transition in other excited states, and k_{ion} is the rate constant of atom ionization, and they rate constants in the helium and argon case were evaluated above. Correspondingly, the electron energy ε_* of Fig. 10.23 from which the tail part of the electron distribution function starts is determined by formula

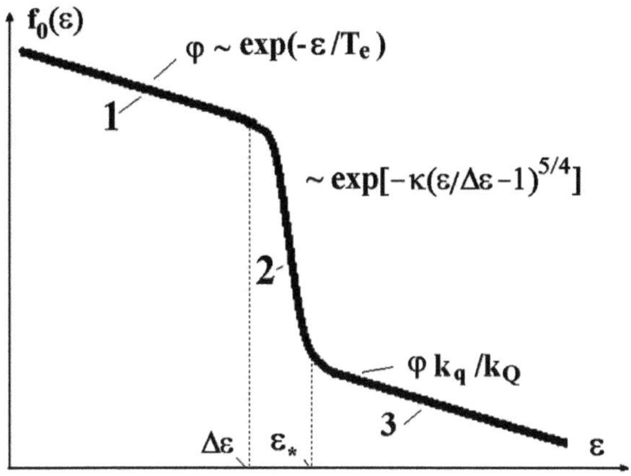

Fig. 10.23 Energy distribution function of electrons in the case of a high electron number density where the trunk of the distribution function in the range 1 corresponds to the Maxwell distribution $\varphi(\varepsilon)$. The range 2 accounts for a drop of the distribution function due to atom excitation, and the distribution function in the range 3 is determined by quenching processes in electron collisions with excited atoms

$$\frac{\varepsilon_* - \Delta\varepsilon}{\Delta\varepsilon} = \frac{1}{\kappa} \ln\left(\frac{k_q}{k_Q}\right) \tag{10.6.2}$$

where we use formulas (6.5.2), (6.5.3) for the distribution function in the range 2 of Fig. 10.23.

It should be noted that the energy electron distribution function in the ranges 2 and 3 of Fig. 10.23 is determined by different processes of electron kinetics. Indeed, the most part of electrons penetrates in the range 2 as a result of diffusion in the energy space from the range of lower energies and accounts for the loss of fast electrons as a result of atom excitation. On contrary, the range 3 of Fig. 10.23 results from quenching of excited atoms by slow electrons. Hence, these range of the electron distribution function are independently, and if the range 2 is narrow, the equilibrium between atoms in the ground and excited states leads to the concentration c_m of excited atoms in lower excited states due to the second mechanism

$$c_m = g_m \frac{k_q}{k_Q} \exp\left(-\frac{\Delta\varepsilon}{T_e}\right) \tag{10.6.3}$$

Evidently, the first mechanism for establishment of the equilibrium between atoms in the ground and excited states due to collisions with electrons results at low electron temperatures, whereas the second mechanism determines this equilibrium for more high electron temperatures. Figure 10.24 contains the comparison of formula (10.6.3) for population of lowest levels $^3P_2$ with experimental data. Figure 10.24 considers two versions of quenching of excited atoms in collisions with electrons. In the second version any transition in more excited states is irreversible, and the rate constant of

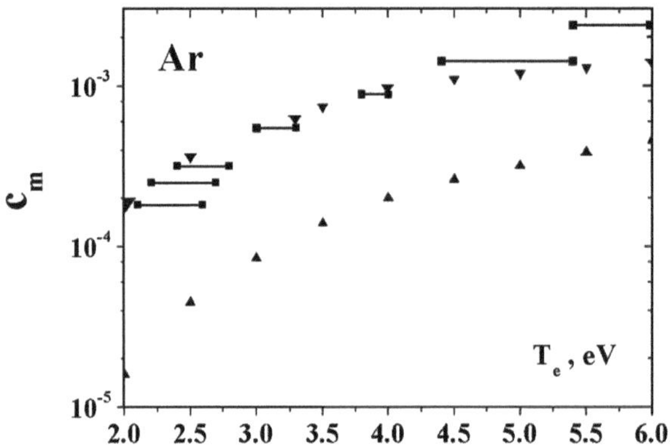

Fig. 10.24 Concentration of argon metastable atoms in the lowest excited state $^3P_2$ via the electron temperature due to the second mechanism of equilibrium in accordance with formula (10.6.3). Experimental results [164, 165] are given between *filled squares*, *triangles* correspond to destruction of metastable atoms by electron impact through subsequent excitation and ionization and *upturned triangles* correspond to quenching of metastable atoms through ionization by electron impact only

this process is given in Fig. 10.20. The second version relates to low electron number densities where after transition in more excited states an atom returns in the initial state as a result of radiation, and hence the irreversible process of quenching of an argon metastable atom in a gas discharge plasma results in atom ionization. Note that in the second version kinetics of excitation may be more complex since mixing of excited states with close excitation energies may be essential.

Chapter 11
Ionization in Helium and Argon Gas Discharge Plasma

Abstract Ionization processes in a helium and argon gas discharge plasma are analyzed for single and stepwise ionization in electron-atom collisions. The rates of ionization processes and the equilibrium reduced electric field strength as a function of the atom number density are determined for helium and argon plasmas.

11.1 Single Ionization of Atoms in Helium Gas Discharge Plasma

Single ionization of atoms in a gas discharge plasma proceeds according to the scheme

$$e + A \to 2e + A^+, \tag{11.1.1}$$

and stepwise ionization proceeds through excited atom states according to the scheme

$$e + A \to e + A^*, \; e + A^* \to 2e + A^+, \tag{11.1.2}$$

It is convenient to describe the rate of single ionization in a gas discharge plasma by the first Townsend coefficient (8.1.1), and the values of the first Townsend coefficient in helium and argon are given in Figs. 8.3 and 8.4. Evidently, stepwise ionization dominates in a gas discharge plasma if the number density of excited atoms is not low.

It is of importance for single ionization that the electron distribution function by energy drops sharply after the threshold of excitation of each atom level. From this it follows that at low electric field strengths stepwise ionization dominates usually. We first determine the dependence of the first Townsend coefficient in helium on the reduced electric field strength at not large electric fields. This allows us to understand the role of the electron distribution fall with an increasing electron energy due to excitation process in gas ionization.

Let us determine the behavior of the energy distribution function electrons above the threshold of excitation that has form (6.5.2) [38, 83] not far from the atom excitation threshold $\Delta\varepsilon$

B.M. Smirnov, *Theory of Gas Discharge Plasma*, Springer Series on Atomic, Optical, and Plasma Physics 84, DOI 10.1007/978-3-319-11065-3_11

$$f_0(\varepsilon) = f_0(\Delta\varepsilon)\exp(-S), \tag{11.1.3}$$

and in the quasiclassical approach is valid if the criterion $(S')^2 >> S''$ holds true. In this case we obtain formula (6.5.3) [83] for the function S that follows from solution of equation (6.5.1) near the excitation threshold

$$S = \kappa\left(\frac{\varepsilon - \Delta\varepsilon}{\Delta\varepsilon}\right)^{5/4}, \ \kappa = \frac{2v_o}{5a}\sqrt{3\frac{g_*}{g_o}\nu_q\nu_{ea}},$$

where v_o is the electron velocity at the excitation threshold, $a = eE/m_e$, the quenching rate is $\nu_q = Nk_q$, where N is the number density of atoms, g_*, g_o are the statistical weights of an atom in the ground and excited states, k_q is the rate constant of quenching of a given state of the helium atom by electron impact with transition in the ground atom state, and $\nu_{ea} = Nv_o\sigma^*_{ea}$ is the rate of elastic electron-atom scattering, so that σ^*_{ea} is the diffusion cross section of electron-atom elastic scattering. As is seen, this quasiclassical solution is valid, if $\kappa \gg 1$.

The first Townsend coefficient α, as well as the ionization rate constant of atoms by electron impact in an electric field, is proportional to the electron distribution function at the atom ionization potential. The electron distribution function drops with an increasing electron energy, and on the basis of formula (6.5.2) the isotropic part of the electron distribution function at the ionization threshold $f_0(J)$ may be represented as

$$f_0(J) \sim \exp\left[-\sum_i \kappa_i\left(\frac{J - \Delta\varepsilon_i}{\Delta\varepsilon_i}\right)^{5/4}\right], \tag{11.1.4}$$

where κ_i is given by formula (6.5.2) for a given state, and $\Delta\varepsilon_i$ is the excitation energy of this state. One can convince that highly excited or Rydberg states with large principal quantum numbers n do not give not a contribution to this sum because it is $\sim n^{-15/4}$. Hence, because mostly s and p states of excited electron in the helium atoms partake in this coupling, highly excited states do not give a contribution to the drop of the electron distribution function.

As it follows from the above analysis, the drop of the electron distribution function in an electron energy space is determined at the atom ionization potential by a restricted number of excited atom states. We below remain in the sum of formula (11.1.4) only 2^3S, 2^1S and 2^1P states, and the parameters of this and previous formulas for the drop of the electron distribution function above the corresponding excitation threshold are given in Table 11.1. Assuming that quenching of metastable states 2^3S and 2^1S by electron impact in the ground atom state results from exchange collisions, i.e. an incident electron replaces the bound excited electron, we obtain almost identical quenching rate constants of 2^1S state for a slow electron $k_q \approx 3\cdot 10^{-9}\,\text{cm}^3/\text{s}$ as it take place for 2^3S-state [198]. Quenching of the resonantly excited state 2^1P by electron impact is determined by the dipole-dipole interaction that gives for the quenching rate constant with transition in the ground state $k_q = 8\cdot 10^{-8}\,\text{cm}^3/\text{s}$ [40].

Table 11.1 The parameters of a sharp decrease of the energy distribution function of electrons due to three excited states of the helium atom

State	2^3S	2^1S	2^1P
$\Delta\varepsilon_i$ (eV)	19.82	20.62	21.22
σ_{ea} ($Å^2$)	2.7	2.7	2.6
k_q ($10^{-9}\,cm^3/s$)	3	3	8
κ_i	270/x	150/x	130/x
$S_i(J)$	45/x	20/x	14/x

Here $\Delta\varepsilon_i$ is the excitation energy of a given state (the ionization potential of the helium atom is $J = 24.59\,eV$), σ_{ea} is the diffusion cross section of elastic electron-atom scattering at the threshold of an indicated excitation, k_q is the rate constant of quenching of this state in collisions with a slow electron with the atom transition in the ground state, $x = E/N$ is the reduced electric field strength given in Td ($1\,Td = 10^{-17}\,V \cdot cm^2$), the quantities κ_i and $S_i(J)$are given by formulas (11.1.3) and (6.5.2) respectively

These quenching rate constants are used for evaluation the parameters represented in Table 11.1.

On the basis of the data of Table 11.1, assuming the first Townsend coefficient to be proportional to the distribution function at the atom ionization potential $\alpha(x) \sim f_0(J)$, we obtain the dependence of the first Townsend coefficient α on the reduced electric field strength in the form

$$\frac{\alpha}{N_a} = C \exp\left(-\frac{80}{x}\right) \tag{11.1.5}$$

Figure 11.1 gives comparison of the measured values of the reduced first Townsend coefficient [43, 345–347] with calculated ones on the basis of this formula (11.1.5). From this comparison one can find the appropriate constant in formula (11.1.5) $C = 0.2\,Å^2$.

In evaluating the ionization rate for helium atoms by electron impact in a gas discharge plasma in the range 10–100 Td, we assume that the dependence of the first Townsend coefficient α on the reduced electric field strength $x = E/N$ is determined by jumps of the logarithm of the electron distribution function near corresponding thresholds of atom excitation. As it follows from the data of Fig. 11.1, this mechanism leads to a more or less right this dependence. Nevertheless, a more deep analysis of the electron distribution function testifies about its specific dependence on the electron energy. Indeed, let us use the approximation (3.2.8) for the diffusion cross section of electron scattering on the helium atom that is compared with the experimental cross sections [134] in Fig. 3.1. We below use the approximation (3.2.8) in the kinetic equation (6.5.1) for the electron distribution function.

The limit of low electron number densities corresponds to the Druyvesteyn case [273, 289] for the electron distribution function, where the diffusion cross section of electron-atom is independent of the electron energy. In this limiting case we have for the electron distribution function

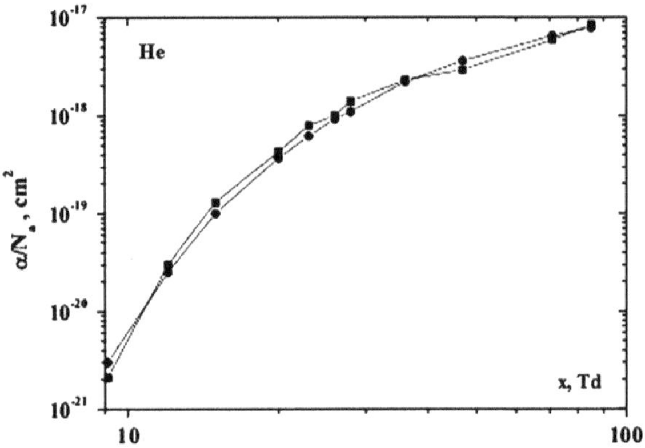

Fig. 11.1 The reduced first Townsend coefficient as a function of the reduced electric field strength in the range of low strengths. *Filled squares* correspond to measured data, which collected in [43] and are averaged now, and *filled circles* relate to values calculated on the basis of formula (11.1.5) with $C = 0.2\,\text{Å}^2$

$$f_0 = B\exp\left(-\frac{\varepsilon^2}{\varepsilon_o^2}\right), \; \varepsilon_o = \sqrt{\frac{M}{3m_e}}eE\lambda, \; \varepsilon \le \varepsilon_c, \; \varepsilon_c = 10\,\text{eV}, \tag{11.1.6}$$

where B is the normalization coefficient, $\lambda = 1/(N_a\sigma_{ea}^*)$, and in the helium case

$$\varepsilon_o = 0.82x, \; \varepsilon \le \varepsilon_c, \tag{11.1.7}$$

where the characteristic energy ε_o is measured in eV, and the reduced electric field strength $x = E/N_a$ is expressed in Td. Next, at more high electron energies $\varepsilon \ge \varepsilon_c$ the electron distribution function has the form

$$f(\varepsilon) = f(\varepsilon_c)\left(\frac{\varepsilon}{\varepsilon}\right)^{-\gamma}, \; \varepsilon \ge \varepsilon_c, \; \gamma = \frac{6A^2m_e}{e^2x^2M} \approx \frac{300}{x^2} \tag{11.1.8}$$

The distribution function at the energy which is equal to the atom ionization potential is given in Fig. 11.2.

We consider the ionization process in a range $10\,\text{Td} < x < 100\,\text{Td}$. One can see that according to formulas (11.1.6), (11.1.7), (11.1.8) the variation of the electron distribution function below the excitation threshold of the helium atom is essential on the lower edge of the indicated range of reduced electric field strengths. As a result, we obtain for the reduced first Townsend coefficient

$$\frac{\alpha}{N_a} = \sigma_o \exp\left(-\frac{80}{x} - \frac{150}{x^2}\right) \cdot 2.46^{-\gamma}, \; \gamma = \frac{300}{x^2}, \tag{11.1.9}$$

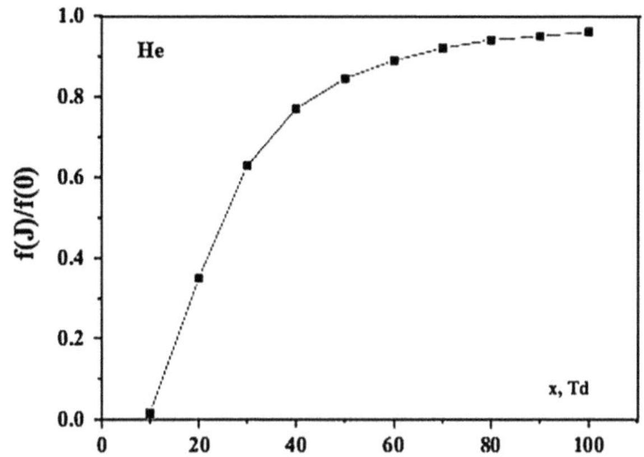

Fig. 11.2 The reduced first Townsend coefficient as a function of the reduced electric field strength in the range of low strengths. *Filled squares* correspond to measured data, which collected in [43] and are averaged now, and *filled circles* relate to values calculated on the basis of formula (11.1.5) with $C = 0.2\,\text{Å}^2$

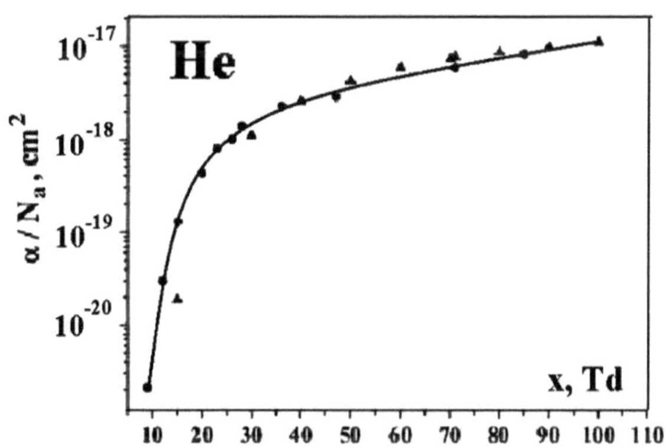

Fig. 11.3 The reduced first Townsend coefficient as a function of the reduced electric field strength in the range of low strengths. *Filled circles* correspond to measured data [43], and *filled triangles* relate to values calculated on the basis of formula (11.1.9)

where $\sigma_o = 2.5\,\text{Å}^2$. The results on the basis of this formula are compared with experimental data in Fig. 11.3. As is seen, formula (11.1.9) is found in a rough accordance with experimental data.

We above consider ionization of helium in an external electric field in the range of the electric field strengths $10\,\text{Td} < x < 100\,\text{Td}$, where the ionization process proceeds at the tail of the energy distribution function of electrons. The ionization process corresponds to direct atom ionization by electron impact according to the scheme

$$e + He \to 2e + He^{+}, \tag{11.1.10}$$

and atomic ions He^{+} result from this process.

11.2 Stepwise Ionization of Atoms in Helium Gas Discharge Plasma

The main contribution to stepwise ionization is determined by the metastable state $He(2^3S)$ because of higher concentration for such excited atoms. To convince in this, we find below the contribution to atom ionization by electron impact through the metastable state 2^1S in the regime of high electron number densities.

Let us represent the balance equation for the rate of formation of electrons in the form

$$\frac{dN_e}{dt} = K_{ion}N_eN_a,$$

and this is the definition of the rate constant K_{ion} of stepwise ionization. Next, expressing this equation through the number density of metastable atoms N_m in the 2^3S state, we obtain for this balance equation

$$\frac{dN_e}{dt} = k_{ion}N_eN_m,$$

where k_{ion} is the rate constant of ionization of the metastable atom in the 2^3S state by electron impact. Comparing these balance equations, we have for the rate constant of stepwise ionization

$$K_{ion} = c_m k_{ion} = \frac{k_{ex}k_{ion}}{k_Q + k_{ion}}, \tag{11.2.1}$$

where we use formula (11.2.6) for the concentration of metastable atoms. Figure 11.4 gives the values of this rate constant as a function of the electron temperature.

We now find the criterion of stepwise ionization in a helium gas discharge plasma restricted by the metastable state $He(2^3S)$ only. Let us introduce a typical time τ of decay of metastable atoms through other decay channels, and the criterion of stepwise ionization has the form

$$\tau \gg \frac{1}{(k_Q + k_{ion})N_e},$$

where k_Q and k_{ion} are the rate constants of quenching and ionization of the metastable state by electron impact correspondingly. From the above formula we have for the

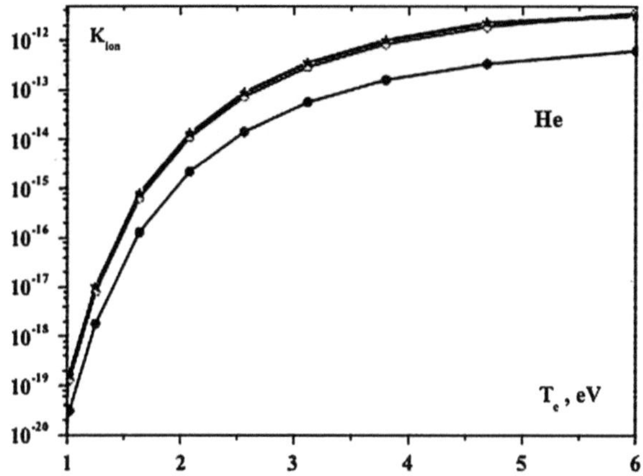

Fig. 11.4 Stepwise ionization rate constant of helium atoms by electron impact depending on the electron temperature T_e according to formula (11.2.1) at the electron concentration $c_e = 10^{-6}$ if ionized helium is located in a constant electric field. The partial rate constant of ionization through the metastable 2^3S state is given by *open circles*, the partial rate constant of ionization through the metastable 2^1S state is marked by *filled circles*, the total rate constant of stepwise ionization is represented by *stars*

criterion of stepwise ionization

$$(N_e\tau)^{-1} \gg k_Q + k_{ion}, \tag{11.2.2}$$

Let us consider firstly the regime of low electron concentrations. Basing on the Druyvesteyn distribution function (6.1.9) of electrons by energy and using the rate constant of atom ionization by electron impact (3.6.16) near the ionization threshold we have for the averaged rate constant of atom ionization

$$k_{ion} = \int \sqrt{\frac{2\varepsilon}{m_e}} \cdot \frac{10e^4(\varepsilon - J)}{9\varepsilon J^2} f_0(\varepsilon)d\varepsilon,$$

where the distribution function $f_0(\varepsilon)$ is given by formula (6.1.9) and is normalized to one

$$f_0(\varepsilon)d\varepsilon = \frac{2}{\varepsilon_o^{3/2}\Gamma(3/4)} \exp\left(-\frac{\varepsilon^2}{\varepsilon_o^2}\right) \varepsilon^{1/2}d\varepsilon,\ \varepsilon_o = eE\lambda\sqrt{\frac{M}{3m_e}}$$

Assuming that ionization proceeds mostly near the threshold, i.e. the integral converges at $\varepsilon_o \ll J$, and taking the integral at this assumption, we obtain

$$k_{ion} = 0.64\sqrt{\frac{\varepsilon_o}{m_e}}\frac{e^4}{J^2} \exp\left(-\frac{J^2}{\varepsilon_o^2}\right) \tag{11.2.3}$$

This formula holds true at $\varepsilon_o \ll J$ where the electron energies near the atom ionization threshold give the main contribution to the above integral. In addition, this formula corresponds to ionization of excited atoms for which the atom ionization energy is small compared to the excitation energy of the ground atom state where the electron distribution function of electrons by energy drops sharply with an increasing electron energy.

In particular, considering ionization of the metastable helium atom $He(2^3S)$, we have $J = 4.77$ eV. Taking the average cross section of electron-atom diffusion cross section $\sigma^*_{ea} = 6\,\text{Å}^2$, we have $\varepsilon_o = 0.82x$, where ε_o is given in eV, and the reduced electric field strength x is measured in Td. In this case formula (11.2.3) takes the form

$$k_{ion} = k_o\sqrt{x}\exp\left(-\frac{33.8}{x^2}\right), \tag{11.2.4}$$

where $k_o = 2.2\cdot10^{-8}\,\text{cm}^3/\text{s}$. In this consideration we are based on the approximation (3.2.8) for the diffusion cross section of electron-atom scattering and take into account that at energies which determine ionization of helium atoms by electron impact, the cross section of elastic electron-atom collision, is independent of the collision energy.

We now consider the case of a high electron number density with the Maxwell distribution function of electrons above the threshold of atom ionization, being guided by ionization of excited atoms. Applying formula (3.6.16) for the ionization rate constant k_{ion} of metastable atoms $He(2^3S)$ and $He(2^1S)$ by electron impact, using the data of Table 3.5 and taking the electron temperature $T_e < 5$ eV that is typical for gas discharge, we obtain for the ionization rates of these atoms

$$k_{ion}(2^3S) = 1.5\cdot10^{-7}\sqrt{T_e}\exp\left(-\frac{4.77}{T_e}\right), \quad k_{ion}(2^1S) = 2.2\cdot10^{-7}\sqrt{T_e}\exp\left(-\frac{3.97}{T_e}\right) \tag{11.2.5}$$

where the rate constant is expressed in cm^3/s, and the electron temperature T_e is given in eV. The rate constant of quenching of the metastable atom $He(2^3S)$ by electron impact in cm^3/s in accordance with formula (10.3.13) takes the form

$$k_Q = 3\cdot10^{-9} + \frac{5\cdot10^{-7}\exp(-1.398/T_e)}{1+5\exp(-0.602/T_e)}$$

The dependencies on the electron temperature for the rate constants k_Q and k_{ion} are given in Fig. 11.5 in accordance with formulas (10.3.13) and (11.2.5).

Since according to Fig. 11.5 the rate constant of ionization of the metastable atom $He(2^3S)$ by electron impact is comparable with the rate constant of quenching of this state in other channels, ionization of metastable atoms may give a contribution to the population of these excited atoms in a gas discharge plasma. Indeed, transforming formula (10.3.9) for the concentration c_m of metastable atoms in the state (2^3S), we have

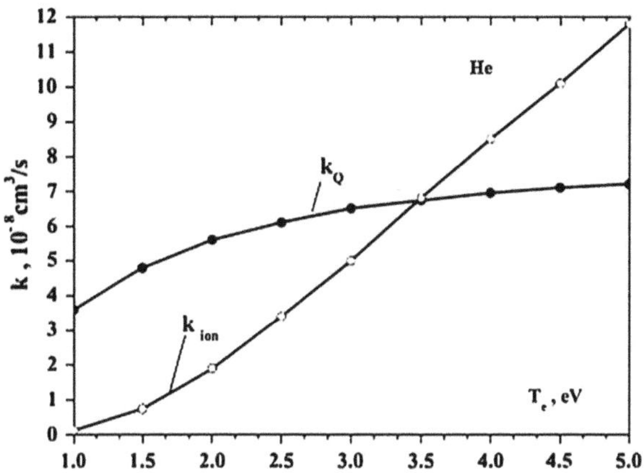

Fig. 11.5 The rate constant k_Q of quenching of the metastable atom $He(2^3S)$ by electron impact in a helium gas discharge plasma and the rate constant k_{ion} of ionization of the metastable atom $He(2^3S)$ by electron impact as a function of the electron temperature

$$c_m = \frac{k_{ex}}{k_Q + k_{ion}} \tag{11.2.6}$$

Figure 11.6 gives the concentration c_m of metastable atoms $He(2^3S)$ calculated on the basis of formula (11.2.6) with using formula (10.2.3) for the excitation rate constant, formula (10.3.13) for quenching with transition in other bound states and formula (11.2.5) for ionization of metastable atoms. As is seen, ionization of metastable atoms does not change significantly the concentration of metastable atoms under typical parameters of a helium gas discharge plasma.

In consideration stepwise ionization of a helium gas discharge plasma we assume that the main contribution to stepwise ionization is determined by the metastable state $He(2^3S)$. In order to ascertain the validity of this, we determine the contribution of the state $He(2^1S)$ in stepwise ionization of helium. Let us introduce the ratio of the ionization rates from the states $He(2^1S)$ and $He(2^3S)$ as

$$y = \frac{N(2^1S)k_{ion}(2^1S)}{N(2^3S)k_{ion}(2^3S)}, \tag{11.2.7}$$

and according to formula (10.3.9)

$$\frac{N(2^1S)}{N(2^3S)} = \frac{\exp(-0.796/T_e)}{3 + 5\exp(-0.602/T_e)}, \tag{11.2.8}$$

where we consider the regime of a high electron number density, and the electron temperature is expressed in eV. The rate constants are taken according to the scheme of Fig. 10.8 and the principle of detailed balance (3.3.7). In addition, we use formula (11.2.5) for the ionization rate constants of atoms in the states $He(2^3S)$ and $He(2^1S)$

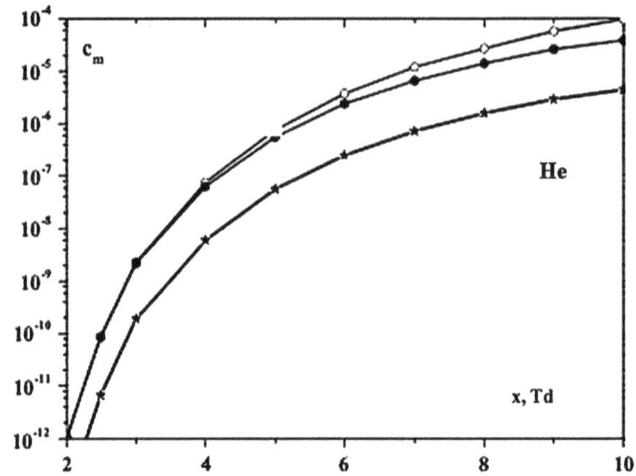

Fig. 11.6 Concentration of metastable helium atoms in a gas discharge plasma of helium in an external electric field as a function of the reduced electric field at the electron concentration $c_e = 10^{-6}$. The concentrations of metastable atoms $He(2^3S)$ are labelled by *filled circles*, the concentrations of metastable atoms $He(2^1S)$ are marked by *stars*. For comparison the concentrations of metastable atoms $He(2^3S)$ in neglecting the ionization decay of these atoms are represented as *open circles*

which are obtained on the basis of formula (3.6.16) and are valid at $T_e < 5\,\mathrm{eV}$. From this we find for the parameter

$$y = \frac{N(2^1S)k_{ion}(2^1S)}{N(2^3S)k_{ion}(2^3S)} = \frac{0.48}{1 + 1.67\exp(-0.602/T_e)} \qquad (11.2.9)$$

Figure 11.7 gives the temperature dependence for the contribution $y/(1+y)$ of ionization of metastable atoms $He(2^1S)$ to stepwise ionization of a helium gas discharge plasma for the regime of a high electron number density. From this it follows that the main contribution to stepwise ionization of a helium gas discharge plasma follows from excitation of the metastable state $He(2^3S)$.

In the case of the Maxwell distribution function of electrons if highly excited atoms are in equilibrium with electrons, the asymptotic rate constant of ionization is given by formula (3.8.8)

$$k_{as} = 2\frac{g_i}{g_o}\frac{m_e e^{10}}{\hbar^3 T_e^3}\exp\left(-\frac{J}{T_e}\right)$$

according to the equilibrium for weakly bound and free electrons. As it is given by (3.8.4)

$$C = 4 \cdot 10^{\pm 0.2}$$

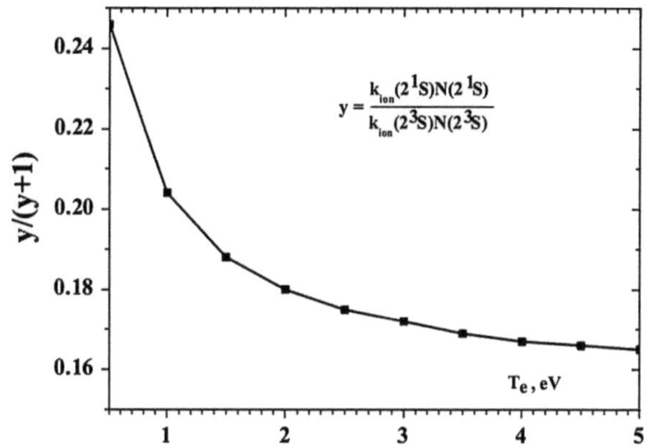

Fig. 11.7 The ratio of the rates of ionization of helium atoms by electron impact in a helium gas discharge plasma through formation of metastable atoms in states 2^1S and 2^3S depending on the electron temperature T_e

This accounts for a hydrogenlike spectrum for excited states. In addition, the asymptotic formula ignores radiative processes. This is fulfilled partially for metastable helium atoms, where in a wide range of electron number densities their decay results from transitions in the $He(2^1P)$ state with their subsequent radiation. Formula (3.8.8) gives for the ionization rate constant κ_{ion} in the case of ionization the metastable atom $He(2^3S)$ by electron impact in this limiting case ($C = 4,\ g_o = 3,\ g_i = 2$)

$$\kappa_{ion} = \frac{6.6 \cdot 10^{-4}}{T_e^3} \exp\left(-\frac{4.77}{T_e}\right), \tag{11.2.10}$$

where κ_{ion} is expressed in cm^3/s, and T_e is given in eV.

Using formula (11.2.5) for the ionization rate constant of the metastable atom $He(2^3S)$ by electron impact and formula (11.2.10) for the ionization rate constant in the asymptotic limit, we find for their ratio

$$\Delta = \frac{k_{ion}(2^3S)}{\kappa_{ion}} = \frac{6.5 \cdot 10^{-4} T_e^{7/2}}{4.77 + T_e} \tag{11.2.11}$$

Figure 11.8 gives the temperature dependence for this ratio in accordance with formula (11.2.11). As is seen, in the range of real electron temperatures the limiting case of the metastable atom $He(2^3S)$ by electron impact is not realized.

Being guided by a gas discharge plasma of glow discharge with $N_e \ll 10^{12}\,\mathrm{cm}^{-3}$, we obtain the role of radiation in the helium gas discharge plasma by resonantly excited atoms $He(2^1P)$. As a result, the main contribution to ionization of this plasma follows from metastable helium atoms $He(2^3S)$ and the ionization rate constant is small compared to the asymptotic ionization rate constant (3.8.8) that we have in the case of thermodynamic equilibrium between excited states and ions. Let us

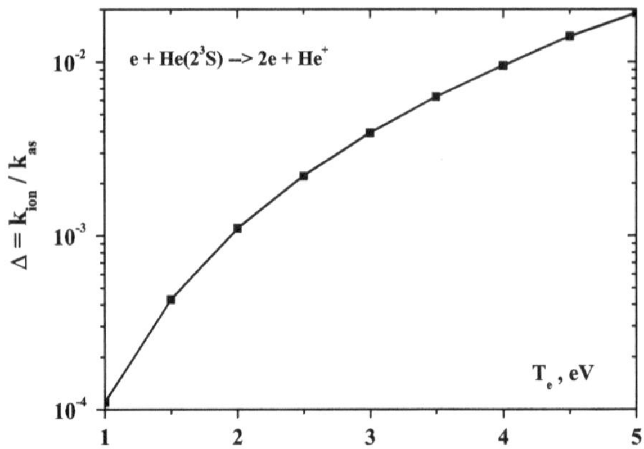

Fig. 11.8 The ratio of the ionization rates the metastable atom $He(2^3S)$ by electron impact through single collisions and in participation of more excited atoms according to formula (11.2.11)

consider another limiting case when one can ignore radiative processes compared to collision ones for excited helium atoms in $He(2^1P)$ and $He(2^3P)$ states. This takes place at high electron number density $N_e \gg 10^{13}\,\text{cm}^{-3}$ and corresponds to arc discharge. Then thermodynamic equilibrium is established for excited atom states and their population is determined by the Boltzmann formula (2.1.5). Correspondingly, formula (3.8.8) for the ionization rate constant is valid in this case. We below ascertain which excited states give the main contribution to this ionization rate constant.

Using the Boltzmann formula (2.1.5) for the population of excited states of helium atoms and formula (3.6.16) for the ionization rate constant from each state, we obtain for the total ionization rate constant k_{ion}

$$k_{ion} = k_{ion}(2^3S) \cdot \sum_i \frac{g_i}{3}\left(\frac{4.77}{J}\right)^2, \tag{11.2.12}$$

where $k_{ion}(2^3S)$ is the rate constant of atom ionization by electron impact from the state 2^3S, J_i is the ionization potential for ith state that is expressed in eV, g_i is the statistical weight of this states 4.77 eV and 3 are the ionization potential and statistical weight for the 2^3S state. From this one can find that taking into account the states 2^3S, 2^1S, 2^3P and 2^1P increases the ionization rate constant in 9 times compared with the case of ionization of the 2^3S state. The contribution of the next group of levels with the principal quantum number $n = 3$ for a valence electrons (the ionization potentials for these states ranges from 1.5 up to 1.87 eV) is 113 compared to that from the 2^3S state. Note that according to formula (11.2.11) the contribution of the 2^3S state to ionization of the equilibrium plasma is $3 \cdot 10^{-4}$ at the temperature $T_e = 1\,\text{eV}$, $1 \cdot 10^{-3}$ at the temperature 1.5 eV, and $4 \cdot 10^{-3}$ at the electron temperature 2 eV. One can conclude from this that excited states with $n > 3$ give the main contribution to ionization of the equilibrium plasma. Since these states are not important for ionization processes in a nonequilibrium plasma, we have that the formula (3.8.8) leads to overvaluation of the ionization rate constant of atoms by electron impact in a nonequilibrium plasma.

11.3 Single and Stepwise Ionization of Atoms in Argon Gas Discharge Plasma

In considering ionization in an argon gas discharge plasma, we analyze two limiting cases for the number density of electrons, as well as in the case of the helium gas discharge plasma. Namely, in the limit of a low electron number density, where electron-electron collisions are negligible, the ionization process proceeds from the ground atom state, and in the other limit of a high electron number density the ionization process has a stepwise character and proceeds through excited states. We first find a drop of the energy distribution function of electrons $f_0(\varepsilon)$ above the excitation threshold $\Delta\varepsilon$ due to excitation of argon atom in states with the electron shell $3p^5 4s$, that according to formulas (6.5.2) and (6.5.3) is given by

$$f_0(\varepsilon) = f(\Delta\varepsilon)\exp(-S), \; S = \kappa\left(\frac{\varepsilon - \Delta\varepsilon}{\Delta\varepsilon}\right)^{5/4}, \; \kappa = \frac{2v_o N_a}{5a}\sqrt{3\frac{\sum_i g_i k_q^i}{g_o}\nu_q k_{ea}}, \tag{11.3.1}$$

where k_q^i is the rate constant of quenching of ith state with transition in the ground state. Restricting by states with the electron shell $3p^5 4s$, we have on the basis of the Table 12.1 data $\sum_i g_i k_q^i = 8.0 \cdot 10^{-9}\,\text{cm}^3/\text{s}$. In addition, we have $v_o = 2.0 \cdot 10^8\,\text{cm/s}$, the rate constant of electron-atom elastic collision at the excitation threshold is $k_{ea} = 2.9 \cdot 10^{-7}\,\text{cm}^3/\text{s}$, that gives in this case

$$\kappa = \frac{407}{x}, \tag{11.3.2}$$

where the reduced electric field strength x is expressed in Td. This gives at the ionization threshold

$$S(J) = \kappa \cdot \left(\frac{J - \Delta\varepsilon}{\Delta\varepsilon}\right)^{5/4} = \frac{110}{x}, \tag{11.3.3}$$

where J is the atom ionization potential, and we use the average excitation energy $\Delta\varepsilon = 11.65\,\text{eV}$ for the state group with the electron shell $3p^5 4s$. Taking the threshold dependence for the ionization rate constant by electron impact $k_{ion}(\varepsilon) \sim \varepsilon - J$, we have for the total ionization rate constant

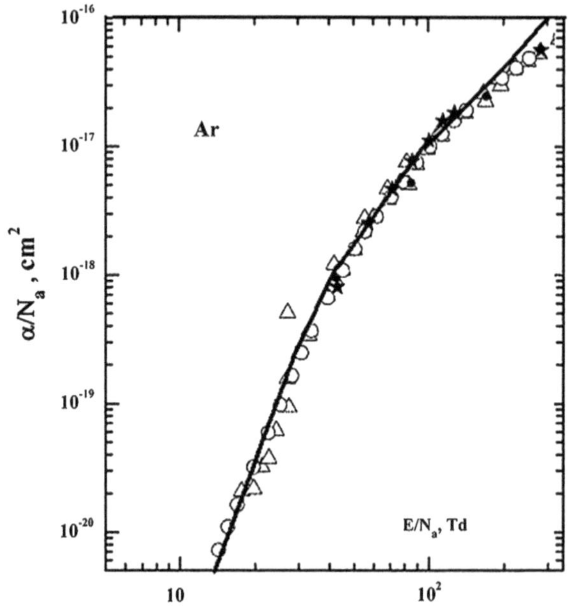

Fig. 11.9 The reduced first Townsend coefficient α/N_a as a function of the reduced electric field strength $x = E/N_a$. *Signs*-measured values given in Fig. 5.2, *solid curve*—evaluation on the basis of formula (11.3.5) with $C = 5 \cdot 10^{-19}\,\text{cm}^2$

$$k_{ion} \sim \int_J^{\infty} \exp[-S(\varepsilon)](\varepsilon - J)d\varepsilon \sim x^2 \exp[-S(J)] = x^2 \exp(-110/x) \quad (11.3.4)$$

We also determine the reduced average energy $\overline{\varepsilon}$ of a released electron as

$$\frac{\overline{\varepsilon} - J}{J} = \frac{\int_J^{\infty} (\varepsilon - J)^2 d\varepsilon \exp[-S(\varepsilon)]}{\int_J^{\infty} (\varepsilon - J) d\varepsilon \exp[-S(\varepsilon)]} = \frac{x}{270}$$

From this it follows that formula (11.3.4) is valid at $x \ll 300\,\text{Td}$. Taking in the range $20\,\text{Td} < x < 300\,\text{Td}$ the dependence of the electron drift velocity in argon $w_e \sim x$ on the reduced electric field strength x according to Fig. 5.2, we obtain

$$\alpha(x) = \frac{k_{ion}}{w_e} = Cx \exp(-110/x) \quad (11.3.5)$$

Figure 11.9 compares measured values of the reduced first Townsend coefficient in accordance with Fig. 5.2 and evaluated values on the basis of formula (11.3.5) with $C = 5 \cdot 10^{-19}\,\text{cm}^2$.

The stepwise character of ionization in an argon gas discharge plasma proceeds through ionization of excited atoms and takes place at not very small number density of excited atoms. Let us estimate the lower limit for the concentration of excited argon atoms being guided by the electric field strength 8–10 Td. We have according to Table 8.1 the rate constant of single ionization to be $6 \cdot 10^{-20}\,\mathrm{cm}^3/\mathrm{s}$ at the reduced electric field strength 10 Td. Next, according to Fig. 6.9 the electron temperature $T_e = 5\,\mathrm{eV}$ corresponds to the reduced electric field strength $x = 4\,\mathrm{Td}$ in the regime of high electron number densities. At this temperature formula (3.6.16) gives $k_{ion} = 1.5 \cdot 10^{-7}\,\mathrm{cm}^3/\mathrm{s}$ for the rate constant of atom ionization in the state $^3P_2$ by electron impact. As it follows from Fig. 5.4, stepwise ionization is realized in an argon gas discharge plasma with the concentration of electrons above $1 \cdot 10^{-9}$, while according to Fig. 10.20 the equilibrium concentration of metastable atoms is $6 \cdot 10^{-5}$ at this electron temperature. Therefore one can expect a compared electron concentration that justifies the regime of a high electron number densities used for the above estimation.

In determination the rate of stepwise ionization, we reduce the ionization rate constant K_{ion} to the ground state, so that the balance equation term accounted for this process has the form

$$\frac{dN_e}{dt} = K_{ion} N_e N_a,$$

where N_a is the number density of atoms in the ground state. According to this definition, the ionization rate constant is given by

$$K_{ion} = \sum_i c_i k_{ion}^i, \qquad (11.3.6)$$

where c_i is the concentration of excited atoms in ith state, and k_{ion}^i is the rate constant of ionization of atoms in this state.

Let us consider the case of not high electron number density $N_e \ll 10^{12}\,\mathrm{cm}^{-3}$ which corresponds to a gas discharge plasma of glow discharge. In this case the concentrations of resonantly excites states $^3P_1$ and $^1P_1$ of argon atoms are small compared with that of metastable atoms, and these states do not give the contribution to stepwise ionization in this plasma. Applying formula (3.6.16) for ionization of excited state, we note that this formula describes liberation of a classical s-electron, and this electron is found now in $4s$ state. Let us apply formula (3.6.16) to each state of the group $4s$ and accounts for $c(^3P_0) = 0.2c(^3P_2)$. In the case under consideration we assume for simplicity $k_{ion}(^3P_2) = k_{ion}(^3P_0)$, and then formula (11.3.6) takes the form

$$K_{ion} = 1.2c(^3P_2)k_{ion}(^3P_2), \qquad (11.3.7)$$

where the argument indicates the state to which a given quantity relates. Figure 11.10 contains the dependence of the stepwise ionization rate constant (11.3.7) on the

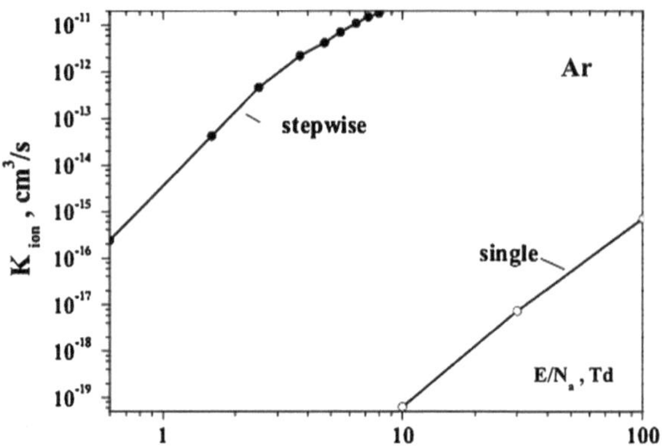

Fig. 11.10 The ionization rate constant of ionization of argon atoms by electron impact under the action of an electric field. The rate constant of stepwise ionization is evaluated on the basis of formula (11.3.7) and the electron concentration $c_e = 10^{-6}$, values of the rate constant of single ionization are taken from Table 8.1

reduced electric field strength for an argon gas discharge plasma with using the connection between the reduced electric field strength and the electron temperature in an argon gas discharge plasma given in Fig. 10.3. In addition, the single ionization rate constant is given in this Fig. 11.10 in accordance with Table 8.1 data. As is seen, stepwise ionization is realized at lower electric field strengths. The reason of this results from a not low number density of excited atoms that optimizes ionization of atoms by electron impact.

We now consider ionization equilibrium in an argon gas discharge plasma with molecular ions as a basic ion type, so that the channel of electron destruction corresponds electron-ion dissociative recombination according to the scheme

$$e + Ar_2^+ \to A + A^*, \qquad (11.3.8)$$

and the ionization balance equation takes the form in the regime of low electron number densities

$$k_{ion} N_a = \alpha N_e \qquad (11.3.9)$$

Here k_{ion} is the rate constant of single ionization of argon atoms by electron impact which values are given in Table 8.1, and α is the coefficient of dissociative recombination. Statistical averaging of measured data [214, 309, 424, 425, 427–431] gives for the dissociative recombination coefficient at room temperature

$$\alpha(Ar_2^+) = (7 \pm 2) \cdot 10^{-7}\,\mathrm{cm}^3/\mathrm{s}$$

The ionization balance equation (11.3.9) gives for the equilibrium electron concentration

$$c_e = \frac{k_{ion}}{\alpha(Ar_2^+)},$$

and in determination the equilibrium electron concentration c_e it is important the temperature dependence of the dissociative recombination coefficient $\alpha(Ar_2^+)$. An experience in measurement of this dependence [429, 430, 432] shows that this dependence is different for cases, where the electron and gas temperatures are identical or different. Roughly, this dependence has the form

$$\alpha \sim T_e^{-1.5},\ T_e = T;\ \alpha \sim T_e^{-0.5},\ T_e \gg T, \qquad (11.3.10)$$

where T is the gas temperature. Basing on the second case and using the data of Table 8.1 for k_{ion} and Table 7.4 for the effective electron temperature, we obtain $c_e = 4 \cdot 10^{-8}$ at the electric field strength $E/N_a = 10\,\mathrm{Td}$ and higher values at higher electric field strengths. Comparing this with data of Fig. 10.1, we obtain that this ionization equilibrium with dissociative electron-ion recombination corresponds to the regime of high electron number densities.

11.4 Thermodynamic of Stepwise Ionization in Gas Discharge Plasma

The above analysis shows violation of the thermodynamic equilibrium in a gas discharge plasma between atoms in the ground and excited states if this equilibrium is determined by electron-atom collisions in the regime of a high electron number density where the Maxwell distribution function of electrons holds true for the most part of electrons. This results from violation of the Maxwell form of the electron distribution function at its tail. The thermodynamic equilibrium may violate also between excited states if emission of excited atoms is a strong process. We now analyze the equilibrium for free electrons in the regime of a high electron number density if we assume stepwise ionization through the metastable state. Then equilibrium for free electrons is established according to the scheme

$$e + He(2^3S) \to 2e + He^+,\ 2e + He^+ \to e + He^* \qquad (11.4.1)$$

Because a formed excited state He^* may differ from the state $He(2^3S)$, the equilibrium number density of electrons differs from that according to the Saha formula (2.1.8). The rate constants of processes for the scheme (11.4.1) are given by formulas (11.2.5), (3.8.3) and (3.8.4) which have the form

$$K_{ei} = \frac{1.0 * 10^{-26}}{T_e^{9/2}},\ k_{ion} = \frac{2.1 \cdot 10^{-7}\sqrt{T_e}}{1 + 0.023T_e} \exp\left(-\frac{4.77}{T_e}\right), \qquad (11.4.2)$$

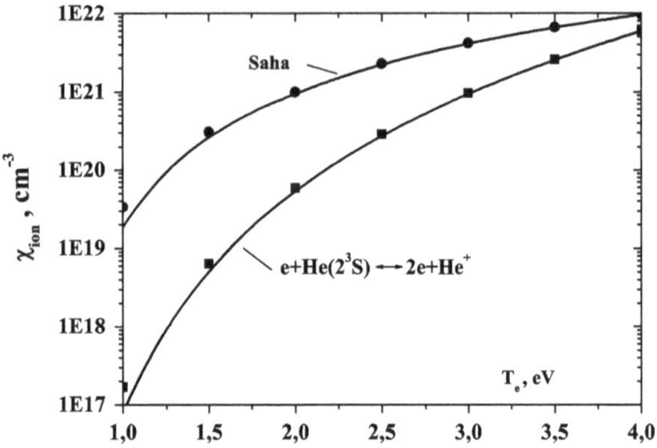

Fig. 11.11 The equilibrium ionization constant for equilibrium between free electrons and metastable atoms $He(2^3S)$ according to formulas (11.4.3) and (11.4.4)

where K_{ei} is expressed in cm^6/s, k_{ion} is given in cm^3/s and is determined by formula (11.2.5), the electron temperature is expressed in eV. From this we find the equilibrium constant between free electrons and metastable atoms that follows from the balance equation for equilibrium (11.4.1) in a helium quasineutral gas discharge plasma

$$\frac{dN_e}{dt} = -K_{ei} N_e^3 + k_{ion} N_e N_m$$

this gives for the equilibrium constant χ between electrons and metastable atoms $He(2^3S)$

$$\chi_{ion} \equiv \frac{N_e^2}{N_m} = \frac{k_{ion}}{K_{ei}} = \frac{2.1 \cdot 10^{19} T_e^5}{1 + 0.023 T_e} \exp\left(-\frac{4.77}{T_e}\right), \tag{11.4.3}$$

One can compare this equilibrium constant with that for the ionization equilibrium according to the Saha formula (2.1.8)

$$\chi_{Saha} \equiv \frac{g_e g_i}{g_m} \left(\frac{m_e T_e}{2\pi \hbar^2}\right)^{3/2} \exp\left(-\frac{J}{T_e}\right) = 4.0 \cdot 10^{21} T_e^{3/2} \exp\left(-\frac{4.77}{T_e}\right), \tag{11.4.4}$$

where the electron temperature is given in eV. Figure 11.11 gives the dependence on the electron temperature for the ionization equilibrium constant between free electrons and metastable atoms $He(2^3S)$ for equilibrium (11.4.1) and the Saha equilibrium in accordance with formulas (11.4.3) and (11.4.4). In addition we give in Fig. 11.12 the electron concentration under thermodynamic equilibrium between free electrons and metastable helium atoms $He(2^3S)$ in accordance with equilibrium (11.4.1). We assume the Maxwell distribution function of electrons which cause

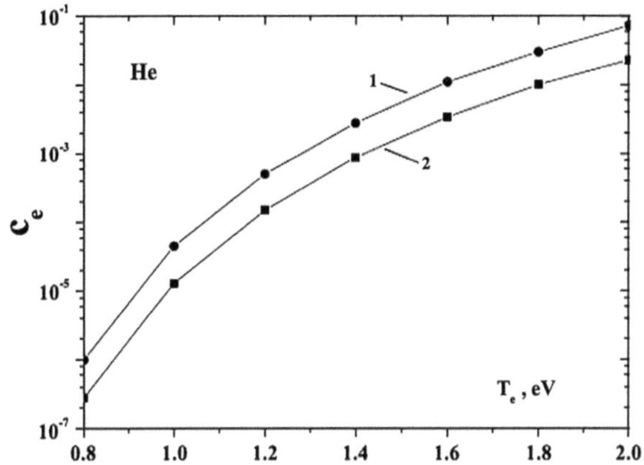

Fig. 11.12 The equilibrium concentration of free electrons in helium in the case of the Maxwell distribution function of electrons and the scheme (11.4.1) for electron equilibrium. The number density of helium atoms is : *1*—$N_a = 10^{16}\,\mathrm{cm}^{-3}$, 2—$N_a = 10^{17}\,\mathrm{cm}^{-3}$

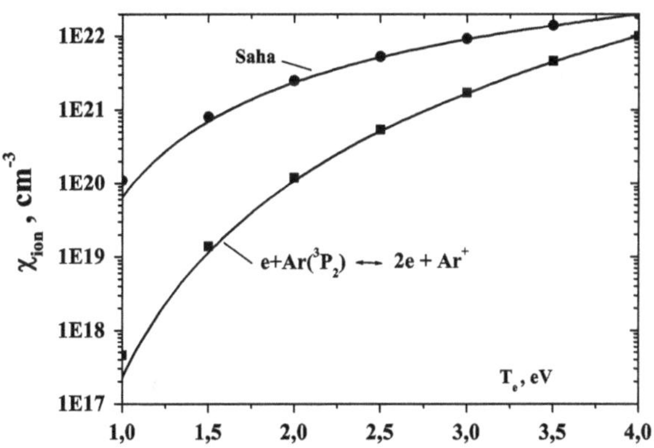

Fig. 11.13 The ionization equilibrium constant in an argon gas discharge plasma of a high electron number density for equilibrium between free electrons and metastable argon atoms $Ar(^3P_2)$ in accordance with formula (11.4.7) for an equilibrium plasma and formula (11.4.9) for a nonequilibrium plasma

formation of free electrons as a result of electron collisions with metastable atoms, and the number density of metastable atoms is determined by formula (10.3.9). Though according to this formula the number density of metastable atoms depends on the electron concentration, this dependence is weak. Next, the concentration of metastable atoms in the range of the electron temperatures of Fig. 11.12 ($T_e = 0.8 - 2\,\mathrm{eV}$) is small compared to the electron concentration and ranges from $c_m = 6 \cdot 10^{-13}$ at $T_e = 0.8\,\mathrm{eV}$ up to $c_m = 1 \cdot 10^{-6}$ at $T_e = 2\,\mathrm{eV}$.

We give in Fig. 11.13 the electron concentration under thermodynamic equilibrium between free electrons and metastable helium atoms $He(2^3S)$ in accordance with equilibrium (11.4.1). We assume the Maxwell distribution function of electrons which cause formation of free electrons as a result of electron collisions with metastable atoms, and the number density of metastable atoms is determined by formula (10.3.9). Though according to this formula the number density of metastable atoms depends on the electron concentration, this dependence is weak. Next, the concentration of metastable atoms in the range of the electron temperatures of Fig. 11.13 ($T_e = 0.8 - 2\,\text{eV}$) is small compared to the electron concentration and ranges from $c_m = 6 \cdot 10^{-13}$ at $T_e = 0.8\,\text{eV}$ up to $c_m = 1 \cdot 10^{-6}$ at $T_e = 2\,\text{eV}$. We also note that the scheme (11.4.1) holds true if the criterion

$$K_{ei} N_e^2 \tau \gg 1 \qquad (11.4.5)$$

is fulfilled, where τ is a typical time of electron loss due to other channels. Usually this criterion is violated.

We also consider the ionization equilibrium for excited argon atoms in an argon gas discharge plasma that by analogy with the helium case (11.4.1) has the form

$$e + Ar(3p^5 4s) \to 2e + Ar^+,\; 2e + Ar^+ \to e + Ar^* \qquad (11.4.6)$$

By analogy with formula (11.4.4), the ionization equilibrium constant in an argon gas discharge plasma according to the Saha formula (2.1.8) has the following form with respect to the excited argon atom state $3p^5 4s(^3P_2)$

$$\chi_{Saha} = 7.2 \cdot 10^{21} T_e^{3/2} \exp\left(-\frac{4.21}{T_e}\right), \qquad (11.4.7)$$

where index "m" corresponds to the atom state $3p^5 4s(^3P_2)$, and the electron temperature T_e is expressed in eV. If we assume that atom ionization in an argon gas discharge plasma is determined by step ionization of atoms in excited states with the electron shell $3p^5 4s$, the equilibrium constant for ionization has the form

$$\chi_{ion} \equiv \frac{N_e^2}{N_m} = \frac{\sum\limits_i c_i k_{ion}^i}{c_m K_{ei}}, \qquad (11.4.8)$$

where c_i is the concentration of excited atoms in ith state, c_m is the concentration of $^3P_2$ atoms, k_{ion}^i is the rate constant of ionization by electron impact for excited atoms in ith state. In particular, in a rare argon gas discharge plasma, where the concentration of excited atoms in resonantly excited states $^3P_1$ and $^1P_1$ is small compared with that in the state $^3P_2$, formula (11.4.8) takes the form $[c(^3P_0) = 0.2c(^3P_2)]$

$$\chi_{ion} \equiv \frac{N_e^2}{N_m} = \frac{k_{ion}}{K_{ei}} = \frac{3.2 \cdot 10^{19} T_e^5}{1 + 0.026 T_e} \exp\left(-\frac{4.21}{T_e}\right), \qquad (11.4.9)$$

Figure 11.13 contains the dependence of the ionization equilibrium constant between free electrons and metastable atoms $Ar(^3P_2)$ in an argon gas discharge

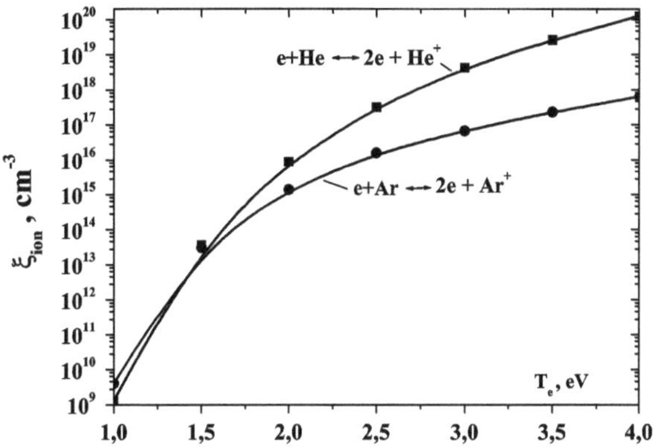

Fig. 11.14 The ionization equilibrium constant (11.4.10) in a helium and argon gas discharge plasma of a high electron number density for equilibrium between free electrons and atoms of helium and argon in the ground state. The data of Figs. 11.11, 10.12, 11.13 and 10.21 are used

plasma on the electron temperature according to formula (11.4.7) for an argon equilibrium plasma and on the basis of formula (11.4.9) for a nonequilibrium argon plasma. Note that under equilibrium conditions the ionization rate constant of argon atoms by electron impact is given by formula (3.8.8) in a gas discharge plasma of a high electron number density. As is seen from Figs. 11.11 and 11.13, in a real gas discharge plasma of helium and argon equilibrium conditions are not fulfilled.

We also give in Fig. 11.14 the ionization equilibrium constant

$$\xi_{ion} \equiv \frac{N_e^2}{N_a} = c\chi_{ion} \qquad (11.4.10)$$

related to the number density N_a of atoms in the ground state.

11.5 Ionization in Argon Gas Discharge Plasma Involving Excited Atoms

In consideration the ionization processes in a gas discharge plasma we above assume that these processes are caused by collisions with electrons. In reality, there are other mechanisms of ionization, and we consider them briefly for a helium gas discharge plasma due to presence of metastable atoms in this plasma. In particular, if an admixture of argon is located in helium, the Penning process (3.7.3) proceeds according to the scheme

$$He(2^3S) + Ar(^1S) \to e + Ar^+(2^2P) + He(1^1S) + 4.06\,\text{eV}, \qquad (11.5.1)$$

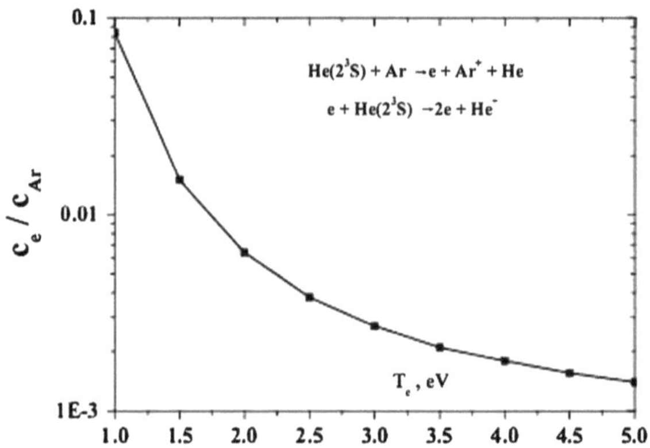

Fig. 11.15 The ratio of the electron and argon atom concentrations for the identical contribution of stepwise ionization and the Penning process (11.5.1) to ionization of a helium gas discharge plasma

and the rate constant of this process is equal to $k_P = 8 \cdot 10^{-11}\,\mathrm{cm}^3/\mathrm{s}$. The identical contribution of the Penning process and stepwise ionization takes place at the relation between the electron c_e and argon c_{Ar} concentrations

$$\frac{c_e}{c_{Ar}} = \frac{k_P}{k_{ion}}, \tag{11.5.2}$$

where the rate of ionization of metastable atoms by electron impact is given by formula (11.2.5). Figure 11.15 represents the dependence of this quantity on the electron temperature.

In the same manner, the Penning process

$$2He(2^3S) \rightarrow e + He^+(1^2S) + He(1^1S) + 14.4\,\mathrm{eV} \tag{11.5.3}$$

may be of importance both for ionization in a helium gas discharge plasma and for formation of fast electrons which increase the rate of excitation of helium atoms. The rate constant of this process in thermal atomic collisions is [198] $k(2^3S) = (9 \pm 1) \cdot 10^{-10}\,\mathrm{cm}^3/\mathrm{s}$ that is larger than the rate constant of the process (11.5.2) by one order of magnitude. Correspondingly, comparing the rate of this process and stepwise ionization of atoms by electron impact, we obtain that they become equal at the ratio of the electron and metastable atom $He(2^3S)$ concentration which is by one order of magnitude higher than that for the ratio of electron and argon atom concentrations given in Fig. 11.14. Hence, if the electron concentration exceeds the concentration of metastable atoms $He(2^3S)$ in a helium gas discharge plasma, the Penning process (11.5.3) gives a small contribution to plasma ionization.

One more channel of ionization in a gas discharge plasma is associative ionization. In the helium case it proceeds according to the scheme

$$e + He(1^1S) \to e + He(3^3D), \ He(3^3D) + He \to He_2^+ + e, \qquad (11.5.4)$$

and the process of associative ionization, the last process in this scheme, is characterized by the cross section $\sigma_{as} \approx 2 \cdot 10^{-15}\,\mathrm{cm}^2$ [433–435]. This channel of ionization is hampered in a gas discharge plasma because of the character of atom excitation in this. Note that because of formation of molecular ions the ionization threshold decreases from 24.59 to 23.07 eV. Nevertheless, associative ionization is not of importance for helium and argon gas discharge plasmas.

Chapter 12
Helium and Argon Plasma in Positive Column of Gas Discharge

Abstract Various types of ionization equilibrium in the positive column of gas discharge are considered. Parameters of ionization equilibrium for helium and argon gas discharge plasmas are found. Properties of a plasma of capillary discharge are analyzed as an example of ionization equilibrium.

12.1 Schottky Regime for Ionization Equilibrium in Positive Column

The Schottky regime of ionization equilibrium corresponds to the regime of a gas discharge plasma located in a cylinder tube where formation of electrons results from electron-atom collisions and their loss is due to motion of a plasma as a whole to walls by ambipolar diffusion. The number of electrons assumes to be not large, so that electric current does not create plasma non-uniformity including gas heating. Under these conditions equation (8.1.1) of ionization balance takes place, and this leads to the relation (8.2.6) between parameters of ionization balance

$$N_a k_{ion} = \alpha w_e = \frac{5.78 D_a}{\rho_o^2} \qquad (12.1.1)$$

Here k_{ion} is the rate of atom ionization by electron impact, and the first Townsend coefficient is $\alpha = N_a k_{ion}/w_e$, where w_e is the electron drift velocity, k_{ion} is the ionization rate constant by electron impact, D_a is the ambipolar diffusion coefficient. Solution of this equation determines the electric field strength that provides the ionization equilibrium under given conditions. Since $D_a \sim 1/N_a$, the reduced electric field strength that follows from solution of this equation on the tube radius ρ_o or the atom number density N_a are included in the balance equation in the form of the combination $N_a \rho_o$.

Applying this formula to the helium case, we use the electron drift velocity in helium [260, 261] given in Fig. 5.1, the values of the first Townsend coefficient in helium [341, 342] from Fig. 8.3 and the values of the ambipolar diffusion coefficient from Fig. 7.15. Assuming that the basic sort of ions is He^+, we give in Fig. 12.1 the

B.M. Smirnov, *Theory of Gas Discharge Plasma*, Springer Series on Atomic, Optical, and Plasma Physics 84, DOI 10.1007/978-3-319-11065-3_12

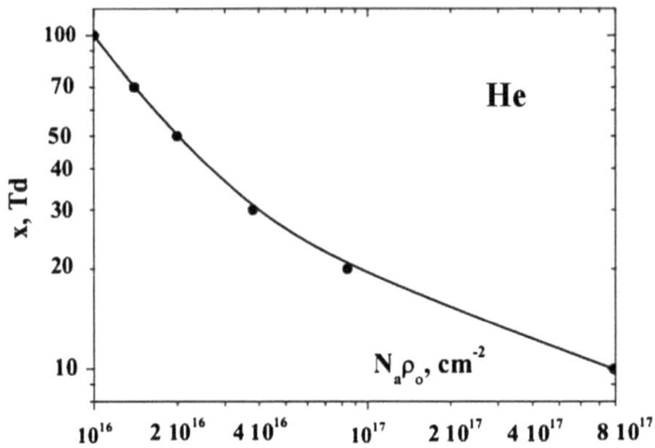

Fig. 12.1 Reduced electric field strength for the Townsend mechanism of atom ionization (the regime of low electron number densities) and basic atomic ions He^+ in a plasma as a function of the reduced number the density of helium atoms according to formula (12.1.1)

dependence of the reduced electric field strength $x = E/N_a$ on the reduced number density of atoms $N_a\rho_o$ in a helium gas discharge plasma in accordance with formula (12.1.1) for the regime of a low number density of electrons.

In the same manner one can analyze the ionization balance equation for argon in the case of the Townsend mechanism of ionization and in the regime of low electron number density. We are based on the dependencies on the reduced electric field strength for the electron drift velocity in argon from Fig. 5.2 and for the first Townsend coefficient in argon from Fig. 8.4, and use the values of the ambipolar diffusion coefficient in argon from Fig. 7.16, assuming that Ar^+ is the basic ion sort in an argon gas discharge plasma. On the basis of these data and formula (12.1.1) we

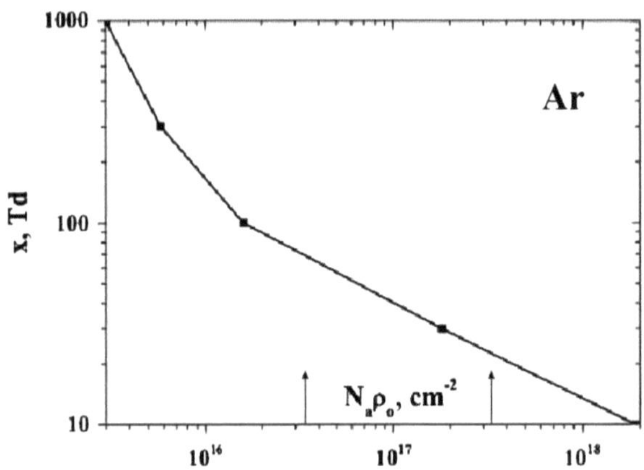

Fig. 12.2 Dependence of the reduced electric field strength on the reduced tube radius $N_a\rho_o$ for Townsend discharge in argon with basic ions Ar^+. *Arrows* correspond to pressures 1 Torr and 10 Torr at the tube radius $\rho_o = 1$ cm

represent in Fig. 12.2 the dependence of the reduced electric field strength $x = E/N_a$ on the reduced number density of atoms $N_a\rho_o$ in an argon gas discharge plasma in the regime of a low number density of electrons at $T = 300$ K.

It should be noted that the drift character of electron motion along the discharge tube axis in the positive column requires the validity of the criterion (8.1.8)

$$\xi = \frac{\alpha T_{ef}}{eE} \ll 1$$

This allows one to neglect by diffusion of electron motion for the gas discharge plasma in comparison with drift motion of an electron. Using the values of parameters of this formula in a helium gas discharge plasma, we find the dependence on the electric field strength for of this parameter, that is given in Fig. 12.3. One can see that the criterion (8.1.8) is fulfilled in the range of electric field strengths under consideration.

The value $1/\alpha$ is the mean free path of electrons with respect to ionization. Formula (12.1.1) for ionization equilibrium allows one to compare this mean free path of electrons with the tube radius, if the Schottky character of ionization equilibrium holds true. According to formula (12.1.1) this parameter is

$$\alpha\rho_o = \frac{5.78D_a}{\rho_o w_e}$$

Figure 12.4 gives the dependence of the parameter $\alpha\rho_o$ on the electric field strength in a helium gas discharge plasma if the ionization equilibrium holds true and the basic ion sort is He_2^+. As is seen, the mean free path of electrons with respect to ionization is large compared to the discharge tube radius.

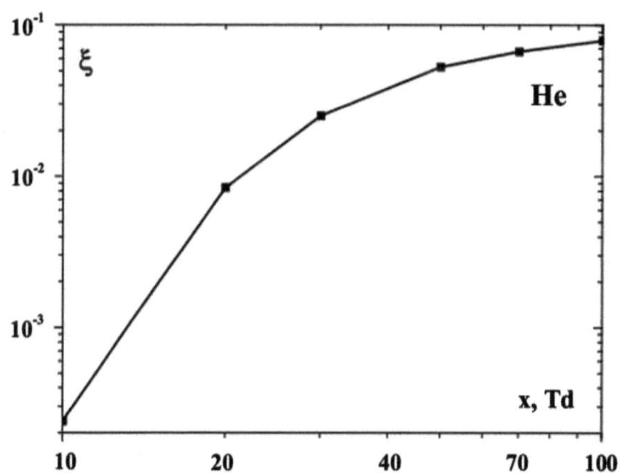

Fig. 12.3 The dependence of the parameter $\xi = \alpha T_{ef}/eE$, defined in accordance with formula (8.1.8), on the reduced electric field strength for Townsend discharge in helium

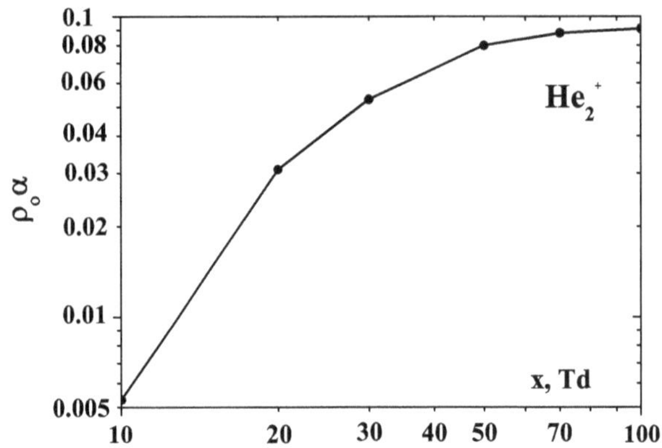

Fig. 12.4 The ratio of the discharge tube radius to the electron mean free path with respect to atom ionization at the Schottky character of ionization equilibrium in a helium gas discharge plasma

The Schottky regime corresponds to a loss of a plasma as a result of attachment of electrons and ions to walls. This can be violated if molecular ions are present in a gas discharge plasma and a loss of electrons and ions results from dissociative recombination. In the helium case this possibility is problematic because dissociative recombination of electrons and molecular ions He_2^+ is effective for vibrationally excited molecular ions. But it is possible for an argon gas discharge plasma where the basic sort of ions is Ar_2^+. Then the Schottky regime of ionization equilibrium according to formula (12.1.1) is realized if the rate of dissociative recombination involving electrons and molecular ions Ar_2^+ is less than the rate of plasma departure to walls as a result of ambipolar diffusion, that is

$$\alpha_{rec} N_e \ll \frac{5.78 D_a}{\rho_o^2} \equiv \alpha w_e \tag{12.1.2}$$

Here α_{rec} is the coefficient of dissociative recombination, N_e is the electron number density. From this criterion it follows that the Schottky regime is realized at low electron number densities. We are based on the dependencies on the reduced electric field strength, and use the values of the ambipolar diffusion coefficient in argon from Fig. 7.15, assuming that Ar^+ is the basic ion sort in an argon gas discharge plasma.

In order to obtain numerical parameters of the criterion (12.1.2) under given conditions, we use the right hand side of this criterion taking the electron drift velocity in argon from Fig. 5.2 and for the first Townsend coefficient in argon from Fig. 8.4. Considering weak discharge currents, we ignore the heating of the gas discharge plasma. The coefficient of dissociative recombination is $\alpha_{rec} = 7 \times 10^{-7}\,\mathrm{cm}^3/\mathrm{s}$ at room temperature according to a sum of measurements [190]. Though the coefficient of dissociative recombination drops with an increasing of an electron energy, we neglect this fact. As a result we have for the criterion of the Schottky ionization equilibrium

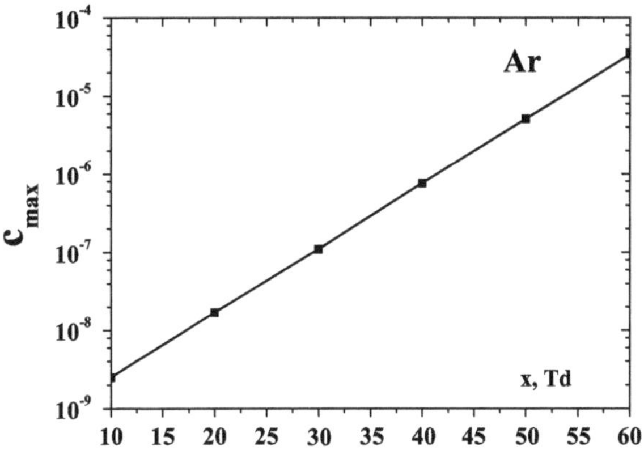

Fig. 12.5 The upper limit for the electron concentration in an argon gas discharge plasma with the basic sort ion Ar_2^+. The electron concentration below this line corresponds to the Schottky regime of ionization equilibrium

$$c_e = \frac{N_e}{N_a} \ll c_{max} \equiv \frac{\alpha}{N_a} \frac{w_e}{\alpha_{rec}} \tag{12.1.3}$$

Fig. 12.5 represents the dependence of the parameter c_{max} on the reduced electric field strength.

According to definition of the positive column of gas discharge, the ionization balance is supported inside each cross section of the tube, and passage of charged particles to tube walls is compensated by atom ionization inside the tube. Therefore a distance between electrodes L exceeds an average way L_o that an electron passes during its lifetime τ inside the tube, and this way equals to

$$L_o = w_e \tau = \frac{w_e}{N_a k_{ion}} = \frac{1}{\alpha},$$

where w_e is the electron drift velocity, and α is the first Townsend coefficient. On the basis of the ionization equilibrium (12.1.1) one can express this value through the ambipolar diffusion coefficient D_a of a gas discharge plasma because this parameter does not depend sharply on the discharge regime. In these terms a length of the positive column satisfies to the criterion

$$L \gg L_o = \alpha \frac{w_e \rho_o^2}{6D_a} \tag{12.1.4}$$

If this criterion holds true, the basic part of a distance between electrodes is occupied by the positive column. Then a subsequent increase of a distance between electrodes will not lead to a change of electrode regions, but gives an increase of the positive column length. Figure 12.6 gives the minimum length of the positive column L_o with a helium gas discharge plasma of a low electron number density. Values of L_o are

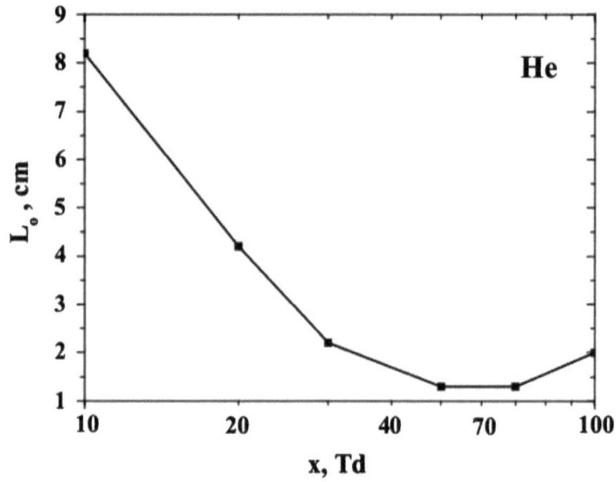

Fig. 12.6 The lowest limit for the positive column length in accordance with formula (12.1.4)

obtained on the basis of formula (12.1.4) with parameters of this formula according to Figs. 7.15 and 12.1, Tables 7.2 and 8.2.

Let us determine the difference of electric potentials between the axis and walls of the positive column of a cylinder discharge tube for a regime of a high density of electrons and for the Schottky regime of ionization balance. We consider an argon gas discharge plasma located in a cylinder tube. For definiteness, we take the tube radius to be $\rho_o = 1\,\text{cm}$, the argon pressure to be $p = 1\,\text{Torr}$, that corresponds to the number density of argon atoms $N_a = 3.3 \times 10^{16}\,\text{cm}^{-3}$ at room temperature, and the electron temperature to be $T_e = 2\,\text{eV}$. According to data of Table 3.6 the cross section of resonant charge exchange in collision $Ar^+ - Ar$ is $\sigma_{res} \approx 8 \times 10^{-15}\,\text{cm}^2$ that gives the mean free path of argon atomic ions in argon to be $\lambda = 1/(2N_a\sigma_{res}) \approx 2 \times 10^{-3}\,\text{cm}$. Assuming that the mean free path of electrons in argon has the same order of magnitude, we obtain from this on the basis of formula (8.2.8) that the difference plasma electric potentials at the axis and near walls to be $\Delta U \approx 17\,\text{eV}$. As is seen, this value exceeds remarkably a thermal electron energy.

We above are guided by the mechanism of atom ionization in a gas discharge plasma by electron impact. We now consider one more mechanism of atom ionization that results from the Penning process proceeds according to the scheme

$$A^* + B \to A + B^+,$$

Let us consider a helium gas discharge plasma with a small addition of argon and include the Penning process in equation of ionization balance. Then equation for the rate of electron formation takes the form

$$\frac{dN_e}{dt} = N_e N_a k_{ion} + N_m N_{Ar} k_P$$

Here N_a, N_{Ar} are the number densities of helium and argon atoms in the ground states, N_m is the number density of metastable helium atoms, k_{ion} is the effective rate constant of atom ionization by electron impact in pure helium, k_P is the rate constant of the Penning process that is equal to $8 \times 10^{-11}\,\text{cm}^3/\text{s}$ for the state $He(2^3S)$ of the metastable helium atom, and is $3 \times 10^{-10}\,\text{cm}^3/\text{s}$ for the state $He(2^1S)$ of the metastable helium atom.

For simplicity we assume the number density of metastable atoms $He(2^3S)$ to be of the order of the electron number density. Then formula (12.1.1) gives the following estimation for the boundary number density of admixture atoms that can influence on the ionization rate

$$N_{Ar} \sim \frac{6D_a}{k_P \rho_o^2}$$

We have the following estimation for the ambipolar diffusion coefficient $D_a \sim 10^4\,\text{cm}^2/\text{s}$ at the helium pressure of 10 Torr. This gives for the boundary number density of argon atoms to helium that can change the ionization balance under given conditions

$$N_{Ar} \sim 10^{14}\,\text{cm}^{-3},$$

that corresponds to the concentration of argon atoms in helium of $\sim$0.03 %.

12.2 Stepwise Ionization in Helium Positive Column

At high electron number densities ionization in the positive column may have a stepwise character where electrons and ions are formed as a result of collisions of electrons with excited atoms. Because the energy distribution function of electrons drops sharply above the threshold of excitation of lowest excited states, only lowest excited states give a basic contribution to the ionization balance. In particular, in the helium case this state is $He(2^3S)$. When ionization in helium proceeds through this state, the ionization balance equation has the following form instead of (12.1.1)

$$N_m k_{ion}^m = \frac{5.78 D_a}{\rho_o^2}, \qquad (12.2.1)$$

where N_m is the number density of metastable atoms in the 2^3S state, and k_{ion}^m in (12.2.1) is the rate constant of ionization of the 2^3S metastable atom by electron impact.

Because the electron temperature T_e exceeds significantly the ion temperature T_i, which coincides with the gaseous temperature at electric field strengths under consideration, we have on the basis of formula (7.7.5)

$$D_a = \frac{T_e}{T} D_i,$$

where D_i is the diffusion coefficient of helium ions in helium. Assuming atomic ions He^+ to be present in a gas discharge plasma of the positive column of gas discharge, we have for the ion mobility $K_i = 10.4\,\mathrm{cm}^2/(\mathrm{V}\cdot\mathrm{s})$ [8] under normal number density of atoms. Transferring to the ion diffusion coefficient D_i on the basis of the Einstein relation, we obtain $D_i N_a = 7.7 \times 10^{18}\,\mathrm{cm}^{-1}\mathrm{s}^{-1}$.

Let us rewrite (12.2.1) in the form

$$N_a K_{ion} = \frac{5.78 D_i T_e}{T \rho_o^2}, \qquad (12.2.2)$$

where we introduce the effective rate constant K_{ion} of stepwise ionization from the ground state on the basis of formula

$$K_{ion} = \frac{k_{ex} k_{ion}^m}{k_Q + k_{ion}^m}, \qquad (12.2.3)$$

so that

$$N_a K_{ion} = N_m k_{ion}^m$$

According to Fig. 11.4 stepwise ionization through the metastable atom state $He(2^3S)$ gives the main contribution to ionization of a helium gas discharge plasma. Figure 12.7 gives the dependence of this quantity on the reduced electric field strength for different values of the electron concentration c_e. As is seen, the dependence of the effective ionization rate constant K_{ion} on the electron concentration is weak. This dependence of the electron concentration in the regime of stepwise ionization occurs due to competition of electron-electron collisions and electron-atom collisions in formation of fast electrons which are able to excite helium atoms in the metastable

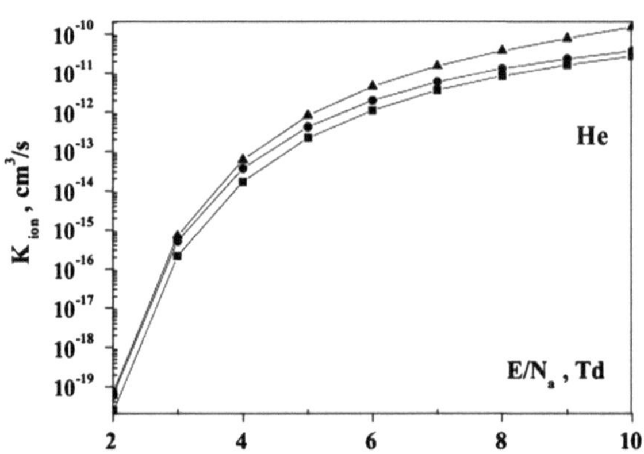

Fig. 12.7 Rate constant K_{ion} of helium atom ionization by electron impact through the metastable state $He(2^3S)$ as a function of the reduced electric field strength. *Squares* correspond to the electron concentration $c_e = 10^{-7}$ in a helium gas discharge plasma, *circles* relate to the electron concentration $c_e = 10^{-5}$, and *triangles* correspond to the electron concentration $c_e = 10^{-3}$

state $He(2^3S)$. Because of an unique dependence of the electron temperature T_e and the electric field strength (6.2.7), one can transfer to the dependence of the effective ionization rate constant K_{ion} on the electric field strength given in Fig. 12.7 in its dependence on the electron temperature that is represented in Fig 12.8.

One can see that equation (12.2.2) of the ionization balance contains the number density of atoms N_a and a discharge tube radius ρ_o in the combination $N_a\rho_o$. Hence, this equation allows us to evaluate the electron temperature as a function of the parameter $N_a\rho_o$. Expressing T_e in eV, K_{ion} in cm^3/s, N_a in cm^{-3}, and ρ_o in cm, we rewrite (12.2.2) as

$$(N_a\rho_o)^2 = \frac{1.7 \times 10^{21} T_e}{K_{ion}(T_e)} \qquad (12.2.4)$$

if atomic ions He^+ are present in a gas discharge plasma.

Figure 12.9 gives this dependence, where we use the dependence $K_{ion}(T_e)$ in accordance with data of Fig. 12.7 for the rate constant of stepwise ionization and formula (12.2.4). In addition, Fig. 12.10 contains the dependence of the reduced tube radius (or the number density of helium atoms) on the reduced electric field strength at different electron concentrations. One can see a weak dependence of $N_a\rho_o$ on the electron concentration. We use here the dependence (6.2.14) of the electron temperature T_e on the reduced electric field strength $x = E/N_a$. We assume in Fig. 12.10 that destruction of metastable atoms results from their quenching by electron impact, rather than their collisions with walls.

One can compare the dependence of the reduced electric strength and the atom number density in the positive column for the regimes of low and high electron number densities, i.e. for the Townsend and stepwise character of ionization equilibrium. In the case of the Townsend mechanism of atom ionization by electron impact we use formula (8.1.7) for the ionization rate constant and take values of the first Townsend

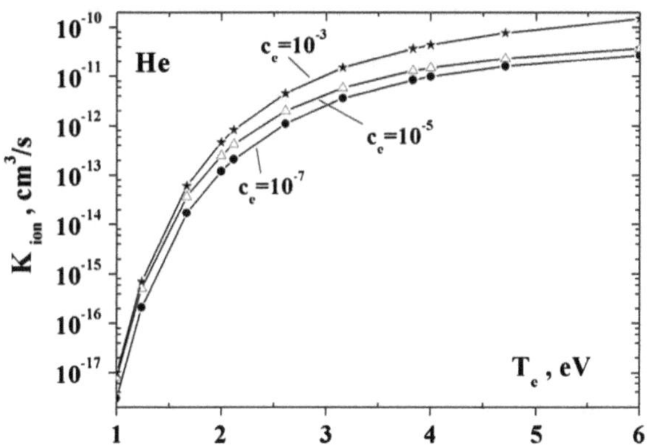

Fig. 12.8 Rate constant K_{ion} of stepwise ionization of a helium atom by electron impact through the metastable state $He(2^3S)$ as a function of the electron temperature in the regime of high electron number densities

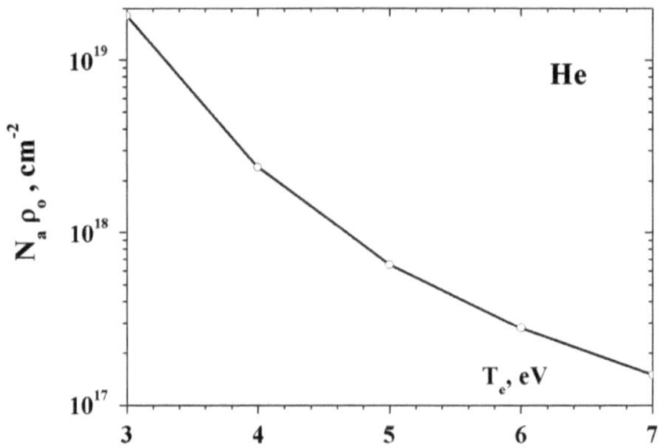

Fig. 12.9 Dependence of the reduced number density of helium atoms on the electron temperature in the positive column of glow discharge for the stepwise character of atom ionization through the metastable state $He(2^3S)$ at the electron concentration $c_e = 10^{-7}$

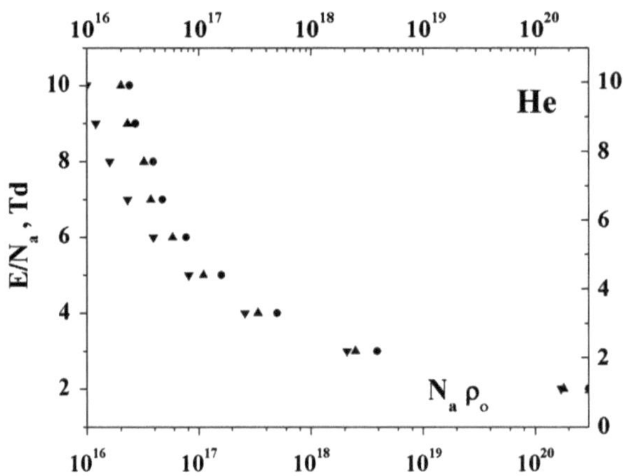

Fig. 12.10 Dependence of the reduced number density of helium atoms (or the radius of the discharge tube) on the reduced electric field strength if atom ionization proceeds through the metastable state $He(2^3S)$. The electron concentration is : *1* $c_e = 10^{-7}$, *2* $c_e = 10^{-5}$, *3* $c_e = 10^{-3}$

coefficient and the ionization rate constant from Table 8.1. We approximate the first Townsend coefficient by formulas (8.1.4) and (8.1.5) with parameters in each range of the electric field strengths given in Table 8.2. The electron drift velocity w_e in helium in a constant electric field strength we approximate on the basis of experimental data [43] (Table 7.3) by the expression $w_e = 2.4 \times 10^5\,\mathrm{cm/s} \cdot \mathrm{x}$, where $x = E/N_a$ is the reduced electric field strength. As a result, one can approximate the ionization rate constant in the range of the reduced electric field strengths $x = 10$–$100\,\mathrm{Td}$ for the regime of low electron number densities as

$$\kappa_{ion} = 4.8 \times 10^{-12} x \exp(-80/x),$$

where the ionization rate constant κ_{ion} is measured in cm^3/s.

Basing on the ionization balance equation given by formula (8.2.6)

$$\alpha w_e = \frac{5.78 D_a}{\rho_o^2},$$

and using the above approximation for the ionization rate constant, we find the connection between the reduced electric field strength $x = E/N_a$ and the reduced tube radius $N_a \rho_o$ that is represented in Fig. 12.11.

As is seen, the ionization balance equations for the Townsend and stepwise mechanisms of ionization equilibrium have an identical form, and hence the mechanism with a larger rate constant of ionization is realized under equilibrium conditions. Let us calculate the rate constant K_{ion}, assuming that the ionization process proceeds through formation of metastable atoms $He(2^3S)$ in the regime of low electron number densities. Using formula (10.1.6) for the rate constant of excitation of the metastable state 2^3S by electron impact, formulas (10.3.5) and (10.3.12) for quenching of the metastable state by electron impact with electron transition in other bound states, and formula (12.2.4) for ionization of this metastable atom by electron impact, we find the total rate constant of ionization K_{ion} in this case. Its values are represented in Fig. 12.11.

One can determine the range of electric field strengths $\kappa_{ion} > K_{ion}$, where the Townsend mechanism of ionization is realized. The first range is $x < 4$ Td, but the rate constant there is very small. For example, $\kappa_{ion} = 4.0 \times 10^{-20}\,cm^3/s$ and $K_{ion} = 1.3 \times 10^{-20}\,cm^3/s$ at the reduced electric field strength $x = 4$ Td. At the helium pressure $p = 1$ Torr and the gas temperature $T = 300$ K a typical time of electron doubling is of the order of a hour. A subsequent decrease of the electric field strength leads to increasing times which are not of interest in reality. Another range of

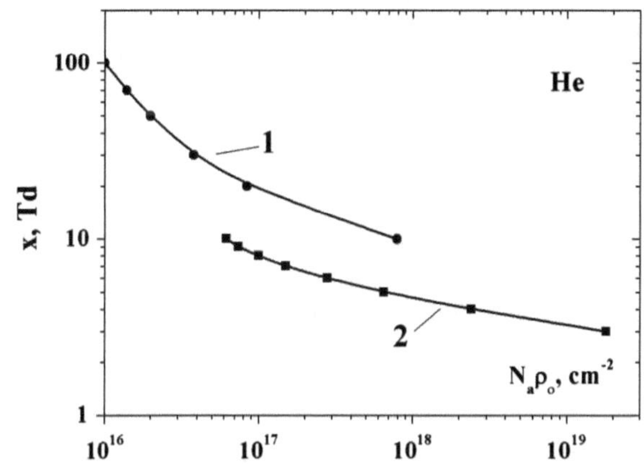

Fig. 12.11 Dependence of the reduced electric field strength on the number density of helium atoms in the positive column of glow discharge for the Townsend (*1*) and stepwise (*2*) character of ionization

the electric field strengths, where the Townsend mechanism of ionization is realized ($\kappa_{ion} > K_{ion}$) takes place at $x > 20\,\mathrm{Td}$ if we use formally the above expressions for the rate constants of atom ionization. But the above evaluation of K_{ion} is valid at $x < 10\,\mathrm{Td}$ because we assume in these evaluations the rate constant of quenching of excited atom states by electron impact to be independent of the electron energy. Hence, one can state that the Townsend mechanism of ionization holds true at large reduced electric field strengths $x \sim 20\,\mathrm{Td}$, but the used approach does not allow one to determine more precisely the boundary between these mechanisms of ionization.

Let us determine the criterion for the glow and Townsend regime of ionization equilibrium of a gas discharge plasma located in the positive column of helium glow discharge. Indeed, the Townsend mechanism of ionization in the positive column of gas discharge is realized at not low gas pressures that corresponds to the criterion of smallness of the mean free path of electrons in a gas λ compared to the tube radius $\lambda \ll \rho_o$. Since the gas-kinetic cross section of an electron in helium is $\sigma_g = 1.3 \times 10^{-15}\,\mathrm{cm}^2$. Hence in considering the Townsend mechanism of ionization in the positive column of gas discharge, we require fulfilling of the criterion

$$N_a \rho_o \gg 8 \times 10^{14}\mathrm{cm}^{-2} \cdot \exp\left(\frac{80}{x}\right) \tag{12.2.5}$$

This dependence of the parameter $N_a \rho_o$ on the electric field strength that is the solution of equation (12.1.1) of ionization equilibrium and is represented in Fig. 12.11, and relation (12.1.1) is valid, at least, at the reduced electric field strengths $x \ll 100$ Td.

In considering the Schottky character of ionization equilibrium, we assume that metastable atoms are destructed in collisions with electrons, while loss of electrons results from their recombination on walls. Along with this, another version of stepwise ionization is possible where destruction of metastable atoms proceed on walls. Thus, we have three regimes of ionization, and they are represented in Fig. 12.12 under assumption that stepwise ionization proceeds only through the metastable state 2^3S, and destruction of metastable atoms proceeds at walls or in collisions with electrons. Ionization equilibrium in the case of stepwise ionization is described by equations (8.3.7) or (8.3.4) respectively depending on the character of destruction of metastable atoms. In the case of the Townsend character of ionization equilibrium in the regime of low electron number densities for the positive column of gas discharge the ionization balance is given by formula (8.2.6). In this formula we use measured values of the first Townsend coefficient for the reduced electric field strengths $x > 10$ Td given in Table 8.2 and continue this dependence to lower electric field strengths. Comparison of the ionization balance for the above three cases is represented in Fig. 12.12.

We now analyze the criterion if the Schottky ionization equilibrium dominates in the regime of high electron number density, and this criterion has the form

$$(k_{ion}^m + k_Q)N_e \gg \frac{5.78 D_m}{\rho_o^2},$$

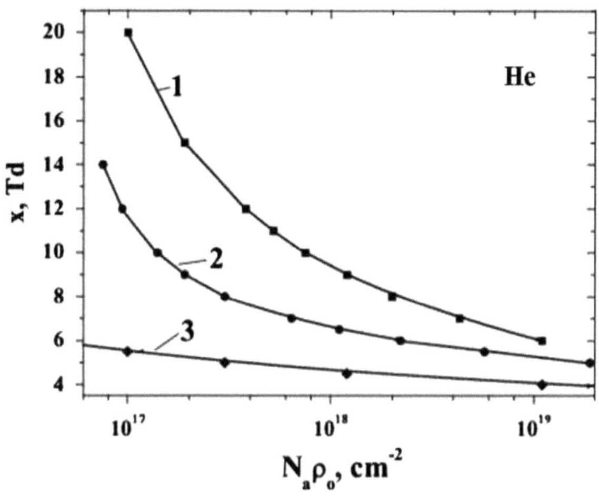

Fig. 12.12 Dependence the reduced radius $N_a \rho_o$ of a discharge tube in the positive column of helium discharge for the reduced electric field strength $x = E/N_a$ in the regime of low electron number density. *1*—Townsend discharge,where the ionization balance is given by formula (8.2.6), *2*, *3*—glow discharge with stepwise ionization of atoms through the state 2^3S, where the ionization balance is determined by formulas (8.3.7) and (8.3.4) correspondingly

where D_m is the diffusion coefficient of metastable atoms in a gas, and this inequality means that metastable atoms are destroyed by electron impact. Comparing this inequality with equation of ionization equilibrium (12.2.1), we obtain the above criterion in the form

$$\frac{c_e}{c_m} \gg \left(\frac{c_e}{c_m}\right)_{min} = \frac{k_{ion}^m}{k_{ion}^m + k_Q} \frac{D_m}{D_a} \tag{12.2.6}$$

Figure 12.13 gives the minimum values of the ratio between the electron and metastable atom concentrations in accordance with formula (12.2.6). We assume that molecular ions He_2^+ are located in ionized helium, the temperature of atoms and ions is room one, and take the diffusion coefficient of metastable atoms in helium to be $D_m \approx 0.59\,\text{cm}^2/\text{s}$ at the normal number density of atoms, and from the sum of experiments [167, 304–315, 317] we have $D_m N_a = (1.6 \pm 0.1) \times 10^{19}\,\text{cm}^{-1}\text{s}^{-1}$. For the diffusion coefficient of molecular ions He_2^+ in helium we have $D_i \approx 0.46\,\text{cm}^2/\text{s}$ [41, 190] at the normal number density of atoms. This gives

$$\frac{D_m}{D_i} = 1.3 \pm 0.1$$

and is used in Fig. 12.13.

As it follows from the data of Fig. 12.13, ionization equilibrium with the stepwise character of atom ionization is realized at not low electron concentrations. In addition, Fig. 12.4 gives the values of the parameter $\alpha\rho_o$, where ρ_o is the tube radius and He_2^+

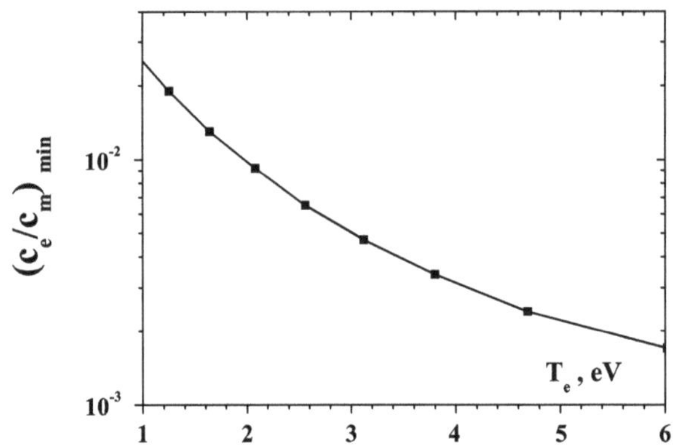

Fig. 12.13 The ratio of electron and metastable atom concentrations. The Schottky character of ionization equilibrium is realized if this ratio is located above a given *curve*

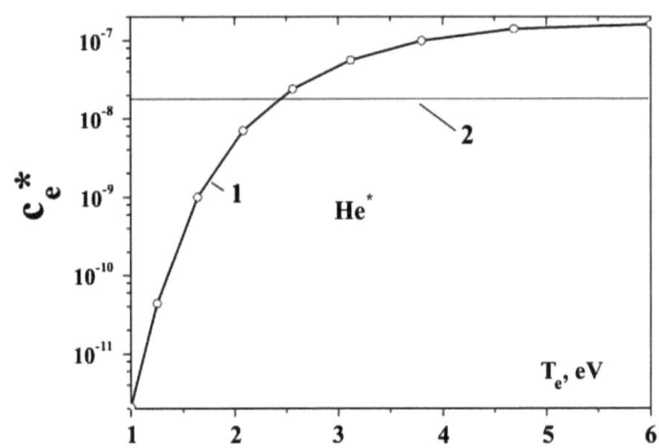

Fig. 12.14 Given by *curve 1* the boundary between two types of ionization equilibrium in a helium gas discharge plasma for the regime of high electron number densities. Above this curve the equilibrium for metastable atoms is established by collisions with electrons, below this curve destruction of metastable atoms results from their attachment to walls. The regime of high electron number densities is realized for electron concentrations above the *line 2*

as the basic ion. In the case of the Townsend character of atom ionization and the Schottky regime of ionization equilibrium in accordance with (12.1.1) the reduced radius of the discharge tube is given by

$$N_a\rho_o = \sqrt{\frac{5.78D_aN_a^2}{\alpha w_e}},$$

that is used in Fig. 12.4.

As is seen, the criterion of the Schottky character of ionization equilibrium (12.2.6) for the regime of a high electron number density includes the concentration c_m of metastable atoms that is contained in Fig. 10.14 for this regime of the gas discharge plasma existence. From this one can find the boundary between two types of ionization equilibrium for the regime of a high number density of electrons by analogy with the regime of low densities of electrons represented in Fig. 12.12. We give in Fig. 12.14 the boundary between these two types of ionization equilibrium, namely, between the Schottky equilibrium where the number density of the metastable helium atoms results from electron-atom collisions and an electron loss is determined by their attachment to walls, and the ionization equilibrium where destruction of metastable atoms is connected with their travelling to walls. Note that the line 2 in Fig. 12.14 characterizes the boundary between the regimes of a low and high electron number densities in accordance with formula (5.6.2) and Fig. 5.3. One can see that the second mechanism of ionization equilibrium results in a narrow range of electron concentrations above line 2 and below curve 1 of Fig. 12.14.

Thus, the analysis of ionization equilibrium in a helium gas discharge of the positive column in the cylinder discharge tube shows the validity of the Schottky character of this ionization equilibrium where a loss of electrons and ions proceeds at tube walls. In the regime of low electron number densities the origin of electrons and ions results from single atom ionization by electron impact, while in the regime of high electron number densities atom ionization has the stepwise character and proceeds mostly through the metastable atom state $He(2^3S)$. Destruction of metastable atoms in this equilibrium results mainly in collisions with electrons.

Properties of a gas discharge plasma are determined by several external and internal parameters. Being guided by a cylinder discharge tube, where the electric field strength in the positive column of stationary gas discharge is constant both along the tube axis and in the transversal direction, we consider above the dependence of the reduced electric field strength on the tube radius and the number density of helium atoms in different regimes of gas discharge. Let us analyze now the voltage–current dependence for the positive column of gas discharge in helium in ignoring the gas heating. In the regime of low electron number density, where the ionization balance equation is given by formula (8.2.6), all terms of this equation are independent of the electron number density and, correspondingly, on the electric current density. Hence, at low current densities the voltage–current characteristic is described by straightforward line, as it is given in Fig. 12.15. In the regime of high electron number densities we have a weak dependence of the reduced electric field strength on the electron current density in accordance with Fig. 12.10, as it is given in Fig. 12.15. The jump from one line to another one proceeds in accordance with Fig. 5.3 for each $N_a\rho_o$, and values for the upper and lower parts of the voltage-current characteristic at a given $N_a\rho_o$ are determined in accordance with data of Figs. 12.10 and 12.11.

In considering the ionization equilibrium in a helium gas discharge plasma of the positive column of gas discharge, we take the electron concentration c_e as an external parameter, i.e. this parameter is determined by an external source of gas discharge. This means that the discharge current and its voltage are determined by parameters of a general system which joins gas discharge and external source. In this case electrons from an external source which are introduced in gas discharge give a certain contribution to formation of electrons and ions in the positive column. Let us consider the case where external electrons do not give a contribution to ionization, i.e.

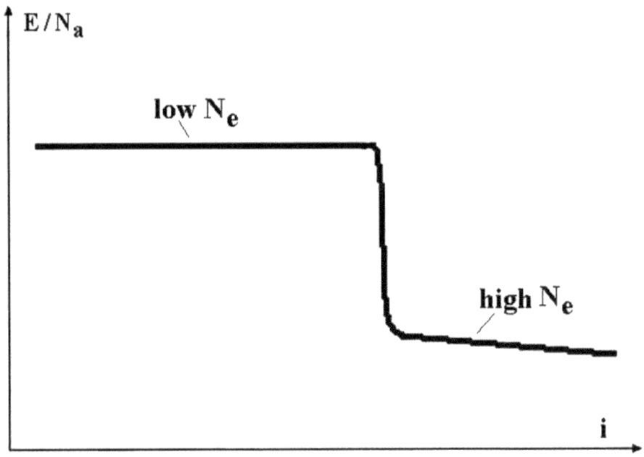

Fig. 12.15 General form of the voltage-current characteristic for a gas discharge plasma in the positive column of gas discharge

the ionization balance is realized due to processes inside a gas discharge plasma. In particular, in the regime of high electron number densities this equilibrium proceeds according to the following scheme

$$e + He(2^3S) \rightarrow 2e + He^+, \; 2e + He^+ \rightarrow e + He^* \tag{12.2.7}$$

In the regime of high electron number densities we use formula (3.8.3) for the rate constant K_{ei} of three body recombination. The ionization balance equation in this case has the form

$$N_e N_m k_{ion}^m \equiv N_e N_a K_{ion} = N_e^2 N_i K_{ei} \tag{12.2.8}$$

Since $N_e = N_i$, we obtain the equilibrium constant for this equilibrium and for the thermodynamic equilibrium described by Saha formula (2.1.8), and these equilibrium constants are given in Fig. 11.11.

Analyzing the positive column with a gas discharge plasma located in a cylinder discharge tube, we include in the ionization balance a loss of electrons and ions at walls. Then equation of the ionization balance takes the following form instead of (12.2.8)

$$N_e N_a K_{ion} = N_e^2 N_i K_{ei} + \frac{5.78 D_a N_e}{\rho_o^2} \tag{12.2.9}$$

If two terms of the right hand side of (12.2.9) are equal, the equilibrium constant

$$\chi = \frac{N_e^2}{N_a}$$

decreases twice compared to its equilibrium value which is given in Fig. 11.11. Thus, an increase of the role for the process of ambipolar diffusion compared three body recombination in a gas discharge plasma, the electron number density decreases.

12.3 Ionization in Argon Gas Discharge Plasma of Positive Column

In analyzing the ionization equilibrium in an argon gas discharge plasma of the positive column of gas discharge, we first evaluate the rate constant of stepwise ionization (12.2.3) by analogy with the helium case. This value is given in Fig. 11.10 as a function of the electric field strength for an optically thin plasma where resonantly excited atoms do not give contribution to the ionization process has the form

$$K_{ion} = \sum_i \frac{k^i_{ex} k^i_{ion}}{k^i_{exci}} = \sum_i k^i_{ion} c_i, \tag{12.3.1}$$

where index i corresponds to each excited state of the argon atom with the electron shell $3p^5 4s$, k^i_{ex} is the excitation rate constant of this state by electron impact, c_i is the concentration of excited atoms in a given excited state, k^i_{ion} is the ionization rate constant of an atom in this state, k^i_{exci} is the rate constant of quenching of this state which results in excitation from this state in states with the electron shell $3p^5 4p$. We give in Fig. 12.16 the ionization rate constants for excited argon atoms of the electron shell $3p^5 4s$ by electron impact which are calculated on the basis of formula (3.6.16). We neglect here the destruction of excited states as a result of ionization and transition in other excited states except indicated ones. Figure 12.16 gives the

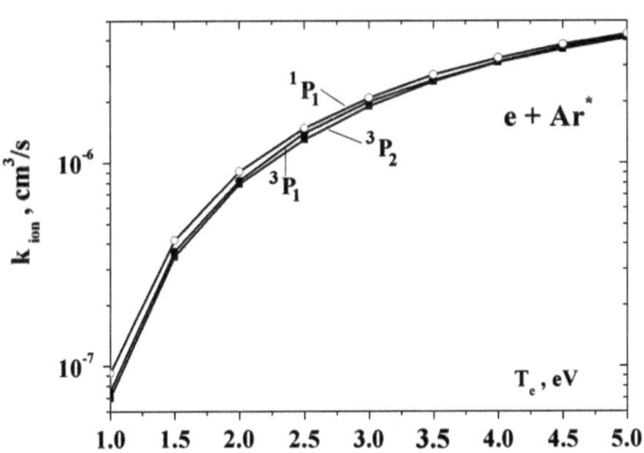

Fig. 12.16 Rate constant of stepwise ionization of an argon atom by electron impact from lower excited atom states of the electron shell $3p^5 4s$ at the electron concentration $c_e = 1 \times 10^{-6}$

dependence on the electron temperature T_e for the effective rate constant of stepwise ionization in two limiting cases $N_e \gg N_o$ and $N_e \ll N_o$. Note that for a cylinder discharge tube of a radius $\rho_o = 1\,\mathrm{cm}$ filled by argon is $N_o = 1.4 \times 10^{12}\,\mathrm{cm}^{-3}$ for the state $Ar(^3P_1)$ and $N_o = 7.2 \times 10^{11}\,\mathrm{cm}^{-3}$ for the state $Ar(^1P_1)$ (Fig. 12.17).

In considering a helium gas discharge plasma of the positive column of gas discharge, we were guided mostly by molecular ions He_2^+, whereas atomic helium ions He^+ may be present in a gas discharge plasma. Note that from the sum of measurements [190] the mobility of molecular ions He_2^+ at the temperature $T = 300\,\mathrm{K}$ is $18 \pm 2\ \mathrm{cm}^2/(\mathrm{V}\cdot\mathrm{s})$, whereas the mobility of atomic ions He^+ at this temperature is $10.4 \pm 0.3\ \mathrm{cm}^2/(\mathrm{V}\cdot\mathrm{s})$. In addition, the mobility dependence on the electric field strength E at large electric field strengths is $E^{-1/2}$ for atomic ions and weak dependence on the high electric field strength corresponds to molecular ions. Hence transition from atomic to molecular ions leads to a change of the ion mobility in two–three times. It should be noted that pair recombination of electrons and molecular helium ions proceeds effectively for excited vibration states only and is not of principle for a helium gas discharge plasma.

In the case of an argon gas discharge plasma with molecular ions the process of dissociative recombination of electrons and molecular ions Ar_2^+ is effective including lower vibration states, and hence formation of molecular ions Ar_2^+ in a gas discharge plasma may influence dramatically on its existence. Figure 10.5 gives the temperature at which the number densities of atomic and molecular ions are equal at a given argon pressure. According to data of this Figure, molecular ions dominate at temperatures below 1,000K in argon of a pressure $p \sim 1$ Torr if an equilibrium between atomic and molecular ions is realized. Let us consider the ionization equilibrium in an argon gas discharge plasma where a loss of charged particles results

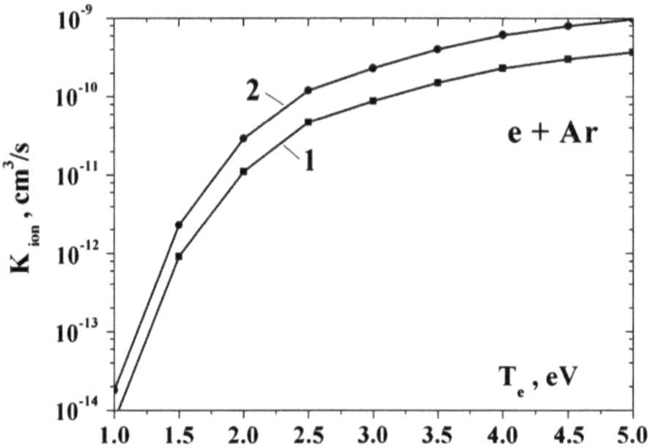

Fig. 12.17 Rate constant of ionization of an argon atom in the ground state by electron impact if the ionization process proceeds through excited atom states of the electron shell $3p^54s$ at the electron concentration $c_e = 1 \times 10^{-6}$. The rate constant is determined by formula (12.3.1) and *curve 1* corresponds to an optically thin plasma where resonantly excited states do not partake in atom ionization, whereas *curve 2* relates to an optically thick gas discharge plasma

from dissociative recombination of electrons and ions. We considered this equilibrium above (Sect. 11.3) and convinced that this ionization equilibrium corresponds to the regime of high electron number densities, so that atom ionization in a gas discharge plasma has the stepwise character. In this case the ionization balance equation has the form

$$K_{ion}N_a = \alpha N_e, \tag{12.3.2}$$

and we use the dependence of the dissociative recombination coefficient α on the electron temperature T_e as

$$\alpha(T_e) \sim T_e^{-0.5}, \tag{12.3.3}$$

and $\alpha(300\mathrm{K}) = (7 \pm 2) \times 10^{-7}\,\mathrm{cm}^3/\mathrm{s}$. We give in Fig. 12.18 the dependence on the electron temperature T_e for the electron or ion concentrations in the case of ionization equilibrium (12.3.2). Let us indicate the conditions where this ionization equilibrium is realized. First, we assume that destruction of excited atoms, which partake in ionization equilibrium, is determined by electron impact rather than by diffusion of excited argon atoms to walls. Second, the lifetime of electrons and ions in this gas discharge plasma is connected with dissociative recombination of electrons and ions.

We now consider the Schottky mechanism of ionization equilibrium where the lifetime of electrons and ions in the positive column of gas discharge is determined by their diffusion to walls, and in accordance with formula (8.2.4) has the form

$$K_{ion}N_a = \frac{5.78 D_a}{\rho_o^2}, \tag{12.3.4}$$

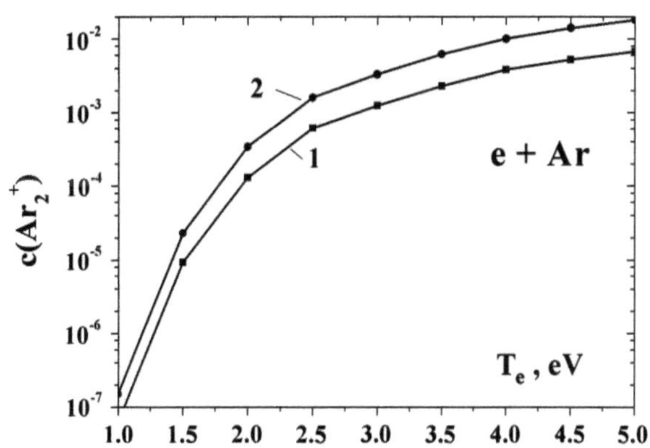

Fig. 12.18 Equilibrium electron and ion concentration for a stepwise formation of electrons and ions in an argon gas discharge plasma in the regime of a high electron number density and dissociative recombination for destruction of charged atomic particles: *1* optically thin plasma, *2* optically thick plasma

and the ambipolar diffusion coefficient D_a is connected with the ion diffusion coefficient D_i by formula (7.7.5)

$$D_a = D_i \left(1 + \frac{T_e}{T}\right)$$

and the ion diffusion coefficient under normal conditions according to Table 7.6 is equal $D_i = 0.048\,\mathrm{cm}^2/\mathrm{s}$. This ionization equilibrium leads to dependence of the reduced tube radius (or reduced number density of atoms) on the electron temperature that according to (12.3.3) has the form

$$(N_a \rho_o)^2 = \frac{5.78 D_i N_a}{K_{ion}} \left(1 + \frac{T_e}{T}\right) \tag{12.3.5}$$

Figure 12.19 represents this dependence in the argon case.

One can combine both mechanisms of ionization equilibrium and then the ionization balance equation takes the form

$$K_{ion} N_a = \frac{5.78 D_a}{\rho_o^2} + \alpha N_e \tag{12.3.6}$$

If we denote the equilibrium electron concentration of Fig. 12.18 as c_* and the reduced tube radius (or the reduced atom number density) of Fig. 12.19 as $(N_a \rho_o)_*$. One can represent the ionization balance equation in these notations in the form

$$\frac{c_e}{c_*} + \frac{(N_a \rho_o)_*^2}{(N_a \rho_o)^2} = 1 \tag{12.3.7}$$

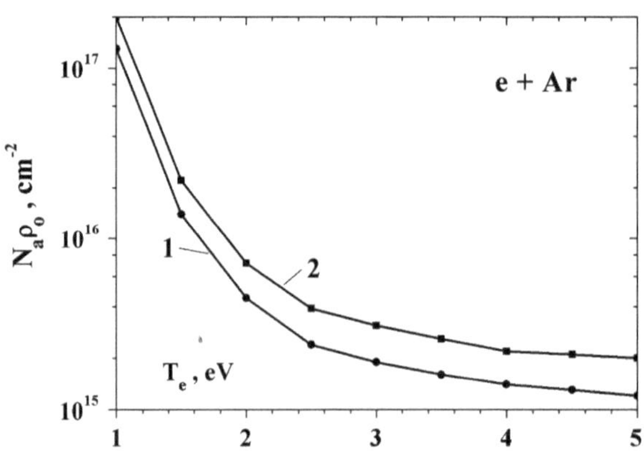

Fig. 12.19 Parameters of an argon gas discharge plasma of the positive column of gas discharge for the Schottky mechanism of plasma destruction and stepwise ionization of atoms by electron impact in accordance with (12.3.4): *1* optically thin plasma, *2* optically thick plasma

From this equation it follows that the electron concentration at ionization equilibrium in a general case is lower than that follows from the ionization balance (12.3.2). Equation (12.3.6) and its solution (12.3.7) allow one to find the connection between parameters of an argon gas discharge plasma of the positive column in a multidimensional space of plasma parameters. In the limit of low electron number densities this equation is converted into (12.3.4) with its solution (12.3.5) (Fig. 12.19), whereas in the limit of high atom density this leads to (12.3.2) of the ionization balance, and its solution is given in Fig. 12.18.

Above we assume the gas temperature T for a gas discharge plasma to be small in accordance with Fig. 10.5 data. In this case atomic argon ions are formed at the first stage and then they are converted into molecular ones according to the scheme

$$Ar^{+} + 2Ar \rightarrow Ar_2^{+} + Ar \qquad (12.3.8)$$

We above assume the rate of this process is enough high, so that molecular ions are located in this gas discharge plasma. Let us include in consideration the process (12.3.8) and then under stationary conditions the balance equations for ions the ionization equilibrium (12.3.6) take the form

$$K_{ion} N_e N_a = \frac{5.78 D_a(Ar^{+}) N(Ar^{+})}{\rho_o^2} + K_{as} N_a^2 N(Ar^{+}),$$
$$K_{as} N_a^2 N(Ar^{+}) = \frac{5.78 D_a(Ar_2^{+}) N(Ar_2^{+})}{\rho_o^2} + \alpha N_e N(Ar_2^{+}) \qquad (12.3.9)$$

Here K_{as} is the rate constant of the three body process and $N(Ar^{+})$, $N(Ar_2^{+})$ are the number densities for atomic and molecular ions correspondingly, and

$$N_e = N(Ar^{+}) + N(Ar_2^{+}) \qquad (12.3.10)$$

because of the plasma quasineutrality.

Note that the diffusion coefficients of the ambipolar diffusion for a plasma with atomic and molecular ions is expressed through the diffusion coefficients of ions through formula (7.7.5), and the diffusion coefficients of atomic and molecular ions which are the basis for the ambipolar diffusion coefficients are close. Namely, according to Table 7.6 data the diffusion coefficients for atomic and molecular ions in argon at room temperatures are $D_i(Ar^{+}) = 0.040\,\mathrm{cm^2/s}$ and $D_i(Ar_2^{+}) = 0.048\,\mathrm{cm^2/s}$ respectively under normal conditions. Let us simplify the problem taking these diffusion coefficients to be equal, namely, $D_i(Ar^{+}) = D_i(Ar_2^{+}) = 0.044\,\mathrm{cm^2/s}$. Then summarizing two equation of the set (12.3.9), we obtain the following equation of the ionization balance

$$K_{ion} N_a = \frac{5.78 D_a}{\rho_o^2} + \alpha N(Ar_2^{+}) \qquad (12.3.11)$$

If molecular ions dominate in this plasma, (12.3.11) coincides with equation (12.3.6). This transition corresponds to a high rate of ion conversion.

Let us construct the parameter with the pressure dimensionality

$$\zeta = \frac{5.78 D_a N_a}{K_{as}(N_a \rho_o)^2} \cdot T \qquad (12.3.12)$$

that is responsible for high rate of ion conversion at large atom number densities or gas pressures. Keeping ionization equilibrium (12.3.4) with the balance equation (12.3.5), we reduce this parameter to the form

$$\zeta = \frac{K_{ion}}{K_{as}} \cdot T \qquad (12.3.13)$$

Statistical averaging gives for the rate constant of the process (12.3.10) $K_{as} = (2.2 \pm 0.7) \times 10^{-31}\,\text{cm}^6/\text{s}$ on the basis of measurements [209, 216, 220, 432] at room temperature. The temperature dependence for this rate constant is [433, 190]

$$K_{as} \sim T^{-3/4}$$

in the case of the polarization interaction of an ion with atoms. If we use the rate constant of stepwise ionization given in Fig. 12.17, we obtain the dependence $\zeta(T_e)$ represented in Fig. 12.20 at the gas temperature $T = 300$ K. Note that the ionization equilibrium in an argon gas discharge plasma of the positive column of gas discharge with destruction of a plasma due to ambipolar diffusion that is described by the balance (12.3.4) is realized at gas pressures $p \ll \zeta$.

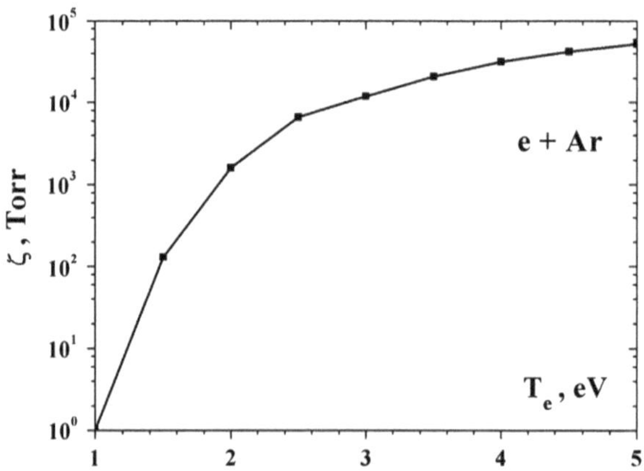

Fig. 12.20 Dependence on the electron temperature T_e for the parameter ζ defined by formula (12.3.12). This parameters corresponds to an optically thick argon gas discharge plasma at room gas temperature, and the ionization equilibrium in this plasma is described by formula (12.3.4). This relates to large number densities of atoms where this parameter is given by formula (12.3.13)

Because of a low rate of the process (12.3.8), from data of Fig. 12.20 it follows that the number density of molecular ions $N(Ar_2^+)$ in the ionization balance (12.3.11) is small compared to the electron number density N_e in the basic range of electron temperatures excluding low ones where the ionization rate is small. Therefore we consider the case where this relation holds true, i.e. $N(Ar_2^+) \ll N_e$, $N(Ar^+) \approx N_e$. Moreover, we consider the case where the Schottky regime of ionization equilibrium holds true, and for an argon gas discharge plasma this corresponds to the criterion

$$\frac{5.78 D_a}{\rho_o^2} \gg \alpha N_e$$

In this case from the second equation of the set (12.3.9) we have

$$\frac{N(Ar_2^+)}{N(Ar^+)} = \frac{N(Ar_2^+)}{N_e} = \frac{K_{as} N_a^2}{5.78 D_a/\rho_o^2},$$

and the ionization balance (12.3.11) takes the form

$$K_{ion} N_a = \frac{5.78 D_a}{\rho_o^2}\left(1 + \frac{N_e}{\chi}\right), \quad \chi = \frac{[5.78 D_a/\rho_o^2]^2}{\alpha \cdot K_{as} N_a^2} \tag{12.3.14}$$

One can see that the Schottky mechanism for ionization equilibrium with the balance (12.3.4) holds true if

$$N_e \ll \chi \tag{12.3.15}$$

Near the Schottky ionization equilibrium (12.3.4) we have

$$\chi = \frac{K_{ion}^2}{\alpha \cdot K_{as}} \tag{12.3.16}$$

Fig. 12.21 represents the dependence $\chi(T_e)$ near this ionization equilibrium. As shows this Figure, the criterion (12.3.15) holds true for the basic range of electron temperatures. This means the realization of the Schottky mechanism for ionization equilibrium with the balance equation (12.3.4).

Note that in this analysis we assume room temperature of argon, i.e. this analysis relates to low powers injected in a gas discharge plasma. In addition we assume the diffusion character of electron and ion transport in this plasma, i.e. the gas pressure is not small, and a not small number density of electrons that leads to stepwise ionization. But even in a restricted range of these parameters one can obtain several versions of ionization equilibrium that depends on rates of processes in this plasma. In an argon gas discharge plasma under consideration we use certain parameters for processes which influence on the ionization equilibrium that decreases a number of regimes for this gas discharge plasma.

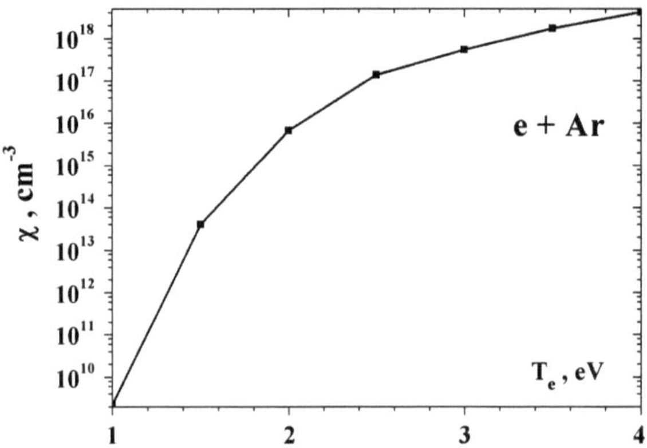

Fig. 12.21 Dependence on the electron temperature T_e for the parameter χ defined by formula (12.3.16). This parameters corresponds to an optically thick argon gas discharge plasma at room gas temperature, and the ionization equilibrium in this plasma is described by formula (12.3.4)

12.4 Hot Gas Discharge Plasma of Positive Column

Above in considering properties of a gas discharge plasma of gas discharge we ignore gas heating under the action of the discharge electric current. Therefore these results are valid at low currents or at low number densities of electrons and ions where gas heating is small. We now determine the temperature at the axis of a discharge tube of the positive column where a gas discharge plasma is located. Denoting the temperature at the axis as T_o and taking the temperature at walls $T_w = 300\,\mathrm{K}$, we are based on equation (8.5.6) of thermal balance

$$\int_{T_w}^{T_o} \kappa(T)dT = 0.14P = 0.14EI, \qquad (12.4.1)$$

where we used the temperature dependence for the thermal conductivity coefficient $\kappa(T) \sim T^{0.7}$ for helium and argon in accordance with data of Table 7.1, P is the power per unit length of a discharge tube, E is the electric field strength that is independent of the coordinate both in transversal and longitudinal direction for a gas discharge plasma of the positive column, I is the discharge current. Let us use the space distribution (8.2.3) for the number density of electrons (and ions) inside the tube, that gives

$$I = 1.36eN_o w_e \rho_o^2,$$

so that N_o is the number density of electrons at the tube axis, w_e is the electron drift velocity, and ρ_o is a discharge tube radius. This reduces the balance equation (12.4.1) to the form

$$\frac{T_o\kappa(T_o)}{1.7}\left[1-\left(\frac{T_w}{T_o}\right)^{1.7}\right]=0.19c_e\frac{eE}{N_a}(N_a\rho_o)^2 w_e, \qquad (12.4.2)$$

where c_e is the electron concentration at the tube axis. If we assume weak heating of a gas inside a tube, one can use the Schottky mechanism for ionization equilibrium in accordance with (8.2.4). Then one can rewrite (12.4.2) of the thermal balance in the form

$$1-\left(\frac{T_o}{T_w}\right)^{1.7}=1.9c_e\frac{eE}{N_a}\frac{D_a N_a}{k_{ion}T_w\kappa(T_w)}w_e, \qquad (12.4.3)$$

In the regime of low electron number densities the ionization rate constant in accordance with formula (8.1.1) is

$$k_{ion}=\frac{\alpha}{N_a}w_e$$

In the regime of high electron number densities it is necessary to replace the ionization rate constant k_{ion} of single ionization in formula (12.4.3) by the effective rate constant K_{ion} of stepwise ionization in accordance with formula (12.2.3). Figure 12.22 represents gas heating for a helium gas discharge plasma in the regime of low electron number densities. According to Fig. 10.1 data, this regime corresponds to the electron concentration below $2*10^{-8}$, and Fig. 12.22 relates to the electron concentration $c_e = 1\times10^{-9}$. The basic ion in a gas discharge plasma is the atomic one He^+, and the coefficient of ambipolar diffusion for this ion is given in Fig. 7.15. The drift velocities of electrons w_e are taken from Table 7.2, and the rate constants of single ionization of helium atoms by electron impact are given in Table 8.3. Note that a range of reduced electric field strengths is taken in accordance

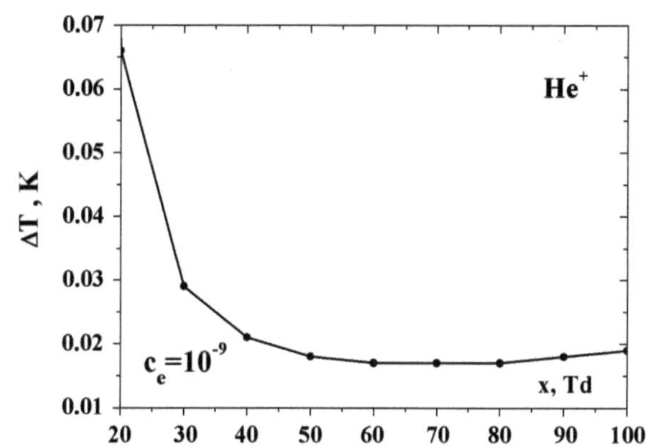

Fig. 12.22 Dependence of gas heating $\Delta T = T_o - T_w$ for a helium gas discharge plasma located in a cylinder tube for the positive column of gas discharge in the regime of low electron number densities at the electron concentration $c_e = 10^{-9}$. Gas heating is given by formula (12.4.3) and He^+ is the basis ion type in a gas discharge plasma

with ionization equilibrium for typical pressures of gas discharge (see Figs. 12.11 and 12.12). As is seen, gas heating in the regime of low electron number densities is weak because of a small electron concentration, and hence one can ignore thermal effects in the analysis of a gas discharge plasma in this regime.

In considering the regime of high electron number densities, we use again formula (12.4.3) for gas heating with use the rate constant K_{ion} of stepwise ionization (12.2.3) and ionization equilibrium that is described by formula (12.2.4). Figure 12.23 represents gas heating for a helium gas discharge plasma in the regime of high electron number densities at the electron concentration $c_e = 1 \times 10^{-7}$. We use the rate constant K_{ion} of stepwise ionization according to formula (12.2.3) and an atomic ion He^+ as the basic ion sort in a gas discharge plasma. One can see from results given in Figs. 12.22 and 12.23 that the gas heating is weak at not large electron concentrations if the ionization process proceeds effectively.

We also consider gas heating in an argon gas discharge plasma of the positive column of gas discharge for stepwise mechanism of atom ionization by electron impact. Figure 12.24 gives argon heating under the action of the discharge current in accordance with formula (12.4.3) at the electron concentration $c_e = 1 \times 10^{-6}$, where the rate constant K_{ion} of stepwise ionization is taken from Fig. 12.17 under conditions if the basic ion sort is Ar^+ and resonant radiation from the states $Ar(^3P_1)$ and $Ar(^1P_1)$ leaves this plasma freely, i.e. stepwise ionization of argon atoms by electron impact is determined by the states $Ar(^3P_2)$ and $Ar(^3P_0)$. Note that the discharge electric currents I under these conditions correspond to the μA scale. In particular, we have $IN_a = 2.1 \times 10^{12}\,\mathrm{A/cm^3}$ at the electron temperature $T_e = 3\,\mathrm{eV}$ and electron concentration $c_e = 1 \times 10^{-6}$. This corresponds to the power per unit length of the tube $IE = 250\,\mu\mathrm{W/cm}$ ($E/N_a = 12\,\mathrm{Td}$). Since the tube radius ρ_o in this regime is large compared to the mean free path of atoms, we obtain in this case

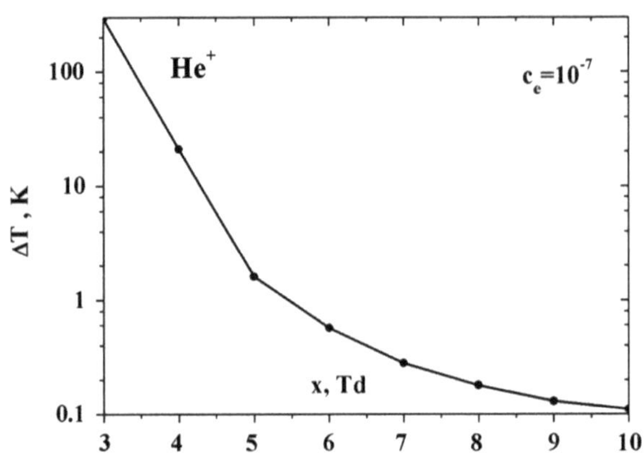

Fig. 12.23 Dependence of gas heating $\Delta T = T_o - T_w$ for a helium gas discharge plasma located in a cylinder tube for the positive column of gas discharge in the regime of high electron number densities at the electron concentration $c_e = 10^{-7}$. Gas heating is given by formula (12.4.3) and He^+ is the basic ion type in a gas discharge plasma

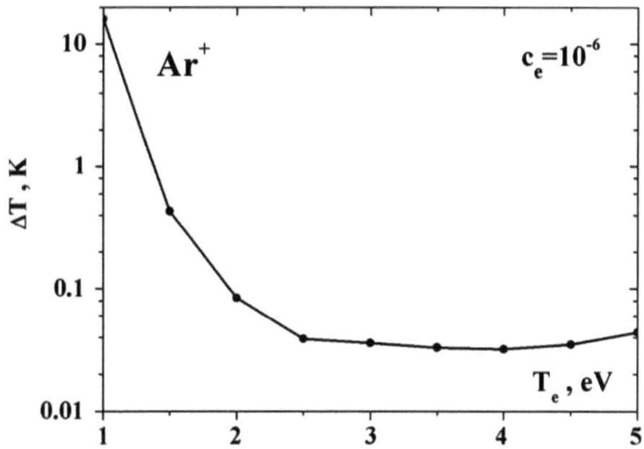

Fig. 12.24 Dependence of gas heating $\Delta T = T_o - T_w$ for an argon gas discharge plasma located in a cylinder tube for the positive column of gas discharge in the regime of high electron number densities at the electron concentration $c_e = 10^{-6}$. Gas heating is given by formula (12.4.3) and Ar^+ is the basis ion type in a gas discharge plasma

$$\frac{I}{\rho_o} \gg 10^{-3}\,\mathrm{A/cm}$$

Heating of a gas discharge plasma in the positive column of gas discharge is determined by a power which is transferred to a gas through electrons. The power per unit length of a discharge tube evolved in a gas discharge plasma is equal

$$P = EI = 1.36 N_o \rho_o^2 \cdot eEw_e = 1.36 c_e (N_a \rho_o)^2 \cdot \frac{eEw_e}{N_a} = 7.9 c_e \frac{eEw_e}{N_a} \frac{D_a N_a}{K_{ion}}, \tag{12.4.4}$$

where N_o is the electron number density at the tube axis, c_e is the electron concentration there. We give in Fig. 12.25 the specific power for an argon gas discharge plasma on the basis of formula (8.57) under the same conditions which are used in Fig. 12.24.

It should be noted that in the above analysis we choose the reduced electric field strength or the electron temperature as a parameter which characterizes a gas discharge plasma under consideration and restrict by low electron concentrations. This means that the electron concentration is a free parameter here that is given by an external source which provides an electric current passed through a gas. We also assume gas heating to be weak that allows us to use the ionization balance equation (8.2.4) where the ionization rate is proportional to the electron number density. If the temperature at the tube axis differs significantly from the wall temperature, the ionization balance equation becomes more complex. In order to understand the scale of powers where this influence becomes strong, we represent in Fig. 12.26 the dependence of the gas temperature T_o at the tube axis on the specific power that is given to electrons of a gas discharge plasma and subsequently it is consumed on gas heating.

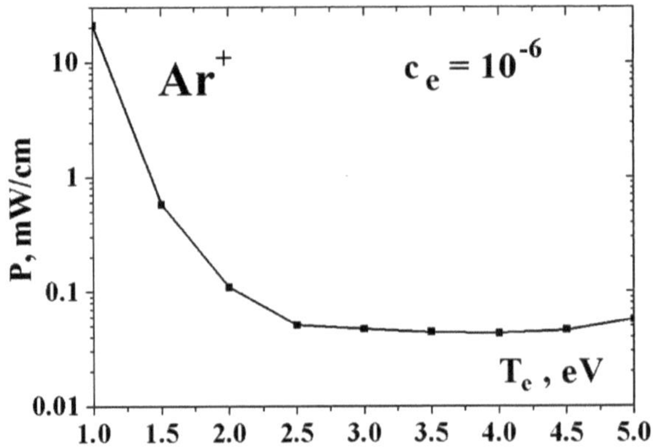

Fig. 12.25 Specific power P transferred from an external electric field to a gas through electrons of an argon gas discharge plasma of the positive column of gas discharge in accordance with formula (12.4.4). These results relate to the electron concentration $c_e = 10^{-6}$ and to ion Ar^+ as the basic ion type in a gas discharge plasma

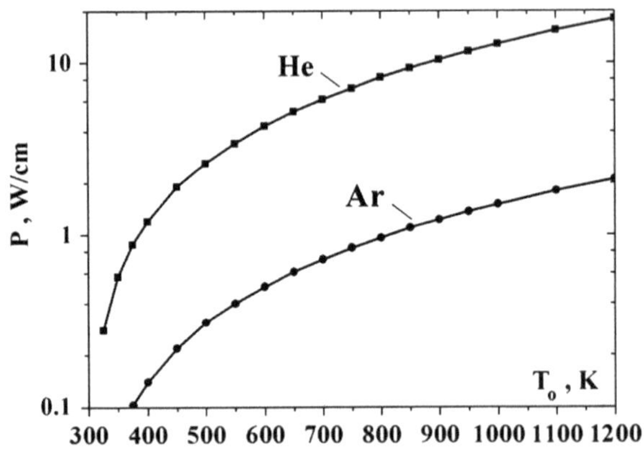

Fig. 12.26 Connection between the gas temperature T_o at the axis of a discharge tube and the specific power P transferred from an external electric field to a gas through electrons in a helium and argon gas discharge plasma of the positive column of gas discharge according to formula (8.5.12)

In conclusion we note that variation of the gas temperature in a gas discharge plasma is relatively small at low electron concentrations both in the regime of low and high electron number densities. This means that the above consideration of properties of a helium and argon gas discharge plasmas which is located in the positive column of gas discharge holds true at not large electron concentrations.

We now consider a gas discharge plasma of arc of high pressure where the concentration of charged atomic particles is not small and thermal processes are important for its properties. Due to a high electron number density, ionization equilibrium at each point is established in an arc plasma due to collisions with electrons. We then encounter with a specific character of ionization processes in a discharge tube. Indeed, the electric field strength is constant in the transversal direction¡ but the ionization rate is determined by the reduced electric field strength E/N_a, and because the gas pressure is constant over the tube cross section, the number density of atoms is higher at the tube axis, and a more strong electric field strength acts in this region. As a result, the main contribution to ionization gives a region near the tube axis. We first consider this problem for the case where electron-atom elastic collisions may be considered within the framework of the hard sphere model (see Sect. 7.3). This model is suitable for a helium gas discharge plasma at electron energies $\varepsilon < 10\,\text{eV}$. In this case the relation between the electron T_e and gas T temperature is given by formula (8.7.1)

$$\frac{T_e}{T}\left(\frac{T_e}{T} - 1\right) = const$$

This consideration is valid if a typical distance, where these temperatures vary remarkably, is large compared to the mean free path of electrons. In this case the rate of atom ionization is nonuniform and ionization may proceed in a narrow region near the tube axis. We analyze in detail this case which holds true for a helium gas discharge plasma of not large temperatures. A simple connection between electron and atomic temperatures simplifies the analysis of the ionization balance. Note we insert now in consideration an additional parameter, the gas temperature T, in comparison with the above analysis of ionization equilibrium. A nonuniform temperature space distribution complicates this analysis.

Let us first ascertain the hierarchy of times for processes which establish an equilibrium in the helium gas discharge plasma under consideration. For definiteness we take the electron temperature at the axis to be $T_e = 3\,\text{eV}$, the helium pressure to be $p = 10$ Torr and the gas temperature at the axis to be $T_o = 600$–$2,000\,\text{K}$ that differs from the wall temperature $T_w = 300\,\text{K}$. We now determine the character of equilibrium in such a system. The connection (6.2.7) between the electron T_e and gas T temperatures is established in elastic electron-atom collisions and leads to the above relation (8.7.1) in helium where the electron-atom diffusion cross section is independent on the collision energy. A typical time for this equilibrium is

$$\tau_{ea} = \frac{1}{N_a v_T \sigma_{ea}} \sim 10^{-10} s$$

Here $N_a \sim 10^{17}\,\text{cm}^{-3}$ is the number density of helium atoms, $v_T \sim 10^8\,\text{cm/s}$ is a thermal electron velocity, $\sigma_{ea} \approx 6\text{Å}^2$ is the diffusion cross section of electron scattering on an helium atom. At this equilibrium formula (6.2.7) connects the electron temperature T_e and the reduced electric field strength $x = E/N_a$. At the chosen electron temperature we have $x \approx 8\,\text{Td}$ or $E \approx 10\,\text{V/cm}$.

A chosen range of the gas temperatures at the axis $T_o = 600$–$2{,}000\,\text{K}$ corresponds to a definite range of powers in accordance with formula (8.5.12). Taking the temperature dependence for the thermal conductivity coefficient $\kappa(T) \sim T^{0.7}$ and its value from Table 7.1, we obtain from this a range of the powers P per unit tube length to be

$P \sim 10$–$100\,\mathrm{W/cm}$ that relates for typical discharge currents $I = P/E = 1$–$10\,\mathrm{A}$. For typical laboratory tube radii $\rho_o \sim 1\,\mathrm{cm}$ this corresponds to the current density $i = eN_e w_e = 1$–$10\,\mathrm{A/cm^2}$. Because a typical electron drift velocity in helium is $w_e \sim 10^6\,\mathrm{cm/s}$ at the above electric field strengths, we estimate the electron number density in this gas discharge plasma as $N_e \sim 10^{13}$–$10^{14}\,\mathrm{cm^{-3}}$. This corresponds to a high electron concentration $c_e \sim 10^{-4}$–10^{-3} for this plasma. These parameters we use for subsequent estimations. Evidently ionization equilibrium is determined by stepwise ionization now. Taking according to the analysis of Sect. 12.3 a typical rate constant of destruction of the metastable state $He(2^3S)$ by electron impact to be $k_Q \sim 10^{-7}\,\mathrm{cm^3/s}$, we have a typical time τ_m for equilibrium establishment for this state

$$\tau_m = \frac{1}{N_e k_Q} \sim 10^{-7} - 10^{-6} s$$

Loss of electrons and ions in the ionization equilibrium is determined by two processes, three body electron-ion recombination (3.8.1) and by pass of electrons and ions to walls as a result of ambipolar diffusion. The rate constant of three body electron–ion recombination at the electron temperature $T_e = 3\,\mathrm{eV}$ according to formula (3.8.5) is equal $K_{ei} \approx 7 \times 10^{-29}\,\mathrm{cm^6/s}$ that leads to a recombination time

$$\tau_{rec} = \frac{1}{N_e^2 K_{ei}} \sim 1\text{–}100\, s$$

The ion diffusion coefficient at the temperature $T = 1{,}000\,\mathrm{K}$ [110] $D_i \approx 0.7\,\mathrm{cm^2/s}$ under normal atom density. Then formula (7.7.5) gives for the ambipolar diffusion coefficient $D_a \approx 6 \times 10^3\,\mathrm{cm^2/s}$ at the electron temperature $T_e = 3\,\mathrm{eV}$ and the above number density of helium atoms $N_a = 1 \times 10^{17}\,\mathrm{cm^{-3}}$. From this we have for a typical time of electron and ion travelling to walls according to formula (8.2.4)

$$\tau_D = \frac{\rho_o^2}{5.78 D_a} \sim 3 \times 10^{-5} s$$

As is seen, the Schottky mechanism of ionization equilibrium is realized now, and the electron number density is less than that according to Saha formula (2.1.8).

As a result of a nonuniform distribution of the electron temperature over the tube cross section, the ionization rate drops remarkably at removing from the tube axis. We take the dependence on a distance ρ from the axis according to formulas (8.5.14) and (8.5.15) as $T(\rho) = T_o - \beta\rho^2$, where T_o is the gas temperature at the axis. Because the temperature dependence on the electron temperature T_e for the rate constant (11.2.1) of stepwise ionization as $K_{ion} \sim \exp(-J/T_e)$, where J is the ionization atom potential, we obtain the following dependence for the ionization rate constant

$$K_{ion}(\rho) = K_{ion}(0) \exp\left(-\frac{\rho^2}{a^2}\right), \quad a^2 = \frac{T_o T_e(0)}{J\beta}, \tag{12.4.5}$$

and we assume $a < \rho_o$. On the basis of formula (8.5.14) we obtain $a^2/\rho_o^2 \approx 0.08$ and $a^2/\rho_o^2 \approx 0.11$ on the basis of formula (8.5.15) at temperatures T_o under consideration. As a result we obtain that formation of free electrons and ions in a gas discharge plasma proceeds in a narrow region of the tube cross section. This leads to the total rate of electron and ion formation per unit tube length

$$J_1 = N_a N_o \int_0^{\rho_o} 2\pi\rho d\rho K_{ion}(\rho) = N_a N_o \pi a^2 K_{ion}(0) \tag{12.4.6}$$

In order to rewrite the ionization balance equation, we analyze the electron and ion distribution outside the ionization region that has the following form by analogy with heat balance equation (8.5.2)

$$\frac{d}{dx}\left(D_a x \frac{dN_e}{dx}\right) = 0,$$

where $x = \rho^2/\rho_o^2$. We have that the ambipolar diffusion coefficient $D_a \sim 1/N_a \sim T$ and according to formula (7.7.5) it is proportional to a factor T_e/T. Hence if we neglect by a more weak temperature dependencies for the ion velocity and cross section of ion-atom collision, the ambipolar diffusion coefficient becomes independent of the gas temperature. Solving the above equation for the electron number density $N_e(x)$, we assume $N_e = N_o$ in the region $\rho < a$. Then we have for the electron number density N_e in a region outside a region of their formation according to formula (8.7.3)

$$N_e = N_o \frac{\ln(\rho_o/\rho)}{\ln(\rho_o/a)}, \quad \rho > a$$

From this we find for the rate of attachment of electrons and ions to walls

$$J_2 = -2\pi\rho_o D_a \frac{\partial N_e}{\partial\rho(\rho = \rho_o)} = \frac{2\pi D_a N_o}{\ln(\rho_o/a)}$$

Equalizing this and the rate (12.4.6) of formation of electrons and ions, we obtain the ionization balance equation $J_1 = J_2$ in this case

$$N_a(0) K_{ion}(0) = \frac{2D_a(\rho_o)}{a^2 \ln(\rho_o/a)} \tag{12.4.7}$$

Taking the number density of atoms at walls $N_a \equiv N_a(\rho_o)$, one can reduce this equation to the form

$$(N_a \rho_o)^2 = \frac{4D_i N_a}{K_{ion}(T_e)\ln(J/T_e)} \cdot \frac{J T_o}{T_w^2} \tag{12.4.8}$$

In particular, for a helium gas discharge plasma ($J = 24.56\,\mathrm{eV}$) with atomic ions ($4D_i N_a = 3.1 \times 10^{19}\,\mathrm{cm}^{-1}\mathrm{s}^{-1}$) at the wall temperature $T_w = 300\,\mathrm{K}$ this formula takes the form

$$(N_a \rho_o)^2 = \frac{4D_i N_a}{K_{ion}(T_e)\ln(J/T_e)} \cdot \frac{J T_o}{T_w^2} \tag{12.4.9}$$

This coincides with (8.7.6) if we replace the ionization rate constant by the rate constant of stepwise ionization at the tube axis. We use in this derivation formula (8.7.1) which connects the electron and gas temperatures, so that formula (12.4.7) holds true if the diffusion cross section of electron-atom scattering is independent on the electron energy.

Note that according to its structure the ionization balance (12.4.7) is analogous to equation (12.2.2) for stepwise ionization at a low power, so that the temperature at the axis coincides with the wall temperature. In the case under consideration equation (12.4.7) of the ionization balance must be added to equation (8.5.12) of the thermal balance. Figure 12.27 compares the reduced gas pressure $p\rho_o$ in helium as a function of the electron temperature T_e at low and high electron concentrations. In the first case the gas temperature at the axis coincides with that at walls, i.e. $T_o = T_w = 300\,\mathrm{K}$, while in the second case the gas temperature at the axis is $T_o = 2{,}000\,\mathrm{K}$, and at walls $T_w = 300\,\mathrm{K}$. As is seen, parameters of a gas discharge plasma depends on several external parameters. Figure 12.28 gives the electron concentration depending on the electron temperature at a certain gas temperature at the discharge tube axis.

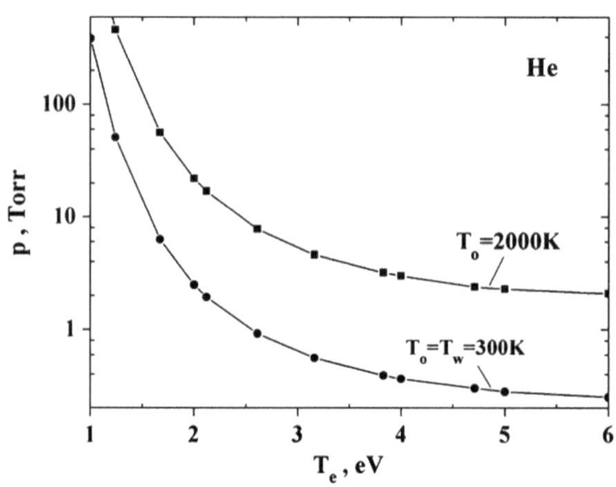

Fig. 12.27 Dependence of the helium pressure on the electron temperature T_e at the axis for the axis temperature $T_o = T_w = 300\,\mathrm{K}$ of helium atoms according to equation of ionization balance (12.2.2) and the gas temperature at the axis $T_o = 2{,}000\,\mathrm{K}$ at the same temperature $T_w = 300\,\mathrm{K}$ at walls in accordance with equation of ionization balance (12.4.7)

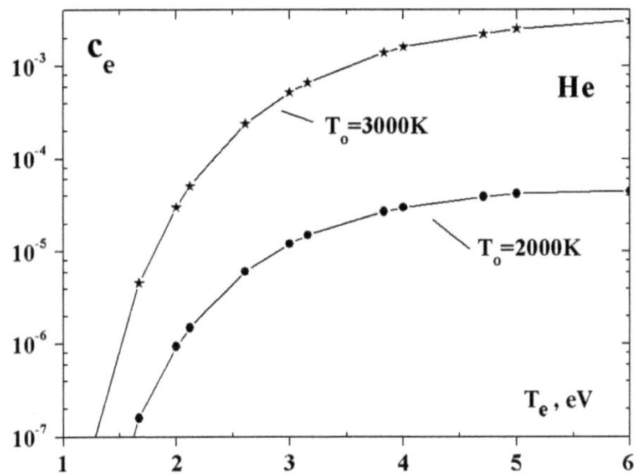

Fig. 12.28 Dependence of the electron concentration c_e at the axis on the electron temperature T_e at the injected specific power $P = 8\,\mathrm{W/cm}$ and $P = 180\,\mathrm{W/cm}$, so that the gas temperature at the axis is $T_o = 2{,}000$ K and $T_o = 3{,}000$ K respectively

12.5 Capillary Discharge

Capillary gas discharge is characterized by a small radius of a discharge tube that provides a high energy of electrons. As a result, the drift velocity of electrons is high and electrons can pass outside discharge tube. A high energy of electrons in a plasma torch outside the tube allows one to use this plasma as a source of radiation in VUV (vacuum ultraviolet) range. In particular, the intensity of radiation near resonance line HeI ($\lambda = 58.43$ nm) is 10^{11}–$10^{12}\,\mathrm{s}^{-1}$ [434–440]. Because of a high electron temperature, capillary discharge may be a source of X-rays [441–443] and even a basis of a laser. In particular, capillary discharge in argon allows one to generate laser radiation on the transition of Ne-like argon multicharge ions at the wavelength 46.9 nm [50–52]. Since the length of a plasma torch outside the tube is up to 5 cm [444, 445], this plasma may have medical applications both for sterilization of microbes and modification of biological materials with specific schemes of discharge depending on applications [72, 446, 448]. Though some schemes of capillary discharge are analyzed theoretically and codes for these discharges are elaborated (for example, [449–451]), a more detailed analysis is required in order to establish connections between processes in this gas discharge. This will help to understand the sensibility of output discharge parameters to the rates of some processes in this gas discharge plasma.

We below analyze capillary gas discharge of low pressure where the mean free path of ions is large compared to the tube radius. Because of a small residence time of ions in a capillary plasma, a high electric field strength is required to support this plasma. Electrons move with a high drift velocity under the action of such fields. As a result, electrons located near the tube edge shoot outside the tube and form a plasma torch with fast electrons. An excess negative charge of the plasma torch returns back along a conductive external side of the tube or is compensated by positive ions which are drawn in the plasma torch. This provides a stationary or quasistationary regime of the capillary plasma.

Let us consider first capillary gas discharge of low pressure, where a tube radius is small compared to the mean free path of ions. A general scheme of capillary discharge is represented in Fig. 12.29. An ionized gas is created in a specific chamber and flows through a capillary tube where it is supported by an external electric field. Because of the rate of ion attachment to walls is large for a small tube radius, a strong field is required to support this plasma in a capillary tube. Therefore fast electrons are located in this plasma and electrons have a high drift velocity. As a result, an electric field creates an additional gas pressure through electrons, and this pressure compels an ionized gas to flow towards an open tube edge. Thus, a plasma torch with fast electrons arises in a space outside the capillary tube. For capillary discharge of low pressure this plasma torch with fast electrons is a source of vacuum ultraviolet radiation.

In determination the residence time of electrons and ions in the capillary plasma, we will reduce the current of ions to walls to formula (8.4.8) obtained on the based of the Firsov method [38, 350]

$$j_i = 0.272 N_o \sqrt{\frac{2T_e}{M}},$$

where N_o is the number density of electrons and ions at the axis, T_e is the electron temperature. Because the electron temperature $T_e = \varepsilon_\perp \sim N_e$ and the electron number density varies over the tube cross section, we introduce the average electron temperature $T_o = \langle T_e \rangle$, where an average is made over the tube cross section. Taking the dependence of the plasma electric potential $U(\rho)$ for electrons with respect to the tube axis as $U(\rho) = \beta\rho^2$, we obtain

$$\langle \sqrt{T_e} \rangle = \sqrt{T_o} \frac{\int_0^{\rho_o} \exp[-1.5U(\rho)/T_o] \cdot 2\pi\rho d\rho}{\int_0^{\rho_o} \exp[-U(\rho)/T_o] \cdot 2\pi\rho d\rho} = 0.8\sqrt{T_o},$$

where $U(\rho_o) = 1.145 T_o$ [38]. This gives according to formula (8.4.8) for the ion (and electron) flux to walls [38]

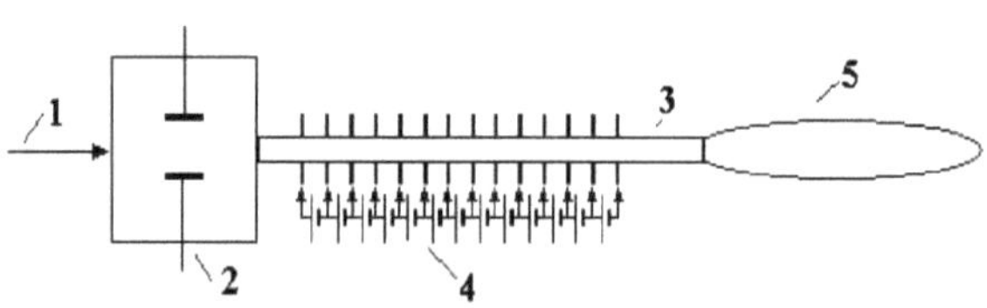

Fig. 12.29 Scheme of capillary discharge. *1* gas flow, *2* container with an ionized gas, *3* capillary tube, *4* battery created an electric field in the tube, *5* plasma torch

$$j_i = 0.22 N_o \sqrt{\frac{2T_o}{M}} \qquad (12.5.1)$$

From this we find the average time τ_w of ion travelling to walls taking for simplicity the parabolic dependence of the self-consistent electric potential $U(\rho)$ on a distance ρ from the axis

$$\tau_i = \frac{N_o \int_0^{\rho_o} 2\pi\rho d\rho \exp[-U(\rho)/T_e])}{2\pi\rho_o j_i} \approx 2\rho_o \sqrt{\frac{M}{T_o}} \qquad (12.5.2)$$

Because for discharge of low pressure the ion residence time inside a tube is small compared to a typical time of the charge exchange process with this ion, the criterion of low pressure discharge has the form

$$\xi = N_a \sigma_{res} \rho_o \ll 1, \qquad (12.5.3)$$

where N_a is the number density of atoms, σ_{res} is the cross section of resonant charge exchange. In particular, taking the tube radius $\rho_o = 0.5\,\text{mm}$, the number density of gas atoms $N_a = 1 \times 10^{15}\,\text{cm}^{-3}$ and being guided by the ion temperature $T_i = 1{,}000\,\text{K}$, we use the cross sections of resonant charge exchange $\sigma_{res} = 34\,\text{Å}^2$ in the helium case and $\sigma_{res} = 70\,\text{Å}^2$ in the argon case [110]. This gives for the parameter ξ characterized the gas rareness for ions the values $\xi = 0.17$ in the helium case and $\xi = 0.35$ in the argon case. In addition, according to formula (12.5.3) the average times of ion residence inside the tube are $\tau_i = 20\,\text{ns}$ in the helium case and $\tau_i = 64ns$ in the argon case.

Due to a small lifetime of electrons and ions in a gas discharge plasma, a strong electric field strength is required to provide ionization equilibrium that is described by

$$\frac{1}{\tau_i} = \alpha w_e \qquad (12.5.4)$$

Taking experimental values of the first Townsend coefficient α from [43] and the electron drift velocity w_e according to formula (6.6.14), one can find from equation (12.5.4) of ionization balance in the helium case $x = E/N_a = 1{,}400\,\text{Td}$ ($\alpha/N_a = 2.8 \times 10^{-16}\,\text{cm}^2$), while in the argon case formula (12.5.4) gives $x = 700\,\text{Td}$ ($\alpha/N_a = 1.4 \times 10^{-16}\,\text{cm}^2$).

We collect in Table 12.1 parameters of a capillary plasma in helium and argon at certain parameters. For definiteness we take, as above, a tube radius $\rho_o = 0.5\,\text{mm}$, the number density of helium and argon atoms $N_a = 1 \times 10^{15}\,\text{cm}^{-3}$, and the gas temperature $T_i = 1{,}000\,\text{K}$. The values of the reduced electric field strengths E/N_a and the residence time for ions τ_i which are determined above are represented this Table. The electron drift velocities in helium and argon are evaluated on the basis of formula (6.6.14); formula (6.6.11) is used for determination of the electron way Δx

Table 12.1 Parameters of helium and argon gas discharge plasmas of capillary discharge of low pressure that is burnt in a capillary tube of a radius $\rho_o = 0.5$ mm, the number density of atoms $N_a = 10^{15}$ cm^{-3} and gas temperature $T = 1,000$ K

Plasma sort	He	Ar
E/N_a (Td)	1,400	700
E (V/cm)	14	7
τ_i (ns)	20	64
w_e (cm/s)	1.7×10^8	1.0×10^8
$m_e w_e^2/2$ (eV)	22	3.1
Δx (cm)	1.4	1.7
k_{ex} (cm^3/s)	1.2×10^{-7}	6.4×10^{-8}
l (cm)	5.6	6.7
τ_{rel} (μs)	0.41	0.22
p_e (Pa)	71	10
N_e/N_*	1.4	7.0
$\varepsilon_o/\Delta\varepsilon$	2.7	1.5

per one atom excitation, and formulas (6.6.17), (6.6.18) are used for calculation of the atom excitation rate constants k_{ex} in the lowest excited states.

Electrons located close to the tube edge do not attach to tube walls and go out the tube forming a plasma torch. Ions attached to tube walls charge it positively near a tube edge; hence dielectric walls near the tube edge repel ions and they follow to the plasma torch also. The length of the region l from which electrons and ions go outside the tube is given by

$$l = w_e \tau_i \tag{12.5.5}$$

Numerical values of this length l of a capillary plasma region from which electrons and ions leave the tube are given in Table 12.1 for helium and argon gas discharge plasmas. Electron drift under the action of an electric field creates a pressure along the tube that is

$$p_e = N_e m_e w_e^2 \tag{12.5.6}$$

The values of p_e for helium and argon capillary plasmas are given in Table under the above conditions and the electron number density $N_e = 10^{13}$ cm^{-3} (the ionization degree of a capillary plasma is 1 %). Since this pressure is small compared to atmospheric one, this capillary plasma of low pressure cannot be injected in atmospheric air and may be used as a source of vacuum ultraviolet radiation only. For comparison, the gas pressure in a capillary tube under considering conditions ($N_a = 10^{15}$ cm^{-3}, $T = 1{,}000$ K) is 14Pa.

Let us analyze the balance of atoms in the lowest excited states for a capillary plasma of low pressure. These atoms are formed in electron-atom collisions and are destructed as a result of attachment to walls, so that the balance equation for the number density N_* of excited atoms has the form

$$N_e N_a k_{ex} = N_* v_T \left\langle \frac{1}{R} \right\rangle, \quad (12.5.7)$$

where $v_T = \sqrt{8T/(\pi M)}$ is the average atom velocity, T is the gas temperature, M is the atom mass, k_{ex} is the excitation rate constant averaged over energies of incident electrons, R is a distance from a point of formation of an excited atom to the surface point on the atom trajectory, and triangle parentheses mean an average over trajectories of excited atoms towards walls. After an average over atom trajectories we obtain for the boundary electron number density

$$\frac{N_*}{N_e} = \frac{N_a k_{ex} \rho_o}{v_T} \quad (12.5.8)$$

where k_{ex} is given by formulas (6.6.17), (6.6.18). Note that we deal with atoms in the lowest excited states because of a low rate constant of atom excitation in higher states. Values of this ratio given in Table 12.1 show that under considering conditions the number density of electrons in the capillary plasma is lower than the number density of excited atoms.

Note that under conditions of a plasma of glow discharge the criterion (6.5.4) holds true and the self-consistency of the excitation process is of importance [83, 287]. Then the energy distribution function of electrons drops sharply above the excitation threshold, and this influences in turn on the excitation rate. In the helium and argon cases it is valid at the reduced electric field strengths $E/N_a \ll 100\,\text{Td}$. In a capillary plasma because of high electric field strengths this effect is absent. Next, let us determine at which parameters the contribution of stepwise atom ionization becomes comparable to single ionization. Evidently, equation of the ionization balance in the case of stepwise ionization has the form

$$k_{ion}^* N_* = \frac{1}{\tau_i} \quad (12.5.9)$$

Taking the rate constants of ionization of excited atoms k_{ion}^* by electron impact according to formula (6.6.29) and the above values of the residence time τ_i for ions, we obtain for the number densities of excited atoms in a capillary plasma which provide stepwise ionization $N_* = 3 \times 10^{14}\,\text{cm}^{-3}$ for a helium gas discharge plasma and $N_* = 1 \times 10^{15}\,\text{cm}^{-3}$ for an argon gas discharge plasma. Comparing these values with the number density of atoms in the ground state $N_a = 10^{15}\,\text{cm}^{-3}$, one can conclude that stepwise ionization is not realized in a capillary plasma.

The main part of capillary discharge is a plasma torch with fast electrons, and a region of the capillary tube of a size $\sim l$ is responsible for formation and existence of the plasma torch. Hence, the scheme of Fig. 12.30 is more profitable with the length l of a capillary tube. This scheme uses the fact that in principle a capillary plasma is not effective because of a low residence time for electrons and ions. In this scheme the regions where a gas discharge plasma is formed and where this plasma is used for origin of the plasma torch. An intermediate region between these ones may have the

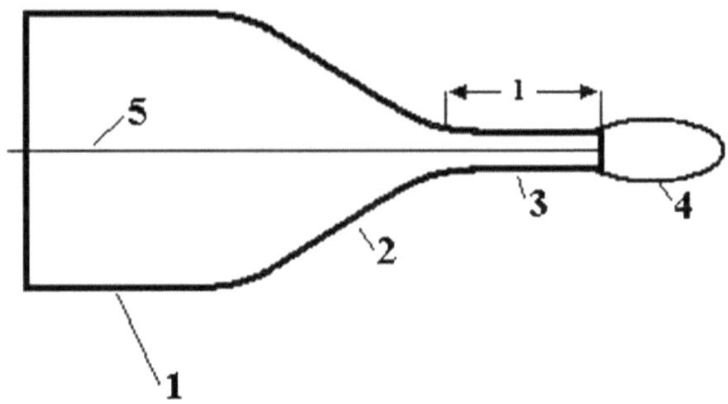

Fig. 12.30 Shape of a gas discharge chamber for formation of the plasma torch. *1* region of a gas discharge plasma of high pressure, *2* intermediate region, *3* region of a capillary plasma of low pressure, *4* plasma torch, *5* axis

conic shape. Next, according to data of Table 12.1 the plasma torch of low pressure is characterized by a low pressure of electrons and hence this plasma torch may be used as a source of vacuum UV radiation. A plasma of high pressure is required for capillary discharge which allows one to obtain a plasma torch in atmospheric air, and this plasma may be used for medical goals. Therefore, we below consider capillary discharge of high pressure.

Being guided by a high pressure of electrons which push out a plasma from capillary tube, we consider this object from another standpoint and determine parameters of capillary discharge which allows one to obtain a plasma torch in atmospheric air. Let us consider a plasma flow inside a capillary tube as laminar motion with small Reynolds numbers [452, 453]. If a gas flows in a cylinder tube of a radius R and length l, and the pressure difference for the beginning and end tube is Δp, the flow velocity $v(r)$ at a distance r from the tube axis is given by [454]

$$v(r) = \frac{\Delta p}{4\eta l}(R^2 - r^2) \tag{12.5.10}$$

This gives the following distribution of the flow velocity over the tube cross section that was obtained in study of blood propagation in a blood vessel [455–457] and is called as the Poiseuille formula

$$v(r) = v_o\left(1 - \frac{r^2}{R^2}\right), \tag{12.5.11}$$

where v_o is the flow velocity at the axis that is given by [454, 458]

$$v_o = \frac{\Delta p R^2}{4\eta l} \tag{12.5.12}$$

The Poiseuille formula leads to the following gas expense Q (the gas mass per unit time) in the course of its flow through a tube [454, 458]

$$Q = \frac{\pi m_a \Delta p R^4}{8\eta l} \tag{12.5.13}$$

In considering capillary discharge of high pressure, we restrict for definiteness by argon as a working gas of pressure $p = 1\,\mathrm{atm}$ and a pressure variation in the tube $\Delta p = 0.1\,\mathrm{atm}$. Taking a tube radius $R = 0.5\,\mathrm{mm}$, a tube length $l = 5\,\mathrm{cm}$ and the argon temperature $T = 1,000\,\mathrm{K}$, we have for the argon expense $Q \approx 60\,\mathrm{mg/s}$ that corresponds to the gas expense approximately $3l/s$ under normal conditions. In addition, formula (12.5.12) gives the flow velocity $v_o = 2.3 \times 10^4\,\mathrm{cm/s}$ at the tube axis under given conditions. On the basis of the ionization balance equation (12.1.1), one can find the reduced electric field strength E/N_a and the number density of electrons N_e which provides these flow parameters.

Note that electron drift cannot provide an addition pressure which causes flow of a gas through a tube. Indeed, let us estimate the electron number density N_e which provides the above pressure variation $\Delta p = p_e = 0.1\,\mathrm{atm}$ at the atmospheric argon pressure in accordance with formula (12.5.6) at the reduced electric field strength $E/N_a = 300\,\mathrm{Td}$. This gives $N_e \approx 5 \times 10^{17}\,\mathrm{cm}^{-3}$ that is a remarkable part of the number density of argon atoms $N_a \approx 7 \times 10^{18}\,\mathrm{cm}^{-3}$ at $T = 1{,}000\,\mathrm{K}$. But this creates very high current and leads to a strong gas heating. From this we conclude that in contrast to the case of capillary discharge of low pressure, it is impossible to create a pressure excess for formation the plasma torch due to an electron drift, and this results from a higher pressure in a container before the capillary tube. In this case it is not necessary attachment of ions to walls and absence this attachment for electrons in the course of this process, i.e. the gas flow due to a pressure excess in the container captures simultaneously electrons and ions. On the basis of formula (12.4.1) we find that the gas temperature at the tube axis $T = 1{,}000\,\mathrm{K}$ at the reduced electric field strength $E/N_a = 300\,\mathrm{Td}$ is provided by the electric current $I = P/E = 0.1\,\mathrm{mA}$. For the above radius of the capillary tube $R = 0.5\,\mathrm{mm}$ and the average flow velocity $v \approx 2 \times 10^4\,\mathrm{cm/s}$ this corresponds to the excess electron number density $N_e \sim 4 \times 10^{12}\,\mathrm{cm}^{-3}$.

In order to decrease the argon expense in this version it is convenient to decrease the radius of a capillary orifice and change the tube shape. Let us take a conic shape of the capillary tube which is ended by a nozzle which radius is large compared to the mean free path of atoms and small compared to the tube radius at the beginning. Assuming the pressure variation in this tube to be relatively small, one can solve the Poiseuille equation (12.5.10) at the flow axis assuming the gas temperature and pressure along the axis to be constant

$$\frac{dp}{dz} = \frac{4\eta v_o}{R^2}, \tag{12.5.14}$$

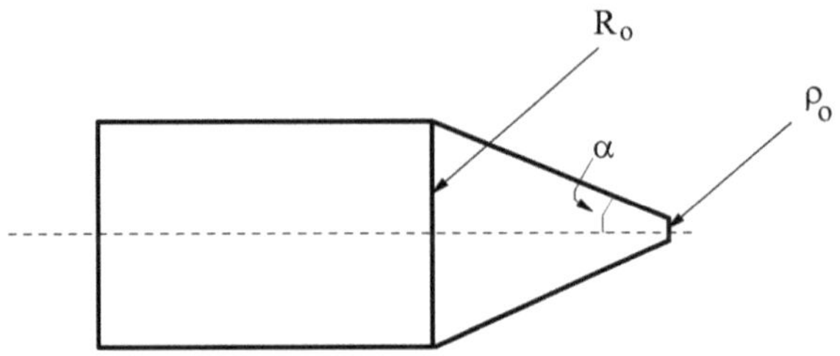

Fig. 12.31 Geometry and notations for parameters of a conic capillary tube

where z is the direction along the axis, R is a current tube radius. Using notations of Fig. 12.31 and solving this equation, we obtain (for example [459])

$$\Delta p = \frac{4\eta v_o}{3r_o \tan\alpha}, \tag{12.5.15}$$

where we assume $r_o \ll R_o$ and v_o is the flow velocity at the tube axis. From this we obtain for the flow velocity at the orifice

$$v_o = \frac{3r_o \Delta p \tan\alpha}{4\eta}, \tag{12.5.16}$$

i.e. transition from a cylinder tube to conic one leads to a change of a cylinder tube length l in formula (12.5.13) by $r_o/3$. In determination the gas expense Q of a gas flow, we assume the gas temperature T_o at the axis is large compared to that T_w at walls, and the gas temperature at a distance r from the axis is

$$T(r) = T_o\left(1 - r^2/R^2\right),$$

where R is a radius for a current cross section. Integration over the cross section gives for a gas expense

$$Q = \pi r_o^2 v_o N_o m_a = \frac{3\pi}{4}\frac{r_o^3 \Delta p \tan\alpha}{\eta}\rho_o, \tag{12.5.17}$$

where N_o is the number density of atoms at the orifice axis, and $\rho_o = N_o m_a$ is the mass density of the flow there.

We now consider the heat regime of the flow. Taking the temperature distribution over the cross section as early

$$T(r) = T_o - \Delta T \frac{r^2}{R^2},$$

where $\Delta T = T_o - T_w$ is the temperature difference between the axis and walls, we find the average temperature change to be $\Delta T/2$ in this process. Averaging the flow velocity and the atom number density over the cross section as early, we obtain for the power Π that is required for heating of a flowing gas, if we model a conic tube near an orifice by a cylinder tube

$$\Pi = \frac{c_p \Delta T}{2m_a} Q = \frac{\pi}{2} c_p \Delta T v_o N_o r_o^2 = \frac{15\pi}{16\eta} \Delta T v_o N_o \Delta p \tan \alpha r_o^3, \qquad (12.5.18)$$

where $c_p = 5/2$ is the atom heat capacity at constant pressure. In particular, if we take for definiteness $\alpha = 20°$, $r_o = 50\,\mu\text{m}$ and $\Delta p = 0.1$ atm, we obtain for argon at atmospheric pressure $v_o = 3.1 \times 10^5\,\text{cm/s}$, $\Pi = 2.1\,\text{W}$, $Q = 400\,\text{sccm}$ (standard cubic cm per second) or 12 mg/s.

Next, this temperature according to formula (8.5.6) corresponds to the power P per unit length $P = 1.8\,\text{W/cm}$ injected in this plasma if we model a flow near an orifice to have the cylindrical shape. As is seen, an external field heats a plasma flow in a volume $\sim 0.01\,\text{cm}^3$. Based on the reduced electric field strength $E/N_a =$ 300 Td (the electric field strength $E = 20\,\text{kV/cm}$ is comparable to the breakdown strength for dry air), we find a current $I = P/E \approx 80\,\mu\text{A}$, and the current density $i \approx 1\,\text{A/cm}^2$. Because the electron drift velocity at the above reduced electric field strength is $w_e = 1.4 \times 10^7\,\text{cm/s}$ (Table 7.4), this corresponds to the excess electron number density $\Delta N_e = 5 \times 10^{11}\,\text{cm}^{-3}$. Note that this is the difference between the number density of plasma electrons and ions, while the total number density of electrons (and ions) exceeds this value, but is less than the number density of atoms ($N_a \approx 10^{19}\,\text{cm}^{-3}$) by several orders of magnitude. A distance L at which an external electric field must support this excess electron number density is given by

$$L \approx \frac{E}{4\pi e \Delta N_e} \approx 8\,\text{cm},$$

i.e. this field is enough to have the above excess electron number density in a region with plasma heating.

Let us consider the ionization equilibrium in the plasma flow under consideration. Taking the ionization equilibrium in the form (8.2.4), we take the transversal diffusion coefficient of argon ions in argon to be $D_i = 0.06\,\text{cm}^2/\text{s}$ at $E/N_a = 300$ Td according to Fig. 7.5. Using data of Fig. 7.5, we obtain on the basis of formula (7.7.5) for the reduced ambipolar diffusion coefficient $D_a \approx 5\,\text{cm}^2/\text{s}$ under given conditions. This leads to a typical time τ_{at} of electron and ion attachment to walls

$$\tau_{at} = \frac{r_o^2}{5.78 D_a} = 9 \times 10^{-7}\,s$$

This corresponds to the ionization rate constant of argon atoms by electron impact $k_{ion} = N_a/\tau_{at} = 1.6 \times 10^{-13}\,\text{cm}^3/\text{s}$ and according to data of Table 8.1 ionization equilibrium may be fulfilled at $E/N_a \approx 10$ Td, i.e. under considering conditions additional ionization proceeds in the plasma flow. One can see that charge separation due to ion attachment to walls and electron drift outside the capillary tube takes place at a distance l from an orifice

$$l = v_o \tau_{at} \sim 0.2\,\mathrm{cm}$$

It is clear that yield parameters of a flow capillary plasma depends on external and geometric parameters of this capillary discharge. Nevertheless, the above analysis with using certain discharge parameters shows the reality of such a system.

Chapter 13
Processes at Boundaries of Gas Discharge Plasma

Abstract Cathode processes in the normal and abnormal regimes of glow discharge in helium and argon are analyzed. They determine parameters of the cathode plasma in helium and argon. The character of plasma processes in helium and argon near walls is represented. Properties of a magnetron plasma in helium and argon are considered.

13.1 Cathode Plasma of Glow Discharge Near Equilibrium

In the classical consideration [4], we assume that reproduction of electrons at the cathode in the normal regime of glow discharge proceeds owing to three above mechanisms, bombardment by ions, quenching by excited atoms and photoionization of the cathode material (see Fig. 9.2) that is described by the second Townsend coefficient γ given by the following ratio between the electron i_e and ion i_i current densities at the cathode

$$i_e = \gamma i_i \tag{13.1.1}$$

The classical theory of the cathode plasma of normal glow discharge is represented in §9.3, where from the minimum of the cathode voltage it follows that the current density is independent of the total discharge current and is given by formula (9.3.11) in a dense gas.

The classical theory of the cathode layer [4] assumes the cylinder shape of the current, so that the depth l of the cathode layer is small compared to a current radius r, that is

$$l \ll r \tag{13.1.2}$$

This leads to an one-dimensional model of the cathode plasma with an excess of the electrons number density N_i compared to the electron one N_e. In this model electrons are formed at the cathode layer boundary, and electrons from the positive column cannot penetrate in the cathode layer because of the potential barrier at is

B.M. Smirnov, *Theory of Gas Discharge Plasma*, Springer Series on Atomic, Optical, and Plasma Physics 84, DOI 10.1007/978-3-319-11065-3_13

boundary (see Fig. 8.1). But when the ion flux propagates from the cathode layer boundary towards the cathode, it expands due to a transverse electric field, and the transversal electric field strength at the cathode plasma boundary $E_\perp$ exceeds the cathode electric field strength E_c, so that a transversal plasma expansion Δr may be large compared to the cathode layer width l. Indeed, because the drift velocity w_i of atomic ions in a parent atomic gas depends on the electric field strength as $w_i \sim \sqrt{E}$, we have $\Delta r/l \sim \sqrt{E_\perp/E_c} \gg 1$. From this we have

$$\frac{\Delta r}{r} \sim \frac{l}{r}\sqrt{\frac{E_\perp}{E_c}} \sim \sqrt{\frac{l}{r}} \ll 1, \tag{13.1.3}$$

and this justifies the validity of the classical theory of the cathode layer.

Following to the classical theory of the cathode layer [4], we require the cathode voltage U_c according to formula (9.3.13) to be minimal U_{min} among its possible values. Let us denote by E_c the electric field strength at the cathode, by l_o the cathode layer depth, by r_o the radius of a region occupied by an electric current. Introducing the number density N_a of atoms and the discharge electric current I, we consider optimal conditions where the cathode voltage is minimal in accordance with formula (9.3.13). Variations of the cathode voltage U, the cathode layer depth l, and the radius r of a plasma region near this equilibrium as a function of the reduced electric field strength E_c/E_n are given in Fig. 13.1.

From this analysis it follows that the dynamics of propagation of the ion flux towards the cathode creates an electric lens that tends to shrink the ion flux in order to keep the cathode plasma in a certain region. From Fig. 13.1 it follows that the quantity $u(x)$ is quadratic with respect to deviation of the specific electric field strength $x = E_c/E_n$ from one, while other quantities of Fig. 13.1 are linear with respect to this deviation from the equilibrium. Let us construct the quantity

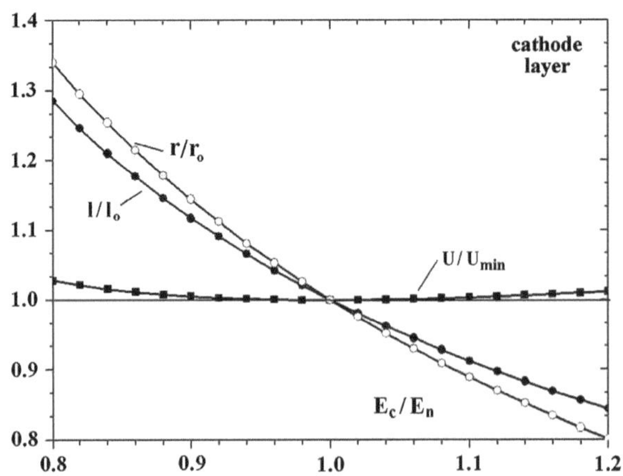

Fig. 13.1 Parameters of the cathode layer near the equilibrium

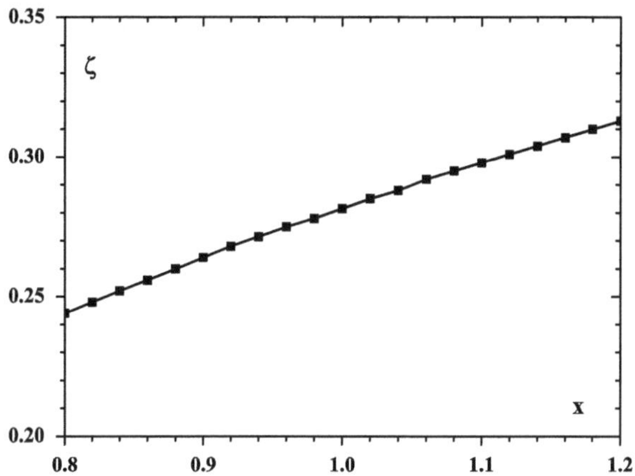

Fig. 13.2 Dependence for $\zeta(x)$ according to formula (9.3.1)

$$\zeta(x) = \frac{U/U_{min} - 1}{(r/r_o - 1)^2} \tag{13.1.4}$$

Figure 13.2 gives this dependence on the parameter $x = E_c/E_n$.

Though the parameter $\zeta(x)$ varies with a varying x, one can introduce an average value of this parameter $\zeta = 0.28$ near the equilibrium with an accuracy 10 %. One can understand the role of this parameter. If some ion transfers outside the current radius r_o, it is returned by a force F which is proportional to $r - r_o$, where r is the ion position. The motion equation for this ion is

$$M\frac{d^2r}{dt^2} = -\frac{\partial^2 U}{\partial r^2}_{|r_o}(r - r_o), \tag{13.1.5}$$

where M is the ion mass. This equation describes oscillation motion of an ion near the equilibrium with the frequency

$$\omega = \sqrt{\frac{0.28U_{min}}{Mr_o^2}} \tag{13.1.6}$$

This character of ion motion testifies about an equilibrium for the cathode layer that is realized near the minimal cathode voltage $U \approx U_{min}$ and is described by the classical theory [4] of the cathode layer for glow discharge.

This effect may lead to stabilization of the cathode plasma if $\omega\tau \gg 1$, where τ is the lifetime of the cathode plasma as a time of the ion pass through the cathode layer

$$\tau = \int \frac{dx}{w_i} = \frac{3l}{2w_i(E_c)} \tag{13.1.7}$$

From this we have on the basis of formula (13.1.6) for the parameter $\omega\tau$ because $U_{min} = 3E_c l/5$

$$\omega\tau = 0.8\frac{l}{r_o}\sqrt{\frac{l}{\lambda_i}} \tag{13.1.8}$$

Since we use an one-dimensional model of the cathode layer, $l \ll r_o$, but $l \gg \lambda_i$. Hence, the stability of the cathode layer, if it is described by the classical theory of the cathode layer [4], is valid in a restricted range of discharge currents and gas pressures. In particular, in the case of a copper cathode in argon the criterion $\omega\tau \gg 1$ gives $N_a r_o \ll 5 \times 10^{16}\,\mathrm{cm}^{-2}$. Note that according to Table 9.5 in this case $N_a l = 9.6 \times 10^{16}\,\mathrm{cm}^{-2}$. As a result we obtain the criterion $\omega\tau \gg 1$ in the case of a copper cathode in argon in the form

$$I \ll I_o, \ I_o = 60\,\mathrm{mA} \tag{13.1.9}$$

Nevertheless, some features of normal glow discharge are conserved outside this criterion.

We below consider another way to violate the classical theory of the cathode layer of normal glow discharge if ionization results through formation of excited atoms. In these case an electric field in the cathode region does not act on excited atoms and hence stepwise ionization may increase a region where the electric current to the cathode is concentrated. In order to account for the contribution of stepwise ionization to the total rate constant of atom ionization by electron impact, we represent the first Townsend coefficient in the form

$$\alpha = \alpha_T + \alpha_{st}, \tag{13.1.10}$$

where α_T is the first Townsend coefficient due to single ionization, and α_{st} is the first Townsend coefficient owing to stepwise ionization. In this consideration we evaluate α_{st} and find the criterion where $\alpha_{st} \ll \alpha_T$, i.e. the Townsend ionization regime is realized.

Assuming that the ionization coefficient α_{st} proceeds due to stepwise ionization, we use the balance equation for the number density N_* of excited atoms that has the form

$$\frac{dN_*}{dt} = -\frac{N_*}{\tau} + N_e N_a k_{ex}, \tag{13.1.11}$$

where k_{ex} is the rate constant of atom excitation by electron impact, τ is the lifetime of excited atoms. This gives for the number density of excited atoms

$$N_* = N_e N_a k_{ex}\tau, \tag{13.1.12}$$

We assume that the process

$$e + Ar(3p^6) \to e + Ar(3p^5 4s) \tag{13.1.13}$$

gives the main contribution to the loss of the electron energy. This leads to the energy balance equation

$$N_a k_{ex} \Delta\varepsilon = eEw_e, \tag{13.1.14}$$

where k_{ex} is the rate constant of atom excitation, w_e is the electron drift velocity, $\Delta\varepsilon$ is the atom excitation energy, w_e is the electron drift velocity. From this mechanism of atom excitation it follows

$$w_e \sim \frac{E}{N_a}, \; k_{ex} \sim \left(\frac{E}{N_a}\right)^2, \tag{13.1.15}$$

and at the cathode these parameters are $k_{ex}^* = 4 \times 10^{-7}\,\mathrm{cm}^3/\mathrm{s}$, $w_e = 1.4 \times 10^8\,\mathrm{cm/s}$.

One can divide the states of an excited argon atom with the electron shell $3p^5 4s$ in metastable $({}^3P_2, {}^3P_0)$ states and resonantly excited ones $({}^3P_1, {}^1P_1)$. Based on the statistical character of excitation of atom sublevels, we assume the rate constant of excitation of a given sublevel to be proportional to its statistical weight. Then we obtain the rate constant of excitation of metastable argon states $({}^3P_2, {}^3P_0)$ at the cathode to be $k_{ex}^m = 2.7 \times 10^{-7}\,\mathrm{cm}^3/\mathrm{s}$ and for resonantly excited states $({}^3P_1, {}^1P_1)$ it is equal $k_{ex}^r = 1.3 \times 10^{-7}\,\mathrm{cm}^3/\mathrm{s}$. We restrict ourselves by contribution of metastable states in stepwise ionization of atoms only since the population of resonantly excited atoms is small. Then the lifetime of excited atoms τ is equal by their quenching on the cathode owing to their diffusion

$$\tau = \frac{l^2}{2D_*}, \; N_* = \frac{N_e N_a k_{ex} l^2}{2D_*}, \tag{13.1.16}$$

where D_* is the diffusion coefficient of excited atoms, and we consider the depth of the cathode layer l to be small compared with a current radius that corresponds to the normal regime of glow discharge. The reduced diffusion coefficient for metastable helium atoms $He(2^3S)$ in helium is $D_* N_a = (1.5 \pm 0.1) \times 10^{19}\,\mathrm{cm}^{-1}s^{-1}$ according to experimental data [306–316], for excited argon atoms in states ${}^3P_2$ and ${}^3P_o$ the reduced diffusion coefficient in argon is $D_* N_a = (1.9 \pm 0.3) \times 10^{18}\,\mathrm{cm}^{-1}s^{-1}$ according to measurements [321–326], and the above data corresponds to the statistical average of indicated measurements. In addition, as it is obtained in §6.6, the rate constant of ionization of metastable helium atoms $He(2^3S)$ by electron impact in strong electric fields is $k_{ion}^*(He) = 2.5 \times 10^{-7}\,\mathrm{cm}^3/\mathrm{s}$, and the rate constant of ionization of metastable argon atoms $Ar(3_2^P)$ by electron impact in strong electric fields is $k_{ion}^*(Ar) = 3.6 \times 10^{-7}\,\mathrm{cm}^3/\mathrm{s}$. Next, according to formulas (6.6.17) and (6.6.18) the rate constant of helium atom excitation in the metastable state $He(2^3S)$ by electron impact near the cathode is $k_{ex}(He) = 1.7 \times 10^{-7}\,\mathrm{cm}^3/\mathrm{s}$.

For metastable state $Ar(^3P_2)$ this rate constant is $k_{ex}(Ar) = 6.9 \times 10^{-7}\,\mathrm{cm^3/s}$. We add to this that at the cathode the drift electron velocity in a helium cathode plasma is $w_e(He) = 3.3 \times 10^8\,\mathrm{cm/s}$, and in the argon case the electron drift velocity at the cathode is $w_e(Ar) = 1.1 \times 10^8\,\mathrm{cm/s}$. These values follow from Tables 7.3, 7.4 or from Figs. 5.1 and 13.2, if we continue their drift velocities for larger electric field strengths.

According to the definition of the reduced first Townsend coefficient due to stepwise ionization is equal in accordance with formula (13.1.16)

$$\frac{\alpha_{st}}{N_e} = \frac{1}{N_e}\frac{k_{ion}^* N_*}{w_e} = \frac{k_{ion}^* k_{ex}(N_a l)^2}{2w_e(D_* N_a)}, \tag{13.1.17}$$

where notations and values of parameters in this formula are given above. Representing formula (13.1.16) in the form

$$\frac{\alpha_{st}}{N_e} = a(N_a l)^2, \tag{13.1.18}$$

and the proportionality coefficient is $a = 4.3 \times 10^{-42}\,\mathrm{cm^6}$ in the helium case and $a = 5.9 \times 10^{-40}\,\mathrm{cm^6}$ in the argon case. Tables 13.1 and 13.2 contain the values of the

Table 13.1 Parameters of the helium cathode plasma near the cathode, namely, a part of the reduced first Townsend coefficient α_{st}/N_e due to stepwise ionization according to formulas (13.1.17) and (13.1.18), the boundary concentration of electrons c_{max}, below which stepwise ionization gives a small contribution to ionization near the cathode, the reduced number density of ions N_i/N_a^2 according to formula (13.1.21)

Cathode	$\alpha_{st}/N_e\,\mathrm{cm^2}$	c_{max}	$N_i/N_a^2\ 10^{-25}\,\mathrm{cm^3}$	$c_e/N_a\ 10^{-29}\,\mathrm{cm^3}$
Mg	1.6×10^{-9}	6.9×10^{-8}	3.2	2.7
Al	1.9×10^{-9}	5.8×10^{-8}	2.9	2.0
K	3.5×10^{-10}	3.1×10^{-7}	6.8	18
Ca	7.3×10^{-10}	1.5×10^{-7}	4.7	7.0
Fe	2.3×10^{-9}	4.8×10^{-8}	2.7	1.6
Ni	2.5×10^{-9}	4.4×10^{-8}	2.5	1.3
Cu	3.1×10^{-9}	3.6×10^{-8}	2.3	1.5
Zn	2.1×10^{-9}	5.2×10^{-8}	2.8	1.9
Sr	7.3×10^{-10}	1.5×10^{-7}	4.7	7.0
Ag	2.7×10^{-9}	4.1×10^{-8}	2.4	1.2
Cd	2.7×10^{-9}	4.1×10^{-8}	2.4	1.2
Ba	7.3×10^{-10}	1.5×10^{-7}	4.7	7.0
Pt	2.7×10^{-9}	4.1×10^{-8}	2.4	1.2
Au	2.7×10^{-9}	4.1×10^{-8}	2.4	1.2
Pb	3.1×10^{-9}	3.6×10^{-8}	2.3	0.97
Bi	1.9×10^{-9}	5.8×10^{-8}	2.9	2.0

Table 13.2 Parameters of the argon cathode plasma near the cathode, which include a part of the reduced first Townsend coefficient α_{st}/N_e due to stepwise ionization according to formulas (13.1.17) and (13.1.18), the boundary concentration of electrons c_{max}, below which the stepwise ionization gives a small contribution to ionization near the cathode, the reduced number density of ions N_i/N_a^2 according to formula (13.1.21)

Cathode	α_{st}/N_e cm^2	c_{max}	N_i/N_a^2 10^{-25} cm^3	c_e/N_a 10^{-29} cm^3
Mg	4.4×10^{-6}	1.1×10^{-10}	2.4	0.53
Al	3.2×10^{-6}	1.5×10^{-10}	2.9	1.0
K	1.4×10^{-6}	3.4×10^{-10}	1.9	1.6
Ca	2.8×10^{-6}	1.7×10^{-10}	2.7	1.1
Fe	8.6×10^{-6}	5.5×10^{-11}	5.0	0.42
Ni	8.6×10^{-6}	5.5×10^{-11}	3.9	0.68
Cu	5.5×10^{-6}	8.5×10^{-11}	3.9	0.69
Zn	4.6×10^{-6}	1.0×10^{-10}	3.5	0.77
Sr	2.8×10^{-6}	1.7×10^{-10}	2.7	1.1
Ag	5.5×10^{-6}	8.5×10^{-11}	3.9	0.69
Cd	4.6×10^{-6}	1.0×10^{-10}	3.5	0.77
Ba	2.8×10^{-6}	1.7×10^{-10}	2.7	1.1
Pt	5.6×10^{-6}	8.4×10^{-11}	3.9	0.68
Au	5.6×10^{-6}	8.4×10^{-11}	3.9	0.68
Pb	5.0×10^{-6}	9.4×10^{-11}	3.7	0.74
Bi	5.9×10^{-6}	8.0×10^{-11}	4.0	0.62

reduced first Townsend coefficient for the helium and argon gas discharge plasma at the cathode.

For the normal regime of glow discharge, where the cathode layer depth l to be small compared to its radius in accordance with (13.1.2), the reduced electric field strength at the cathode is independent of a cathode material and according to formula (9.3.15) is equal $E_n = 1650$ Td in the helium case and $E_n = 2300$ Td in the argon case. Correspondingly, as it follows from formula (9.3.15), the reduced first Townsend coefficient at the cathode is

$$\frac{\alpha_T}{N_a} = A\exp(-0.665) = 0.51\,\mathrm{A}, \tag{13.1.19}$$

if we use approximations (8.1.4) and (8.1.5) for this quantity. From this on the basis of the Table 8.2 data we obtain for the reduced first Townsend coefficient at the cathode $\alpha_T/N_a = 1.1\text{Å}^2$ in the helium case and $\alpha_T/N_a = 4.7\text{Å}^2$ in the argon case. Hence, the criterion of a small contribution to ionization of the cathode plasma from stepwise ionization $\alpha_T \gg \alpha_{st}$ may be represented in the form $c_e \ll c_{max}$, where

$$c_{max} = \frac{\alpha_T/N_a}{\alpha_{st}/N_e} \tag{13.1.20}$$

Values of this parameter for the helium and argon gas discharge plasmas at the cathode are given in Tables 13.1 and 13.2.

We now determine the ion and electron concentration at the cathode in a real cathode plasma that allows us to estimate the contribution of stepwise ionization to the total ionization. The number density of ions at the cathode follows from the Poisson equation (9.3.2) and formula (9.3.12) is given by

$$N_i = \frac{1}{4\pi e}\frac{dE}{dz} = \frac{E_n}{6\pi el} \tag{13.1.21}$$

Tables 13.1 and 13.2 contain the values of the reduced ion velocity in accordance with formula (13.1.21). In this case the ion drift velocity w_i and the average ion energy $\overline{\varepsilon}$ are determine by formula (7.6.12) for a dense gas, where $l \gg \lambda_i$, and $\lambda_i = 1/(N_a\sigma_{res})$, σ_{res} is the cross section of resonant charge exchange process. In the helium case at the cathode $E_n/N_a = 1650$ Td we have from formula (7.6.12) for the average ion energy $\overline{\varepsilon} = 1.8$ eV, for the ion drift velocity $w_i = 9.3 \times 10^5$ cm/s, and the cross section of resonant charge exchange $\sigma_{res} = 26\text{Å}^2$ corresponds to this average energy. In the argon case at the cathode, where $E_n/N_a = 2300$ Td, these parameters are equal $\overline{\varepsilon} = 1.3$ eV, $w_i = 2.4 \times 10^5$ cm/s, and $\sigma_{res} = 57\text{Å}^2$.

For determination the electron number density at the cathode, use formula of Sect. 9.3 $i_e = \gamma i_i$, where i_e, i_i are the current densities at the cathode for electrons and ions. From this we obtain

$$\frac{N_e}{N_i} = \gamma\frac{w_i}{w_e}, \tag{13.1.22}$$

Using the above drift velocities of electrons and ions near the cathode, we rewrite formula (13.1.22) in the form

$$\frac{N_e}{N_i} = 2.2 \times 10^{-3}\gamma, \tag{13.1.23}$$

and the proportionality coefficient in formula (13.1.22) is identical for helium and argon cases. Tables 13.1 and 13.2 contain the values of the reduced electron concentrations at the cathode in a helium and argon cathode plasma obtained on the basis of formulas (13.1.21) and (13.1.23). From the results given in Tables 13.1 and 13.2 it follows that at the gas pressures $p \ll 1$ atm related to glow discharge we have $c_e \ll c_{max}$, i.e. stepwise ionization gives a small contribution to the ionization process.

13.2 Cathode Layer of Abnormal Glow Discharge

Until the criterion (9.3.5) is fulfilled, the normal regime of the cathode region is realized. Then the discharge current occupies only a part cathode region and the current density i_n is independent of the total discharge current and provides the optimal cathode voltage U_c that is independent of the discharge current as well as the electric field strength E_c near the cathode. We below consider the abnormal regime of glow gas discharge, where the character of electron formation at the cathode is determined by emission of secondary electrons due to ion impact and the current density i exceeds that i_n in the case of the normal regime of the cathode region. We below consider the abnormal regime of the cathode region $i > i_n$ for helium and argon if atomic ions are located in the cathode region.

In this analysis we use as early the Poisson equation for the electric field strength in the cathode region whose solution is given by formula (9.3.10) that gives by analogy with formulas (9.3.11) and (9.3.12)

$$i = \frac{E_n w_n}{6\pi L} = \frac{E_n^{3/2}(e\lambda)^{1/2}}{3\pi L(2\pi M)^{1/2}}, \quad E(z) = E_n\left(\frac{z}{L}\right)^{2/3}, \tag{13.2.1}$$

where i is the current density, w_n is the ion drift velocity near the cathode, where is E_n is the electric fields strength at the cathode, L is the depth of the cathode layer, z is the distance from the boundary of the cathode region.

We add to this the equation of ionization balance in the cathode region of the abnormal regime that has the form in accordance with (9.3.6)

$$\int_0^1 \exp\left(-\frac{b}{z^{2/3}}\right) dz = \frac{1}{AN_a L}\ln(1 + 1/\gamma),$$

where $b = BN_a/E_c$, A and B are the parameters of formula (8.1.4) for the first Townsend coefficient. Let us introduce the function

$$f(b) = \int_0^1 \exp\left(-\frac{b}{z^{2/3}}\right) dz = \frac{1}{AN_a L}\ln(1 + 1/\gamma), \tag{13.2.2}$$

We give in Fig. 13.3 the dependence (13.2.2).

On the basis of the above equations, one can reduce the cathode region parameters in the abnormal regime to those in the normal regime. Let i_n be the current density for the cathode region in the normal regime, E_n is the electric field strength at the cathode, L_n is the thickness of the cathode region; i, E_c, L are the same parameters of the cathode region in the abnormal regime. Figure 13.4 gives the reduced electric field strength E_c at the cathode and the reduced voltage drop in the cathode region. Figure 13.5 contains the dependence of the reduced depth of the cathode region for abnormal glow discharge as a function of the reduced current density.

We have the following relations between these parameters

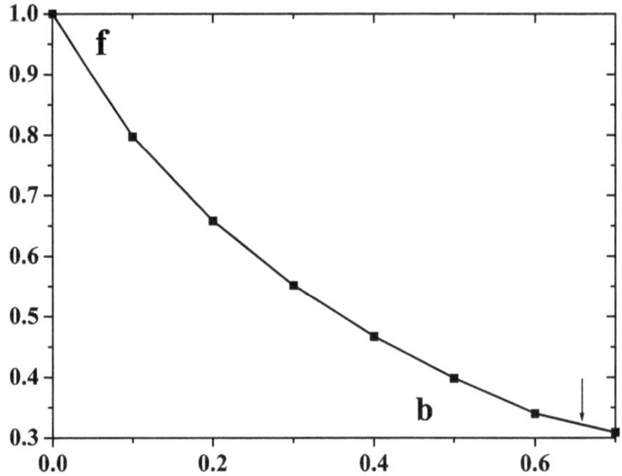

Fig. 13.3 Dependence $f(b)$ according to formula (13.2.2). The *arrow* corresponds to the normal regime of the cathode region if atomic ions are located in the cathode region

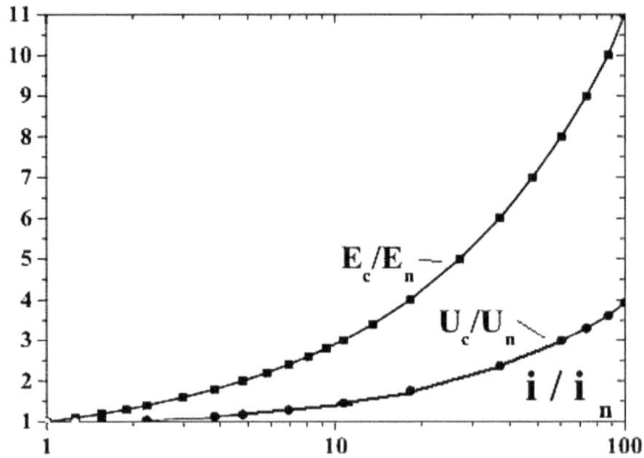

Fig. 13.4 Reduced electric field strength and the cathode voltage for the cathode region in argon in the abnormal regime according to formulas (13.2.3) and (13.2.4)

$$\frac{i}{i_n} = \left(\frac{E_c}{E_n}\right)^{3/2} \frac{L_n}{L}, \quad \frac{L}{L_n} = \frac{0.665}{f\left(\frac{E_n}{1.5E_c}\right)} \tag{13.2.3}$$

We obtain also on the basis of formula (9.3.13) for the voltage of the cathode region U_c in the abnormal regime

$$\frac{U_c}{U_n} = \frac{E_c}{E_n} \frac{1}{1.5 f\left(\frac{E_n}{1.5E_c}\right)} \tag{13.2.4}$$

where U_n is the cathode voltage in the normal regime.

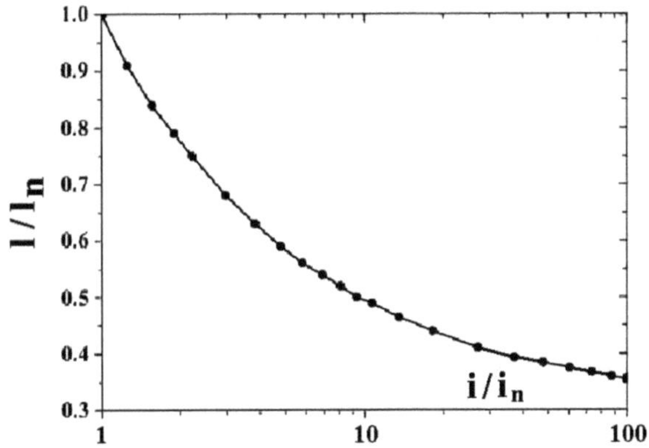

Fig. 13.5 Reduced thickness of the cathode region in argon for the abnormal regime according to formula (13.2.4)

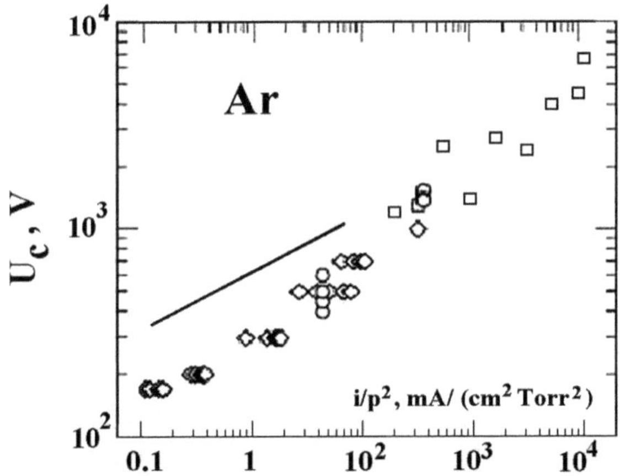

Fig. 13.6 Voltage-current characteristic of the cathode region of abnormal glow discharge in argon. Signs—experimental data [392], *solid curve*—formulas (13.2.3) and (13.2.4) for $\gamma = 0.1$

The peculiarities of these evaluations follow from the following assumptions. We assume that the electric current of gas discharge does not influence on the heat balance of the cathode region and its temperature is equal to $T = 300$ K. This may be violated at high electric currents. Next, we used the approximation (8.1.5) for the first Townsend coefficient on the basis of experimental data in accordance with data of Figs. 8.5 and 8.6. But these measurements are made for restricted values of the electric field strengths, at least, $E/N_a < 10^4$ Td.

Figures 13.6 and 13.7 contain parameters of the cathode region in the abnormal regime on the basis of experimental data which are collected in a review [392]. Note that these data include high reduced electric field strengths which correspond to high electron and ion energies. We give in Figs. 13.6 and 13.7 the results of evaluations for

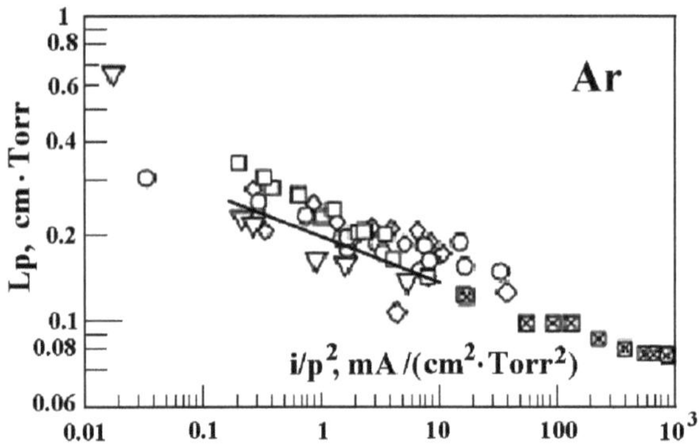

Fig. 13.7 Reduced thickness of the cathode region of abnormal glow discharge in argon as a function of the reduced current density. Signs—experimental data [392], *solid curve*—formulas (13.2.3) for $\gamma = 0.1$

restricted electric field strengths where the used theory holds true. Other mechanisms of ionization in the cathode region are possible at higher energies.

13.3 Transition Region in Glow Discharge

The transition region occurs between the cathode layer and positive column of normal glow discharge provides the transition from the cathode layer of normal glow discharge where a discharge electric currents occupies a part of the cross section and the positive column where a gas discharge plasma fills all the space inside a discharge tube. Hence the transition region which we call a dark region ensures expansion of the electric current at removal from the cathode (see Fig. 13.8).

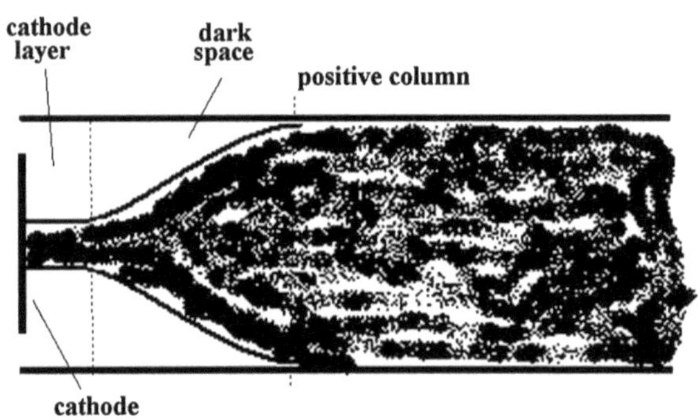

Fig. 13.8 Schematic distribution of the electric current over the cross section of glow discharge

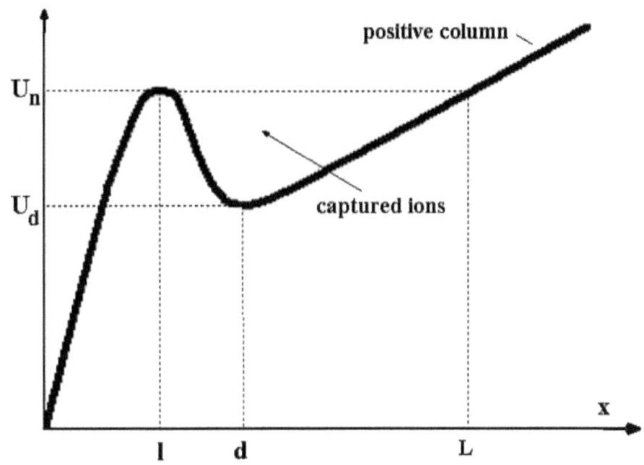

Fig. 13.9 Character of distribution of the electric voltage in the transition region between cathode layer and positive column

Defining the depth l of the cathode layer as $E(l) = 0$, we found at its boundary $N_i \gg N_e$ in the helium and argon case. Figs. 8.1 and 8.10 give the character of distribution of the electric field strength and electric voltage at the axis of a cylinder discharge tube, and the electric potential for ions in the transition region is given additionally in Fig. 13.9. It is important the voltage maximum at the boundary of the cathode region. This creates a partition for ions at the cathode layer boundary, so that ions cannot transfer between the cathode layer and dark region and and vice versa. This prevents from expansion the current region of the cathode layer.

Note that the potential well for ions exists only at the axis. There is a potential well in the cathode region in the radial direction, and this well prohibits for the cathode plasma to expand and shrink. On contrary, a radial potential well is absent in the dark region, and a dark region plasma propagates to the walls.

An electric current in the cathode region of normal glow discharge occupies a small part of the cathode area. If glow discharge is burnt in a long tube, the positive column takes usually all the tube length and the electric current in it propagates over all the tube cross section. Hence, there is an intermediate region between the cathode region and positive column, and different names are used for this region depending on a property under consideration. We below call this region as the dark region of glow discharge and will consider it from the standpoint of enlargement of the electric current as it is given schematically in Fig. 13.8.

In order to estimate a size of the dark region, we use three values of the length dimension, namely, l_n, the depth of the cathode layer, $1/\alpha$, a way where single ionization proceeds, and ρ_o, a tube radius. One can exclude from this list the value $1/\alpha$, since $\alpha(E_p)\rho_o \ll 1$ according to Fig. 12.4 and $E_d < E_p$, where E_p is the electric field strength in the positive column, and E_d is a typical electric field strength in a dark region. Usually $l_n \ll \rho_o$, and hence a size of the dark region is of the order of the tube radius, and electrons penetrate in the positive column from the cathode layer as a result of diffusion.

The character of ionization processes is of importance for formation of the transition region between the cathode layer and positive column in glow discharge. Using the relation (8.1.3) for the ionization balance, we assume the local character of this process where the first Townsend coefficient α is determined by the electric field strength at a point where the ionization process proceeds. This means that the energy distribution function of electrons is established at each point in accordance with the electric field strength under the local ionization equilibrium. In reality in accordance with (5.5.6) we have two relaxation time if electrons travel in a gas in a nonuniform electric field.

The electric field strengths drops in the cathode region as we remove from the cathode. Because rates of atom excitation and ionization by electron impact depend strongly on the electron energy, these processes in cathode layer regions with a small electric fields has a nonlocal character. In other words, electrons which excite an ionize atoms in these regions, go there from regions with a large electric field strengths. This is possible because of a high time of electron energy relaxation in atomic gases. The rate constants of relaxation of the electron momentum and energy in helium and argon are given in Table 7.5, as well as in Figs. 7.6 and 7.7. One can see that relaxation of the electron energy proceeds slower than relaxation of the electron momentum, and proceeds at a distance Λ from a point that is given by

$$\Lambda \sim \lambda k_P / k_\varepsilon \tag{13.3.1}$$

Because we assume here that the drift character of electron motion dominates, this corresponds to the criterion

$$\Lambda \gg \frac{D_L}{w} \tag{13.3.2}$$

13.4 The Wall Plasma Sheath for Positive Column in Helium and Argon

We now consider the transition layer between walls and the positive column of glow discharge in helium and argon. Because the thickness of this layer Δ is small compared to the wall curvature, this layer may be considered as flat. In the limiting case of a dense gas

$$\Delta \gg \lambda_i, \tag{13.4.1}$$

where $\lambda_i = (N_a \sigma_{ia}^*)^{-1}$ is the mean free path of ions in a gas, so that σ_{ia}^* is the diffusion cross section of ion-atom scattering. In particular, if atomic ions are present in the positive column of gas discharge, we represent the criterion (13.4.1) as

$$N_a \sigma_{res} \Delta \gg 1, \tag{13.4.2}$$

where σ_{res} is the cross section of resonant charge exchange.

The thickness of the plasma sheath layer Δ and the electric field strength at the wall surface E_o follows from (9.5.7) to (9.5.10) which give

$$\Delta = \left(\frac{a_i T_i \rho_o}{2.5\pi N_o e^2}\right)^{1/3}, \quad eE_o = (a_i T_i)^{2/3}\left(\frac{2.5\pi N_o e^2}{\rho_o}\right)^{1/3}, \tag{13.4.3}$$

where N_o is the number density of electrons and ions at the center of a cylinder discharge tube, ρ_o is a radius of this tube, parameters a_i and a_e follow from (9.5.10)

$$\frac{D_e(T_e)}{D_i(T_i)} = \frac{f(a_i)}{F(a_e)},$$

and $a_i T_i = a_e T_e$. We use expressions (9.5.10) for functions $f(a)$ and $F(a)$ which are given in Fig. 8.9 and their asymptotic expressions follow from formulas (9.5.11).

Applying these formulas to a glow gas discharge plasma in helium and argon, we take into account that the mean free path of electrons with respect to variation of their energy is large compared to the layer thickness Δ, and electron parameters in the double layer coincides with those in the positive column. We use the data [43] of Tables 7.3 and 7.4 for parameters of electron drift in helium and argon, and the electron mobility in the transversal direction is given by

$$K_e = \sqrt{\frac{\partial w_e}{\partial E}\frac{w_e}{E}}$$

where $w_e(E)$ is the electron drift velocity at an indicated electric field strength. This allows us to evaluate the voltage in the positive column of glow discharge in helium and argon between the center and walls of a cylinder discharge tube. One can see that this exceeds significantly the voltage drop near walls.

Assuming atomic ions to be the basic ion sort in helium and argon, we take into account that typical values of the reduced electric field strengths in the positive column do not change a thermal distribution function of ions. Therefore according to the Table 7.6 data we have for the reduced diffusion coefficients of ions $N_a D(He^+ - He) = 7.3 \times 10^{18}\,\text{cm}^{-1}s^{-1}$ and $N_a D(Ar^+ - Ar) = 1.1 \times 10^{18}\,\text{cm}^{-1}s^{-1}$. Figure 13.10 contains the double layer thickness Δ between the positive column of glow gas discharge as a function of the effective temperature, and these values are based on the data of Tables 7.3 and 7.4. According to formula (13.4.3) this quantity depends on the electron and ion number density N_o at the axis and a tube radius ρ_o as

$$\Delta \sim \left(\frac{\rho_o}{N_o}\right)^{1/3},$$

and according to Fig. 13.10 this value depends weakly on the effective electron temperature. Figure 13.11 contains values of the electric field strength E_o at walls of glow gas discharge as a function of the effective temperature obtained on the basis of

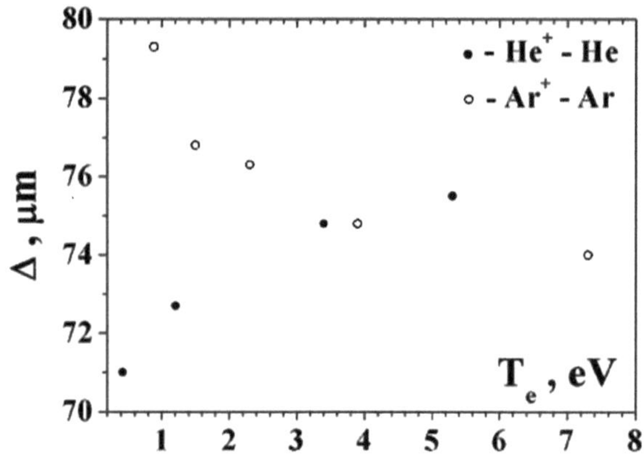

Fig. 13.10 Thickness of the double layer near walls of glow discharge in dense helium and argon at the number densities of electrons and ions $N_o = 10^{12}\,\mathrm{cm}^{-3}$ at the axis of a discharge tube of a radius $\rho_o = 1\,\mathrm{cm}$. The ion temperature $T_i = 300\,\mathrm{K}$ is equal to the gas and wall temperature

the data of Tables 7.3 and 7.4. According to formula (13.4.3) this quantity depends on the electron and ion number density N_o at the axis and a tube radius ρ_o as

$$E_o \sim \left(\frac{N_o}{\rho_o}\right)^{1/3},$$

and according to Fig. 13.11 this value depends weakly on the effective electron temperature. Figure 13.12 represents the voltage U_o of the double layer of glow gas discharge in helium and argon as a function of the effective electron temperature, and these values are based on the data of Tables 7.3 and 7.4. According to formula (13.4.3) this quantity is independent of both the electron number density N_o at the axis of a gas discharge tube and a tube radius ρ_o. Again we have a weak dependence of this quantity on the electric field strength in the positive column and the same order of its values for helium and argon. In addition, the voltage drop in this region of the tube cross section is small compared with the total voltage drop from the tube axis up to its walls in this regime of a high gas pressure.

The double layer thickness given in Fig. 13.10 allows us to determine the boundary of the regime of a high gas density in accordance with the criterion (13.4.2). Let us introduce the boundary gas pressure $p_b = N_b T$, so that according to criterion (13.4.2) the boundary number density of atoms $N_b = 1/(\Delta\sigma_{res})$ separates ranges of high and low densities of gases. Taking the gas temperature to be $T = 300\,\mathrm{K}$, we have that at an average collision energy about 0.1 eV the cross section of resonant charge exchange is $\sigma_{res}(He^+ - He) = 3.5 \times 10^{-15}\,\mathrm{cm}^2$ and $\sigma_{res}(Ar^+ - Ar) = 7.0 \times 10^{-15}\,\mathrm{cm}^2$ according to the data of Table 3.4. On the basis of these cross sections and the data of Fig. 13.10 we obtain for the boundary pressure $p_b \approx 1.2\,\mathrm{Torr}$ in the helium case and $p_b \approx 0.6\,\mathrm{Torr}$ in the argon case.

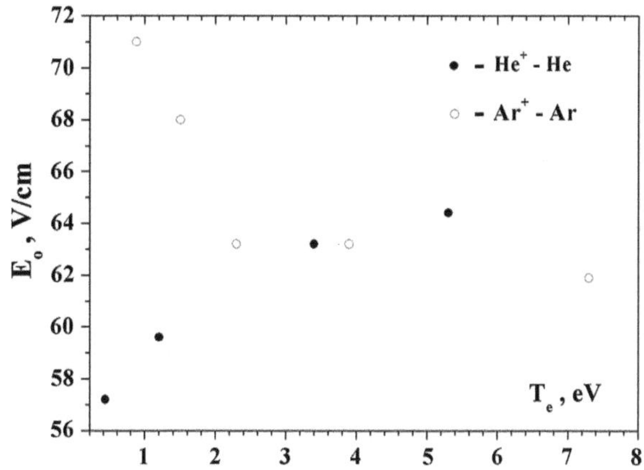

Fig. 13.11 Electric field strength of the double layer E_o between the positive column and walls for glow discharge in dense helium and argon in a discharge tube of a radius $\rho_o = 1$ cm and the number densities of electrons and ions $N_o = 10^{12}$ cm^{-3} at its axis. The ion temperature $T_i = 300$ K is equal to the gas and wall temperatures

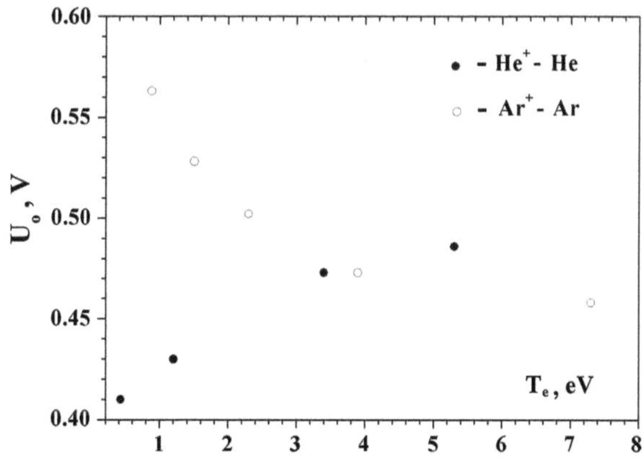

Fig. 13.12 Voltage drop of the double layer separated the positive column and walls of glow discharge in dense helium and argon at the number densities of electrons and ions $N_o = 10^{12}$ cm^{-3} at the axis of a discharge tube of a radius $\rho_o = 1$ cm. The ion temperature $T_i = 300$ K is equal to the gas and wall temperatures

Let us consider now the limit of low atom number densities where the double layer thickness is small compared to the ion mean free path. In this limit the voltage of the double layer becomes larger than that in the limit of a high gas density and exceeds the value T_e/e because just this electric potential determines the separation of charges in the plasma sheath region. If the number densities of electrons and ions are equal, the voltage drop in the plasma sheath region is determined by formula (9.5.14)

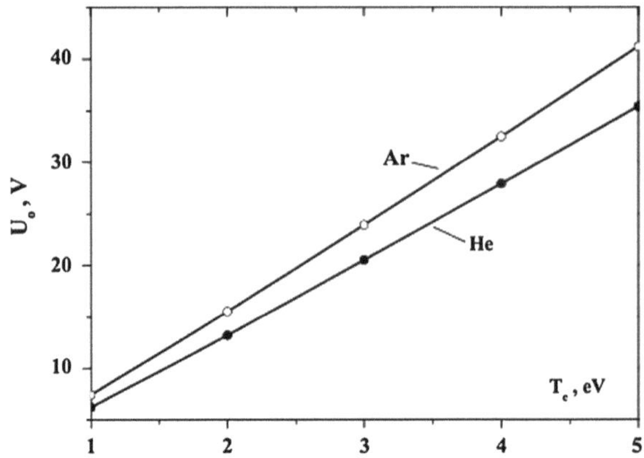

Fig. 13.13 Voltage between the tube center and walls for glow discharge in rareness helium and argon as a function of the electron temperature T_e at the center if the ion temperature $T_i = 300\,\mathrm{K}$ is equal to the gas and wall temperatures

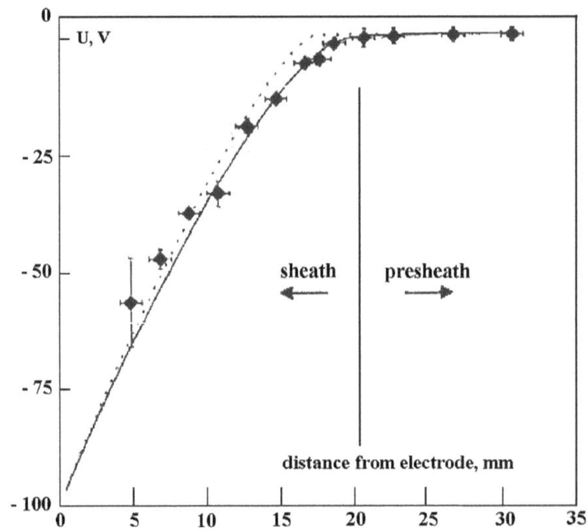

Fig. 13.14 Voltage drop of the plasma sheath for a gas discharge plasma located between two parallel electrodes [462]. Signs—experiment, *solid curve*—numerical solution of the Poisson equation (9.5.2), *dotted curve*—the Child model [380–383]

$$U_o = \frac{T_e}{2e} \ln \left(\frac{T_e m_i}{T_i m_e} \right)$$

in the regime of a high number density of electrons where the electron T_e and ion T_i temperatures may be introduced. Figure 13.14 represents the voltage drop of the plasma sheath region for this regime.

In the limit of a low gas density where the double layer thickness is small compared to the mean free path of ions. One can use the Child laws (9.3.6), (9.3.7) and (9.3.8)

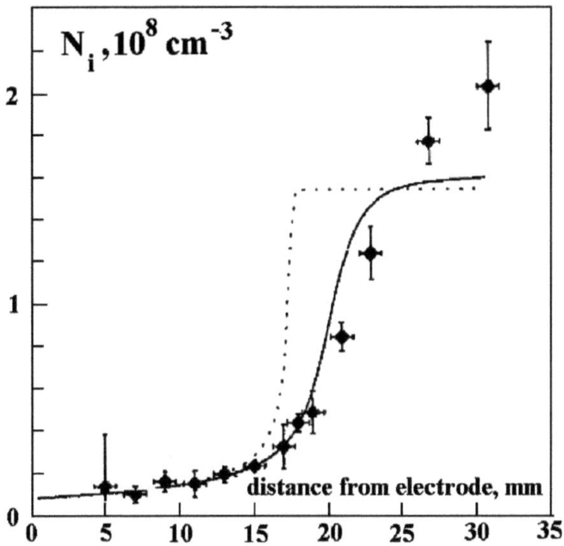

Fig. 13.15 The ion number density in the plasma sheath region for a gas discharge plasma located between two parallel electrodes [462]. Signs—experiment, *solid curve*—numerical solution of the Poisson equation (9.5.2), *dotted curve*—the Child model [380–383]

under the assumption that one can ignore the presence of electrons in the plasma sheath. In the argon case, the results of this model are compared with experimental data [462] in Figs. 13.14, 13.15 and 13.16. In this experiment the argon pressure is $0.05Pa$ (the number density of argon atoms at the temperature $T = 300\,\mathrm{K}$ is $1.3 \times 10^{13}\,\mathrm{cm}^{-3}$) that corresponds to the ion mean free path $\lambda \approx 11\,\mathrm{cm}$. A gas discharge plasma is located between two parallel electrodes, and outside the plasma sheath the electron temperature is $T_e = 0.53\,\mathrm{eV}$ and the number density of electrons and ions is $N_o = 9 \times 10^7\,\mathrm{cm}^{-3}$ that corresponds to the Debye radius $r_D = 0.57\,\mathrm{mm}$, while the sheath thickness ia approximately 2 cm. Figure 13.14 compares the potential distribution between plane electrodes according to experiment (closed circles), to numerical solution of (9.5.2) (solid curve) and its solution in neglecting the electron number density in the plasma sheath (dotted curve). The same notations are used in Fig. 13.15, where the ion number density is given in a gap, and in Fig. 13.16, where the ion velocity in the plasma sheath is represented.

It should be noted that strong separation of the regions of the positive column and plasma sheath near walls in gas discharge that is burnt in a cylinder discharge tube takes place in the limit of a high gas density because there is a simultaneous separation of the region of ionization equilibrium and the near-wall region that corresponds to criterion (13.4.1)

$$\lambda_i \ll (r_D^2 \rho_o)^{1/3}, \qquad (13.4.4)$$

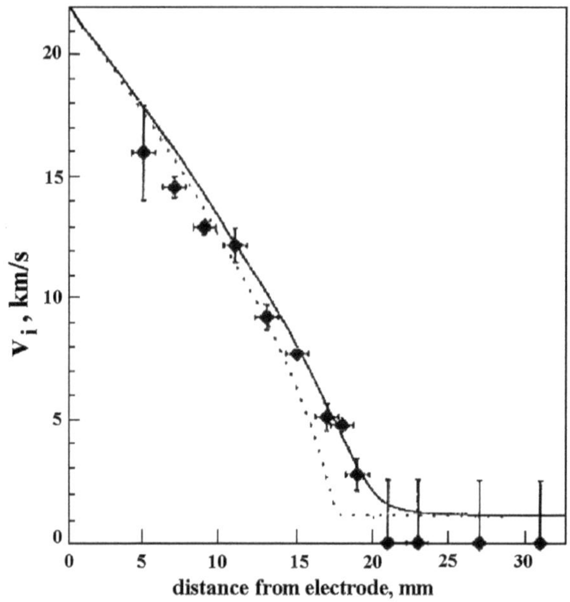

Fig. 13.16 Ion velocity in the plasma sheath region for gas discharge between two parallel electrodes [462]. Signs—experiment, *solid curve*—numerical solution of the Poisson equation (9.5.2), *dotted curve*—the Child model ([380–383]

where the Debye-Hückel radius r_D corresponds to a gas discharge plasma in the region of ionization equilibrium. In the case of a rareness gas the regions of ionization equilibrium and near-wall layer are not separated as it takes place in arc of low pressure with criterion (8.4.10). Therefore the plasma sheath layer exists in a plane geometry where its properties may be connected with the character of plasma generation.

13.5 Magnetron Plasma in Helium and Argon

We now consider various aspects of a helium and argon gas discharge plasma in magnetron discharge. A general scheme of magnetron discharge is given in Fig. 9.10 and we determine parameters of each region of this gas discharge in helium and argon. Figure 13.17 indicates principal regions of magnetron discharge. Since electrons are magnetized in the region of a magnetic trap, electrons do not penetrate in the cathode region, so that ions bombard the cathode and generate secondary electrons with some γ probability which depends on the ion energy. When the secondary electron passes the cathode layer, it acquires an energy eU_o, where U_o is the cathode voltage. We first determine a number of electrons resulted from one fast secondary electron emitted from the cathode.

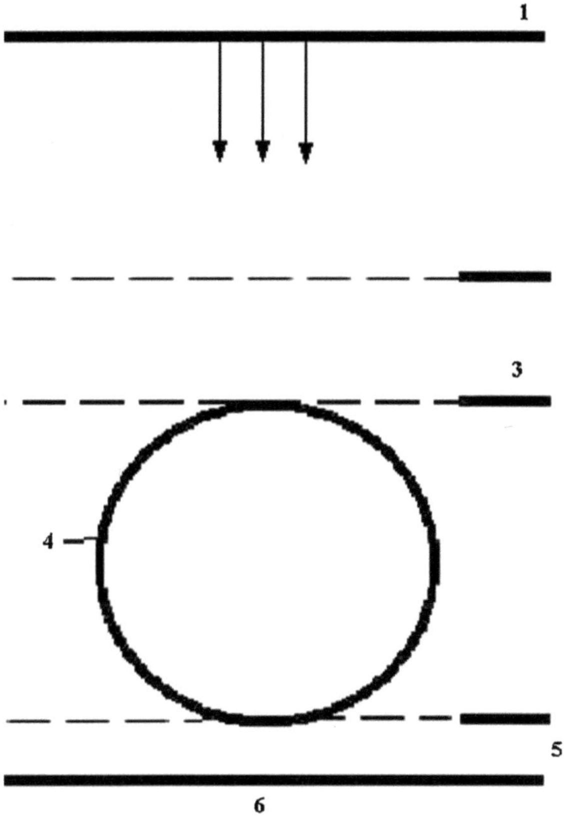

Fig. 13.17 Radial cross section of magnetron gas discharge. *1* anode, *2* ion current, *3* the region with a steady plasma, *4* magnetic trap, *5* cathode layer, *6* cathode

Let us use the Thomson formula (3.6.4) according to which the average energy ϵ consumed per a released electron in the ionization process with ε—the energy of an incident electron is given by

$$\epsilon = \frac{\varepsilon J}{\varepsilon - J} \cdot \ln \frac{\varepsilon}{J} \, ,$$

where J is the atom ionization potential. In the limit $\varepsilon \gg J$ the number n of released electrons is large and is determined by

$$n = \int_J^{U_o} \frac{d\varepsilon}{\epsilon} = \int_1^{U_o/J} \frac{dx(x-1)}{x \ln x} = \int_0^{\ln U_o/J} \frac{dt(e^t - 1)}{t},$$

where $x = \varepsilon / J$ and $t = \ln x$.

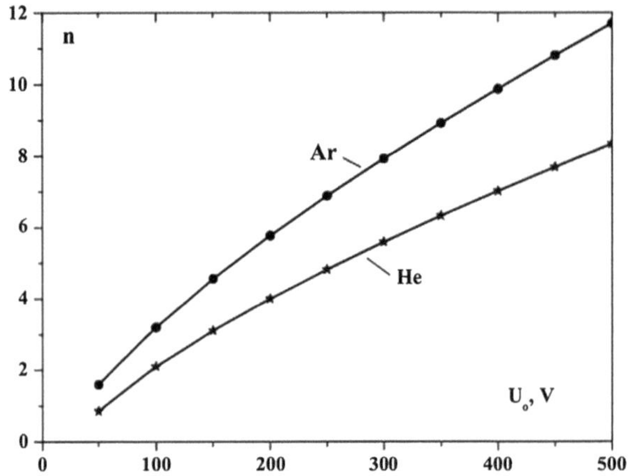

Fig. 13.18 Average number of electrons results from collision of an electron of energy eU_o with helium and argon atoms

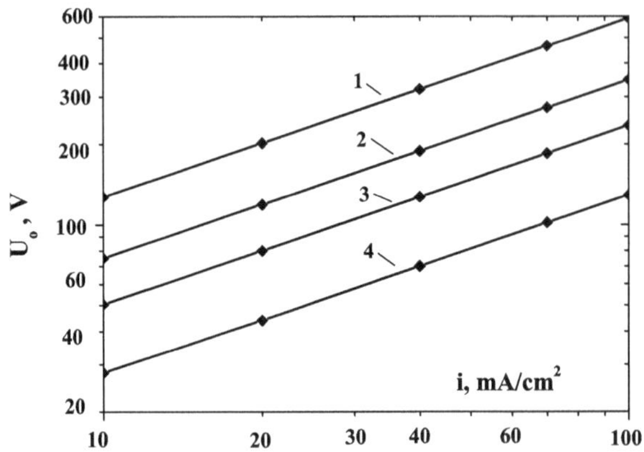

Fig. 13.19 Cathode voltage U_o depending on the current density i in magnetron discharge in argon ($1 - L = 0.4$ mm, $3 - L = 0.2$ mm) and in helium ($2 - L = 0.4$ mm, $4 - L = 0.2$ mm)

Let us apply this formula for ionization of helium and argon atoms to magnetron discharge. Taking $J = 24.59$ eV for the ionization potential of the helium atom and $J = 15.76$ eV for that of the argon atom, we find on the basis of the above formula the average number of released electrons n depending on the initial energy eU_o of a secondary electron after passage the cathode region. This dependence is represented in Fig. 13.18. The criterion of self-maintaining of magnetron discharge is given by formula (9.6.3) $n = 1/\gamma - 1$, where γ is the probability of formation of a secondary electron as a result of cathode bombardment by ions with energy eU_o.

Another aspect of self-maintaining of magnetron discharge is connected with the character of screening of the discharge voltage in the cathode region where electrons are absent. We use for this goal the Child law (9.3.6), (9.3.8) and (9.3.22). It should

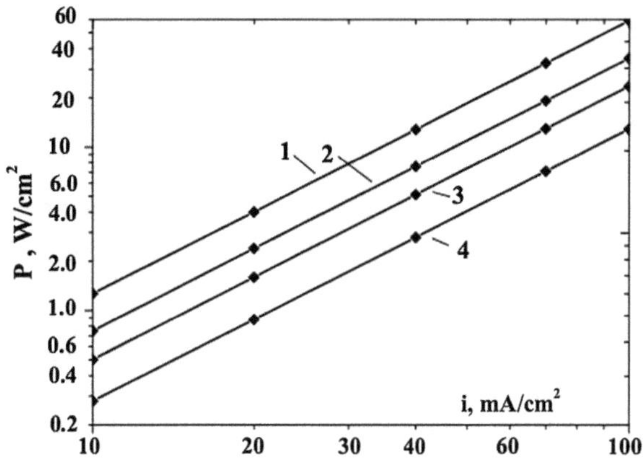

Fig. 13.20 Power P released per cathode unit area as a function of the current density i for magnetron discharge in argon ($1 - L = 0.4\,\mathrm{mm}$, $3 - L = 0.2\,\mathrm{mm}$) and in helium ($2 - L = 0.4\,\mathrm{mm}$, $4 - L = 0.2\,\mathrm{mm}$)

be noted that a gap between the magnetic trap and cathode of magnetron discharge is given by the discharge geometry, and it is profitable to place the magnetic trap so close to the cathode as it is possible. The connection between the cathode voltage U_o and the current density i is given now by formula (9.3.7)

$$U_o = \left(9\pi i \sqrt{\frac{M}{2e}}\right)^{2/3} L^{4/3},$$

where M is the ion mass. Fig. 13.19 gives the dependence of the cathode voltage on the electric current density in accordance with this formula, and Fig. 13.20 represents the reduced power at the cathode $P = iU_o$ that follows from these data. It should be noted that the cathode layer voltage is the basic part of voltage for magnetron discharge. Moreover, according to an empirical formula [413], the cathode layer voltage is approximately $0.73U$, where U is the total voltage of magnetron discharge in a cylinder chamber.

We also represent the criterion for magnetron discharge that requires that the thickness of the cathode layer L is small compared with the mean free path of ions λ_i. Let us introduce the probability ζ to survive of ions

$$\zeta = \exp(-\xi), \ \xi = \int N_a \sigma_{res}(\varepsilon) dx,$$

where the integral is taken over the cathode layer depth. For definiteness, we take the cathode temperature $T = 1000\,\mathrm{K}$. Using the cross sections of resonant charge exchange [230, 231] $\sigma_{res}(He^+ - He) = 35\,\text{Å}$ at the collision energy $\varepsilon = 0.1\,\mathrm{eV}$

and $\sigma_{res}(Ar^+ - Ar) = 70$ Å at this collision energy and the dependence of these cross sections on the collision energy, we obtain

$$\xi \approx 0.9 N_a \sigma_{res}(\varepsilon_o) L,$$

where L is the cathode layer thickness, $\varepsilon_o \approx 0.1\,\text{eV}$. Assuming that magnetron discharge exists if the probability of ion surviving is $\xi \leq 0.9$, we obtain from this that magnetron discharge is realized in helium at the pressure $p < 0.2\,\text{Torr}(20\,\text{Pa})$ and in argon at $p < 0.1\,\text{Torr}\ (10\,\text{Pa})$.

In considering the magnetic trap region of magnetron discharge in helium and argon, we take for definiteness the maximum magnetic field strength to be $H = 140\,\text{Gs}$ that can be obtained on the basis of simple magnetic materials, and the gas pressure to be $p = 10\,\text{Pa}$ that does not exceed the maximum pressure provided the existence of magnetron discharge. In addition, we will be guided by the gas temperature in the region of the magnetic trap to be $T = 500\,\text{K}$ (i.e. the number density of atoms is $N_a = 1.6 \times 10^{15}\,\text{cm}^{-3}$) and a typical electron temperature in this region to be $T_e = 3\,\text{eV}$. These parameters more or less correspond to real parameters of magnetron discharge. The Larmor frequency at this magnetic field strength is $\omega_H = 2.5 \times 10^9\,\text{s}^{-1}$, and in the helium case the rate of electron-atom collision is $\nu = N_a v_T \sigma_{ea}^* = 1 \times 10^8\,\text{s}^{-1}$, $v_T = \sqrt{8T_e/(\pi m_e)} = 1.2 \times 10^8\,\text{cm/s}$ is the average electron velocity at a given electron temperature $T_e = 3\,\text{eV}$, and $\sigma_{ea}^* = 6\text{Å}^2$ is the diffusion cross section for electron scattering on the helium atom. As is seen, the criterion (7.9.4) of a magnetized plasma

$$\omega_H \gg \nu,$$

holds true. Thus, electrons are magnetized in the magnetic trap region. Electrons remain to be magnetized also at the magnetic trap periphery, because the relation $\omega_H = \nu$ corresponds to the magnetic field strength $H = 8\,\text{Gs}$, i.e. in the region of the magnetic trap and near it where $H > 8\,\text{Gs}$ plasma electrons are magnetized.

Let us consider the validity of the criterion (7.9.4) for atomic ions in a helium and argon magnetron plasma under the above conditions. The Larmor frequencies are equal for helium and argon atomic ions correspondingly at the magnetic field strength $H = 140\,\text{Gs}$

$$\omega_{He} = \frac{eH}{M_{He}c} = 3.4 \cdot 10^5\,\text{s}^{-1}, \quad \omega_{Ar} = \frac{eH}{M_{Ar}c} = 3.4 \cdot 10^4\,\text{s}^{-1},$$

where M_{He} is the mass of the helium atom, M_{Ar} is the mass of the argon atom. The rate of ion-atom collisions is given by

$$\nu_{ia} = N_a v_T \sigma_{ia}^* = N_a \sqrt{\frac{8T}{\pi M}} \cdot 2\sigma_{res},$$

where $N_a = 1.6 \times 10^{15}\,\mathrm{cm}^{-3}$ is the number density of atoms, M is the atom mass, T is the gas temperature, σ^* is the cross section ion-atom scattering, σ_{res} is the cross section of resonant charge exchange, and since the cross section of resonant charge exchange exceeds the elastic ion-atom cross section, we have [337] $\sigma^* = 2\sigma_{res}$. Since $N_a = 2.6 \times 10^{15}\,\mathrm{cm}^{-3}$, and in the helium case we have a thermal velocity of ions $v_T = 1.6 \times 10^5\,\mathrm{cm/s}$, and the cross section of resonant charge exchange is $\sigma_{res} = 35\text{Å}^2$. This gives for the rate of ion-atom collisions in the helium case under the above conditions $\nu_{ia} = 3 \times 10^6\,s^{-1}$. In the same manner we obtain in the argon case $v_T = 5 \times 10^4\,\mathrm{cm/s}$, $\sigma_{res} = 70\text{Å}^2$, that gives in the argon case $\nu_{ia} = 1.9 \times 10^6\,\mathrm{s}^{-1}$. Comparing these rates with the Larmor frequencies, we find that ions in the helium plasma are magnetized under given conditions, if $H \gg 1200\,\mathrm{Gs}$. The same criterion in the case of the argon plasma gives $H \gg 8 \times 10^3\,\mathrm{Gs}$. As is seen, ions are not magnetized in the magnetron plasma under the above conditions if stationary magnets are used.

Note that a size of the magnetic trap must exceed the Larmor radius of electrons. In particular, for a secondary electron of energy $\varepsilon = 200\,\mathrm{eV}$ that corresponds to the electron velocity of $v_e = 8.4 \times 10^8\,\mathrm{cm/s}$, the Larmor radius is $r_L = 0.3\,\mathrm{cm}$. If it is small compared to a size of the magnetic trap region, electrons penetrated in the region of the magnetic trap are reflected from the magnetic trap and delay in this region. In this case a strong interaction between a secondary electron with a magnetic trap leads to electron delay in the magnetic region, and ionization owing to secondary electrons proceeds just in this region. The Larmor radius for thermal electrons is lower than that for secondary electrons. Hence a magnetic trap dimension $\sim 1\,\mathrm{cm}$ is suitable for action on electrons.

We have for the electron temperature T_e of a magnetized plasma located in perpendicular electric and magnetic fields according to formula (7.9.10)

$$T_e - T = \frac{M v_\tau^2}{3} = \frac{M c^2 E^2}{3H^2}, \quad \omega_H \gg \nu$$

where M is the atom mass, c is the light velocity, E, H are the electric and magnetic field strengths respectively. Assuming the electron temperature to be $T_e = 3\,\mathrm{eV}$, so that $T_e \gg T$, and the magnetic field strength to be $H = 140\,\mathrm{Gs}$, we obtain from this for the electric field strength $E = 2.1\,\mathrm{V/cm}$ that provides these parameters of a magnetized plasma in helium and $E = 0.65\,\mathrm{V/cm}$ in argon. Correspondingly, the electron drift velocity in crossed electric and magnetic fields under these conditions is $v_\tau = 1.5 \times 10^6\,\mathrm{cm/s}$ in the helium case and $v_\tau = 4.7 \times 10^5\,\mathrm{cm/s}$ in the argon case (Fig. 13.21).

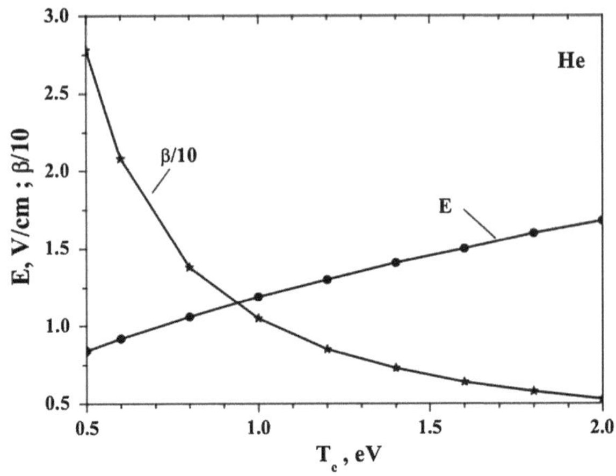

Fig. 13.21 Parameters of a helium magnetron plasma in the magnetic trap. T_e is the electron temperature in accordance with formula (7.9.10), β is the ratio of the electron number density in the magnetic trap to that outside trap according to formulas (13.5.1) and (13.5.2). The magnetic field strength is $H = 140\,\mathrm{Gs}$, the magnetic trap radius is $r = 4\,\mathrm{cm}$

Let us consider the behavior of a magnetron plasma in the magnetic trap. Being guided by experimental conditions, we take the circle radius with the maximum magnetic field to be $r = 4\,\mathrm{cm}$ that according to formula (9.6.1) gives the potential depth of the magnetic trap $U_{max} = eEr/2 = 4.2\,\mathrm{eV}$ in the helium case and $U_{max} = 1.3\,\mathrm{eV}$ in the argon case. From this it follows

$$\chi = \frac{U_{max}}{T_e} = \frac{3erH^2}{2Mc^2E} \qquad (13.5.1)$$

This parameter is equal $\chi = 1.4$ in the helium case and $\chi = 0.44$ in the argon case. Let us assume that the space distribution of electrons is established quickly. Then the ratio of the number density of electrons inside the magnetic trap N_{max} and far from it N_o is given by

$$\beta = \frac{N_{max}}{N_o} = \exp\left(\frac{U_{max}}{T_e}\right), \qquad (13.5.2)$$

and under the above conditions these values are equal to 4.0 and 1.6 in the helium and argon cases correspondingly. The parameter β characterizes an increase of the electron number density in magnetic trap. Because this plasma is quasineutral, the number density of ions increases in the same manner. Ions are not magnetized in the magnetic trap, and hence the ion current from the magnetic trap region to the cathode exceeds in β times the ion current from regions with a low magnetic field if we assume the electric field to be uniform in a space.

Figure 13.21 gives the dependencies of the electric field strength E and an increase of the electron number density in the magnetic trap β on the electron temperature. One can expect that large electric field strengths are more favorable for the magnetic trap action, while according Fig. 13.21 the magnetic trap acts at not high electric fields at which the parameter β characterized electron capture in the magnetic trap becomes small. Evidently, large values of β cause an instability of the magnetic trap, and its optimal values are $\beta \sim 10$ that are used in Fig. 13.21. In order to transfer from helium to argon, it is convenient to rewrite formula (13.5.2) in the form

$$\beta = \frac{N_{max}}{N_o} = \exp\left(\frac{U_{max}}{T_e}\right) = \exp\left(\frac{E_o}{E}\right), \quad E_o = \frac{3eH^2 r}{2Mc^2} \tag{13.5.3}$$

One can see that $E_o \sim M$, and for parameters under consideration ($H = 140Gs,\ r = 4\,\text{cm}$) we have $E_o = 2.8\,\text{V/cm}$ in the helium case and $E_o = 0.28\,\text{V/cm}$ in the argon case. Correspondingly, according to formula (7.9.10) we have $T_e(E_o) = 5.5\,\text{eV}$ in the helium case and $T_e(E_o) = 0.55\,\text{eV}$ in the argon case.

One can connect the current parameters for the cathode layer and the magnetic trap region. We require the current lines near the cathode to be straightforward lines, i.e. the current density i does not vary in transition from the magnetic trap to the cathode layer. Let us determine the conductivity Σ of a plasma inside the magnetic trap that is the sum of the electron Σ_e and ion Σ_i conductivities which ratio is proportional to the ratio of the electron w_e drift velocity in the direction of the electric field and the ion drift velocity. The electron velocity along the electric field in a plasma with magnetized electrons according to formula (7.9.8) is given by

$$w_e = \frac{4eE}{3m_e\lambda_e\omega_H^2}v_T, \quad v_T = \sqrt{\frac{8T_e}{\pi m_e}}$$

This formula relates to the helium case where the cross section of electron-atom scattering is independent of the collision velocity.

In determination the electron drift velocity for magnetron discharge in argon, we take for definiteness in the center of the magnetic trap $E = 0.9\,\text{V/cm}$ that corresponds to the Fig. 13.19 data $T_e = 0.6\,\text{eV}$ and the ratio of the plasma density in the magnetic trap and far from it $\beta = 21$. Next, $v_T = 4.6 \times 10^7\,\text{cm/s}$, $\omega_H = 2.5 \times 10^9 s^{-1}$, $\lambda_e \approx 1\,\text{cm}$ (the number density of helium atoms is $1.6 \times 10^{15}\,\text{cm}^{-3}$ at the temperature $T = 500\,\text{K}$ and pressure $p = 10\,\text{Pa}$), and according to the above formula the electron drift velocity along the electric field is $w_e = 1.4 \times 10^4\,\text{cm/s}$. Using the above parameters of the helium gas discharge plasma in the magnetic trap, we have for ions $eE\lambda_i/T = 4$, and the ion drift velocity according to Fig. 7.12 data is $w_i = 2.1 \times 10^5\,\text{cm/s}$. Though the above operations has the character of estimations rather than evaluations, they exhibit that the plasma conductivity is determined by ions. We note also that the electron mean free path in this gas discharge plasma $\lambda_e \approx 1\,\text{cm}$ is larger than the mean free path of ions $\lambda_i = 1/(N_a\sigma_{res}) \approx 0.2\,\text{cm}$.

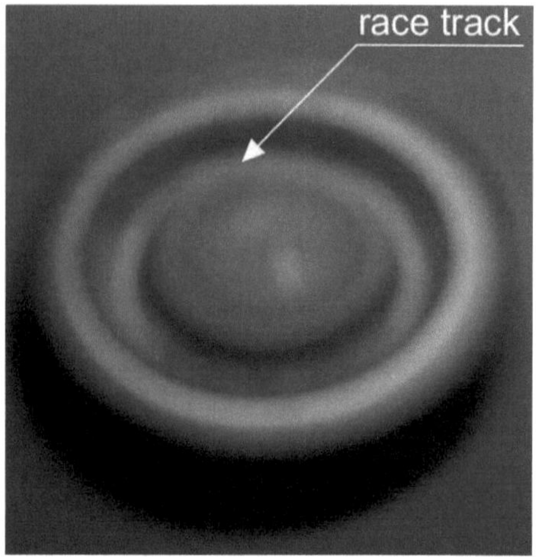

Fig. 13.22 Track on the cathode surface formed as a result of cathode erosion under the action of ion current on the cathode [409]

If the conductivity of a magnetron plasma is determined by ions and a magnetic trap is approached to the cathode, the electric current density in the region of the magnetic trap center exceeds significantly that for the cathode region far from the magnetic trap center. One more action of the magnetron current is cathode sputtering. As a result, a track is formed on the cathode as it is shown in Fig. 13.22. Of course, track formation changes characteristics of magnetron discharge as well as the space distribution of the electric current to the cathode.

Let us take the distribution of the density current $i(\rho)$ to the cathode as

$$i(\rho) = i_o \exp\left[-\frac{\beta(\rho - r)^2}{T_e}\right] \tag{13.5.4}$$

where i_o is the maximum current density that passes through the magnetic field maximum $\rho - r = 0$, where r is the radius of the magnetic trap. As is seen, the electric current of magnetron discharge is concentrated mostly near the magnetic field maximum. From this the connection between the total current I and the current density i at the center of the magnetic trap takes the form

$$I = 2\pi^{3/2} r \beta^{-1/2} i,$$

In particular, for conditions under consideration ($p = 10\,\mathrm{Pa}$, $r = 4\,\mathrm{cm}$) it follows from Fig. 13.22 $\beta = 2\,\mathrm{cm}^{-2}$. Then under optimal parameters of magnetron discharge with the current $I = 0.5\,\mathrm{A}$ we obtain for the electric current density $i = 16\,\mathrm{mA/cm}^2$ if this current passes through the magnetic trap center. Taking the cathode temperature $T = 1000\,\mathrm{K}$, we have the number density of argon atoms near the cathode $N_a = 7.5 \times 10^{14}\,\mathrm{cm}^{-3}$, and the mean free path for argon ions is equal $\lambda_i = 1/(N_a \sigma_{res}) \approx 3\,\mathrm{mm}$. Under these conditions the dimension of the cathode region where ions are accelerated, is comparable with the mean free path of ions in the cathode region.

Note that though the above evaluations relate to the helium case under indicated parameters of magnetron discharge, general conclusions about the role of ion charge transport are valid for other parameters in a helium plasma and also for magnetron discharge in argon. From this one can find the number density of electrons and ions in a magnetron helium plasma under consideration according to the relation

$$N_e = \frac{i}{ew_i}$$

This gives the number density of electrons and ions at the magnetic trap center is $N_e = 4.5 \times 10^{11}\,\mathrm{cm}^{-3}$, and far from it is equal $N_e = 2.1 \times 10^{10}\,\mathrm{cm}^{-3}$.

Let us connect the magnetic trap region of magnetron discharge with the region of steady discharge where magnetic field is absent and the plasma conductivity is created by electrons. In this consideration we are based on the continuity of the electric current density and the straightforward form of electric current lines. The electron drift velocity in this region is equal $w_e = \beta w_i \approx 4 \times 10^6\,\mathrm{cm/s}$. According to Fig. 5.2 this corresponds to the reduced electric field strength $E/N_a \approx 20\,\mathrm{Td}$, or if the helium pressure is $p = 10\,\mathrm{Pa}$ and its temperature is $T = 400\,\mathrm{K}$, this corresponds to the electric field strength $E = 0.4\,\mathrm{V/cm}$. Thus, the electric field strength in regions of the magnetic trap and steady form of discharge are comparable. This is valid both for magnetron discharge in helium and argon.

We now determine parameters of the magnetron plasma in the anode region. In accordance with the scheme of magnetron discharge represented in Fig. 13.17, electrons move to the anode and attach to it, and the electron reproduction takes place in the anode region. Hence each electron propagated through the anode region creates one electron-ion pair, and the condition of ion reproduction in the anode region has the following form

$$\int_0^L \alpha dx = 1,$$

instead of (8.1.3). Here α is the first Townsend coefficient, and L is a size of the anode region. Take for definiteness the helium pressure $p = 10\,\mathrm{Pa}$ and the temperature $T = 400\,\mathrm{K}$ in the anode region that corresponds to number density of atoms $N_a = 1.8 \times 10^{15}\,\mathrm{cm}^{-3}$. We use the approximation (8.1.5) for the first Townsend coefficients with parameters $A = 2.5\,\text{Å}^2$ and $B = 1,100\,\mathrm{Td}$ for large reduced electric field strength $E/N_a > 10^3\,\mathrm{Td}$. Denoting by E_a the electric field strength in the anode region, we have now $E_a/N_a = 1550Td$ on the basis of formula (9.3.24) that corresponds to $E_a = 13\,\mathrm{V/cm}$. Next, from formula (9.3.25) we have $AN_aL = 3.06\ln 2$ that gives $L = 0.5\,\mathrm{cm}$. From formula (9.3.23) we have for the anode voltage $U_a = 2E_aL/3 = 4\,\mathrm{V}$. We note that used formulas correspond to the criterion $\lambda_e \ll L$, and in this case $\lambda_e \approx 1\,\mathrm{cm}$, i.e. this criterion is not fulfilled. This

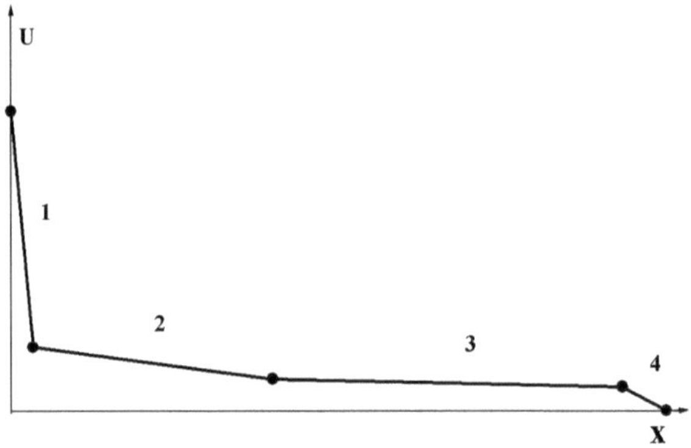

Fig. 13.23 Space distribution of the electric potential in magnetron discharge : *1* cathode layer, *2* magnetic trap region, *3* region of steady discharge, *4* anode layer region

leads to a higher length of the anode region L and larger voltage U_a. Figure 13.23 gives the schematic space distribution of voltage for magnetron discharge. Note that though our estimations correspond to the helium case under certain parameters, a general character of properties of magnetron discharge are valid also in the argon case.

Chapter 14
Principles of Gas Discharge Plasma

Abstract Principal properties of a gas discharge plasma and models for their description are the object of this analysis. It is shown that the tau-approximation that is a basis of the gasdynamical plasma model leads to an error in the evaluation of plasma parameters. Self-consistent character of atom excitation in a gas discharge plasma is of importance for its analysis. Many regimes of a gas discharge plasma may be realized depending on plasma parameters and character of plasma processes.

14.1 Gas Discharge Plasma as Complex Physical Object

In summing up the results of the above analysis, one can conclude that a gas discharge plasma is a complex physical object, and we repeat briefly its peculiarities. The nature of this object is connected with the character of energy injection in a gas by energy transfer from an electric field to gas atoms through electrons. Therefore a gas discharge plasma is an nonequilibrium system which properties are determined by some processes, and the behavior of electrons is described by electron energy distribution function (EEDF). In addition, because gas discharge is supported by an electric voltage between two electrodes, a space plasma distribution corresponds to the minimum of this voltage.

Since a gas discharge plasma consists, at least, of three components, atoms (molecules), electrons and ions, and processes involving each of these components are responsible for certain plasma properties, a gas discharge plasma is characterized by a large number of regimes which differ from each other in principle. Each regime requires a specific description, and hence we have no an universal description of a gas discharge plasma. To decrease the number of these regimes, we restrict ourselves by helium and argon gas discharge plasmas, but even in this case the regimes are manifold.

Ionization processes are important for properties and evolution of a gas discharge plasma, and ionization equilibrium is supported basically by electron-atom collisions. Usually the average electron energy in a gas discharge plasma is small compared to the atom ionization potential or the atom excitation energy, and subsequently the rate of atom ionization in a gas discharge plasma is less than the rate of elastic

B.M. Smirnov, *Theory of Gas Discharge Plasma*, Springer Series on Atomic, Optical, and Plasma Physics 84, DOI 10.1007/978-3-319-11065-3_14

electron-atom collisions. Therefore processes of atom excitation and ionization in a gas discharge plasma are determined by fast electrons, i.e. by a tail of the EEDF that includes a small part of electrons. From this it follows that processes of atom excitation and ionization by electron impact in a gas discharge plasma has a self-consistent character. Indeed, these processes lead to a decrease of fast electrons that in turn causes a decrease of the rate of these processes. Along with the self-consistency of excitation and ionization processes, from this it follows the role of excited atoms in a gas discharge plasma. Excited atoms facilitate ionization processes, and in some regimes the ionization process has a stepwise character, which first stage is formation of excited atoms and the second stage is ionization of excited atoms.

The complexity of a gas discharge plasma complicates its analysis. To simplify this analysis, simple models are used which allow one to change the kinetic character of processes in a gas discharge plasma by hydrodynamic one or apply to a gas discharge plasma a thermodynamic description. Being guided as early by the helium and argon cases, we check the accuracy of such models on simple examples.

Let us consider first the validity of the hydrodynamic model for drift of electrons and ions in gases in an external electric field. Within the framework of this model the friction force F is introduced for motion of electrons in a gas that is independent of an electron velocity. Equation for the average electron velocity w_e, that is the electron drift velocity, is analogous to the equation of motion for electrons

$$m_e \frac{dw_e}{dt} = F, \; F = \frac{m_e w_e}{\tau}, \tag{14.1.1}$$

where m_e is the electron mass, and the rate of electron-atom collisions is $1/\tau = N_a k_{el}$, so that N_a is the number density of gas atoms, k_{el} is the rate constant of elastic collisions of a test electron with gas atoms, and it assumes to be independent of the electron velocity. In this approximation we describe the friction force by one parameter τ, and this corresponds to tau-approximation (5.1.3) that gives usually a qualitative description of drift of atomic particles in a gas.

Let us use this approximation for electron drift in gases under the action of an external electric field. In the stationary case solution of the motion equation (14.1.1) has the form

$$w_e = \frac{eE}{m_e N_a k_e} = cx, \; x = \frac{eE}{N_a}, \; c = \frac{e}{m_e k_{el}} \tag{14.1.2}$$

According to this formula, the electron drift velocity is proportional to the reduced electric field strength E/N_a. If we take the rate constant of elastic electron-atom scattering at the electron energy $\varepsilon = 1\,\mathrm{eV}$, that is $k_{el} = v_e \sigma^*_{ea} = 4.1 \times 10^{-8}\,\mathrm{cm}^3/\mathrm{s}$ at the collision energy of 1 eV for helium and $k_{el} = 8 \times 10^{-9}\,\mathrm{cm}^3/\mathrm{s}$ for argon if we take the cross sections of electron-atom collisions taken from Table 3.1. This gives for the proportionality coefficient in formula (14.1.2) $c = 4 \times 10^5\,\mathrm{cm/s}$ in the helium case and $c = 2 \times 10^6\,\mathrm{cm/s}$ in the argon case. Comparison of formula (14.1.2) with experimental results is given in Fig. 14.1 in the helium case and in Fig. 14.2 in the

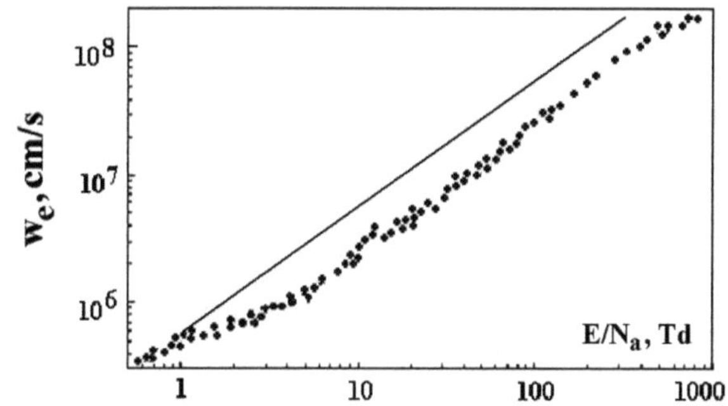

Fig. 14.1 Electron drift velocity in helium as a function of the electric field strength in the regime of a low electron number density. Signs–experimental data [43], a *solid line*–formula (14.1.2)

argon case. The experimental data are taken from review [43] in the helium case and from old Townsend and Bailey measurements [7] of the electron drift velocity in argon that accord to contemporary data.

Comparison of the hydrodynamic approach (14.1.2) with experimental results in Figs. 14.1 and 14.2 shows that this approach corresponds more or less to experimental data in the helium case and differs from this in principle in the argon case. But it is necessary to note the principal lack of this consideration. The proportionality of the electron drift velocity to the electric field strength takes place at low electric field strengths [formula (2.3.2)] where an external electric field does not perturb the energy electron distribution and at large electric fields [formula (6.6.6)] where electron acceleration in an external electric field is limited by inelastic processes. But the later does not relate to hydrodynamic approach in principle where the friction force is determined by elastic electron-atom collisions. At lower electric field strengths,

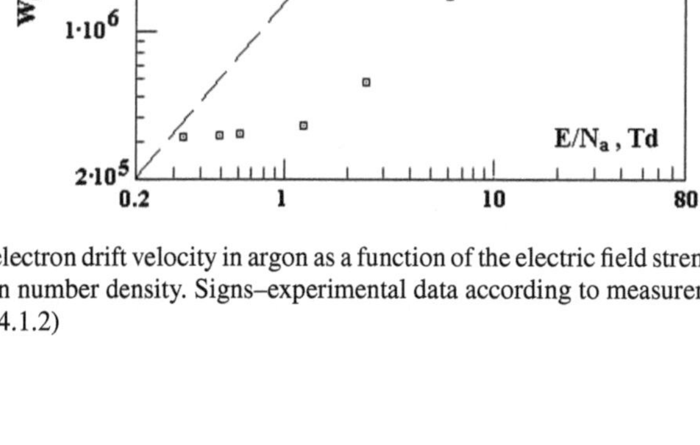

Fig. 14.2 The electron drift velocity in argon as a function of the electric field strength in the regime of a low electron number density. Signs–experimental data according to measurements [7], a *dash line*–formula (14.1.2)

where the average electron energy exceeds significantly a thermal energy of gas atoms and the electron drift velocity is determined by elastic electron-atom collisions, the proportionality of the electron drift velocity to the electric field strength is not fulfilled in helium and argon cases.

Thus, from this comparison we obtain that the kinetic analysis of the electron behavior in a gas located in a strong electric field gives additional aspects of this behavior which cannot be taken into account in the hydrodynamic or tau-approximation. Hence, though the hydrodynamic approximation may be used for the qualitative analysis of a gas discharge plasma, this approach is valid only at low electric field strengths (2.3.4) where an external field does not influence on distributions in this plasma. As is seen, the kinetic consideration gives additional aspects of a gas discharge plasma compared with the hydrodynamic one.

Usually in consideration complex plasma properties, such as plasma oscillations, plasma structures and plasma instabilities, the number density of electrons N_e and electron temperature T_e usually are taken as plasma parameters (for example, [465–474]). Note that the electron temperature T_e may be used as a plasma parameter if an energy exchange in electron-electron collisions is stronger than that in electron-atom collisions, and then an electron subsystem is separated from an atomic system thermodynamically. This possibility is fulfilled strictly if the criterion (6.1.2) holds true (see Fig. 5.4), and the collision integral of electron-electron collisions exceeds that for electron-atom collisions in the kinetic equation (6.1.1). In particular, at the electron energy $\varepsilon = 3\,\mathrm{eV}$ the regime of low electron number densities takes place at the electron concentrations $c_e = N_e/N_a$ below 2×10^{-8} for helium, and below 1×10^{-9} for argon. Correspondingly, at larger electron concentrations electrons may be considered as an independent thermodynamic subsystem, and the velocity distribution function of electrons has the Maxwell form (6.2.4).

In this context it should be noted that the electron temperature as the parameter of the EEDF may be used in a restricted range of electric field strengths and corresponds to the range 2 of Fig. 5.2 where elastic electron-electron collisions dominate in formation of the EEDF compared with electron-atom collisions and the action of the electric field. But this relates only to the skeleton of the energy distribution function, whereas the tail of distribution function differs from the Maxwell distribution at not large electric field strengths, as it is shown above.

14.2 Self-consistent Character of Processes Involving Excited Atom States

Since usually a typical electron energy is small compared with the atom excitation energy and the atom ionization potential, the processes of atom excitation and ionization are determined by the tail of EEDF. Though the tail of the distribution function is distorted due to processes of atom excitation, we consider first the case where the Maxwell distribution is valid at the tail of EEDF. The excitation rate constant k_{ex} for a given atom state $*$ from a state o is connected with the rate constant k_q of quenching of an excited state $*$ by electron impact by the principle of detailed balance which is given by formula (3.3.8)

$$k_{th} = k_q \frac{g_*}{g_o} \exp\left(-\frac{\Delta\varepsilon}{T_e}\right), \quad T_e \ll \Delta\varepsilon \qquad (14.2.1)$$

which corresponds to thermodynamic equilibrium (2.1.5) between these states under the action of collisions with electrons and is convenient for comparison.

We now construct the EEDF with taking into account the self-consistent character of the excitation process and the EEDF for fast electrons. Then we have the following kinetics equation for EEDF as an enlarged stationary kinetic equation (5.1.1)

$$-\frac{a^2}{3v^2}\frac{d}{dv}\left(\frac{v^2}{\nu_{ea}}\frac{df_0}{dv}\right) = I_{ea}(f_0) + I_{ee}(f_0) - N_a \int\limits_{\Delta\varepsilon}^{\infty} k_{ex}(\varepsilon) f_0(\varepsilon) d\varepsilon$$

$$+ N_m \int\limits_0^{\infty} k_q(\varepsilon - \Delta\varepsilon) f_0(\varepsilon - \Delta\varepsilon) d\varepsilon - \frac{f_0(\varepsilon)}{\tau}, \qquad (14.2.2)$$

where $a = eE/m_e$, and the left hand side of equation (14.2.2) describes action of an external electric field on the electron distribution function, the integral of electron-atom collisions is given by formula (5.3.9), the integral of electron-electron collisions is determined by formula (5.5.5) with formula (5.5.4) for the diffusion coefficient of a fast electron in the electron energy space. The third and forth terms of the right hand side of equation (14.2.2) correspond to processes

$$e + A \leftrightarrow e + A^*, \qquad (14.2.3)$$

and the last term takes into account other processes (without participation of electrons) for loss of electrons with a given energy. Kinetic equation (14.2.2) is coupled with the balance equation for the number density N_m of excited atoms

$$\frac{dN_m}{dt} = N_a \int\limits_{\Delta\varepsilon}^{\infty} k_{ex}(\varepsilon) f_0(\varepsilon) d\varepsilon - N_m \int\limits_0^{\infty} k_q(\varepsilon - \Delta\varepsilon) f_0(\varepsilon - \Delta\varepsilon) d\varepsilon - \frac{N_m}{\tau_m}, \qquad (14.2.4)$$

where the first and second processes are given by equation (14.2.3), and the third term accounts for other processes involving excited atoms. Set of equations (14.2.2) and (14.2.4) allows one to consider the behavior of the electron distribution function for various electron energies.

Let us divide the energy dependence for the energy distribution function of electrons in parts. At low energies far from the excitation threshold the energy

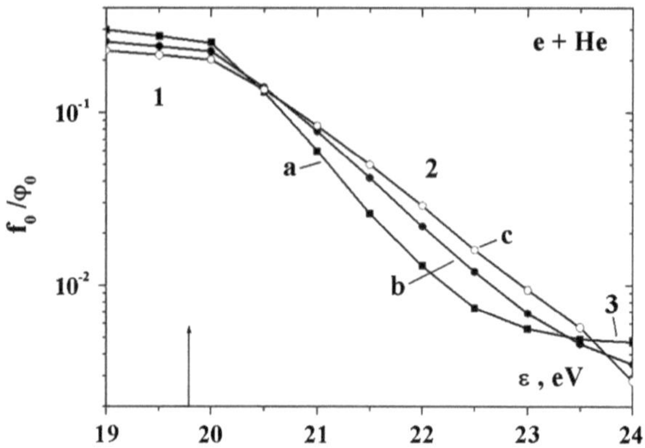

Fig. 14.3 Reduced EEDF, where ε is the electron energy, f_0 is the electron distribution function, and φ_0 is the Maxwell distribution function (6.2.4). The *arrow* indicates the atom excitation threshold $\Delta\varepsilon$ and the atom ionization potential $J = 24.59eV$. *Curve a* corresponds to the electron temperature $T_e = 2\,\text{eV}$, *curve b* relates to $T_e = 3\,\text{eV}$, and *curve c* corresponds to $T_e = 4\,\text{eV}$

distribution function of electrons $f_0(\varepsilon)$ coincides with the Maxwell distribution function $\varphi_0(\varepsilon)$ given by formula (6.2.4). At the excitation threshold we have

$$\frac{f_0(\varepsilon)}{\varphi_0(\varepsilon)} = \frac{k_<}{k_< + k_>}$$

according to data of Fig. 6.11 and formula (6.17). This is represented as the range 1 in Fig. 14.3. The following range 2 of Fig. 14.3 is given by formulas (6.5.2) and (6.5.3) according to which the electron distribution function drops sharply due to the atom excitation process. This range occurs at the simultaneous solution of equations (14.2.2) and (14.2.4), if we conserve the third term in the right hand side of equation (14.2.2) and the first term of the right hand side of equation (14.2.4).

In the range 3 of Fig. 14.3 the equilibrium of electrons with atoms in the ground and excited states proceeds according to the scheme (14.2.3) that gives the electron distribution function as

$$f_0(\varepsilon) = \frac{N_m}{N_a} f_0(\varepsilon - \Delta\varepsilon) \qquad (14.2.5)$$

Assuming the electron distribution function corresponds to the Maxwell one (6.2.4) $f_0(\varepsilon) = \varphi(\varepsilon)$, we obtain in the range 3 of Fig. 14.3

$$f_0(\varepsilon) = \frac{N_m}{N_m^B} \varphi_0(\varepsilon), \qquad (14.2.6)$$

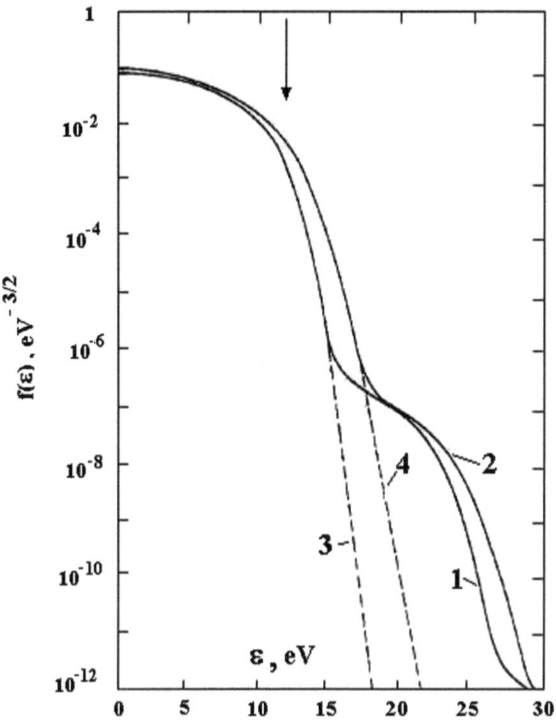

Fig. 14.4 EEDF in argon at the electric field strengths of 10 Td (*curves 1* and *3*) and 20 Td (*curves 2* and *4*) under stationary conditions according to calculations [45]. In the case of *curves 1* and *2* the concentration of metastable argon atoms is $c_m = 6 \times 10^{-5}$, and in the case of *curves 3* and *4* metastable atoms are absent in ionized argon. The *arrow* indicates the excitation energy of the metastable state

where the number density N_m^B of excited atoms is given by Boltzmann formula (2.1.5)

$$N_m^B = N_a \frac{g_m}{g_o} \exp\left(-\frac{\Delta\varepsilon}{T_e}\right) \quad (14.2.7)$$

One can see the self-consistent character of processes of atom excitation and quenching in collisions with electrons influences both on the EEDF and the rate of atom excitation in a gas discharge plasma. The form of the connection between these quantities depends on the electron energy. One more example of the behavior of the EEDF is given in Fig. 14.4 [45] for an argon gas discharge plasma. As is seen, we have the same behavior of the EEDF as in the helium case in three electron energy ranges, and the presence of excited atoms in a gas discharge plasma is responsible for a slightly sloping part of the EEDF at large electron energies. A sharp decrease of the EEDF above the excitation threshold leads to the character of formation of following excited scheme in accordance with Fig. 14.5.

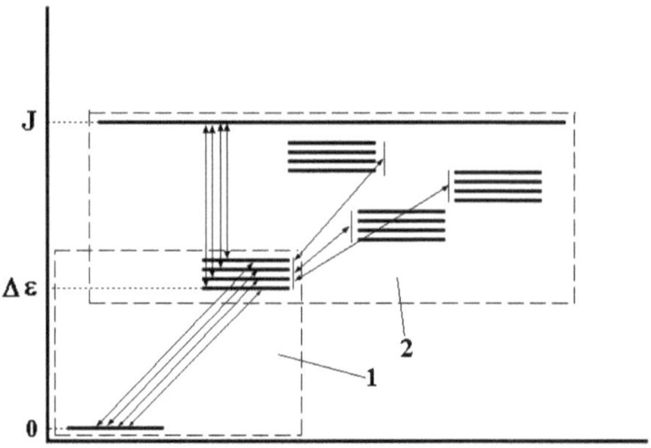

Fig. 14.5 Scheme of inelastic processes in electron-atom collisions in a gas discharge plasma where electron-atom inelastic processes are separated in two blocks, so that the ground and lowest excited atom states are included in the first block, whereas the second block includes other excited states and ionized states. Here the ordinate axis indicates the excitation energy for atomic levels, and 0 relates to the atom ground state, $\Delta\varepsilon$ is the excitation energy for the lowest excited states, J is the atom ionization potential

Since excited atoms influence on the properties of a gas discharge plasma, one can consider the latter to be consisted of four components, atoms, electrons, ions and excited atoms, so that each component is responsible for certain its properties. In particular, processes involving excited atoms are determined so called dark phase in evolution of the positive column of glow discharge [475–479]. This stage of gas discharge characterizes transition to a stationary plasma state with respect to its ionization or radiative properties after initiation of gas discharge or its pulse excitation. The presence of excited atoms in a gas discharge plasma is determined its radiative properties. Spectrum of plasma radiation due to excited atoms has a discrete character and determines a basic radiative flux of the plasma at not large gas pressures. Spectral lines due to excited atom states are broaden significantly, and spectroscopy of a gas discharge plasma [39, 480, 481] is the simplest method for its diagnostics.

In considering the self-consistent character of atom excitation in a gas discharge plasma, we assume only one excited state to be responsible for the EEDF for fast electrons. In reality, excited states may be combined in groups by their excitation energies. In particular, one can divide excited states of atoms first in accordance with an electron shell of these atoms. In the case of an argon gas discharge plasma the first group of excited states belongs to the electron shell $3p^5 4s$, the second group of states corresponds to the electron shell $3p^5 4p$ as it is shown in Fig. 2.10. We have two limiting cases in consideration of transitions involving excited atom states. If the rate of mixing of excited states by electron impact, as the process (3.5.4) in the argon case, is large compared with the rate of quenching of these states, one can combine the states of the same electron shell in one state with appropriate parameters that corresponds to the block model [15, 482, 483] in the kinetic equation for electrons and excited atoms. If a reciprocal relation is fulfilled for the rates of the above processes, it is necessary to consider the excited states of the electron shell $3p^54s$ independently.

Excitation of lowest excited states by electron impact leads to a sharp decrease in the EEDF with an increasing electron energy, as it is shown in Figs. 14.3 and 14.4. Excitation of the lowest excited atom states has a self-consistent character, i.e. excitation of each state is accompanied by a subsequent decrease of the number density of fast electrons which are able to excite the states of a subsequent group. As a result, excitation and ionization of atoms by electron impact has a stepwise character, as it is shown in Fig. 14.5, so that the rate of formation of not low excited states is determined by collisions involving atoms in the lowest excited states, rather than collisions of electrons with atoms in the ground state. The stepwise character of atom excitation takes place at low electron temperatures

$$T_e \ll \Delta\varepsilon,$$

where $\Delta\varepsilon$ is the atom excitation energy, and this criterion is fulfilled usually.

It should be noted that computer models for description of a gas discharge plasma are disseminated and in the argon case were used in [77–82]. Computer models allow one to use a large number of processes. For example, 64 levels are included in the Vlsek model [82] which is a basis of subsequent computer calculations [78–80]. In this book we develop some algorithms to transfer step by step from elementary processes in a plasma and plasma kinetics to some numerical parameters of helium and argon gas discharge plasmas. Computer models promise to transfer from elementary processes to some parameters of a gas discharge plasma in one step with using various information about processes in this plasma. In principle, it may be so if all the elements of computer simulation including information used are reliable. Unfortunately, it is not fulfilled for indicated papers; in particular, the self-consistent character of atom excitation, the basis of Figs. 14.3 and 14.4, is ignored in the above computer models. This lack of computer models is a result of the complexity of the analysis of a gas discharge plasma and reflects the contemporary state of this problem. This means that the analysis of a certain gas discharge plasma requires a careful preparation to it, and the most part of this book (parts I, II, III) are devoted to this stage of the analysis.

14.3 Regimes of Gas Discharge Plasma

A self-consistent character of atom excitation in a gas discharge plasma influences on parameters of this plasma. Summation of results of Sect. 6 for the electron drift velocity w_e and an average electron energy $\overline{\varepsilon}$ is given in Fig. 14.6. A range 1 of this figure corresponds to low electric fields, elastic electron-atom scattering determines electron parameters in a range 2, in a range 3 a test electron consumes a remarkably part of the energy obtained from an external field on atom excitation. In other words,

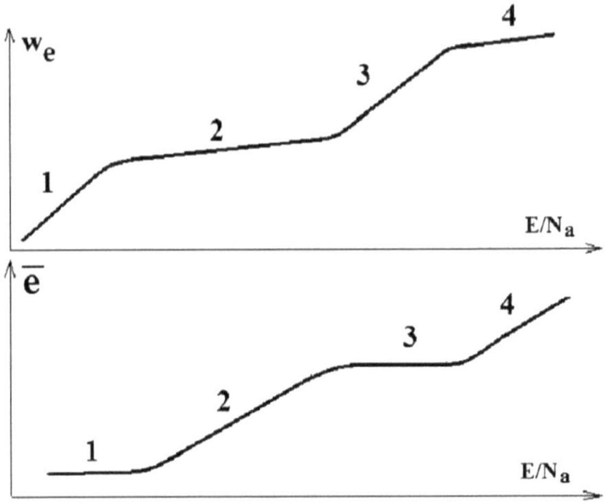

Fig. 14.6 Character of dependencies of the electron drift velocity and average electron energy on the reduced electric field strength E/N_a

if an electron energy exceeds the atom excitation energy, an atom is excited, and a formed slow electron increases its energy as a result of interaction with an external field and scattering on atoms. In region 4 the rate of energy loss in inelastic electron-atom collisions is less than the rate of an energy increase under the action of the electric field if the electron energy exceeds the atom excitation energy.

An important feature of a gas discharge plasma results in a variety of regimes of its existence and evolution. As an example, let us consider ionization equilibrium in an argon gas discharge plasma where a sharp decrease of the energy distribution function of electrons takes place with an increasing electron energy, as it is represented in Figs. 14.3 and 14.4. Then if excited atoms are accumulated in a plasma, the scheme of Fig. 14.5 is realized in excitation and ionization of a gas. But if excited atoms do not give a contribution to atom ionization, as it takes place at low electron number densities, ionization equilibrium shifts to higher electric field strengths. As a result, we obtain two regimes of ionization equilibrium depending on the electron number density and the electric field strength, as it is shown in Fig. 14.7.

Presence of atomic or molecular ions in helium and argon gas discharge plasmas lead to different plasma properties and regimes of its evolution. The processes involving ions determine transport of a plasma as a whole and recombination of electrons and ions. Thus, a gas discharge plasma is governed by various processes and parameters, and as a result many regimes of existence of a gas discharge plasma are possible. Table 14.1 lists processes and plasma parameters which may lead to different regimes of its existence and evolution.

Production of electrons and ions in a gas is an important element of ionization equilibrium in a gas discharge plasma. Though we consider ionization processes as a result of electron-atom collisions as a wide spread channel of atom ionization,

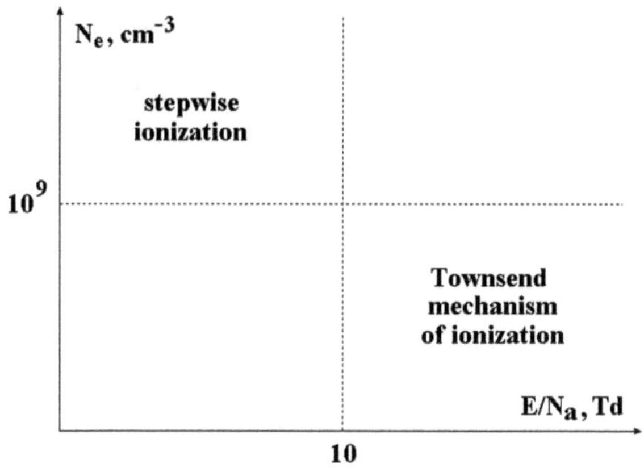

Fig. 14.7 Regimes of ionization of an argon gas discharge plasma depending on the electric field strength and electron number density at the argon pressure of the order of 1 Torr

Table 14.1 Factors which determine different regimes of a gas discharge plasma

	Factor	Possibilities
1.	Energy distribution function for thermal electrons	The electron distribution function may be determined by electron-atom collisions or electron-electron collisions depending on the number density of electrons
2.	Single or stepwise ionization	Depending on the number density of electrons, the ionization process of gas atoms by electron impact results from single ionization of atoms in the ground state or proceeds through excited states
3.	Radiative transitions between atom states	Excited atoms are quenched in collisions with electrons or as a result of photon emission
4.	Atomic or molecular ions in a plasma	The rate of plasma decay proceeds through different processes depending on an ion sort
5.	Ionization through excited atom states	Associative ionization or Penning process may influence on plasma parameters
6.	Heat processes	Diffusive and constricted forms of the positive column may depend on heat release processes
7.	Cathode processes	Emission electrons from the cathode may proceed through cathode bombardment by ion impact or thermoemission

processes of collisions of excited and nonexcited atoms, as associative ionization and Penning process, may be significant in ionization equilibrium. In general, new elementary processes and new combination of elementary processes may lead to new regimes of plasma existence. Along with elementary and transport processes in a plasma, heat processes may be significant in this problem. In particular, the diffusive and constricted regimes of the positive column of gas discharge are possible depending on a discharge power.

In order to reduce a number of plasma regimes, we restrict ourselves in this book by a stationary gas discharge plasma of helium and argon located between plane electrodes or in a cylinder discharge tube. One can expect that removal of these restrictions causes an increase of possible regimes of plasma existence because of new elementary processes in a gas discharge plasma and new conditions of its supporting. In particular, new possibilities occur if a plasma is supported by an alternative field, and a typical field frequency ω is large compared to a typical rates of collisions processes. Elementary processes in a plasma depend on a gas sort. Excitation of vibration states by electron impact is of importance for the energy balance and ionization equilibrium in molecular gases. Formation of negative ions is possible in an electronegative gas. A variety of regimes for plasma existence and its evolution does not allow us to use universal methods for plasma description.

14.4 Conclusion

Summarizing the above analysis of a gas discharge plasma, where we try to transfer from understanding of the problem to its numerical simulation, note that this analysis cannotbe fulfilled in a universal form due to many situations and regimes for this object. From this book we conclude on examples of helium and argon that determination of numerical parameters of a gas discharge plasma requires information about some processes which dominate for this system under given external conditions. In addition, the understanding of physical principles of a gas discharge plasma is a basis of such an analysis and the description of these fundamentals occupies the most part of this book.

Appendix

A.1 Fundamental Physical Constants

Electron mass	$m_e = 9.10939 \times 10^{-28}$ g
Proton mass	$m_p = 1.67262 \times 10^{-24}$ g
Atomic unit of mass	$m_a = \frac{1}{12} m(^{12}C) = 1.66054 \times 10^{-24}$ g
Ratio of masses	$m_p/m_e = 1836.15$, $m_a/m_e = 1822.89$
Electron charge	$e = 1.602177 \times 10^{-19}$C $= 4.8032 \times 10^{-10}$ CGSE
	$e^2 = 2.3071 \times 10^{-19}$ $erg \cdot$ cm
Planck constant	$h = 6.62619 \times 10^{-27}$ $erg \cdot$ s, $\hbar = 1.05457 \times 10^{-27}$ $erg \cdot$ s
Light velocity	$c = 2.99792 \times 10^{10}$ cm/s
Fine-structure constant	$\alpha = e^2/(\hbar c) = 0.07295$
Inverse fine-structure constant	$1/\alpha = \hbar c/e^2 = 137.03599$
Bohr radius	$a_o = \hbar^2/(m_e e^2) = 0.529177$ Å
Rydberg constant	$R = m_e e^4/(2\hbar^2) = 13.6057$ eV $= 2.17987 \times 10^{-18}$ J
Bohr magneton	$\mu_B = e\hbar/(2m_e c) = 9.27402 \times 10^{-24} J/T = 9.27402 \times 10^{-21}$ erg/Gs
Avogadro number	$N_A = 6.02214 \times 10^{23}$ mol^{-1}
Stephan-Boltzmann constant	$\sigma = \pi^2/(60\hbar^3 c^2) = 5.669 \times 10^{-12}$ W/(cm^{2}K^4)
Molar volume	$R = 22.414$ l/mol
Loschmidt number	$L = N_A/R = 2.6867 \times 10^{19}$ cm^{-3}
Faraday constant	$F = N_A e = 96485.3$ C/mol

A.2 Conversional Factors for Units

See Tables A.1, A.2, A.3, A.4, A.5, A.6, A.7, A.8, A.9, A.10, and A.11.

B.M. Smirnov, *Theory of Gas Discharge Plasma*, Springer Series on Atomic, Optical, and Plasma Physics 84, DOI 10.1007/978-3-319-11065-3

Table A.1 Units of energy

	1 J	1 *erg*	1 eV	1 K	1 cm^{-1}	1 MHz	1 kcal/mol	1 kJ/mol
1 J	1	10^7	6.2415×10^{18}	7.2429×10^{22}	5.0341×10^{22}	1.5092×10^{27}	1.4393×10^{20}	6.0221×10^{20}
1 *erg*	10^{-7}	1	6.2415×10^{11}	7.2429×10^{15}	5.0341×10^{15}	1.5092×10^{20}	1.4393×10^{13}	6.0221×10^{13}
1 eV	1.6022×10^{-19}	1.6022×10^{-12}	1	11604	8065.5	2.4180×10^8	23.045	96.485
1 K	1.3807×10^{-23}	1.3807×10^{-16}	8.6174×10^{-5}	1	0.69504	2.0837×10^4	1.9872×10^{-3}	8.3145×10^{-3}
1 cm^{-1}	1.9864×10^{-23}	1.9864×10^{-16}	1.2398×10^{-4}	1.4388	1	2.9979×10^4	2.8591×10^{-3}	1.1963×10^{-2}
1 MHz	6.6261×10^{-28}	6.6261×10^{-21}	4.1357×10^{-9}	4.7992×10^{-5}	3.3356×10^{-5}	1	9.5371×10^{-9}	3.9903×10^{-7}
1 kcal/mol	6.9477×10^{-21}	6.9477×10^{-28}	4.3364×10^{-2}	503.22	349.76	1.0485×10^7	1	4.184
1 kJ/mol	1.6605×10^{-21}	1.6605×10^{-28}	1.0364×10^{-2}	120.27	83.594	2.5061×10^6	0.23901	1

Table A.2 Units of pressure

	1 Pa = 1 N/m^2	1 dyn/cm^2	1 Torr	1 atm[a]	1 at[b]	1 bar
1 Pa = 1 N/m^2	1	10	7.5001×10^{-3}	9.8693×10^{-6}	1.0197×10^{-5}	10^{-5}
1 dyn/cm^2	0.1	1	7.5001×10^{-4}	9.8693×10^{-7}	1.0197×10^{-6}	10^{-6}
1 Torr	133.332	1333.32	1	1.3158×10^{-3}	1.3595×10^{-3}	1.33332×10^{-3}
1 atm[a]	1.01325×10^5	1.01325×10^6	760	1	1.01332	1.01325
1 at[b]	9.80665×10^4	9.80665×10^5	735.56	0.96785	1	0.980665
1 bar	10^5	10^6	750.01	0.98693	1.0197	1

[a]atm—physical atmosphere
[b]at = kg/cm^2—technical atmosphere

Table A.3 Units of electric charge

	1 e	1 CGSE	1 C
1 e	1	4.8032×10^{-10}	1.60218×10^{-19}
1 CGSE	2.0819×10^{9}	1	3.33564×10^{-10}
1 C	6.2415×10^{18}	2.99792×10^{9}	1

Table A.4 Units of electric voltage

	1 V	1 CGSE	1 CGSM
1 V	1	3.33564×10^{-3}	10^{8}
1 CGSE	299.792	1	2.99792×10^{10}
1 CGSM	10^{-8}	3.33564×10^{-11}	1

Table A.5 Units of electric field strength

	1 V/cm	1 CGSE	1 CGSM
1 V/cm	1	3.33564×10^{-3}	10^{8}
1 CGSE	299.792	1	2.99792×10^{10}
1 CGSM	10^{-8}	3.33564×10^{-11}	1

Table A.6 Units of specific electric field strength

	1 Td[a]	$1\,V/(cm \cdot Torr)$
1 Td[a]	1	2.829
$1V/(cm \cdot Torr)$	0.3535	1

[a] $1\,Td = 1 \times 10^{-17}\,V \cdot cm^2$

Table A.7 Units of conductivity

	$1/(\Omega \cdot cm)$	$1/s$
$1/(\Omega \cdot cm)$	1	1.11265×10^{-12}
$1/s$	8.98755×10^{11}	1

Table A.8 Units of electric resistance

	$1\,\Omega$	1 CGSE	1 CGSM
$1\,\Omega$	1	1.11265×10^{-12}	10^{9}
1 CGSE	8.98755×10^{11}	1	8.98755×10^{20}
1 CGSM	10^{-9}	1.11265×10^{-21}	1

Table A.9 Units of current density

	$1\,e/(cm^2 \cdot s)$	1 CGSE	$1\,A/m^2$
$1\,e/(cm^2 \cdot s)$	1	$2.99792\ 4.8032 \times 10^{-10}$	1.60218×10^{-15}
1 CGSE	2.0819×10^{9}	1	3.3356×10^{-6}
$1\,A/m^2$	6.2415×10^{14}	2.9979×10^{5}	1

Table A.10 Units of magnetic field strength

	1 Oe	1 CGSE	1 A/m
1 Oe	1	2.99792×10^{10}	79.5775
1 CGSE	3.33564×10^{-11}	1	2.65442×10^{-9}
1 A/m	0.012566	1.11265×10^{-21}	1

Table A.11 Units of magnetic induction

	1 CGSE	$1\,T = 1\,\mathrm{Wb/m^2}$	1 Gs
1 CGSE	1	2.99792×10^{6}	2.99792×10^{10}
$1\,T = 1\,\mathrm{Wb/m^2}$	3.33564×10^{-7}	1	10^4
1 Gs	3.33564×10^{-11}	10^{-4}	1

A.3 Conversional Factors in Formulas

See Tables A.12, A.13, A.14, A.15, and A.16.

Table A.12 Formulas of general physics

Number	Formula[a]	Factor C	Units used
1.	$v = C\sqrt{\varepsilon/m}$	5.931×10^7 cm/s	ε in eV, m in m_e[a]
		1.389×10^6 cm/s	ε in eV, m in m_a[a]
		5.506×10^5 cm/s	ε in K, m in m_e[a]
		1.289×10^4 cm/s	ε in K, m in m_a[a]
2.	$v = C\sqrt{T/m}$	6.692×10^7 cm/s	T in eV, m in m_e[a]
		6.212×10^5 cm/s	T in K, m in m_e[a]
		1.567×10^6 cm/s	T in eV, m in m_a[a]
		1.455×10^4 cm/s	T in K, m in m_a[a]
3.	$\varepsilon = Cv^2$	3.299×10^{-12} K	v in cm/s, m in m_e[a]
		6.014×10^{-9} K	v in cm/s, m in m_a[a]
		2.843×10^{-16} eV	v in cm/s, m in m_e[a]
		5.182×10^{-13} eV	v in cm/s, m in m_a[a]
4.	$\omega_H = CH/m$	$1.759 \times 10^7\ \mathrm{s}^{-1}$	H in Gs, m in m_e[a]
		$9648\ \mathrm{s}^{-1}$	H in Gs, m in m_a[a]
5.	$v = CE/H$	1×10^8 cm/s	E in V/cm, H in Gs
6.	$r_H = C\sqrt{\varepsilon m}/H$	3.372 cm	ε in eV, m in m_e[a], H in Gs
		143.9 cm	ε in eV, m in m_a[a], H in Gs
		3.128×10^{-2} cm	ε in K, m in m_e[a], H in Gs
		1.336 cm	ε in K, m in m_a[a], H in Gs
7.	$p = CH^2$	4.000×10^{-3} Pa $= 0.04\,erg/\mathrm{cm}^3$	H in Gs

[a] $m_e = 9.108 \times 10^{-28}$ g is the electron mass, $m_a = 1.6605 \times 10^{-24}$ g is the atomic mass unit

Explanations to Table

1. The particle velocity is $v = \sqrt{2\varepsilon/m}$, where ε is the energy, m is the particle mass.
2. The average particle velocity is $v = \sqrt{8T/(\pi m)}$ with the Maxwell distribution function over velocities; T is the temperature expressed in energetic units, m is the particle mass.
3. The particle energy is $\varepsilon = mv^2/2$, where m is the particle mass, v is the particle velocity.
4. The Larmor frequency is $\omega_H = eH/(mc)$ for a charged particle of a mass m in a magnetic field of a strength H.
5. The drift velocity of a singly charged particle in crossed electric and magnetic fields of strengths E and H correspondingly.
6. The Larmor radius of a charged particle is $r_H = \sqrt{2\varepsilon/m}/\omega_H$, where ε is the energy of a charged particle, m is its mass, ω_H is the Larmor frequency.
7. The magnetic pressure is $p_m = H^2/(8\pi)$.

Table A.13 Formulas of gas and plasma physics

Number	Formula	Proportionality factor C	Units
1.	$\alpha = CN_e/T^3$	2.998×10^{-21}	N_e in cm^{-3}, T in eV
		4.685×10^{-9}	N_e in cm^{-3}, T in K
2.	$f = Cm^{3/2}T^{3/2}$	2.415×10^{15} cm^{-3}	T, m in m_e [a]
		3.019×10^{21} cm^{-3}	T in K, m in m_e [a]
		1.879×10^{20} cm^{-3}	T in K, m in m_a [b]
		2.349×10^{26} cm^{-3}	T in eV, m in m_a [b]
3.	$K = C/T_e^{9/2}$	2.406×10^{-22} cm^6/s	T_e in 1,000 K
		3.894×10^{-27} cm^6/s	T_e in eV
4.	$\omega_p = C\sqrt{N_e/m}$	5.642×10^4 s^{-1}	N_e in cm^{-3}, m in m_e [a]
		1322 s^{-1}	N_e in cm^{-3}, m in m_a [b]
5.	$r_D = C\sqrt{T/N_e}$	525.3 cm	N_e in cm^{-3}, T in eV
		4.876 cm	N_e in cm^{-3}, T in K

[a] $m_e = 9.108 \times 10^{-28}$ g is the electron mass
[b] $m_a = 1.6605 \times 10^{-24}$ g is the atomic mass unit

Explanations to Table

1. The ideality plasma parameter is $\alpha = N_e e^6/T_e^3$, where e is the electron charge, N_e is the number density of electrons, T_e is the electron temperature.
2. Preexponent of the Saha formula is $\xi = [mT/(2\pi\hbar^2)]^{3/2}$.
3. The rate constant of three body electron ion recombination ($2e + A^+ \to e + A$) is $K = 1.5e^{10}/(m_e^{1/2}T_e^{9/2})$.
4. The plasma frequency is $\omega_p = \sqrt{4\pi N_e e^2/m_e}$.
5. The Debye-Hückel radius is $r_D = \sqrt{T/(8\pi N_e e^2)}$.

Table A.14 Radiative transition between atom states

Number	Formula	Conversional factor C	Used units
1.	$\varepsilon = C\omega$	4.1347×10^{-15} eV	ω in s^{-1}
		6.6261×10^{-34} J	ω in s^{-1}
2.	$\omega = C\varepsilon$	$1.519 \times 10^{15}\, s^{-1}$	ε in eV
		$1.309 \times 10^{11}\, s^{-1}$	ε in K
3.	$\omega = C/\lambda$	$1.884 \times 10^{15}\, s^{-1}$	λ in μm
4.	$\varepsilon = C/\lambda$	1.2398 eV	λ in μm
5.	$f_{o*} = C\omega d^2 g_*$	1.6126×10^{-17}	ω in s^{-1}, d in D [a]
		0.02450	$\Delta\varepsilon = \hbar\omega$ in eV, d in D [a]
6.	$f_{o*} = Cd^2 g_*/\lambda$	0.03038	λ in μm, d in D [a]
7.	$1/\tau_{*o} = C\omega^3 d^2 g_o$	$3.0316 \times 10^{-40}\, s^{-1}$	ω, s^{-1}, d, D [a]
		$1.06312 \times 10^6 s^{-1}$	$\Delta\varepsilon = \hbar\omega$ in eV, d in D [a]
8.	$1/\tau_{*o} = Cd^2 g_o/\lambda^3$	$2.0261 \times 10^6\, s^{-1}$	λ in μm, d in D [a]
9.	$1/\tau_{*o} = C\omega^2 g_o f_{o*}/g_*$	$1.8799 \times 10^{-23}\, s^{-1}$	ω in s^{-1}, d in D [a]
		$4.3393 \times 10^7\, s^{-1}$	$\Delta\varepsilon = \hbar\omega$ in eV; d in D [a]
10.	$1/\tau_{*o} = Cf_{o*}g_o/(g_*\lambda^2)$	$6.6703 \times 10^7\, s^{-1}$	λ in μm, d in D [a]

[a] D is Debye, $1\,D = ea_o = 2.5418 \times 10^{-18}$ CGSE

Explanation to Table

1. The photon energy $\varepsilon = \hbar\omega$, where ω is the photon frequency.
2. The photon frequency is $\omega = \varepsilon/\hbar$.
3. The photon frequency is $\omega = 2\pi c/\lambda$, where λ is the wavelength, c is the light speed.
4. The photon energy is $\varepsilon = 2\pi\hbar c/\lambda$.

5, 6. The oscillator strength for a radiative transition from the lower o to the upper $*$ state of an atomic particle that is averaged over lower states o and is summed over upper states $*$ is equal to

$$f_{o*} = \frac{2m_e\omega}{3\hbar e^2}\left|\langle o\left|\mathbf{D}\right|*\rangle\right|^2 g_* = \frac{2m_e\omega}{3\hbar e^2}d^2 g_* = \frac{4\pi c m_e}{3\hbar e^2\lambda}d^2 g_* ,$$

where $\mathbf{d} = \langle o|\mathbf{D}|*\rangle$ is the matrix element for the operator of the dipole moment of an atomic particle taken between transition states. Here m_e, $\hbar$ are atomic parameters, g_* is the statistical weight of the upper state, $\omega = (\varepsilon_* - \varepsilon_o)/\hbar$ is the transition frequency, where ε_o, ε_* are the energies of transition states, λ is the transition wavelength.

7–10. The rate of the radiative transition is

$$\frac{1}{\tau_{*o}} = B_{*o} = \frac{4\omega^3}{3\hbar c^3}d^2 g_o = \frac{32\pi^3}{3\hbar\lambda^3}d^2 g_o = \frac{2\omega^2 e^2 g_o}{m_e c^3 g_*}f_{o*} = \frac{8\pi^2 g_o}{\hbar g_*\lambda^2 c}f_{o*}$$

Here B is the Einstein coefficient; other notations are the same as above.

Table A.15 Transport coefficients

Number	Formula	Coefficient C	Units
1.	$D = CKT$	$8.617 \times 10^{-5}\,cm^2/s$	K in $cm^2/(V \times s)$, T in K
		$1\ cm^2/s$	K in $cm^2/(V \times s)$, T in eV
2.	$K = CD/T$	$11{,}604\,cm^2/(V \times s)$	D in cm^2/s, T in K
		$1\ cm^2/(V \cdot s)$,	D in cm^2/s, T in eV
3.	$D = C\sqrt{T/\mu}/(N\overline{\sigma_1})$	$4.617 \times 10^{21}\,cm^2/s$	$\overline{\sigma_1}$ in $Å^2$, N in cm^{-3}, T in K, μ in m_a [a]
		$1.595\,cm^2/s$	$\overline{\sigma_1}$ in $Å^2$, $N = 2.687 \times 10^{19}\,cm^{-3}$, T in K, μ in m_a [a]
		$171.8\,cm^2/s$	$\overline{\sigma_1}$ in $Å^2$, $N = 2.687 \times 10^{19}\,cm^{-3}$, T in eV μ in m_a [a]
		$68.1115\,cm^2/s$	$\overline{\sigma_1}$ in $Å^2$, $N = 2.687 \times 10^{19}\,cm^{-3}$, T in K, μ in m_e [a]
		$7338\,cm^2/s$	$\overline{\sigma_1}$ in $Å^2$, $N = 2.687 \times 10^{19}\,cm^{-3}$, T in eV, μ in m_e [a]
4.	$K = C(\sqrt{T}\mu N\overline{\sigma_1})^{-1}$	$1.851 \times 10^4\,cm^2/(V \cdot s)$	$\overline{\sigma_1}$ in $Å^2$, $N = 2.687 \times 10^{19}\,cm^{-3}$, T in K, μ in m_e [a]
		$171.8\,cm^2/(V \cdot s)$	$\overline{\sigma_1}$ in $Å^2$, $N = 2.687 \times 10^{19}\,cm^{-3}$, T in eV, μ in m_a [a]
		$7.904 \times 10^5\,cm^2/(V \cdot s)$	$\overline{\sigma_1}$ in $Å^2$, $N = 2.687 \times 10^{19}\,cm^{-3}$, T in K, μ in m_e [a]
		$7338\ cm^2/(V \cdot s)$	$\overline{\sigma_1}$ in $Å^2$, $N = 2.687 \times 10^{19}\,cm^{-3}$, T in eV, μ in m_e [a]
5.	$\kappa = C\sqrt{T/m}/\overline{\sigma_2}$	$1.743 \times 10^4\ W/(cm \cdot K)$	T in K, m in m_a [a], $\overline{\sigma_2}$ in $Å^2$
		$7.443 \times 10^5\ W/(cm \cdot K)$	T in K, m in m_e [a], $\overline{\sigma_2}$ in $Å^2$
6.	$\eta = C\sqrt{Tm}/\overline{\sigma_2}$	$5.591 \times 10^{-5}\,g/(cm \cdot s)$	T in K, m in m_a [a], $\overline{\sigma_2}$ in $Å^2$
7.	$\xi = CE/(TN\sigma)$	1.160×10^{20}	E in V/cm, T in K, σ in $Å^2$, N in cm^{-3}
		1×10^{16}	E in V/cm, T in eV, σ in $Å^2$, N in cm^{-3}

[a] $m_e = 9.108 \times 10^{-28}\ g$ is the electron mass, $m_a = 1.6605 \times 10^{-24}$ g is the atomic mass

Explanation to Table

1. The Einstein relation for the diffusion coefficient of a charged particle in a gas $D = KT/e$, where D, K are the diffusion coefficient and mobility of a charged particle, T is the gas temperature.
2. The Einstein relation for the mobility of a charged particle in a gas $K = eD/T$.
3. The diffusion coefficient of an atomic particle in a gas in the first Chapman-Enskog approximation $D = 3\sqrt{2\pi T/\mu}/(16N\overline{\sigma_1})$, where T is the gas temperature, N is the number density of gas atoms or molecules, μ is the reduced mass of a colliding particle and gas atom or molecule, $\overline{\sigma_1}$ is the average cross section of collision.

4. The mobility of a charged particle in a gas in the first Chapman-Enskog approximation $K = 3e\sqrt{2\pi/(T\mu)}/(16N\overline{\sigma_1})$; notations are the same as above.
5. The gas thermal conductivity in the first Chapman-Enskog approximation $\kappa = 25\sqrt{\pi T}/(32\sqrt{m}\overline{\sigma_2})$, where m is the atom or molecule mass, $\overline{\sigma_2}$ is the average cross section of collision between gas atoms or molecules; other notations are the same as above.
6. The gas viscosity in the first Chapman-Enskog approximation $\eta = 5\sqrt{\pi Tm}/(24\overline{\sigma_2})$; notations are the same as above.
7. The parameter of ion drift in a gas in a constant electric field $\xi = eE/(TN\sigma)$, where E is the electric field strength, T is the gas temperature, N is the number density of atoms or molecules, σ is the cross section of collision.

Table A.16 Clusters or small particles

Number	Formula	Coefficient C	Used units
1.	$r_W = C(m/\rho)^{1/3}$	0.7346 Å	m in m_a [a], ρ in g/cm^3
2.	$n = C(r_o/r_W)^3$	4.189	r_o and r_W in Å
3.	$k_o = Cr_W^2\sqrt{T/m}$	4.5714×10^{-12}cm^3/s	r_W in Å, T in K, m in m_a [a]
4.	$w = C\rho r^2/\eta$	0.2179 cm/s	r in μm, ρ in g/cm^3, η in 10^{-5}g/(cm · s)
	$w = Cr^2$	0.01178 cm/s	r in μm, ρ, η relate to air at $p = 1$ atm, $T = 300$ K
5.	$D = CKT$	8.617×10^{-5} cm^2/s	K in cm^2/(V · s), T in K
		1 cm^2/s	K in cm^2/(V · s), T in eV
6.	$K = CD/T$	11,604 cm^2/(V · s)	D in cm^2/s, T in K
		1 cm^2/(V · s),	D in cm^2/s T in eV
7.	$D_o = C\sqrt{T/m}/(N_a r_W^2)$	1.469×10^{21}cm^2/s	r_W Å, N_a in cm^{-3}, T in K, m in m_a [a]
		0.508 cm^2/s	r_W Å, $N_a = N_o$, T in K, m in m_a [a]
		54.69 cm^2/s	r_W in Å, $N_a = N_o$, T in eV, m in m_a [a]
8.	$K_o = C(\sqrt{Tm}\,Nr_W^2)^{-1}$	1.364×10^{19} cm^2/(V · s)	r_W in Å, N_a in cm^{-3}, T in K, m in m_a [a]
		0.508 cm^2/(V · s)	r_W in Å, $N_a = N_o$, T in K, m in m_a [a]
		54.69 cm^2/(V · s)	r_W in Å, $N = N_o$, T in K, m in m_a [a]
9.	$D_o = CT/(r_W\eta)$	7.32×10^{-5} cm^2/s	r_W in Å, T in K, η in 10^{-5}g/(cm · s)
10.	$K_o = C/(r_W\eta)$	0.085 cm^2/(V · s)	r_W in Å, η in 10^{-5}g/(cm · s)

[a] $m_a = 1.66054\times10^{-24}$ g is the atomic mass

Explanations to Table

1, 2. A number of atoms in a cluster or spherical particle consisting of n atoms and having a radius r is $n = (r/r_W)^3$ within the framework of the liquid drop model, where r_W is the Wigner-Seits radius of a cluster material. A cluster mass is $nm_a = 4\pi r^3\rho/3$, where m_a is the atom mass, ρ is the mass density of a cluster material.

3. The reduced rate constant for collisions involving clusters is $k_o = \pi r^2 \cdot \sqrt{8T/(\pi m_a)}$.
4. The free fall velocity w in the gravitation field for a spherical particle of a radius $r_o : w = 2\rho g r_o^2/(9\eta)$, where g is the free fall acceleration, ρ is the mass density for a particle material, η is the viscosity of a media where the particle moves.
5. The Einstein relation for a charged particle located in a gas $D = KT/e$, where D, K are the diffusion coefficient and mobility of a charged particle, T is the gas temperature.
6. The Einstein relation $K = eD/T$ for the mobility K of a charged particle moved in a gas.
7. The reduced diffusion coefficient D_o of a particle in a gas in the kinetic regime, so that the particle diffusion coefficient D of a particle consisting of n atoms at the normal number density $N_a = N_o = 2.687 \times 10^{19}\,\mathrm{cm}^{-3}$ of gas molecules, that is equal $D_n = D_o/n^{2/3}$, where $D_o = 3\sqrt{2T/\pi m}/(16N_o r_W^2)$, T is the gas temperature, m is the gas atom mass, r_W is the Wigner-Seits radius.
8. The reduced zero-field mobility of a spherical particle K_o in the kinetic regime, so that K is the mobility of a particle consisting of n atoms is $K_n = K_o/n^{2/3}$, where $K_o = 3e/(8N_o r_W^2\sqrt{2\pi mT})$ at the normal number density of gas atoms $N_a = N_o = 2.687 \times 10^{19}\,\mathrm{cm}^{-3}$, and other notations are indicated in the previous point.
9. The reduced diffusion coefficient $d_o = T/(6\pi r_W \eta)$ of a particle in a gas in the diffusion regime at the normal number density $N_a = N_o = 2.687 \times 10^{19}\,\mathrm{cm}^{-3}$ of gas molecules, so that the particle diffusion coefficient D_n of a particle consisting of n atoms is equal to $D_n = d_o/n^{1/3}$. Here, η is the gas viscosity, r_W is the Wigner-Seits radius; other notations are given above.
10. The reduced zero-field mobility of a spherical particle is $K_o = e/(6\pi r_W \eta)$ in the diffusion regime at the normal number density of gas atoms $N_a = N_o = 2.687 \times 10^{19}\,\mathrm{cm}^{-3}$, so that the mobility of a particle consisting of n atoms is equal $K_n = K_o/n^{1/3}$, and the notations used are indicated above.

References

1. J.J. Thomson, *Conduction of Electricity Through Gases* (Cambridge University Press, Cambridge, 1904)
2. J.S. Townsend, *Electricity in Gases* (Claredon Press, Oxford, 1915)
3. J.S. Townsend, *Motion of Electrons in Gases* (Claredon Press, Oxford, 1925)
4. A. Engel, M. Steenbeck, *Electrische Gasenladungen, ihre Physik und Technick* (Springer, Berlin, 1932–1934)
5. J.S. Townsend, *Electrons in Gases* (Hutchinson, London, 1947)
6. J.S. Townsend, V.A. Bailey, Phil. Mag. **43**, 593 (1922)
7. J.S. Townsend, V.A. Bailey, Phil. Mag. **44**, 1033 (1922)
8. A. Tyndall, *The Mobility of Positive Ions in Gases* (Cambridge University Press, New York, 1938)
9. M.F. Penning, *Electrical Discharges in Gases* (Macmillan, New York, 1957)
10. J.D. Cobine, *Gaseous Conductors* (Dover, New York, 1958)
11. F. Llewelyn-Jones, *The Glow Discharge* (Methuen, New York, 1966)
12. S.C. Brown, *Basic Data of Plasma Physics: The Fundamental Data on Electrical Discharges in Gases* (Wiley, New York, 1966)
13. M.F. Hoyaux, *Arc Physics* (Springer, New York, 1968)
14. B. Chapman, *Glow Discharge Processes* (Wiley, New York, 1980)
15. L.M. Biberman, V.S. Vorob'ev, I.T. Yakubov, *Kinetics of Nonequilibrium Low-Temperature Plasmas* (Consultant Bureau, New York, 1987)
16. W. Neuman, *The Mechanism of the Thermoemitting Arc Cathode* (Academic-Verlag, Berlin, 1987)
17. G.A. Mesyats, D.I. Proskurovsky, *Pulsed Electrical Discharge in Vacuum* (Springer, Berlin, 1989)
18. Y.P. Raizer, *Gas Discharge Physics* (Springer, Berlin, 1991)
19. M.A. Lieberman, A.J. Lichtenberger, *Principles of Plasma Discharge and Materials Processing* (Wiley, New York, 1994)
20. Y.P. Raizer, M.N. Schneider, N.A. Yatzenko, *Radio-Frequency Capacitive Discharges* (CRC Press, Boca Raton, 1995)
21. E.M. Bazelyan, Y.P. Raizer, *Spark Gischarge* (CRC Press, Boca Raton, 1997)
22. Yu.D. Korolev, G.A. Mesyats, *Physics of Pulsed Breakdown in Gases* (URO Press, Ekaterinburg, 1998)
23. A.A. Kudryavtzev, A.A. Smirnov, L.D. Tsendin, *Physicas of Gaseous Discharge* (Petersburg, Lan', Russian, 2009)

B.M. Smirnov, *Theory of Gas Discharge Plasma*, Springer Series on Atomic, Optical, and Plasma Physics 84, DOI 10.1007/978-3-319-11065-3

24. L.B. Loeb, *Basic Processes of Gaseous Electronics* (University California Press, Berkeley, 1955)
25. H.S.W. Massey, E.H.S. Burhop, *Electronic and Ionic Impact Phenomena* (Oxford University Press, Oxford, 1969)
26. E. Nasser, *Fundamentals of Gaseous Ionization and Plasma Electronics* (Wiley, New York, 1971)
27. L.G.H. Huxley, R.W. Crompton, *The Diffusion and Drift of Electrons in Gases* (Wiley, New York, 1973)
28. A.M. Howartson, *Introduction to Gas Discharge* (Oxford University Press, Oxford, 1975)
29. R.N. Franklin, *Plasma Phenomena in Gas Discharges* (Pergamon Press, Oxford, 1975)
30. E.E. Son, *Kinetic Theory of Low Temperature Plasma* (MFTI Press, Moscow, 1978)
31. H.S.W. Massey, *Atomic and Molecular Collisions* (Taylor and Francis, New York, 1979)
32. A. von Engel, *Electric Plasmas: Their Nature and Uses* (Taylor and Francis, London, 1983)
33. A.F. Alexandrov, L.S. Bogdankevich, A.A. Rukhadze, *Principles of Plasma Electrodynamics* (Springer, Berlin, 1984)
34. E.W. McDaniel, E.A. Mason, *Transport Properties of Ions in Gases* (Wiley, New York, 1988)
35. P.P.J.M. Schram, *Kinetic Theory of Gases and Plasmas* (Kluwer, Dordrecht, 1991)
36. M.I. Boulos, P. Fauchais, E. Pfender, *Thermal Plasmas* (Plenum, New York, 1994)
37. A.N. Lagarkov, I.M. Rutkevich, *Ionization Waves in Electrical Breakdown in Gases* (Springer, New York, 1994)
38. B.M. Smirnov, *Plasma Processes and Plasma Kinetics* (Wiley, Berlin, 2007)
39. V.N. Ochkin, *Spectroscopy of Low Temperature Plasma* (Wiley, Weinheim, 2009)
40. B.M. Smirnov, *Fundamental of Ionized Gases* (Wiley, Berlin, 2012)
41. B.M. Smirnov, *Properties of Gas Discharge Plasma* (Polytechnical University Press, Petersburg, 2010)
42. J.S. Townsend, Electrifician **50**, 971 (1903)
43. J. Dutton, J. Phys. Chem. Ref. Data **4**, 577 (1975)
44. M. Capitelli, M. Dilonardo, C. Gorse, Chem. Phys. **56**, 29 (1981)
45. J. Bretagne, M. Capitelli, C. Gorse, V. Puec, EPL **3**, 1179 (1987)
46. I. Langmuir, Proc. Nat. Acad. Sci. US **14**, 628 (1928)
47. L. Tonks, I. Langmuir, Phys. Rev. **33**, 195 (1929)
48. I. Langmuir, Science **58**, 290 (1923)
49. http://en.wikipedia.org/wiki/File:Compact-Fluorescent-Bulb.jpg
50. C.D. Macchietto, B.R. Benware, J.J. Rocca, Opt. Lett. **24**, 1115 (1999)
51. J.J. Rocca, M. Frati, B. Benware et al., C. R. Acad. Sci. Paris Ser. IV **1**, 1065 (2000)
52. Y. Lin et al., Phys. Rev. A **63**, 033802 (2001)
53. http://ru.wikipedia.org/wiki/
54. A.A. Kudryavtsev, L.D. Tsendin, Tech. Phys. Lett. **27**, 284 (2001)
55. J.P. Boeuf, J. Phys. **36D**, R53 (2003)
56. L.F. Weber, IEEE Trans. Plasma Sci. **34**, 268 (2006)
57. http://en.wikipedia.org/wiki/Plasma-display
58. P. Friedman, R. Ball, J. Beene et al., J. Soc. Inform. Display **21**, 46 (2013)
59. D.C. Levin, R. Ball, J.R. Been et al., Nucl. Instr. Meth. Phys. Res. A **652**, 315 (2010)
60. F. Stuhlinger, *Ion Propulsion for Space Flight* (McGrow Hill, New York, 1964)
61. R.G. Jahn, *Physics of Electric Propulsion* (McGrow Hill, New York, 1968)
62. G.R. Brewer, *Ion Propulsion Technology and Applications* (Gordon and Breach, New York, 1970)
63. S.D. Grishin, L.V. Leskov, *Electrical Rocket Engines of Space Vehicles* (Mashinostroenie, Moscow, 1989)
64. D.M. Goebel, I. Katz, *Fundamentals of Electric Propulsion* (Wiley, Hoboken, 2008)
65. V.N. Ochkin, N.G. Preobrazhensky, N.Y. Shaparev, *Optohalvanic Effect in Ionized Gas* (Gordon and Breach, London, 1998)
66. http://www.plasmasurgical.com
67. V.A. Nemchinsky, W.S. Severance, J. Phys. **39**, R423 (2006)

68. J.R. Holohan, A.T. Bell, *Techniques and Applications of Plasma Chemistry* (Wiley, New York, 1974)
69. J. Proud et al., *Plasma Processes of Materials: Scientific Opportunities and Technological Challenges* (National Academy Press, Washington, 1991)
70. J. Proud et al., *Plasma Science: from Fundamental Research to Technological Applications* (National Academy Press, Washington, 1995)
71. M.A. Lieberman, A.J. Lichtenberger, *Principles of Plasma Discharge and Materials Processing* (Wiley, Hoboken, 2005)
72. A.A. Fridman, *Plasma Chemistry* (Cambridge University Press, Cambridge, 2008)
73. J. Meichner, H.-E. Wagner, M. Schmidt, R. Schneider (eds.), *Nonthermal Plasma Chemistry and Physics* (CRC Press, Boca Raton, 2013)
74. R. D'Agostino, *Plasma Deposition, Treatment, and Etching of Polymers* (Academy Press, New York, 1990)
75. B.M. Smirnov, Phys. Usp. **52**, 519 (2009)
76. L.D. Tsendin, Phys. Usp. **53**, 133 (2010)
77. M. Ferreira, J. Loureiro, A. Ricard, J. Appl. Phys. **57**, 82 (1985)
78. A. Bogaerts, R. Gijbels, J. Vlcek, J. Appl. Phys. **84**, 121 (1998)
79. A. Bogaerts, R. Gijbels, J. Vlcek, J. Appl. Phys. **86**, 4124 (1999)
80. A. Bogaerts, R. Gijbels, Spectrochim. Acta **55B**, 263 (2000)
81. N.A. Dyatko et al., J. Phys. **41D**, 055204 (2008)
82. J. Vlsek, J. Phys. **22D**, 623 (1989)
83. B.M. Smirnov, *Physics of Ionized Gases* (Wiley, New York, 2001)
84. J.C. Maxwell, *Theory of Heat* (Longmans, Green and Co, London, 1871)
85. M.N. Saha, Proc. Roy. Soc. **99A**, 135 (1921)
86. L.D. Landau, E.M. Lifshitz, *Statistical Physics*, vol. 1 (Pergamon Press, Oxford, 1980)
87. B.M. Smirnov, *Reference Data on Atomic Physics and Atomic Processes* (Springer, Heidelberg, 2008)
88. P. Debye, E. Hückel, Phys. Zs. **24**, 185 (1923)
89. I. Langmuir, L. Tonks, Phys. Rev. **34**, 376 (1927)
90. L. Tonks, I. Langmuir, Phys. Rev. **34**, 876 (1929)
91. B.M. Smirnov, *Introduction to Plasma Physics* (Mir Publishers, Moscow, 1975)
92. A. Einstein, Ann. Phys. **17**, 549 (1905)
93. A. Einstein, Ann. Phys. **19**, 371 (1906)
94. A. Einstein, Zs. für Electrochem. **14**, 235 (1908)
95. W. Nernst, Zs. Phys. Chem. **2**, 613 (1888)
96. J.S. Townsend, V.A. Bailey, Philos. Trans. **A193**, 129 (1899)
97. J.S. Townsend, V.A. Bailey, Philos. Trans. **A195**, 259 (1900)
98. L.G. Huxley, R.W. Crompton, M.T. Elford, Brit. J. Appl. Phys. **17**, 1237 (1966)
99. A. Einstein, *Investigations on Theory of the Brownian Movement* (Dover, New York, 1956)
100. O.W. Richardson, Science **38**, 57 (1913)
101. O.W. Richardson, Proc. Roy. Soc. **89**, 507 (1913)
102. O.W. Richardson, Science **54**, 283 (1921)
103. O.W. Richardson, Phys. Rev. **23**, 153 (1924)
104. O.W. Richardson, Phys. Rev. **33**, 195 (1929)
105. S. Dushman, Phys. Rev. **21**, 623 (1923)
106. W.F. Gale, T.C. Totemeir (eds.), *Smithells Metals Reference Book*, 8th edn. (Elsevier, Oxford, 2004)
107. A.A. Radzig, B.M. Smirnov, *Reference Data on Atoms, Molecules and Ions* (Springer, Berlin, 1985)
108. National Institute of Standards and Technology Atomic Spectra Database http://physics.nist.gov/ PhysRefData/ASD/
109. O. Zatsarinny, K. Bartschat, J. Phys. **39B**, 2145 (2006)
110. L.D. Landau, E.M. Lifshitz, *Mechanics* (Pergamon Press, London, 1980)
111. L.D. Landau, E.M. Lifshitz, *Quantum Mechanics* (Pergamon Press, Oxford, 1980)

112. G.H. Wannier, Bell Syst. Tech. **32**, 170 (1953)
113. A. Dalgarno, M.R.C. McDowell, A. Williams, Philos. Trans. **A250**, 411 (1958)
114. E.A. Mason, H.W. Schamp, Ann. Phys. **4**, 233 (1958)
115. T. Kihara, M.H. Taylor, J.O. Hirschfelder, Phys. Fluids **3**, 715 (1960)
116. L.D. Landau, Zh. Exp. Theor. Fiz. **7**, 203 (1937)
117. Y. Faxen, J. Holtzmark, Zs. Phys. **45**, 307 (1927)
118. G.F. Drukarev, *Collissions of Electrons with Atoms and Molecules* (Plenum Press, New York, 1987)
119. V.S. Lebedev, I.L. Beigman, *Physics of Highly Excited Atoms and Ions* (Springer, Berlin, 1988)
120. V.S. Lebedev, I.L. Beigman, Phys. Rep. **250**, 95 (1995)
121. E.E. Fermi, Nuovo Cimento **11**, 157 (1934)
122. M.Y. Ovchinnikova, Zh. Exp. Theor. Fiz. **49**, 275 (1965)
123. B.M. Smirnov, Theor. Exp. Khim. **7**, 154 (1971)
124. E.E. Nikitin, B.M. Smirnov, *Atomic and Molecular Collisions* (Nauka, Moscow, 1988). (in Russian)
125. T.F. O'Malley, L. Spruch, T. Rosenberg, J. Math. Phys. **2**, 491 (1961)
126. T.F. O'Malley, Phys. Rev. **130**, 1020 (1963)
127. B.M. Smirnov, *Atomic Collisions and Elementary Processes in Plasma* (Atomizdat, Moscow, 1968) (in Russian)
128. C. Ramsauer, Ann. Phys. **72**, 345 (1923)
129. C. Ramsauer, R. Kollath, Ann. Phys. **3**, 536 (1929). (Leipzig)
130. J.L. Pack et al., J. Appl. Phys. **71**, 5363 (1992)
131. V.M. Atrazhev, I.T. Iakubov, High Temp. **18**, 1292 (1980)
132. V.M. Atrazhev, I.T. Iakubov, J. Phys. **C14**, 5139 (1981)
133. B.M. Smirnov, Phys. Usp. **45**, 1251 (2002)
134. L.J. Gurevich, *Grounds of Physical Kinetics* (Leningrad, GITTL, Russian, 1940)
135. E.P. Wigner, Phys. Rev. **73**, 1002 (1948)
136. N.F. Mott, H.S.W. Massey, *The Theory of Atomic Collisions* (Claredon Press, Oxford, 1965)
137. L.T. Specht, S.A. Lawton, T.A. DeTempe, J. Appl. Phys. **51**, 166 (1980)
138. M. Born, Z. Phys. **37** 863 (1926)
139. M. Born, *Problems of Atomic Dynamics* (MIT Press, Cambridge, 1926)
140. M. Born, *Atomic Physics* (Blakie and Son, London, 1935)
141. L.A. Vainstein, I.I. Sobelman, E.A. Yukov, *Excitation of Atoms and Broadening of Spectral Lines* (Nauka, Moscow, 1979) (in Russian)
142. V.P. Schevel'ko, L.A. Vainstein, *Atomic Physics for Hot Plasmas* (Institute of Physics Publishing, Bristol, 1993)
143. V.S. Lisitsa, *Atoms in Plasmas* (Springer, Berlin, 1994)
144. A.V. Eletskii, B.M. Smirnov, Zh. Tekhn. Fiz. **38**, 3 (1968)
145. A.V. Eletskii, B.M. Smirnov, ZhETF **84**, 1639 (1983)
146. A.V. Eletskii, B.M. Smirnov, J. Sov. Laser Res. **7**, 207 (1986)
147. E.U. Condon, G.H. Shortley, *Theory of Atomic Spectra* (Cambridge University Press, Cambridge, 1967)
148. I.I. Sobelman, *Radiative Spectra and Radiative Transitions* (Springer, Berlin, 1979)
149. V.P. Krainov, H. Reiss, B.M. Smirnov, *Radiative Transitions in Atomic Physics* (Wiley, New York, 1997)
150. Y.P. Korchevoi, A.M. Przonskki, Sov. Phys. JETP **24**, 1089 (1967)
151. E.E. Antonov, Y.P. Korchevoi, Ukr. Fiz. Zh. **22**, 1557 (1977)
152. I.P. Zapesochnyi, E.N. Postoi, I.S. Aleksakhin, Sov. Phys. JETP **41**, 865 (1976)
153. D. Leep, A. Gallagher, Phys. Rev. **10A**, 1082 (1974)
154. J.O. Phelps, C.C. Lin, Phys. Rev. **24A**, 1299 (1981)
155. J.O. Phelps, J.E. Solomon, D.F. Korff et al., Phys. Rev. A **20**, 1418 (1979)
156. L.M. Volkova, A.M. Devyatov, Izv. Acad. Nauk SSSR Ser. Fizich. **27**, 1052 (1963)
157. G.A. Piech, M.E. Lagus, L.M. Anderson et al., Phys. Rev. A **55**, 2842 (1997)

158. M.E. Lagus, J.B. Boffard, L.M. Anderson, C.C. Lin, Phys. Rev. A **53**, 1505 (1996)
159. I.P. Bogdanova, S.V. Yurgenson, Opt. Spectrosc. (USSR) **62**, 281 (1987)
160. I.P. Bogdanova, S.V. Yurgenson, Opt. Spectrosc. (USSR) **62**, 424 (1987)
161. I.P. Bogdanova, S.V. Yurgenson, Opt. Spectrosc. (USSR) **63**, 815 (1987)
162. I.P. Bogdanova, S.V. Yurgenson, Opt. Spectrosc. (USSR) **68**, 730 (1990)
163. I.P. Bogdanova, O.V. Veide, V.I. Yakovleva, Opt. Spectrosc. (USSR) **81**, 330 (1996)
164. J.B. Boffard, B. Chiaro, T. Weber, C.C. Lin, Atom Data Nucl. Data Tab. **93**, 831 (2007)
165. J.B. Boffard, R.O. Jung, C.C. Lin, A.E. Wendt, Plasma Sour. Sci. Technol. **18**, 035017 (2009)
166. L.D. Schearer, Phys. Rev. **171**, 81 (1968)
167. A.V. Phelps, Phys. Rev. **99**, 1307 (1955)
168. A.V. Phelps, Phys. Rev. **114**, 1011 (1959)
169. I.P. Bochkova et al., Opt. Spectrosc. **38**, 185 (1975)
170. B.M. Smirnov, *Physics of a Weakly Ionized Plasma* (Nauka, Moscow, 1972)
171. J.J. Thomson, Phil. Mag. **23**, 449 (1912)
172. L. Thomas, Proc. Cambridge Philos. Soc. **28**, 713 (1927)
173. D. Webster, W. Hansen, F. Duveneck, Phys. Rev. **43**, 833 (1933)
174. H. Bethe, Ann. Phys. **4**, 325 (1930)
175. M. Seaton, *Atomic and Molecular Processes* (Academic Press, New York, 1962)
176. A.E. Kingston, Phys. Rev. **135A**, 1537 (1964)
177. A.E. Kingston, Proc. Phys. Soc. **87**, 193 (1966)
178. A.E. Kingston, J. Phys. **1B**, 559 (1968)
179. W.L. Fite, R.T. Brackman, Phys. Rev. **112**, 1141 (1958)
180. D. Rapp, P. Englander-Golden, J. Chem. Phys. **43**, 2462 (1965)
181. K. Dolder, M.F.A. Harrison, P.C. Thonemann, Proc. Roy. Soc. **264**, 367 (1961)
182. L. Vriens, T.F.M. Bonsen, J.A. Smith, Physica **40**, 229 (1968)
183. D.R. Long, R. Geballe, Phys. Rev. **A1**, 260 (1970)
184. W.C. Lineberger, J.W. Hooper, E.W. McDaniel, Phys. Rev. **141**, 151 (1966)
185. J.B. Wareing, K.T. Dolder, Proc. Phys. Soc. **91**, 887 (1967)
186. B. Peart, K.T. Dolder, J. Phys. **B1**, 872 (1968)
187. I.P. Zapesochnyi, I.S. Aleksakhin, ZhETF **55**, 76 (1968)
188. R.H. McFahrland, J.D. Kinney, Phys. Rev. **137A**, 1058 (1965)
189. R.L.F. Boyd, G.W. Green, Proc. Phys. Soc. **71**, 351 (1958)
190. B.M. Smirnov, *Ions and Excited Atoms in Plasma* (Atomizdat, Moscow, 1974) (in Russian)
191. H.N. Kücükarpaci, H.T. Saelee, J. Lucas, J. Phys. **14D**, 9 (1981)
192. A.A. Kruithof, F.M. Penning, Physica **4**, 430 (1937)
193. A.A. Kruithof, M.J. Druyvestein, Physica **4**, 450 (1937)
194. B.M. Smirnov, *Excited Atoms* (Energoatomizdat, Moscow, 1982) (in Russian)
195. A.L. Schmeltekopf, F.C. Fehsenfeld, J. Chem. Phys. **53**, 3173 (1970)
196. N.B. Kolokolov, P.M. Pramatorov, Zh. Tekhn. Fiz. **48**, 294 (1978)
197. A.B. Blagoev, T.K. Popov, Phys. Lett. **70**, 416 (1979)
198. N.B. Kolokolov et al., Zh. Tekhn. Fiz. **53**, 913 (1983)
199. N.B. Kolokolov, in *Chemistry of Plasma*, vol. 12, ed. by B.M. Smirnov (Energoatomizdat, Moscow, 1984), pp. 56–95 (in Russian)
200. R.H. Neynaber, S.Y. Tang, J. Chem. Phys. **68**, 5112 (1978)
201. D.R. Bates, A.E. Kingston, Nature **189**, 652 (1961)
202. D.R. Bates, A.E. Kingston, R.W.P. Mc Whirter, Proc. Roy. Soc. A **267**, 297 (1962)
203. D.R. Bates, A.E. Kingston, R.W.P. Mc Whirter, Proc. Roy. Soc. A **270**, 155 (1962)
204. B.M. Smirnov, *Introduction to Plasma Physics* (Mir, Moscow, 1977)
205. J.J. Thomson, Phil. Mag. **47**, 337 (1924)
206. A.V. Phelps, S.C. Brown, Phys. Rev. **86**, 102 (1952)
207. H.J. Oskam, Phillips Res. Rep. **13**, 401 (1958)
208. R.R. Hackam, J.J. Lennon, Proc. Roy. Soc. **84**, 133 (1953)
209. C.B. Kretschmer, H.L. Peterson, Appl. Phys. **34**, 3209 (1963)
210. H.J. Oskam, V.R. Mittelstadt, Phys. Rev. **132**, 1435 (1963)

211. E.E. Niles, W.W. Robertson, J. Chem. Phys. **40**, 3568 (1964)
212. E.C. Beaty, P.L. Patterson, Phys. Rev. **137A**, 346 (1965)
213. C.G. Morgan, F. Llewellyn-Jones, D.K. Davies, Proc. Phys. Soc. **86**, 403 (1965)
214. G.F. Sauter, R.A. Gerber, H.J. Oskam, Phys. Rev. Lett. **19**, 656 (1966)
215. J.C. Cronin, M.C. Sexton, Br. J. Appl. Phys. **1–2**, 8890 (1967)
216. D. Smith, M.J. Cromey, J. Phys. **B1**, 638 (1968)
217. J.P. Gaur, L.M. Chanin, Phys. Rev **182**, 167 (1969)
218. J.L. Gilkinson, H. Held, L.M. Chanin, J. Appl. Phys. **40**, 2350 (1969)
219. V.B. Brodskii, N.B. Kolokolov, Zh. Tekh. Fiz. **40**, 496 (1970)
220. A.K. Bhattacharya, J. Appl. Phys. **41**, 1707 (1970)
221. R.A. Gerber, M.A. Gusinov, Phys. Rev **4A**, 2027 (1971)
222. B.M. Smirnov, High Temp. **24**, 239 (1986)
223. A. Dalgarno, Philos. Trans. **A250**, 428 (1958)
224. O.B. Firsov, Zh. Exp. Theor. Fiz. **21**, 1001 (1951)
225. L.A. Sena, Zh. Exp. Theor. Fiz. **9**, 1320 (1939) (in Russian)
226. L.A. Sena, Zh. Exp. Theor. Fiz. **16**, 734 (1946)
227. L.A. Sena, *Collisions of Electrons and Ions with Atoms* (Gostekhizdat, Leningrad, 1948)
228. B.M. Smirnov, Phys. Scr. **60**, 111 (2000)
229. B.M. Smirnov, Phys. Usp. **44**, 221 (2001)
230. B.M. Smirnov, JETP **92**, 251 (2001)
231. W. Elenbaas, *High Pressure Mercury Vapor Lamps and their Applications* (Philips Technical Library, Eindhoven, 1965)
232. J.F. Waymouth, *Electric Discharge Lamps* (MIT Press, Cambridge, 1971)
233. M. Scele, *Fundamentals of Light Sources and Lasers* (Wiley, Hoboken, 2004)
234. N.G. Basov, *Pulse Gas-discharge Atomic and Molecular Lasers* (Consultant Bureau, New York, 1976)
235. G.A. Mesyats, V.V. Osipov, V.F. Tarasenko, *Pulse Gas Lasers* (SPIE, Washington, 1995)
236. I.G. Ivanov, E.L. Latush, M.F. Sem, *Metal Vapor Ion Lasers* (Wiley, London, 1996)
237. O. Svelto, *Principles of Lasers* (Plenum Press, New York, 1998)
238. Ch.E. Little, *Metal Vapor Lasers* (Wiley, Chichester, 1999)
239. J. Berkowitz, *Photoabsorption, Photoionization and Photoelectron Spectroscopy* (Academic Press, New York, 1979)
240. A. Sommerfeld, G. Schur, Ann. Phys. **4**, 409 (1930)
241. M. Stobbe, Ann. Phys. **7**, 661 (1930)
242. H. Bethe, E. Salpeter, *Quantum Mechanics of One- and Two-Electron Atoms* (Springer, Berlin, 1975)
243. H.A. Kramers, Phil. Mag. **46**, 836 (1923)
244. L.A. Bureeva, V.S Lisitza, *Perturbating Atom* (Izd. AT, Moscow, 1997) (in Russian)
245. D.R. Bates, A. Damgaard, Philos. Trans. **242**, 101 (1949)
246. A.F. Starace, Theory of Atomic Photoionization, in *Encyclopedia of Physics*, vol. 31 (Springer, Berlin, 1982), pp. 1–121
247. B.M. Smirnov, *Physica of Atoms and Ions* (Springer, New York, 2003)
248. Yu.A. Vdovin, V.M. Galitskii, Sov. Phys. JETP **52**, 1345 (1967)
249. A. Einstein, Verh. Dtsch. Phys. Ges. **18**, 318 (1916)
250. M. Planck, Ann. Phys. **4**, 453 (1901)
251. F. Reif, *Statistical and Thermal Physics* (McGrow Hill, Boston, 1965)
252. B.A. Veklenko, ZhETF **33**, 629 (1957)
253. B.A. Veklenko, ZhETF **33**, 817 (1957)
254. B.M. Smirnov, *Physics of a Weakly Ionized Gas* (Nauka, Moscow, 1978) (in Russian)
255. B.M. Smirnov, *Physics of Weakly Ionized Gases* (Mir, Moscow, 1981)
256. L. Boltzmann, Wien. Ber. **66**, 275 (1872)
257. J.S. Townsend, V.A. Bailey, Phil. Mag. **46**, 657 (1923)
258. R.A. Nielsen, Phys. Rev. **50**, 950 (1936)
259. A.V. Phelps, J.L. Packs, L.S. Frost, Phys. Rev. **117**, 470 (1960)

260. J.L. Packs, A.V. Phelps, Phys. Rev. **121**, 798 (1961)
261. R.W. Warren, J.H. Parker, Phys. Rev. **128**, 2661 (1962)
262. R.W. Crompton, M.T. Elford, R.L. Jory, Aust. J. Phys. **20**, 369 (1967)
263. P. Herrend, C. R. **217**, 75 (1943)
264. G. Jager, W. Otto, Z. Phys. **169**, 517 (1962)
265. R.P. Caren, Phys. Rev. **131**, 1904 (1963)
266. K.H. Wagner, Z. Phys. **178**, 64 (1964)
267. N.E. Levine, M. Uman, J. Appl. Phys. **35**, 2618 (1964)
268. H.A. Lorentz, Proc. Roy. Acad. Amst. **7**, 438, 585, 684 (1905)
269. M.J. Druyvesteyn, Physica **10**, 61 (1930)
270. B.I. Davydov, Phys. Z. Sowjetunion **8**, 59 (1935)
271. P.M. Morse, W.P. Allis, E.S. Lamar, Phys. Rev. **48**, 412 (1935)
272. B.I. Davydov, Phys. Z. Sowjetunion **12**, 269 (1937)
273. W.P. Allis, H.W. Allen, Phys. Rev. **52**, 703 (1937)
274. S. Chapman, T.G. Cowling, *The Mathematical Theory of Non-Uniform Gases* (Cambridge University Press, Cambridge, 1939)
275. W.P. Allis, *Handbuch der Physik*, vol. 21, ed. by S. Flugge (Berlin, Springer, 1956), p. 383
276. E.M. Lifshits, L.P. Pitaevskii, *Physical Kinetics* (Pergamon Press, Oxford, 1981)
277. L.D. Landau, A. Kompaneetz, Phys. Z. Sowjetunion **6**, 163 (1934)
278. L.D. Landau, A. Kompaneetz, Zh. Exp. Theor. Fiz. **5**, 276 (1935)
279. B.I. Davydov, Phys. Z. Sowjetunion **9**, 433 (1936)
280. L.D. Landau, E.M. Lifshitz, Phys. Z. Sowjetunion **9**, 477 (1936)
281. J. Yamashita, M. Watanabe, Prog. Theor. Phys. **12**, 433 (1954)
282. A.D. Fokker, Ann. Phys. (Leipzig) **43**, 810 (1914)
283. M. Planck, Preuss. Akad. Wiss. Phys. Mat. Kl. **5**, 324 (1917)
284. B.M. Smirnov, *Principles of Statistical Physics* (Wiley, Berlin, 2006)
285. M.J. Druyvesteyn, Physica **1**, 1003 (1934)
286. M.J. Druyvesteyn, Physica **3**, 65 (1936)
287. B.M. Smirnov, Zh. Exp. Theor. Fiz. **116**, 58 (2013)
288. A.V. Eletskii, V.D. Kulagin, Phys. Plasma **5**, 98 (1979)
289. G.D. Birkhoff, Proc. Nat. Acad. Sci. USA **17**, 656 (1931)
290. J. Neumann, Proc. Nat. Acad. Sci. USA **18**, 70 (1932)
291. J. Neumann, Proc. Nat. Acad. Sci. USA **18**, 263 (1932)
292. B.M. Smirnov, D.V. Tereshonok, EPL **105**, 35001 (2014)
293. S. Chapman, Philos. Trans. R. Soc. Lond. **A216**, 279 (1916)
294. S. Chapman, Philos. Trans. R. Soc. Lond. **A217**, 115 (1917)
295. D. Enskog, Arkiv Ast. Fys. **16**, 1 (1921)
296. S. Chapman, T.G. Cowling, *The Mathematical Theory of Non-uniform Gases* (Cambridge University Press, Cambridge, 1952)
297. J.H. Ferziger, H.G. Kaper, *Mathematical Theory of Transport Processes in Gases* (North Holland, Amsterdam, 1972)
298. M. Capitelli, D. Bruno, A. Laricchiuta, *Fundamental Aspects of Plasma Chemical Physics: Transport.* (Springer, New York, 2013)
299. J.C. Maxwell, Phil. Mag. **19**, 19 (1860)
300. J.C. Maxwell, Phil. Mag. **20**, 21 (1860)
301. N.B. Vargaftik, *Tables of Thermophysical Properties of Liquids and Gases*, 2nd edn. (Halsted Press, New York, 1975)
302. C.J. Zwakhals, K.W. Reus, Physica **100C**, 231 (1980)
303. N.B. Vargaftik, L.P. Filippov, A.A. Tarismanov, E.E. Totskii, *Reference Book on Thermophysical Properties of Liquids and Gases* (CRC Press, Boca Raton, 1994)
304. E. Ebbinghaus, Ann. Phys. **7**, 267 (1930)
305. M.A. Biondi, Phys. Rev. **82**, 453 (1951)
306. M.A. Biondi, Phys. Rev. **88**, 660 (1952)
307. V.S. Egorov, YuG Kozlov, A.M. Shukhtin, Opt. Spectrosc. **15**, 839 (1963)

308. V.S. Egorov, YuG Kozlov, A.M. Shukhtin, Opt. Spectrosc. **17**, 154 (1964)
309. G. Rogers, M.A. Biondi, Phys. Rev. **134A**, 1215 (1964)
310. E.E. Ferguson, F.C. Fehsenfeld, A.L. Schmeltekopf, Phys. Rev. **138A**, 381 (1965)
311. P.L. Pakhomov, I.Ya. Fugol', Yu.F. Shevchenko, Zh. Tekh. Fiz. **34**, 1312 (1966)
312. P.L. Pakhomov, G.P. Reznikov, I. Ya. Fugol', Opt. Spektrosk. **20**, 10 (1966)
313. I.Ya. Fugol', P.L. Pakhomov, Zh. Exp. Teor. Fiz. **53**, 266 (1967)
314. R.W. Huggins, J.H. Cahn, J. Appl. Phys. **38**, 180 (1967)
315. I.Ya. Fugol', Sov. Phys. Usp. **97**, 429 (1969)
316. M.A. Gusinov, R.A. Gerber, Phys. Rev. **2A**, 1973 (1970)
317. A. Pesnelle, G. Watel, C. R. Acad. Sci. **272B**, 785 (1971)
318. N.P. Penkin, T.P. Red'ko, in Spectroscopy of Gas Discharge plasma, ed. by N.P. Penkin (Leningrad University Press, Russian, 1976), p. 51 (in Russian)
319. R.W. Engstrom, W.S. Huxford, Phys. Rev. **58**, 67 (1940)
320. J.P. Molnar, Phys. Rev. **83**, 940 (1951)
321. A.V. Phelps, J.P. Molnar, Phys. Rev. **83**, 1202 (1953)
322. A.H. Futch, F.A. Grant, Phys. Rev. **104**, 356 (1956)
323. E. Ellis, N.D. Twiddy, J. Phys. **2B**, 1366 (1969)
324. W. Wieme, J. Wieme-Lenaerts, Phys. Lett. **47A**, 37 (1974)
325. J.H. Kolts, D.W. Setser, J. Chem. Phys. **68**, 4848 (1978)
326. L.S. Frost, A.V. Phelps, Phys. Rev. **127**, 1621 (1962)
327. J.S. Townsend, Proc. Roy. Soc. A **80**, 207 (1908) (method for D/w)
328. J.S. Townsend, Proc. Roy. Soc. A **81**, 464 (1908) (method for D/w)
329. G. Cavalleri, Phys. Rev. **179**, 186 (1969) ($D_\perp(He)$)
330. D.R. Nelson, F.J. Davis, J. Chem. Phys. **51**, 2322 (1969) ($D_\perp(He)$)
331. C.S. Lakshminarasimha, J. Lucas, J. Phys. **10D**, 313 (1977) ($\alpha(Ar)$)
332. A.G. Robertson, Aust. J. Phys. **30**, 391 (1977)
333. H.B. Millloy, R.W. Crompton, Aust. J. Phys. **30**, 51 (1977)
334. L. Spitzer, *Physics of Fully Ionized Gases* (Wiley, New York, 1962)
335. T. Holstein, J. Chem. Phys. **56**, 832 (1952)
336. T. Kihara, Rev. Mod. Phys. **25**, 944 (1953)
337. V.I. Perel, Zh. Exp. Teor. Fiz. **32**, 526 (1957)
338. B.M. Smirnov, Dokl. Akad. Nauk SSSR **181**, 61 (1968)
339. B.M. Smirnov, in *Reviews of Plasma Physics*, vol. 23, ed.by V.D. Shafranov (Kluwer Academic, New York, 2003), p. 209
340. E.A. Mason, E.W. McDaniel, *Transport Properties of Ions in Gases* (Wiley, New York, 1988)
341. D.E. Davies, F. Llewellyn-Jones, C.G. Morgan, Proc. Roy. Soc. **80**, 898 (1962)
342. L.M. Chanin, G.D. Rork, Phys. Rev. **133A**, 1005 (1964)
343. T. Sakurai, Y. Ito, T. Ueda, Y. Inoue, H. Hori, J. Phys. **14D**, 9 (1981)
344. A.A. Kruithof, Physica **7**, 519 (1940) ($\alpha(Ar)$)
345. D.E. Davies, J.G.C. Milne, Br. J. Appl. Phys. **10**, 301 (1959)
346. H.N. Kücükarpaci, H.T. Saelee, J. Lucas, J. Phys. **14D**, 2001 (1981)
347. W. Schottky, Phys. Z. **25**, 345 (1924)
348. W. Schottky, Phys .Z. **25**, 635 (1924)
349. I. Langmuir, L. Tonks, Phys. Rev. **34**, 876 (1927)
350. O.B. Firsov, Zh. Exp. Teor. Fiz. **26**, 445 (1956)
351. L.B. Loeb, J.M. Meek, *Mechanism of Electric Spark* (Stanford University Press, Stanford, 1941)
352. J.M. Meek, J.D. Crags, *Electrical Breakdown in Gases* (Claredon Press, Oxford, 1953)
353. F. Llewellyn-Jones, *Ionization and Breakdown in Gases* (Methuen, London, 1957)
354. L.B. Loeb, *Basic Processes of Gaseous Electronics* (University of California Press, Berkeley, 1961)
355. H. Raether, *Electron Avalanches and Breakdown in Gases* (Butterworths, London, 1964)
356. F. Llewellyn-Jones, *Ionization, Avalanches and Breakdown* (Methuen, London, 1967)
357. J.M. Meek, J.D. Crags, *Electrical Breakdown in Gases* (Wiley, New York, 1998)

358. F. Pashen, Ann. Phys. **273**, 69 (1889)
359. F.M. Penning, Physica **8**, 137 (1928)
360. F.M. Penning, Proc. Roy. Soc. Amst. **31**, 14 (1928)
361. F.M. Penning, Proc. Roy. Soc. Amst. **33**, 841 (1930)
362. M.P. Auger, C. R. **180**, 65 (1925)
363. M.P. Auger, J. Phys. Radium **6**, 205 (1925)
364. E.H.S. Burhop, *The Auger Effect and Other Radiationless Transitions* (Cambridge University Press, Cambridge, 1952)
365. D. Chattarji, *The Theory of Auger Transitions* (Academic Press, London, 1976)
366. M.L.E. Oliphant, Proc. Roy. Soc. Lond. **A124**, 227 (1929)
367. M.L.E. Oliphant, P.B. Moon, Proc. Roy. Soc. Lond. **A128**, 388 (1930)
368. H.D. Hagstrum, Phys. Rev. Lett. **43**, 1050 (1979)
369. L.B. Loeb, Br. J. Appl. Phys. **3**, 341 (1952)
370. W. Bartolomeyczyk, Ann. Phys. **42**, 534 (1943)
371. H.D. Hagstrum, Phys. Rev. **89**, 244 (1953)
372. H.D. Hagstrum, Phys. Rev. **91**, 543 (1953)
373. H.D. Hagstrum, Phys. Rev. **96**, 325 (1954)
374. H.D. Hagstrum, Phys. Rev. **104**, 244 (1956)
375. H.D. Hagstrum, Phys. Rev. **104**, 672 (1956)
376. R.N. Varney, Phys. Rev. **93**, 1156 (1953)
377. H.D. Hagstrum, Phys. Rev. **119**, 940 (1960)
378. C.D. Child, Phys. Rev. **32**, 492 (1911)
379. I. Langmuir, Phys. Rev. **2**, 450 (1913)
380. I. Langmuir, Phys. Z. **15**, 348 (1914)
381. I.Langmuir, Phys. Z. **15**, 516 (1914)
382. E. Blechschmidt, Ann. Phys. **81**, 999 (1926)
383. W.R. Grove, Philos. Trans. Roy. Soc. **5**, 87 (1852)
384. J. Plücker, Ann. Phys. **103**, 88 (1858)
385. J. Plücker, Ann. Phys. **104**, 113 (1858)
386. J.P. Gassiot, Philos. Trans. Roy. Soc. **148**, 1 (1858)
387. F.M. Penning, Physica **3**, 873 (1936)
388. E.K. Kay, J. Appl. Phys. **34**, 760 (1963)
389. W.D. Gill, E.K. Kay, Rev. Sci. Instrum. **36**, 277 (1965)
390. A.V. Phelps, Plasma Sources Sci. Technol. **10**, 329 (2001)
391. V.A. Lisovskii, S.D. Yakovin, Tech. Phys. Lett. **26**, 891 (2000)
392. A.S. Pokrovskaya-Soboleva, B.N. Klyarfeld, Sov. Phys. Tech. Phys. **5**, 812 (1957)
393. B.N. Klyarfeld, L.G. Guseva, A.S. Pokrovskaya-Soboleva, Sov. Phys. Tech. Phys. **11**, 520 (1966)
394. P.L. Auer, Phys. Rev. **111**, 671 (1958)
395. Y. Miyoshi1, M. Hayashi1, T. Mori, J. Phys. Soc. Jpn. **18**, 396 (1963)
396. D.D. Sijacic, U.M. Ebert, Phys. Rev. **66E**, 066410 (2002)
397. YuP Raizer, U.M. Ebert, D.D. Sijacic, Phys. Rev. **70E**, 017401 (2004)
398. F.F. Chen, *Introduction to Plasma Physics* (Plenum, New York, 1974)
399. J.A. Bittencourt, *Fundamentals of Plasma Physics* (Springer, New York, 2004)
400. P. Belan, *Fundamentals of Plasma Physics* (Cambridge University Press, Cambridge, 2006)
401. T.F. Sheridan, J.A. Goree, IEEE Trans. Plasma Sci. **17**, 885 (1989)
402. D. Bohm, E.P. Gross, Phys. Rev. **75**, 1851 (1949)
403. D. Bohm, E.P. Gross, Phys. Rev. **75**, 1864 (1949)
404. D. Bohm, E.P. Gross, Phys. Rev. **79**, 992 (1950)
405. K. Wasa, S. Hayakawa, Rev. Sci. Instrum. **40**, 693 (1969)
406. J.R. Mullay, Res. Dev. **22**(2), 40 (1971)
407. P.V. Kashtanov, B.M. Smirnov, R. Hippler, Phys. Usp. **50**, 455 (2007)
408. I. Shyjumon, Ph.D. thesis, University of Greifswald, Greifswald, Germany, 2005
409. http://www.oaresearch.co.uk/cluster.htm

410. S.M. Rossnagel, H.R. Kaufman, J. Vac. Sci. Technol. **A6**, 223 (1988)
411. M.J. Goeckner, J.A. Goree, T.E. Sheridan, IEEE Trans. Plasma Sci. **19**, 301 (1991)
412. J.A. Thornton, Thin Solid Films **54**, 23 (1968)
413. H. Kersten et al., Vacuum **63**, 385 (2001)
414. T.E. Sheridan, M.J. Goeckner, J. Goree, J. Vac. Sci. Technol. **A8**, 30 (1990)
415. M.J. Goeckner, J. Goree, T.E. Sheridan, J. Vac. Sci. Technol. **A8**, 3920 (1990)
416. T.E. Sheridan, M.J. Goeckner, J. Goree, Appl. Phys. Lett **57**, 2080 (1990)
417. J. Goree, T.E. Sheridan, Appl. Phys. Lett **59**, 1052 (1991)
418. N.B. Kolokolov, A.B. Blagoev, Phys. Usp. **163**, 55 (1993)
419. L.D. Landau, E.M. Lifshitz, *Classical Theory of Fields* (Pergamon Press, Oxford, 1979)
420. M.A. Biondi, S.C. Brown, Phys. Rev. **76**, 1697 (1949)
421. A. Redfield, R.B. Holt, Phys. Rev. **82**, 874 (1951)
422. M.A. Biondi, Phys. Rev. **83**, 1079 (1951)
423. S. Takeda, J. Phys. Soc. Jpn. **13**, 112 (1958)
424. M.H. Mentzoni, J. Appl. Phys. **36**, 57 (1965)
425. F.J. Mehr, M.A. Biondi, Bull. Am. Phys. Soc. **13**, 199 (1968)
426. A.J. Cunningham, R.M. Hobson, Phys. Rev. **185**, 98 (1969)
427. D. Smith et al., J. Phys. **3B**, 34 (1970)
428. J.N. Fox, R.M. Hobson, Phys. Rev. Lett. **17**, 161 (1966)
429. M.P. Teter, P.E. Niles, W.W. Robertson, J. Chem. Phys. **44**, 3018 (1966)
430. H.F. Wellenstein, W.W. Robertson, J. Chem. Phys. **56**, 1072 (1972)
431. B. Dubreuil, A. Catherinot, Phys. Rev. A **21**, 188 (1980)
432. J.C. Cronin, M.C. Sexton, Br. J. Appl. Phys. **1–2**, 889 (1967)
433. B.M. Smirnov, JETP **51**, 1747 (1966)
434. J.A. Kinsinger, W.L. Stebbings, R.A. Valenzi, J.W. Taylor, Anal. Chem. **44**, 773 (1972)
435. J.E. Rowe, S.B. Christman, E.E. Chaban, Rev. Sci. Instrum. **44**, 1675 (1973)
436. H. Fellner-feldegg et al., J. Electron Spectrosc. Relat. Phenom. **5**, 643 (1974)
437. T.V. Vorburger, B.J. Waclawski, D.R. Sandstrom, Rev. Sci. Instrum. **47**, 501 (1976)
438. D.C. Mason, D.M. Mintz, A. Kuppermann, Rev. Sci. Instrum. **48**, 926 (1977)
439. E. Pellach, H.Z. Sar-El, J. Electron Spectrosc. Relat. Phenom. **14**, 259 (1978)
440. G. Schönhense, U. Heinzmann, J. Phys. **16E**, 74 (1983)
441. S. Caballero, H. Chuaqui, M. Favre, I. Mitchell, E. Wyndham, J. Appl. Phys. **98**, 023305 (2005)
442. M. Favre, E. Wyndham, A.M. Lenero et al., Plasma Sources Sci. Technol. **17**, 024011 (2008)
443. E.S. Wyndham, M. Favre, M.P. Valdivia et al., Rev. Sci. Instrum. **81**, 093502 (2010)
444. M. Laroussi, X. Lu, Appl. Phys. Lett. **87**, 113902 (2005)
445. D.H. Shin, Y.C. Hong, H.S. Uhm, IEEE Trans. Plasma Sci. **34**, 246 (2006)
446. H.S. Uhm, J.P. Lim, S.Z. Li, Appl. Phys. Lett. **90**, 261501 (2007)
447. G. Fridman, G. Friedman, A. Gutsol, A.B. Shekhter, V.N. Vasilets, A. Fridman, Plasma Processes Polym. **5**, 503 (2008)
448. M.S. Kim, D.G. Jang, H. Suk, H.S. Uhm, J. Korean Phys. Soc. **58**, 39 (2011)
449. N.A. Bobrova et al., Phys. Rev. **65E**, 016407 (2001)
450. B. Li, D. Kwok, J. Plasma Phys. **70**, 379 (2004)
451. B.H.P. Broks et al., Phys. Plasmas **14**, 2 (2007)
452. G. Stokes, Trans. Cambridge Philos. Soc. **9**, 8 (1851)
453. O. Reynolds, Philos. Trans. R. Soc. **174**, 935 (1883)
454. L.D. Landau, E.M. Lifshitz, *Fluid Dynamics* (Pergamon Press, Oxford, 1986)
455. G. Hagen, Poggendorfs Ann. **46**, 423 (1839)
456. J.L.M. Poiseuille, C. R. Acad. Sci. **11**, 1041 (1840)
457. J.L.M. Poiseuille, C. R. Acad. Sci. **15**, 1167 (1842)
458. V.P. Krainov, *Qualitative Methods in Physical Kinetics and Hydrodynamics* (American Institute of Physics, New York, 1992)
459. B.M. Smirnov, *Nanoclusters and Microparticles in Gases and Vapors* (De Gruyter, Berlin, 2012)

460. M.J. Goeckner, J.A. Goree, T.F. Sheridan, Phys. Fluids **B4**, 1663 (1992)
461. V.L. Ginzburg, A.V. Gurevich, Sov. Phys. Usp. **3**, 115 (1960)
462. B.B. Kadomtzev, *Plasma Turbul.* (Academic Press, New York, 1965)
463. R.Z. Sagdeev, A.A. Galeev, *Nonlinear Plasma Theory* (Benjamin, New York, 1969)
464. V.N. Tsytovich, *Nonlinear Effects in Plasmas* (Plenum Press, New York, 1970)
465. D.A. Frank-Kamenetskii, *Plasma—the Forth State of Matter* (Plenum Press, New York, 1972)
466. A. Hasegawa, *Plasma Instabilities and Nonlinear Effects* (Springer, Berlin, 1975)
467. A.B. Mikhailovskii, *Electromagnetic Instabilities in an Inhomogeneous Plasma* (IOP Publishing, Bristol, 1992)
468. A. Sitenko, V. Malnev, *Plasma Physics Theory* (Chapman and Hall, London, 1995)
469. D.G. Swanson, *Plasma Waves* (Institute of Physics Publications, Bristol, 2003)
470. L. Spitzer, *Physics of Fully Ionized Gases* (Dover, New York, 2006)
471. R.Kh. Amirov, E.I. Asinovskii, V.V. Markovets, Teplfiz. Vys. Temp. **19**, 47 (1981)
472. Yu.Z. Ionikh, I.N. Kostyukevich, N.V. Chernysheva, Opt. Spectrosc. **74**, 274 (1993)
473. Yu.Z. Ionikh, Yu.G. Utkin, N.V. Chernysheva, A.S. Evdokimenko, Plasma Phys. Rep. **22**, 289 (1996)
474. R.Kh. Amirov, E.I. Asinovskii, V.V. Markovets, Plasma Phys. Rep. **27**, 424 (2001)
475. N.A. Dyatko, YuZ Ionikh, A.V. Mershchanov, A.P. Napartovich, Plasma Phys. Rep. **31**, 875 (2005)
476. H.R. Grim, *Principles of Plasma Spectroscopy* (Cambridge University Press, New York, 1997)
477. A.G. Zhiglinskii (ed.), *Reference Data for Constants of Elementary Processes Involving Atoms, Electrons, Ions and Photons* (Petersburg University Press, Petersburg, 1994) (in Russian)
478. L.M. Biberman, V.S. Vorob'ev, I.T. Jakubov, Sov. Phys. Usp. **15**, 375 (1973)
479. L.M. Biberman, V.S. Vorob'ev, I.T. Jakubov, Sov. Phys. Usp. **22**, 411 (1979)

Index

A
Abnormal regime of gas discharge, 215, 364
Absorption coefficient, 75
Absorption cross section, 76
Ambipolar diffusion, 184
Argon atom and ion, 25
Associative ionization, 58
Auger effect, 231

B
Balance equations, 41
Bethe formula, 43
Block model, 8, 88
Boltzmann distribution, 14
Boltzmann kinetic equation, 93
Born approximation, 36, 43
Broadening of spectral lines, 69

C
Capillary discharge, 347
Capture cross section, 62
Cathode emission, 231
Cathode layer, 21, 232, 357
Cathode voltage, 236
Chapman-Enskog approximation, 153
Characteristic temperature, 158
Child law, 238, 248, 374
Coeffcient of ambipolar diffusion, 186
Collision integral, 93
Collision rate, 98
Computer plasma simulation, 8
Conductivity of ionized gas, 170
Continuous spectrum of plasma radiation, 285
Conversion of atomic ions into molecular ones, 60
Coulomb logarithm, 33
Cross section of atom excitation, 41, 205
Cross section of atom ionization, 52
Cross section of electron-electron scattering, 36
Cross section of electron-ion scattering, 36
Cross section of resonant charge exchange, 63

D
Dalgarno formula, 176
Dark space, 22, 369
Debye-Huckel radius, 17
Differential cross section, 31
Diffusion coefficient, 31, 151
Diffusion cross section, 37, 38
Diffusive motion of particles, 137, 248
Dimensionality consideration, 60
Diode gas discharge, 239
Discharge lamp, 3
Dissociative equilibrium, 16
Dissociative recombination, 58, 332
Doppler broadening of spectral lines, 70
Double layer, 248, 372
Drift velocity of electrons, 94, 171
Drift velocity of ions, 176
Druyvesteyn distribution function, 115

E
Effective electron temperature, 117
Effective radiative lifetime, 84
Efficiency of atom excitation, 130
Einstein coefficients, 76

B.M. Smirnov, *Theory of Gas Discharge Plasma*, Springer Series on Atomic, Optical, and Plasma Physics 84, DOI 10.1007/978-3-319-11065-3

Einstein relation, 20
Electron temperature, 120
Elenbaas-Heller equation, 220
Energy electron distribution function (EEDF), 387
Equilibrium ionization constant, 308
Exchange process, 48
Excitation cross section, 44, 45
Excitation rate constant, 47, 264
Excitation transfer, 86

F
Fermi formula, 35
Fine-structure constant, 65
Firsov method, 211
First Townsend coefficient, 1, 22, 199, 292
Fock solution, 221
Fokker-Planck equation, 100

G
Gas discharge plasma, 1
Gas-kinetic cross section, 32
Glow discharge, 22, 207, 246
Grotrian diagram, 26

H
Hard sphere model, 154, 155, 169
Heat balance, 215, 338, 345
Heat flux in plasma, 191
Helium atom and ion, 25
Hydrodynamic model, 388

I
Impact broadening of spectral lines, 71
Integral of electron-atom collisions, 99
Integral of electron-electron collisions, 101, 110
Ion diffusion coefficient, 178, 179
Ion drift velocity, 178
Ionization cross section, 52
Ionization equilibrium, 15, 211, 224
Ion mobility, 19

K
Kinetic equation, 20
Kramers formula, 68

L
Lamps, 3, 65
Landau collision integral, 101
Larmor frequency, 193
Lorenz profile of spectral lines, 70

M
Magnetic lines of force, 254
Magnetic trap, 252, 307, 382
Magnetized plasma, 193
Magnetron discharge, 238, 252, 378
Maxwell distribution function, 14
Mobility of ions in gas, 177
Momentum of electron, 95

N
Normal regime of glow discharge, 220, 239

O
Optical thickness, 80
Oscillator strength, 43, 74

P
Pashen curve, 229
Pashen notations, 26
Penning process, 58, 231, 311, 320
Photoionization, 67
Photon distribution function, 69
Photon emission, 278
Photorecombination, 67
Planck formula, 76
Plasma, 247
Plasma display, 4
Plasma engine, 5
Plasma knife, 5
Plasma oscillations, 18
Plasma torch, 6
Poiseuille formula, 353
Polarization capture process, 62
Polarization interaction, 38
Positive column, 22
Principle of detailed balance, 42
Probability of photon surviving, 81, 88
Propagation of resonant radiation, 79

Q
Quasistatic broadening of spectral lines, 71
Quenching cross section, 41
Quenching rate constant, 42, 267, 274, 280

R
Radiative transitions, 65
Ramsauer effect, 36
Rate constant of atom excitation, 45, 136
Rate constant of atom ionization, 204
Rate constant of atom quenching, 44, 267, 274, 279
Rate of atom excitation, 44, 136
Reabsorption of photons, 85
Regimes of electron energy relaxation, 107
Regimes of gas discharge plasma, 396
Resonant charge exchange, 62
Resonantly excited states, 43, 79
Reynolds number, 352
Richardson-Dushman formula, 24
Richardson parameter, 25

S
Saha formula, 15
Scattering amplitude, 34
Scattering length, 35
Scattering phase, 34
Schottky regime, 249, 315
Second Townsend coefficient, 1, 22, 232, 242
Secondary electrons, 245, 254
Self-consistency of gas discharge, 2, 390
Single ionization of atoms, 291, 303
Skin layer, 288
Spectral methods for plasma, 5
Spectrum of released electrons, 55
Spitzer formula, 172
Sputtering, 239, 252
Stepwise ionization, 61, 209, 296, 303, 321, 331

T
Tau-approximation, 94
Thermal conductivity coefficient, 152
Thermal instability, 224
Thermoemission of electrons, 23, 246
Thomson model of atom ionization, 50
Thomson theory for three body processes, 59
Three body processes, 59
Three body recombination of electrons and ions, 60
Threshold law for excitation, 41
Total cross section, 33, 71
Townsend gas discharge, 22, 220, 230, 245
Townsend scheme, 1, 199, 252
Transport coefficients, 19, 156
Transport coefficients for electrons, 161

V
Viscosity coefficient, 152

W
Weiskopf radius, 72
Width of spectral line, 70, 74, 81